Edsel Bryant Ford. (Photo courtesy of Edsel & Eleanor Ford House)

EDSEL

The Story of Henry Ford's Forgotten Son

Other SAE books of interest:

Edsel Ford and E.T. Gregorie:
The Remarkable Design Team and Their Classic Fords of the 1930s and 1940s
By Henry Dominguez
Order No. R-245

Ford: The Dust and The Glory, Volumes 1 and 2
By Leo Levine
Order No. R-292.SET

The Fords in My Past
By Harold L. Brock
Order No. R-270

Henry Ford: A Hearthside Perspective
By Donn P. Werling
Order No. R-266

EDSEL

The Story of Henry Ford's Forgotten Son

Henry Dominguez

SAE INTERNATIONAL®

Society of Automotive Engineers, Inc.
Warrendale, Pa.

 eISBN: 978-0-7680-4187-3

For permission and licensing requests, contact:

SAE Permissions
400 Commonwealth Drive
Warrendale, PA 15096-0001 USA
E-mail: permissions@sae.org
Fax: 724-772-4028
Tel: 724-772-4891

Library of Congress Cataloging-in-Publication Data

Dominguez, Henry L., 1953–
Edsel: the story of Henry Ford's forgotten son / Henry Dominguez.
p. cm.
Includes bibliographical references and index.
ISBN 0-7680-0920-0
1. Ford, Edsel, 1893–1943. 2. Businessmen—United States—Biography. 3. Automobile industry and trade—United States—History. 4. Ford Motor Company—History. I. Society of Automotive Engineers. II. Title.

HD9710.U52 F583 2002
338.7'6292'092—dc21
[B] 2002066956

SAE
400 Commonwealth Drive
Warrendale, PA 15096-0001 USA
E-mail: custsvc@sae.org
Fax: 724-776-0790
Tel: 877-606-7323 (inside USA and Canada)
724-776-4970 (outside USA)

ISBN 0-7680-0920-0

SAE Order No. R-329

Printed in the United States of America.

Front cover: Portrait of Edsel B. Ford, 1932, by Diego M. Rivera. Bequest of Eleanor Clay Ford. Photo © 1994, The Detroit Institute of Arts.

For Sam

Contents

Introduction

Edsel.

To students of business, the word Edsel conjures up only one thing: disaster. It was the name given to an automobile in the late 1950s that failed miserably in the marketplace and lost Ford Motor Company hundreds of millions of dollars. Most of the cause of the failure was timing—the country fell into a recession during the year in which the Edsel was introduced. Its styling also played a significant part, as did its name.

However, Edsel the car is not what sparked my interest in Edsel the man. I have been a Fordophile for the better part of 30 years, and in all of the books written about the Ford Motor Company and the Ford family, precious little has been written about Edsel Ford. I found that curious. Edsel was, after all, the son of Henry Ford and, more importantly, the president of the Ford Motor Company for a quarter of a century. Who was this man? What was he like? And why has he been ignored? After I had written *The Ford Agency*, I decided to pursue a biography on this most elusive of Fords.

I started my biography almost too late. By the time I began my research in the late 1980s, most of the people who had known Edsel were long gone. The one person who could have given the most information about Edsel was his wife, Eleanor, and she had died in 1976, taking a wealth of information to her grave. The story goes that Eleanor had finally consented to give an interview to noted Ford historian Professor David Lewis, but she fell ill shortly before their scheduled interview and the two never had a chance to meet. Edsel and Eleanor's two youngest children, Josephine Ford and William Ford, are both still alive. Unfortunately, neither consented to an interview for this book. I wrote to each of them several times but never received an answer. That is understandable, although disappointing—it seems that they too adhere to Edsel's desire for privacy.

Thus, I was left with only a handful of people and any available written material with which to do my research. Nonetheless, off I went, and, with the help of Richard Folsom—a Fordophile if ever there was one—and the resources of the Ford Archives, I was able to piece together the story of Edsel Bryant Ford.

During the late 1970s and the 1980s, Dick Folsom had tracked down many of the remaining Ford relatives and Ford employees who had known both Henry and Edsel Ford, simply because he wanted to speak to people who actually knew the Fords. He taped many of these conversations, and the transcriptions now reside at the Ford Archives or at the Henry Ford Estate. During my preliminary research, I was told of Dick's activities. I contacted Dick and told him what I was doing, and he graciously introduced me to these same people. I was able to talk with them and ask them specifically about Edsel. These interviews were priceless, particularly those with the Ford relatives, because many of them had known Edsel from the time he was a teenager.

Another fortuitous event during my research occurred when I learned that Bob Gregorie, the first design chief of Ford Motor Company and Edsel's closest confidant during the last years of his life, was still alive. Gregorie is now in his early nineties, but I first met him when he was in his eighties. My good friend Dave Crippen, chief archivist at the Ford Archives during these early years of my research, told me about Gregorie, and noted automotive historian, Mike Lamm, put me in touch with the great designer. Through extensive interviews with Gregorie, I began to understand Edsel as a man, a manager, and a design executive.

The extensive collections of the Ford Archives resposited at the Henry Ford Museum & Greenfield Village in Dearborn, Michigan, filled in the remainder of the information that the interviews and other secondary materials could not provide. Here, the massive collection of material on the Ford family and the Ford company are segregated into Accessions, and Accession 6, comprised of 464 boxes of material, relates to Edsel Ford. Over a period of several months, I gleaned through nearly every box of Accession 6, searching for the tidbits of information pertaining to Edsel's life.

In addition to Accession 6, the Oral History Collection of the Archives was also a rich source of information. Called Reminiscences, this is a collection of more than 250 transcribed interviews of Ford employees and family members conducted in the early 1950s in preparation for the fiftieth anniversary of Ford Motor Company. These Reminiscences—depending on whom was being interviewed—range in length from a few pages to more than a thousand pages. Most are indexed, so it was relatively easy for me to comb through the many interviews, looking for references to Edsel Ford. In some cases, the Reminiscences concentrated on Edsel—one is by the doctor who delivered him, and another is by Edsel's personal secretary.

In the end, I had thousands of note cards, each one containing a tidbit of information on Edsel. My task at that point was to assemble all of this information into a coherent manuscript, which I proceeded to do during the next decade.

Looking back, it is surprising that no one had ever written Edsel's biography, for I indeed had found a lot of information about him. I suppose it was simply one of those serendipitous events that occurs in all of our lives. I decided that Edsel's biography needed to be written, and when I started pursuing it, things started falling into place. And the story turned out more interesting than I could ever have imagined. I am glad it is finally told.

Henry Dominguez
San Francisco, California

Chapter 1

A Man Called Edsel

In celebration of its fiftieth anniversary in 1953, Ford Motor Company commissioned Norman Rockwell to compose a number of paintings depicting the salient events in the history of the company. One painting is of ten-year-old Henry Ford repairing watches in his bedroom as his father looks on. Another shows a mature Henry Ford driving through the streets of Detroit in a Model A Ford, the first car produced by his newly formed company. Another depicts a farmer and his family showing off their new Model T to neighbors.

In Rockwell's detailed scene of Henry Ford in his workshop constructing his first motor carriage, wrenches are lying on the wooden floor, a wood stove gives off radiant heat, and a cat sleeps quietly in the corner of the brick workshop. The quadricycle, almost complete, sits in the middle of the floor as Henry finishes one of the hundreds of parts that he had made for it. His wife, Clara, is sitting next to him and is mending one of his socks, something she continued to do even when the Fords became fabulously wealthy. However, a significant part of this reenacted scene is missing.

During the real event, the wrenches would be lying on the floor. The cat would be sleeping in the corner. Henry would be busy finishing his four-wheeled contraption. And Clara, basking in the warmth of the wood-burning stove, would be watching her inventive husband as he tinkered with his mechanical contraption. In the real scene, however, unlike Rockwell's portrayal, the Fords' infant son, Edsel, would have been there as well—crawling under his father's motorcar, marveling at its shiny brass parts, or sitting in the seat, playing with the steering tiller. At the time, Edsel was two-and-a-half years old.

It would be understandable if Rockwell had omitted both Clara and Edsel from the painting. Clara would have been in the house, tending to Edsel while Henry worked alone in the shop, which he often did. Or, Rockwell could have painted the scene as it actually happened, with all three Fords in the workshop. However, the truth was that Rockwell simply forgot about Edsel.

That is the way it had always been with Edsel. First, there was Edsel's father—vainglorious, self-centered, and autocratic. Then there was Edsel's eldest son—proud, brash, and persevering. Between the two was Edsel—taciturn, introverted, and reticent. In biography after biography, only Henry Ford and his grandson, Henry Ford II, are the centers of attention. Whenever Edsel is mentioned—if he is mentioned at all—it is always parenthetically: he was president in name only…; he emphasized styling of Ford products…; or he and his father never got along… Less important men in the history of the company, such as Harry Bennett, who was Henry Ford's confidant, or Charles Sorensen, who was Henry's right-hand man, have generated much more copy than Edsel.

***

Edsel did not mind being left in the shadows. In fact, he enjoyed it. "He was kind of a loner," said a Ford executive.[1] He preferred anonymity to celebrity, and he deliberately kept himself out of the limelight. When asked by company publicists to pose for pictures or to respond to reporters' questions, Edsel's usual reply was, "See if you can't get father to do that. He likes that sort of thing—I don't."[2]

In the days before television and hounding reporters, that was easy for Edsel to do. So rare were Edsel's public appearances that even at the peak of his power in the 1930s and 1940s, most people outside Detroit did not even know he existed. Even in Detroit, Edsel could drive around town alone without being bothered, as he often did.[3] To maintain his anonymity, Edsel would not allow his photograph to be given to anyone, not even to the press or to Ford dealers.[4] "Please be advised," Edsel wrote to Sorensen, "that my personal photograph is to be withheld from distribution or publication for any purpose."[5]

"Edsel was happiest," reported a close associate, "when he got by without being recognized."[6] He considered it a pleasant triumph, for example, when he managed to visit the Ford exhibit at the 1935 California-Pacific Exposition in San Diego 19 times without being identified.[7]

Once Edsel drove 200 miles north of the Ford factory in Dearborn to road test a modified Model A with Lawrence Sheldrick, the chief engineer at Ford. When they stopped to eat at a hotel where it was required to register for lunch, Edsel's eyes lit up. He grabbed Sheldrick's arm and said, "Let's have some fun. You register under my name, and I'll use yours." For the remainder of the afternoon, Sheldrick was the center of attention because everyone thought he was the Ford heir. Edsel simply sat to the side and grinned.[8]

Edsel preferred staying out of the limelight, allowing his father to take all of the credit for the success and running of the company. "It's his business," said Edsel. "He built it from nothing. He has a right to run it as he pleases."[9]

Plus, Edsel simply did not enjoy the attention. "I don't want to be in the position that father's in," he told Fred Black, the advertising manager at Ford. "He can't go down the street in any city without being recognized and set upon by autograph hunters and people trying to sell him some idea."[10]

If Edsel had outlived his father, he surely would have run the company on his own terms. However, it is unlikely that he would have subjected himself to a bevy of newspaper reporters in the hopes of seeing his quotations in the morning papers. He would have delegated those tasks to his public relations or advertising men. To the extent that he could, Edsel managed by delegation while he was president, pushing decisions as far down the corporate ladder as he could, and company publicity would have been no exception.

***

Edsel was Henry and Clara Ford's only child. Coming from relatively large families themselves—Henry from a family of eight, and Clara from a family of five—the couple probably would have had more children had they been able to do so. Apparently, Clara had physical problems that prevented her from having more children. She did not give birth to Edsel until five years after her marriage, which was a long period of time for those days. In early 1907, she was rushed to the hospital for "a serious operation."[11] The nature of her surgery was never revealed because her grandson, Henry Ford II, had all of her medical records destroyed, as well as those of Henry and Edsel.[12] Speculation suggests that Clara's operation was a hysterectomy. Such an operation would support Clara's long-time malady and explain why she and Henry never had more children.[13]

It was Henry's decision to give his son the curious name of Edsel. Its Germanic root is Attila, but its Old English meaning is "rich man's son."[14] Had it not been for Henry's closest schoolboy friend, Edsel Ruddiman, Henry probably would have named his son something more commonplace or Biblical in the same way that all of the Fords before him had named their sons. Or, Henry might have named his son after himself, for that was common practice for Fords. However, Henry and Edsel Ruddiman had been close friends since first grade.[15] Ruddiman eventually ended up a Ford employee, becoming chief chemist for the company, and a Ford relative when Margaret Ford, Henry's sister, married James Ruddiman, Edsel's brother. Even before these associations, however, Henry had resolved that if he ever had a son, he would name him after his dear friend, Edsel Ruddiman.[16]

Perhaps because of its uniqueness, Edsel Ford never really cared for his Christian name.[17] Even Clara had second thoughts about her son's name and "thought perhaps children should choose their own name after reaching the age of perhaps sixteen or seventeen."[18] In fact, few people ever called him Edsel. His schoolmates began calling him Ed or Eddie, as did his parents.[19] In Edsel's adult years, his wife and close friends called him Ned.[20] Edsel's children called him Daddy or Father when they were young; however, for reasons none of them could later explain, they started calling him Paul when they became older.[21] At work, almost everyone called him Mr. Edsel Ford, to distinguish him from his father, whom they addressed as Mr. Ford. When Henry Ford was talking to his executives, he would refer to his son as Mr. Edsel. Only a few old-time Ford executives, such as Charles Sorensen, who had known Edsel since he was a little boy, called him Edsel.[22]

Nevertheless, Edsel thought enough of his name to christen his second son as Edsel, Jr.[23] However, because of pressure from his wife, who wanted to recognize her side of the family, Edsel changed his son's name to Benson several months later.

For all of its adversity, Edsel's name lives on in the Ford family to this day. After Henry became a wealthy and powerful industrialist, the family started a tradition of alternately naming their firstborn sons after Henry or Edsel. Edsel began the tradition by naming his first son Henry Ford II. In turn, Henry Ford II named his first and only son Edsel Ford II. And Edsel Ford II named his first son Henry Ford III. Perhaps in the future, there will be an Edsel Ford III.

Edsel's middle name, Bryant, was in honor of Clara's father, Melvin Bryant.[24]

Edsel inherited his short stature, dark complexion, brown eyes, brown hair, and quiet demeanor from his mother.[25] His early childhood pictures depict him as a boyish Clara. As he matured, however, Edsel became "the spitting image" of his father, developing his father's angular features and slender build, which made him appear taller than his five-foot four-inch height.[26] This did not become apparent until Edsel was in his twenties. In a photograph taken when he was 21 years old, during his 1915 cross-country trip to San Francisco, he looks exactly like his father did at the same age.

* * *

Although Edsel resembled his father, he behaved like his mother, who was "considerate, unpretentious, and gracious."[27] Clara had always been a reserved and modest person, and she became even more reserved after Henry became famous. The demands that were made on Clara after Henry's success taught her to act reserved while in the public eye. When she was among friends, however, she let down that protective veil and her personality came through. "She was a most entertaining and charming person," recalled a friend of the family. "She was a great deal of fun, and she had a keen sense of humor."[28] Edsel likewise had a humorous side that he exposed only to those whom he knew well. "He liked to joke and get one on you," explained his cousin. "And he'd get a kick out of it."[29]

Ernest Kanzler, Edsel's brother-in-law, was one of the few people who not only worked with Edsel but saw him socially as well. Thus, he witnessed Edsel's humorous side. "Edsel was in a class all by himself," Kanzler recalled. "Nobody ever left him but was completely charmed. He was... a fun fellow to be with. He was great company."[30] One time while Kanzler, Edsel, and some other Ford executives were waiting for Henry Ford, Edsel helped pass the time by telling a story about an elderly man on his deathbed.

"The old man asked his wife to bring in the children so that he might speak to them before he passed on," explained Edsel, barely able to hold back chuckling. "When they lined up before him, with tears in his eyes, the old fellow said, 'Children, I want to make a confession to you. You know, your ma and me never married.'

"A deep silence followed until a boy about fifteen years old, standing in the center, looked up and down the line and suddenly blurted out, 'Well, you bastards, ain't you go'n to say somethin'?' "[31]

Unfortunately, historians and many of Edsel's business associates misinterpreted Edsel's demeanor as a subjugation by his father. Charles Sorensen called him "timid," and Harry Bennett, Henry's henchman, even went so far as to call him "a weak character."[32] However, although he rarely exhibited anger—"There was no shouting or table thumping from him," explained one Ford associate—Edsel was far from timid, as we shall see later.[33] His most notable trait was modesty, and he inherited that from his mother.[34]

* * *

People who knew him well agreed that Edsel's most striking features were his "prepossessing" eyes and genuine smile. His eyes were "bright, observing, and intelligent," they all said.[35] And his smile, when Edsel did smile, "lights everything round about."[36] Similar to everything else about Edsel, his smile was genuine. "There is not a scrap of artificiality about Mr. Edsel Ford's smile," wrote a reporter. "It just happens."[37]

Unfortunately, Edsel rarely smiled before a camera. All of his portraits are formal and serious. Only in those instances where the camera caught him off guard do we see a genuine smile.

Behind the smile was a melodic voice, slightly high-pitched.[38] Edsel spoke clearly, confidently, and concisely—but not loudly. "He talks as if he turns the matter over carefully in his mind before answering questions," wrote one reporter. "But his answers are always to the point and given with little hesitation."[39] If Edsel had the typical Michigan accent of accentuating his r's, he must have lost it along the way, for it was not evident in later years.[40]

In addition to these personal qualities, Edsel was well known for his immaculate appearance. He was always perfectly groomed and fastidiously dressed. There is not a photograph that captured Edsel in tattered clothing. Even when he was a child, Clara made sure that her son was always clean and properly dressed. One photograph shows Edsel at age five, standing on a boat dock on Lake Orion and holding a fishing pole in one hand and a string of fish in the other. Most children fishing at this age would have been found in ragged "go'en fish'en" clothes, but not Edsel. He appears prim and proper in what could be called a fishing outfit: a large straw hat, a well-pressed shirt, and perfectly fitting overalls, rolled up to his knees. Edsel seems to be posing for a children's clothing catalog rather than returning from a day of fishing. Even during his cross-country trip in a Model T touring car, Edsel dressed in what he called "touring tweeds." This decorum continued for all of his life, and it is not surprising that in 1935 Edsel made the list of best-dressed men, along with actor Fred Astaire, designer Walter Dorwin Teague, and master-merchandiser Marshall Field. The critics described Edsel as the "perfectly dressed business executive."[41]

Edsel had clothes for every occasion. He had tobogganing clothes, skiing clothes, and tennis clothes. Golf clothes, hunting clothes, and sailing clothes. In winter, he wore brightly colored sweaters and heavy woolens with stocking caps.[42]

As was the mode of the day, Edsel wore suits to work. Dunn & Company, a prominent Detroit haberdashery, hand-made all of Edsel's suits. Because his time was limited, when he decided to place an order, he purchased in quantity.[43] A retinue of tailors would go to Edsel's office at the Ford Administration Building and do the fitting there, ensuring that "his pants cuffs were tailored at exactly the right height."[44] Edsel's taste in suits was traditional, and he wore them with a nice combination of shirt and tie colors.[45] "All the Ford executives dress neatly, conservatively," wrote a reporter in 1933, "but Edsel is the only one you may notice is wearing the green-and-white tie today, was wearing the brown-and-white tie yesterday."[46] He also kept a tuxedo and tailcoat at the office in the event that he went out in the evening after working late.[47]

All of his shoes were hand-made from top-grain leather by shoemakers in New York and London. Similar to his ties and sweaters, Edsel's shoes were of all colors and styles and for all occasions. His favorite shoes were tan-colored Oxfords with pointy toes and thick soles.[48] Interestingly, it was Edsel's shoes that Ralph Roberts, an automobile designer who worked with Edsel for many years, remembered about him so vividly more than 60 years later. "Edsel's shoes always glistened," he recalled.[49]

***

Unfortunately, one thing Edsel did not inherit from his father or his mother was their robust health. Henry was vigorous almost to the day he died at age 83, and Clara lived well into her eighties. The kindest thing people could say about Edsel's physical condition was that he was "delicate."[50] He seems to have been a healthy baby but, in a time when vaccines did not exist, contracted many of the diseases so prevalent during the era, some of which could have been fatal. During his childhood, Edsel contracted many of the typical childhood diseases but suffered many other maladies as well. He narrowly escaped a smallpox epidemic that swept through Detroit in 1895—Clara took him to stay with her father in Dearborn as soon as she heard of cases in the city.[51] When he was three, Edsel was treated for bronchial pneumonia, and when he was nine, he contracted whooping cough.[52]

Edsel also was afflicted with hives and was constantly scratching his forehead or the side of his face.[53] Years later, Edsel had an appendectomy, which was his first but by no means his last operation.[54] However, Edsel's most pernicious infliction may have been congenital. According to John Dahlinger, purportedly Edsel's half brother, Edsel was "born without a certain acid in his system."[55] Dahlinger may have been correct, or at least partially correct. There is only one acid in the human system, and that is hydrochloric acid in the stomach. It is possible for a person to be born without the ability of producing hydrochloric acid, a condition known as achlorhydria. Although not fatal in and of itself, achlorhydria has been considered to be a precursor to stomach cancer.[56]

We will never know if Edsel had this congenital disease, but what is sorrowfully evident is that no matter what his age, Edsel's health was always a problem to him. "As long as I knew him," recalled Jan Willemse, the chef who made lunch for Henry and Edsel for more than 20 years, "Edsel always looked ill."[57] His docile personality and introversion notwithstanding, Edsel's fragile health was to become his nemesis.

***

Ford Motor Company was an oddity among American corporations during Henry Ford's reign. Henry hated formal organizations and therefore ran his mammoth enterprise loosely. Even in those early days, well-run corporations had formal hierarchies with a chairman, president, vice presidents, and other executives and managers. However, at Ford, there were few titles, and the lines of authority were blurred. Edsel was president, but in reality, his father, who held no official capacity in the company—"I haven't a job at all!" Henry once quipped—held all the power.[58]

During the times that Henry did allow his son to manage the company, the employees were elated. They had grown tired of working for the old man. He was not keeping up with the competition, ignoring all suggestions from his executives to the contrary. It was always Henry's way, or no way. However, with Edsel, it was different. Edsel was well educated, methodical, had his pulse on the marketplace, and, unlike his father, wanted to move the company in a new direction. "There was something sound, stable, and logical about Edsel," explained the Ford sales manager.[59] This is what the employees respected in Edsel. They could talk to him, reason with him, and work together to arrive at decisions that were best for the company because "it was his nature to do it that way."[60]

Many who worked for Edsel over the years believed that if Henry had allowed his son to take over running the company in the late 1920s, Edsel would not have allowed General Motors to assume the Number One position in the industry, the way his father did. "We would have kept them behind us," explained a Ford executive. "We would have been out in front of them all the time."[61] Whether or not that would have happened is pure speculation, but the notion that Edsel Ford was instrumental in saving the Ford company—not only from the quagmire of the Great Depression, but also from his father's stubbornness—is highly plausible. "Without Edsel Ford's lifetime influence in every field of Ford Motor Company, there is suspicion that the company might have found itself in a much rougher position," explained Ernest Kanzler.[62]

This was Kanzler's way of saying, in guarded language that only a lawyer would use, that if it were not for Edsel Ford, Ford Motor Company would not exist today.

Whatever remnants of Edsel's memory survived after his death, they were almost obliterated by Henry Ford II when he began running the company shortly after the end of World War II. He brought in a group of executives from General Motors and proceeded to revamp Ford Motor Company into the image of General Motors. In the process, he destroyed anything and everything associated with his grandfather and his father. "It was as if Henry the Second wanted to wipe the slate clean, to start anew, forget the past," explained a former Ford executive. "He wanted to run the company on his own terms and not be associated at all with the old regime."[63]

Erasing the memory of Henry Ford would be impossible. After all, his name was stamped, painted, or otherwise attached to the radiators, grilles, and hoods of more than 200 million automobiles throughout the world. Although Henry Ford II was less egotistical, he never hesitated to point out that it was *his* name that adorned the company's world headquarters.

* * *

Almost 60 years after his death, Edsel Ford is more obscure than ever. The residents of Grosse Pointe, the luxurious suburb that lies on Lake St. Clair east of Detroit, know who Edsel Ford was—only because he lived there all of his adult life and because his grandson, who is named after him, lives there today. However, outside that affluent community and even in adjacent Detroit, few people recognize Edsel Ford's name. If asked, they would probably reply that he had something to do with a defunct automobile, not that he was Henry Ford's son. Some automobile buffs would know that Edsel was Henry's son, and also probably know that he was president of Ford Motor Company. However, most would not know during what period of time that occurred.

The last time the Ford Motor Company had any association with Edsel was in the late 1950s. With the tremendous sales volume that automobile companies enjoyed after World War II, and consumers' desire to buy bigger and more expensive cars, Ford developed a mid-size car and called it the Edsel.

Ostensibly, the name Edsel was chosen to finally recognize Edsel Ford's contributions to the company. In reality, the name was chosen because nobody could come up with a better name for the new product.

Before the new Edsel automobile was introduced, Ford public relations men were in the trenches, looking for good stories to promote the new product. Only one person, a long-time Ford dealer, brought up the man behind the name. "Put this down if you put nothing else," he told the company man. "It is impossible to visualize the Ford Motor Company today if there had not been an Edsel Ford. When changing times began to pass the company by, Edsel pulled it ahead by sheer force of his personality, against tremendous inertia and resistance."[64]

Those were incredible words about a man who was little known when he was alive, let alone a decade and a half later. However, the old dealer was right. As we shall see, Edsel Ford did indeed push Ford Motor Company toward the modern era, against unbelievable odds, and it is indeed unfortunate that it has taken so long to tell his story.

Chapter 2

A Son for Henry

In 1893, Detroit was a hustling, bustling city of more than 200,000 residents and the fourteenth largest city in the United States.[1] Located in the southeastern corner of Michigan, adjacent to the expansive Detroit straits, which connect the three upper Great Lakes to the lower two, the city was already the center of the U.S. industrial heartland. With its close proximity to abundant natural resources—timber lands and iron ore deposits in the north, and the waterway connecting the city to the larger population base on the East Coast—Detroit was producing lumber, cast iron stoves, chemicals, and railroad equipment for most of the country's needs. Thirty-five miles to the north, in the city of Flint, William Crapo Durant, later founder of General Motors, was producing horse-drawn buggies by the thousands. Henry Ford, chief engineer of the Edison Illuminating Company in Detroit, was preoccupied with a newly developed mechanical marvel—the internal combustion engine.

From the center of the city, located on the banks of the Detroit River, wide boulevards fan outward in a spoke-like fashion for miles in several directions: Gratiot Avenue to the northeast, Woodward Avenue to the north, Grand River Avenue to the northwest, and Michigan Avenue to the west. Less than two miles from downtown, just off Woodward Avenue, the Fords were living in a modest $45-a-month, two-room apartment at 570 Forest Avenue.

Today, thousands of cars a day speed along the Edsel Ford Freeway, which wiped out much of this part of the city in the late 1950s. The Detroit Cultural Center, which includes Edsel's beloved Detroit Institute of Arts, now stands close to where the Fords once lived. However, on the unusually spring-like day of November 6, 1893, Edsel Bryant Ford was born in the once middle-class neighborhood.[2]

***

Although the Fords were living modestly, they had never known a hungry day since the days that Henry's father, William, left County Cork, Ireland, during the great potato famine of the 1830s. William Ford and his family settled in Dearborn Township, a tiny community 10 miles west of the burgeoning city of Detroit in 1832. They took up farming and eventually owned hundreds of acres around the area.

In 1861, William Ford married Mary Litogot O'Hern, the daughter of a Dutch carpenter, William Litogot, and step-daughter of Patrick O'Hern, a resident of Wyandotte, another small community southeast of Dearborn.

Born in 1863, Henry Ford was the second child of William and Mary. Their first was still-born two years earlier, and they subsequently had five others after Henry: John, Margaret, Jane, William, Jr., and Robert.

Clara Bryant Ford was the third child born in a family of ten children. Her oldest brother, Frederick, was five years older than she was; her youngest brother, Edgar, was nineteen years younger.[3] Their father, Melvin S. Bryant, was of Pennsylvania Dutch heritage, and their mother, Martha Bench, was born in Warwickshire, England.[4]

***

About a year before Edsel was born, when the Fords were living in an even smaller apartment on Cass Avenue, Dr. David H. O'Donnell, a young physician fresh from Detroit's College of Medicine and Surgery, met Harry Bryant, Clara's brother, in a local cigar store. Harry was buying a box of cigars in anticipation of the birth of his first child, who was due at any time.[5] Harry and his wife had not yet selected a doctor to help with the delivery, but by the time he left the tobacco emporium, Dr. O'Donnell told Harry to call him when the time arrived.[6]

That night, Mrs. Bryant gave birth to a boy, whom they named Melvin, and Dr. O'Donnell was there to assist as promised. When Clara and Henry went to see their new nephew, Harry referred Clara to Dr. O'Donnell, and she became his patient for about four months.

When Clara went into labor, Henry called the doctor. Dr. O'Donnell strapped his doctor's bag to the handlebars of his bicycle and pedaled as quickly as he could to the Fords' apartment. Henry waited in the kitchen while the doctor attended to Clara in the bedroom. "I didn't run into any difficulty," recalled Dr. O'Donnell. "Mrs. Ford didn't give me any trouble at all. She never complained a bit." Nor did baby Edsel. "He was born without circumstance. He was a quiet, sweet baby. Not a fusser."[7]

Henry went to see his wife and new son, and he offered his finger to the newborn baby. Edsel's little fist opened instantly and curled around it. "You can see he's smart, Callie, by the way he knows me," Henry proudly explained to his wife, calling her by his favorite endearment. "He cost us twice as much as Grandpa Holmes charged to get me safely into this world, but he'll be worth it."[8]

For Edsel's delivery, Henry paid Dr. O'Donnell $10.00, and he hired a live-in nurse at $4.00 a week to help Clara while she recuperated.[9] Clara's mother also came to live with them for a short while to help her daughter.[10] Roy Bryant, Edsel's cousin who was eight years old at the time, also came to visit his new cousin.[11]

***

A few weeks after Edsel was born, Henry again moved his family—this time to a brick duplex at 58 Bagley Avenue, in the heart of the city. With living quarters on the first floor and bedrooms on the second, this residence was almost twice the size of their Forest Avenue apartment, which pleased Clara. What caught Henry's eye, however, was not the extra living space but the small brick coal shed in the back yard. As soon as Henry got his family settled, he cleaned out the coal shed and turned it into a small workshop, where he immediately began building a gasoline engine of his own design.

Henry had first learned of gasoline engines while working in machine shops in Detroit in the early 1880s. One of his coworkers at the James Flower & Brothers machine shop showed Henry an article in *World of Science*, which described in great detail the gasoline engine developed by German engineer Nicholas Otto.

Henry's first engine was simple enough. He made it from a piece of gas pipe, wire, and discarded lathe parts. It was sufficient for Henry to learn the basic principles of the internal combustion engine. He completed his crude model on Christmas Eve in 1893, after working on it for about a week in his coal shed-turned workshop. He was so anxious to see if it would run that he barged into the kitchen where Clara was preparing the next day's dinner. Tiny Edsel, only seven weeks old, was sleeping in his bassinet in the doorway between the kitchen and the living room.[12]

Henry would have started the contraption himself in the workshop, but he could not simultaneously turn the engine's flywheel, adjust the spark, and pour gasoline into the carburetor; therefore, he asked Clara to help him. Henry pushed the dishes and cooking utensils out of the way, clamped the engine to the kitchen sink, and hooked the spark plug wire to the overhead light socket. He instructed Clara to delicately drip gasoline into the carburetor while he spun the flywheel and adjusted the spark.

Henry spun the flywheel once, which sucked air and fuel into the long cylinder. The kitchen light flickered as the wires sparked. The engine coughed and sputtered but did not start. Henry spun it another time, with another flicker. This time, the engine began running on its own, spewing flames, belching smoke, and sending vibrations and shot gun-like blasts throughout the house. It ran for 55 seconds.[13]

Henry was ecstatic with his success, but the commotion startled Clara and baby Edsel. Edsel awoke, screaming and crying as if he hoped someone would comfort him. Clara, shocked by the loud explosions, almost dropped the can of gasoline when the engine first started. She was afraid that the exhaust fumes, which quickly filled the room, would poison her new son. Thus, she quickly snatched Edsel from his bassinet and ran him upstairs to the safety and fresh air of her bedroom. By the time Clara had calmed Edsel and came downstairs again, Henry had returned to his workshop and was making plans for his next machine.[14]

***

Instead of making another one-cylinder engine, Henry started construction of a two-cylinder engine that would be large enough to power a bicycle. He roughed out his plans in a few days and began his new project before the beginning of the New Year.

Because the second engine was more complicated and more refined than the first engine, it took Henry until the fall of 1895 to complete it. The new engine ran well enough, but it proved too heavy for a bicycle frame. Therefore, Henry was forced to abandon that idea to undertake a more ambitious one in which the engine would be mounted into a four-wheeled frame. He called his new contraption a quadricycle.

Henry worked on the quadricycle in every spare moment. Even after working 12 hours at the Edison plant, he would come home for a quick supper and then work until eleven o'clock or midnight, until he completed his horseless carriage.

Before Henry wandered out to the workshop, he would spend some time playing with Edsel. Father and son would roll on the floor together, and Edsel would get a kick out of crawling all over his father.[15] If Clara and Edsel wanted to see Henry further, they visited him in the workshop. After doing the dishes, Clara would take Edsel to the workshop, and the two of them would keep Henry company while he worked. By this time, Edsel was walking and probably gave his father more company than he wanted. The toddler was constantly underfoot, crawling under the contraption or climbing on top of it, exploring the mechanical mysteries of his father's horseless carriage. Henry kept both the shop and the machine immaculately clean, so Clara did not mind her son crawling around it. Her only consternation was that Edsel repeatedly burned his little fingers on two brightly polished brass valves, which protruded from the top of the engine and became hot every time Henry fired up the engine.[16] Although hot, the shiny brass valves were too irresistible for Edsel to ignore.

As she sat in a chair, doing needlepoint, Clara kept a close watch over Edsel to make sure nothing more serious happened to him in the workshop. While Henry was pleased that his wife and son were with him in the workshop, he could not afford to take the time to keep a close eye on Edsel. He was always too engrossed in what he was doing to assume that responsibility. That was proven one evening when Clara left Edsel alone with Henry and his friend, Jim Bishop, while she went into the house for more thread. While Henry and Jim were working intently on the horseless carriage, Edsel managed to sneak to the corner of the shop and began playing with a full can of gasoline that Henry had stored there. As Edsel was lifting off the lid of the can, Bishop noticed Edsel from the corner of his eye. He ran over and grabbed Edsel in horror, expecting an immediate explosion.[17] Fortunately, nothing happened, but after that incident, Henry never allowed Edsel to remain in the shop without Clara.

***

Edsel was sound asleep when his father finally got the quadricycle ready for its maiden run in the early morning hours of June 4, 1896. As Clara related the story years later, "Henry came running into the house to get me and to wake up Edsel so that he too could witness the glorious event."[18] There was a light rain falling in Detroit, and Clara held Edsel in one arm and an umbrella over Henry's head with the other, as Henry pushed the spindly vehicle out of the coal shed. Henry spun the large flywheel, the motor came to life, and he jumped into the seat and drove off.

Henry's first trip in the quadricycle was not long, however. One of the "ignitors" on the engine broke, but Henry was able to make the machine more reliable with a few more trial runs. Soon, Henry was taking Clara and Edsel with him on excursions through Detroit. Edsel sat on Clara's lap for the first few trips, but eventually Henry built a seat in which Edsel could sit—on the left side of the vehicle and between the front wheels and the steering tiller. The seat faced backward, so when Edsel sat in it, he faced his father. Edsel held onto the steering tiller to keep from falling off.[19]

During one Sunday morning trip, Henry drove Clara and Edsel to the western edge of Detroit. It was only another 10 miles or so to the Ford farm in Dearborn, so Henry decided to put his quadricycle to the test and headed for the farm. None of Henry's relatives were expecting him. Margaret Ford was outside gardening when she noticed her brother's contraption coming down the lane. "The wheels on one side were deep in a rut made by farm wagons, while the wheels on the other side were high in the center of the road," she recalled. "Clara and Edsel were with him. Edsel was a small boy in dresses at the time, and he was held tightly by his mother on her lap."[20] (Obviously, Clara was uncomfortable with Edsel sitting by himself in the quadricycle.)

After that, the Fords took many jaunts to the farm in Henry's new motorcar. Their last trip, the one they took before Henry sold the quadricycle to raise money to build his next automobile, was made not so much to show off the durability of the car as it was to show off Edsel's first pair of trousers. The boy was so excited about them that as soon as Henry pulled up to the farmhouse gate, Edsel jumped out of the car and ran to his Aunt Margaret and Aunt Jane, who were standing on the porch, waiting for them. "Maggie, Jane! Maggie, Jane!" shouted Edsel as he ran toward them. "I come all the way out here with my pants on!"[21]

***

Unfortunately, Edsel was too young to remember the explosions of his father's first engine, burning his fingers on the bright brass valves, or taking rides in the quadricycle. "I did not know anything about the first automobile my father made," Edsel recalled in an interview more than 30 years later. "My father told me that I rode in it, but, of course, I now recall nothing about it. But I do remember well that the mayor of Detroit came to see the first machine, because I was standing at the window, watching for him to come. I also recall playing in Bagley Avenue, mostly in the middle of it."[22]

Chapter 3

The Early Years

"The family lived within its means,
and there were no servants
and no luxuries."[1]
Samuel Crowther

A month after Henry Ford had taken his quadricycle on its maiden voyage, the Association of Edison Illuminating Companies held its seventeenth annual convention in New York City. As chief engineer of the Detroit Edison Company, his boss, Alexander Dow, asked Henry to attend the convention with him. Henry had never met Thomas Edison, and he looked forward to meeting him and possibly talking to him about his new motorcar.

It was Henry's first time in New York and his first time away from his family. Clara did not want to stay by herself with Edsel while Henry was away, so she and Edsel visited her parents in Dearborn. Although Henry was gone for only a week, Clara wrote to him every day. They were not long letters—simply notes telling Henry how much she missed him and keeping him up to date on Edsel's antics. "It is 5 P.M. now and the baby is having a glorious time on the lawn," she wrote on August 13th. "They have been cutting the grass and he enjoys it, I can tell you. I asked him if he would like to send Papa a kiss and he said yes paper him over one. Just like one of his speeches, isn't it?"[2]

***

Henry was delighted to receive these notes of endearment from his wife and to be kept up to date on his little boy's antics. However, at this particular time, Clara's letters were less important than what had happened to Henry just before returning to his hotel room. That evening, at a banquet culminating the activities of the week, Henry Ford met Thomas Edison.

Edison was sitting at the foot of the long banquet table, and Henry, a junior employee, was seated several seats away from the great inventor. Toward the end of the evening, discussions turned to electric carriages. At that point, Dow pointed to Henry Ford and told Edison, "That young man there has made a horseless carriage—a gas carriage." Edison was intrigued and asked Henry to sit by him and explain its operation. Henry Ford explained the scene years later[3]:

> *When I had finished explaining it to him—and I remember that I drew my plans out for him on a menu card—he banged his fist on the table and said, "Young man, that's the thing! You have it! Keep at it! Electric cars must keep close to battery stations. The storage battery is too heavy. Steam cars won't do either, for they have to have a boiler and fire. Your car is self-contained—carries its own power plant—no fire, no boiler, no smoke, and no steam. You have the thing! Keep at it!"*

Henry returned to Detroit, full of enthusiasm and determined to build an even better motorcar. "You are not going to see very much of me until I am through building this car," he told Clara.[4]

***

Toward the end of the year, the Detroit Edison Company increased Henry's pay to $160 per month, which was an above-average salary for the time, and he used the extra money to build the second automobile that he had mentioned to Clara. Henry had sold his quadricycle that fall and had already begun working on its successor. The money from the sale of his first motorcar (he had received $200 for it) had already been used, and Henry was spending most of his extra earnings on his new car.[5] However, Clara, ever the vigilant housewife, watched her husband's expenses closely and managed to save a few dollars during this same period.

One of the ways the Fords saved money was by moving to cheaper and cheaper houses. Prior to their Bagley Avenue residence, Henry and Clara had lived in six other houses and apartments, each one being less expensive to rent than the previous. While Henry started work on his second car in the now famous Bagley Avenue coal shed, he continued working on it at their eighth residence at Alexandrine Avenue and finished it at their ninth residence at 192 Second Avenue.

***

Meanwhile, Henry had interested a number of prominent and wealthy businessmen in his new motorcar. Together, they started the Detroit Automobile Company on August 5, 1899. Henry was made superintendent of the fledgling company, and ten days later he quit his job at Edison, devoting full time to the automobile business. It was a risky move, but as Henry recalled many years later, "I had to choose between my job and my automobile. I chose the automobile."[6]

While Henry was busy working in his new company, Clara took Edsel to visit her brother Milton Bryant, who was a salesman in Louisville, Kentucky. Because Clara was the oldest girl of the Bryant family, she kept tabs on the entire family. When she learned that her younger brother was homesick, she decided to visit him in the spring of 1900 to cheer him up.

Although Henry had plenty to keep himself occupied, he missed his family. After working long hours, he would come home and write letters to them. On Detroit Automobile Company letterhead, Henry wrote to his son[7]:

My Dear Little Son

I am well. And hope you are all O.K. Say do they carry whiskey jugs and bandwagons in their blouse in K.Y. I hope you are having a good time and will be back soon for I am lonesome [sic]

From your loving
PaPa

The letter was jovial, in spite of the dire conditions in which Henry Ford had found himself. For more than a year, he had been vacillating over what type of automobile his company should build. The easy route—the one that his investors supported—would have been to simply duplicate his second automobile. This would have put the company into production quickly. Instead, Henry set out to design a delivery van, supposedly hoping to capture the interest of businessmen.

Approximately 20 vehicles were produced, but the they were not well made—at least not according to Henry's exacting standards. They also were too expensive, and consequently they did not sell well. When Henry approached his backers for more money to improve the quality, they refused, and the company went out of business shortly after Clara and Edsel returned from Louisville.[8]

***

With the Detroit Automobile Company now defunct, Henry turned to the idea of building a racing car. This may seem like a diametrical turn because Henry had not yet perfected a motorcar for production. However, even at this early date, the public had become enamored by fast cars, and people began equating speed with quality. It was already well known that the publicity alone after winning a race was worth untold sums in advertising.

From Henry's perspective, he was not confident enough in his designs to put them into the marketplace, and he felt he could improve the quality of his designs by testing his ideas in racing cars. If he could solve the problem of speed—the hardest possible automotive problem at the time—all other design problems would become easier.[9] Additionally, it was the fastest way Henry knew of to earn some sorely needed money.

In the summer of 1901, the Detroit Driving Club announced it would sponsor a race in the fall. The race would be open to all types of motorcars—electric, steam, and gasoline propelled. The prize was a beautiful crystal punchbowl and $1,000. Henry could not have cared less about the punchbowl, but the money certainly would come in handy now that he was without a job and living on his savings.

With financial backing from some of the stockholders in the now defunct Detroit Automobile Company, Henry completed a racecar in time for the October race. Noted automobile racecar drivers were there, many of whom had already gained notoriety in the racing business, including Alexander Winton of Cleveland. Winton had been building and driving racecars for several years, and his current objective was to become the first man in history to travel a mile a minute.

The race of the gasoline-powered cars was slated to run for 25 miles. However, due to long delays caused by the other races during the day, the gasoline-powered car race was shortened to only 10 miles. Of the number of gasoline-powered cars that showed up that day, only two managed to make it to the starting line—Winton's and Ford's.

The race started, and Winton, being the more experienced driver, took an early lead. However, Ford's car proved to be the better machine. Halfway through the race, Ford caught up to Winton. On the eighth lap, Winton's car broke down, allowing Henry to pull ahead and cross the finish line first.[10]

The crowd went wild! Clara and seven-year-old Edsel were jumping up and down at the finish line as Henry raced past them. "Henry has been covering himself with glory and dust," wrote Clara to her brother Milton. "I wish you could have seen him. Also heard the cheering when he passed Winton. One man threw his hat up in the air and when it came back down he stomped on it, he was so excited. Another man had to hit his wife on the head to keep her from going off the handle. She stood up in her seat and screamed 'I'd bet 50 dollars on Ford if I had it!' "[11]

Ford was presented with the lovely punchbowl and the $1,000 purse.

***

To tide them over during these austere times, Henry asked his father, William, and his sister Jane, who still was not married, to live with him and his family. It was a matter of convenience for all of them. William, now in his seventies, preferred being with his extended family rather than living secluded on the Ford farm. For Henry's family, William paid the entire rent while he stayed there, thereby freeing desperately needed funds for Henry to use in his automobile ventures. Although there is no proof, it is likely that William also helped with other expenses.[12]

The financial arrangements worked well, but the living quarters at Second Avenue soon became too cramped for the five of them, so they moved to a larger residence on West Grand Boulevard. All the while, William continued paying the rent.

At this same time, Clara began a diary of the family's daily activities. Unfortunately, she kept up the diary for only a short time, but through her jottings we gain a detailed account of Edsel's early years[13]:

> *January 9, 1901:*
>
> *Worked all day getting settled [in West Grand residence]. Very tired tonight. Edsel found lots of his playthings that he had not seen for some time. Decorated his rocking horse with Xmas-tree trimmings.*
>
> *January 11:*
>
> *Snowed all day. Edsel got soaking wet. He and Grandpa played checkers. Edsel cheated awful and beat every game. Went to bed so full of laughs he could not say his prayers.*
>
> *January 13:*
>
> *Edsel and I went to Sunday School. Came home, had dinner, then Henry fixed Edsel's old sleigh to take him coasting, but Edsel would not go, said sleigh was no good. He was sent up stairs for punishment for his pride. He was sorry.*

January 15:

Edsel started piano lessons for first time this morning.

January 19:

Henry bought Edsel new coaster.

January 30:

Edsel made snow fort and decorated it with flags.

Clara's sister, Katherine, was expecting a baby in early 1901. Thus, in February, Clara took Edsel with her to stay with Aunt Kate to help her with the delivery and afterward. As always, Clara wrote to Henry every day, and she also continued making notations in her diary.

February 2, 1901:

Edsel had a fine time with Uncle Sam feeding cows and lambs and giving pigs chocolate ice cream soda as he calls it.

February 3:

Snowed all day. Wrote a letter to Henry. Felt pretty lonely. Kate said she wished it was not Sunday so we could play cards.

Here is the letter Clara wrote to Henry:

Dear Husband—

I will write you a few lines today to let you know how we are. We got here all right. Sam [Kate's husband] was at depot to meet us got home at nine found Kate in best of health. It is storming hard mostly sleet. Edsel is cleaning the snow off the steps, with the fire shovel. He follows Sam all over. Sam, Edsel and Frank went down in the woods this morning and shot a rabbit. I have thought of you so much today. What are you doing? I suppose it seems strange for you to be staying there without us. Are you lonely? I am. I suppose I will have it bad if I have to stay two weeks.

Good bye dearest—

Clara

Again from Clara's diary:

February 4:

Sam, Frank, and Edsel went for cutter ride to Jasper. Played cards all night. Edsel had a bad attack of chilblains.

And again, another letter to Henry:

> *Dearest:*
>
> *Last night the snow drifted so that they had to hitch up the sleigh and drive it around to make paths. Edsel is as happy as a bird toddling around in the snow. I have taken some snap shots of him. He has been making some caves Sam picks him up and throws him up and lets him come down in a bank. Kate is starting dinner and is as lively as ever, but we don't know how long it will last. She says come if you possible can. I say so too.*
>
> *C.*
>
> *Good bye honey*

Kate had her baby two days later, and Clara and Edsel returned home at the end of the month, as Henry was finishing his latest motorcar. He took Clara and Edsel to see it in mid-March, and about a month later, the three of them were making regular trips in the new machine.[14]

While Henry benefited from his father and sister living with them, Edsel got to know his grandfather Ford and Aunt Jane well during their time together. Mary Litogot Ford, Henry's mother, had died many years earlier, so Edsel never knew her. However, both Bryant grandparents lived well into their eighties, so Edsel got to know them well. Clara was always taking Edsel to the Bryant homestead to visit Grandpa Milton, Grandma Martha, and all nine of Clara's brothers and sisters. He and Clara spent many summer weekends with the Bryants, and whenever Henry went out of town on business, they would stay with the Bryants as well. There are many photographs of Edsel, still in piglets and dresses, frolicking with the Bryant clan on their Dearborn farm.

***

While the Fords were not living from hand to mouth during these early years of Edsel's life and Henry's experiments with automobiles, they were austere. Clara watched every penny Henry earned and did a stellar job in managing a limited budget, especially in the light that Henry wanted every extra cent to go toward his motorcars. Grandpa Ford's rent payments helped considerably, and he probably contributed more financially than what is known. In the summer of 1901, the Fords took a month-long trip with Clara's mother to Niagara Falls—perhaps Mrs. Bryant helped defray the expenses. However, by the time Christmas was approaching, eight-year-old Edsel was concerned about what he might get that year[15]:

> *Dear Santa Claus*
>
> *I havent had any Christmas tree in 4 years and I have broken all my trimmings and I want some more I want a pair of roller skates and a book I cant think of any thing more I want you to think of something more too by. (sic)*
>
> *Edsel Ford*

Regardless of his concern, Edsel had a fine Christmas. Henry got Edsel the roller skates he wanted, not one but several books, new clothes, and lots of toys. Henry even bought Edsel a Christmas tree, although they were in extremely short supply that year. In all of Detroit, only 15,000 trees were available, and Edsel got one of them.[16] Edsel gave his father a set of crystal cups for Christmas to go with the punchbowl he had won in the Winton race.[17]

Edsel's friend, Helen Gore, was also one of the lucky ones to receive a Christmas tree that year, and she and Edsel helped decorate each other's tree. They decorated Edsel's tree with strings of popcorn and cranberries and "loads of tiny little bulbs that resembled world globes." They covered Helen's tree with candy canes and glitter. "I can remember the two of us sat in the front room behind my tree and ate all the candy off of it," Helen recalled.[18] Obviously, she and Edsel did not stop with her tree, thus prompting Edsel's comments in his letter to Santa.

Edsel had met Helen when he and his family moved to Alexandrine Avenue. He and Helen played together almost every day when Clara visited Mrs. Gore in the afternoons. Both children were about four years old when they met, and they remained good friends throughout their lives. In winter, Edsel and Helen sleighed together; in summer, Clara would take them to the Ford farm for picnics.[19] Grandpa Ford would hitch up Edsel's pony cart so the two could take rides around the yard, and, after that, Helen and Edsel would chase the chickens.[20] On rainy days, Helen would come over to Edsel's house, and the two would sit at the top of the stairs and watch the rain patter against the skylight. As it continued to rain and they became bored, they would come downstairs, take Clara's ironing board from the pantry, and improvise a makeshift slide by leaning one end of the ironing board against a chair.[21] At other times, they passed the dreary days by baking Edsel's favorite concoction, which he called *marguerites*. "Edsel liked to cook and fuss around in the kitchen," recalled Helen, "and under Clara's watchful eye, he would make this delicacy from soda crackers, nuts, and sugar."[22] Regardless if the weather were good or bad, when Helen was at Edsel's house during lunch, Clara invariably would make them a batch of alphabet soup. Helen and Edsel would eat the soup at Edsel's own little kitchen table. "We learned our ABC's that way," Helen recalled.[23] If they still needed something to do after lunch, they entertained themselves by sliding down the banister, much to Clara's chagrin.[24] Clara was not afraid of the children getting hurt, but she was concerned that one of them would land in the middle of the crystal punchbowl, which Henry had won at the Grosse Pointe race, located at the foot of the stairs.

When the Fords moved to West Grand Boulevard, Edsel became acquainted with another little girl. Faye Beebe was about a year younger than Edsel, and the two got along splendidly. Faye was "the noisy one," and Edsel was "rather quiet."[25] If Faye was not at Edsel's house, he was at her house. They were always doing something together. When they found themselves with nothing to do one afternoon, Faye coaxed Edsel into climbing one of the apple trees in her back yard. At first, he hesitated because he had never climbed a tree. However, when Faye started climbing, he was right behind her. Edsel should have followed his instincts, because as he climbed down the tree, he lost his grip and fell to the ground, breaking his arm as he tried to cushion his fall.[26]

Before Edsel's arm had a chance to heal, the Fords packed up again and moved to another Bagley Avenue-style double house located on Hendrie Avenue. Edsel was sad to leave Faye, but the Fords and Beebes had become good friends, so they continued seeing each other long after the Fords had moved. Even after Henry became rich in the automobile business, the Fords continued seeing the Beebes. Occasionally on Sunday afternoons, if the weather cooperated, the Fords would climb into a new Ford touring car and pick up the Beebes. Then they all would take a leisurely drive to Grosse Pointe or Mount Clemens. Edsel sat in front between his father and Mr. Beebe, while Faye sat in back with the women.[27]

***

Henry's underdog victory over Winton rekindled the interest of some of the original stockholders in the Detroit Automobile Company. They began questioning their hastiness in not allowing Henry to continue on his path of building a better car. Surely, his new racing victory was proof that he knew how to design and build automobiles, and they were prepared to give him another chance.

With as many stockholders as there were with the Detroit Automobile Company, the Henry Ford Company was established in January 1902. Ford was made chief engineer and given a $10,000 interest in the company, in recognition of his value to the new organization. The firm established a factory on Cass Avenue in Detroit.[28]

Clara was relieved that Henry again had a source of income, although in a letter she wrote to her brother Milton, she seemed weary that it would last long[29]:

> *Henry has worked very hard to get where he is. That race has advertised him far and wide. And the next thing will be to make some money out of it. I am afraid that will be a hard struggle. You know rich men want it all...* "

In light of Henry's prize money and the salary he was earning with the new company, William and Jane returned to the Ford homestead. Clara, who had sacrificed much for her husband, wrote to her brother Milton in Louisville, telling him of the good news[30]:

> *We are keeping house again and very happy to be alone. We have a very nice cozy little house. We did not build on account of Henry building the racer. He could not see anything else. So we have to put up with rented homes a little longer. We got Edsel a bicycle for his birthday. He rides it to school and thinks it is fine. He and Henry both have raglan overcoats. Edsel thinks he is as big as his father. You would laugh to see him imitate Henry.*

However, the Fords' new independence was hardly secure. The racing bug had bitten Henry, and he first wanted to build a bigger, faster racing machine and then turn his attention to designing a vehicle that the new company could manufacture in volume. Unfortunately, his investors were not interested. They brought in expert machinist, Henry M. Leland, as a consultant to keep the project going. Insulted by the investors' move, Henry Ford resigned from the company in March but apparently on good terms. They allowed Henry to take his plans for his new racecar, and they gave him $900 for his efforts. That was no small sum, for it was enough on which Henry's family could live for more than half a year.

As for the Henry Ford Company, it was reorganized under Henry Leland into the Cadillac Motor Car Company and became one of the earliest successful motorcar manufacturers in the country. At the time, Henry did not give much thought to the reorganized company; however, years later, memorics of how Leland forced Henry out of the company bearing his own name came flooding back to Henry. As we shall see later, Henry acted with resentment when Leland asked him for help.

***

For the moment, Henry was undeterred. He had found another backer—one who was interested in getting into automobile racing. Tom Cooper was in the crowd that had watched the Ford-Winton race. He was a champion bicycle racer who had earned tens of thousands of dollars, both from the many races he had won and as a representative of bicycle firms that used his notoriety in promoting their products. After watching the Ford-Winton race, Cooper thought he would like to change professions from that of a bicycle racer to an automobile racer. He looked up Henry Ford and convinced him to build two racecars—one for him, and the other for Henry. Cooper would finance the entire project.

Henry quickly agreed, and the two rented a shop at 81 Park Place in downtown Detroit. Henry began work on two racecars—the *999*, named after a well-known New York Central express train that had made a record run from New York to Chicago, and the *Arrow*, also named for an express train. Henry's goal was to have the cars ready for the next year's Grosse Pointe race.[31]

By October 1902, only the *999* was ready. It was a fast and powerful machine, developing almost 80 horsepower. In preliminary runs, the car proved to be so powerful and so fast that neither Ford nor Cooper wanted to risk their lives driving the contraption. Therefore, they recruited Barney Oldfield, a cyclist friend of Cooper's and an all-around daredevil, to drive the monstrous racer.

While many races were held that day at Grosse Pointe, the crowd was there for one race—the second match between Henry Ford and Alexander Winton.

When the race began, Oldfield immediately gave the *999* full throttle and never looked back. Winton held his own for a short while but eventually gave up. Ford's car won easily.

In celebration of his new victory, Henry took Clara on a long-deserved holiday in January. They left Edsel behind, under the care of Henry's sister, but they kept in touch with each other through telegrams and letters. Edsel was only nine years old at this time, but in a letter he wrote to his parents, we see that he is already adept at writing[32]:

> *Dear papa and moma,*
>
> *I received your telegram this PM. We went to the Temple theater the show was good but I have seen better ones. There was a train wreck last night. Miss Gould of Taylor-Woolfenden was killed in it. The car was crossing the track when a Grand Trunk freight train shot the car a rod down the track. Aunt Jane came and stayed with me until Roy came from school. We made fudge last night. We have had the music box going every night since.*
>
> *I guess will close for tonight. Write Monday your little son ("of a gun")*
>
> *EDSEL*

Henry had learned a lot by building these racecars. With his objective attained, he again returned to the design of a manufacturable vehicle. In fact, he had been working on a production design concurrently with the *999* and the *Arrow*. However, he needed money to develop it and kept his eyes open for new investors.

In the summer of 1902, Alexander Malcomson, a local coal dealer, invited Henry to his office to talk about automobiles. Malcomson had known Henry for years. (He had delivered coal to the Edison plant where Henry worked.) He told Henry that he was thinking about investing in an automobile company, perhaps Winton's. Henry quickly pointed out that Winton had not done too well against him in two races, and that he was working on a new design of his own. Shortly thereafter, Malcomson agreed to forsake Winton and provide funds to Henry to enable him to complete his new design.

During the next year, Henry, with the assistance of several other mechanics, completed his new design, the Model A, and on June 16, 1903, the Ford Motor Company was incorporated. Henry Ford was finally in the automobile business for good. From that moment, the Fords never wanted for money.

***

Early the next year, in an effort to boost publicity for his new company, Henry Ford announced that he would try to break the world speed record in the *Arrow*, which he had finally completed. Unlike the last race, however, Henry Ford would be the driver.

On Saturday morning, January 9, 1904, accompanied by Clara, ten-year-old Edsel, and cousin Roy Bryant, Henry took the *Arrow* to frozen Lake St. Clair. He had the snow scraped off the ice on the lake, and he and fellow mechanic Edward "Spider" Huff, taking their lives into their own hands, took off on the frozen lake to see how fast the car would go. They were astonished. According to their calculations, the monstrous car sped across the frozen lake at more than 100 miles per hour! However, that was an unofficial record. Three days later, with officials from the American Automobile Association present, Ford and Huff made another attempt, this time attaining a speed of slightly more than 90 miles per hour—not as fast as their earlier run, but a new world record nonetheless. Not only had they achieved Winton's goal of a mile a minute, they beat it by a full 20 seconds![33]

The notoriety was helpful to the company, but it was Henry's last foray into racing. He had proven to the world that he could build a powerful racecar. Now he wanted to prove that he could also build a reliable production automobile, taking lessons he had learned from building record-breaking racecars.

***

Years later, when Henry Ford was asked to reminisce about these years of austerity and experimentation before the founding of Ford Motor Company, he always gave credit first and foremost to his wife. Clara was ever the supportive wife, foregoing her own house, nice clothes, and secure income, placing her faith in her husband. Henry gave her the nickname "The Believer."[34] However, Henry likewise did not lack compassion. Although Clara would have allowed him to spend all of their extra money on his automobiles, Henry knew that having some money set aside was prudent, and he allowed the occasional extravagance. They bought a piano for Clara, went on an occasional holiday, and bought a camera. Until this time, photographs were taken primarily in photography studios; however, when portable cameras became available, they became all the rage to those who could afford them. Henry, in particular, realized the importance of capturing images for posterity, and he bought one of these portable cameras in 1893, shortly before Edsel was born.[35] Thus began Henry's affair with the recorded image, which lasted the remainder of his life. The Ford Archives in Dearborn, arguably the most significant industrial archives in the world, is filled with thousands of photographs—initially taken by the Ford family and later by the photography department of the Ford Motor Company—chronicling the history of both the Ford family and the company. Within this precious collection are photographs of Edsel. There is one photograph of Edsel when he was a baby in his crib, the same crib in which he was lying when his father fired up his first experimental engine on Christmas Eve. There are some photographs taken when Edsel was one year old, still wearing dresses. And there are many photographs of Edsel taken at the Ford farm and the Bryant homestead, surrounded by family. Not surprisingly, only a few photographs show Edsel with his father, because Henry was always away, working on his automobiles, while Clara tended to Edsel. However, two endearing studio photographs were taken of Henry and Edsel in 1902, when Henry was 39 and Edsel was 8. Both are dressed "to the nines"—Henry in the latest business fashion of the day, and Edsel in what appears to be a fashionable sailor's jersey. Edsel, with an angelic smile, holds on affectionately to his father, and Henry, with only a hint of a smile, appears proud of his son. The love between the two seems undeniable.

***

Similar to the days when he rode with his father and mother in the quadricycle, Edsel was too young to remember many of these important events in his father's life. When he was a grown man in his thirties, he was asked to recall those times when he was not yet a teenager. He said he could vaguely remember watching his father set the land speed record in the *Arrow.*[36]

> *... and I don't remember much about the Detroit Automobile Company, which later became the Cadillac Company. But I well know the one-story brick shop at 81 Park Place, where my father set up for himself to experiment and to build racing cars. There he built the* 999, *which Barney Oldfield drove. It was an 80-horsepower, four-cylinder machine. We still have it, and I have driven it, although never at full speed, for nobody knows just how fast it can go, and compared with its roar, an airplane engine just purrs! I watched that machine being built piece by piece. On its success hung my father's future as a maker of automobiles, although I did not know this at the time. I only knew that he and Tom Cooper worked on it day and night and that it was the thing principally talked about in our home.*

Chapter 4

From Bicycles to Bugattis

"Edsel never had any ham-on-rye days,
no what-are-we-going-to-do-
when-the-rent-comes-around days."[1]
William C. Richards

The first Ford Motor Company plant was located on Mack Avenue in downtown Detroit, about a mile and a half from the Fords' home on Hendrie Avenue. It was a small, wooden building, and the automobiles that Ford made there were essentially hand-made, one at a time. Production remained at Mack Avenue until late 1904 when the company moved to a newly constructed building specifically designed for the manufacture of motorcars. Located on the corner of Piquette Avenue and Beaubien Street on Detroit's north side, the spacious new factory was to become the birthplace of mass production and Henry's most famous car—the Model T.

During these years of company development, Henry Ford spent many of his waking hours at the plant. When Edsel was not in school, Henry would bring his nine-year-old son to the plant "an average of twice a week, sometimes more," recalled a worker.[2] From that point, "I have always been in and about the company and, in fact, have grown up with it," Edsel said.[3]

Too young to appreciate the workings of the factory, Edsel would bring his wagon and push himself along the factory floor in it as his father met with employees.[4] At other times, Edsel would bring his bicycle and "ride around the factory, dodging the posts."[5] "I remember seeing Edsel at the plant," recalled Richard Kroll, a mechanic at the Piquette plant. "I can see him now, dragging his ice skates along the cement floor. He'd come up to the plant from school and have me sharpen his skates—I'd give them a nice, hollow grind."[6]

As he grew older, Edsel went into the plant and helped the office workers stamp mail, put brass identification tags on machinery, and do other small tasks.[7] He also liked to visit Charles Sorensen, then a lowly pattern maker with the company. (Sorensen would later become the manufacturing guru at the company and Henry Ford's right-hand man.) Edsel would watch Sorensen make wooden models of various parts of the cars. As Edsel watched, he told

Sorensen of the things that he was making at school.[8] One day, Sorensen removed the tiller from Edsel's wagon and installed a steering wheel and a horn, making it resemble a little racecar.[9]

***

Until Edsel was old enough to fend for himself, Henry would not allow Edsel to go onto the factory floor unattended. Automobile factories were treacherous places in the early days. Small electric motors had not yet been invented; therefore, factory floors were crammed with drill presses, lathes, grinders, and other types of machine tools powered by long leather belts attached to overhead pulleys driven by a central power plant. Looking down the aisle of one of these factories, one would see hundreds of these exposed belts whirling from floor to ceiling. The curiosity of a small boy in this environment meant certain disaster, so Henry gave strict orders to all gate watchmen "not to let Edsel in."[10]

Eventually, the Ford factory became a second home to Edsel, as it had become to his father. Edsel spent a lot of time there after school, and he could hardly wait for school to be dismissed so that he could go to the Piquette plant and be with this father. With school books strapped over his shoulder, Edsel would walk into the plant office, throw his books on bookkeeper George Brown's desk, say "Hello, George!" and then head straight for the experimental shop at the rear of the third floor, where Henry was spending much of his time. Henry and his engineers—Joe Galamb, Charles Sorensen, Clarence Avery, and others—were in the midst of developing the Model T.

Expecting Edsel around that same time, Henry would look for his son. "Have you seen Ed?" he'd ask Brown. "Oh, yes, he's here," Henry would say, not giving Brown a chance to reply. "I see his books!" Away Henry would go to catch up with Edsel. Then, the two would remain in the experimental shop for hours, many times not leaving until everyone else had gone home.[11]

Galamb, Sorensen, and the other engineers, as well as Henry, spent the next six years teaching Edsel all they knew about running an automobile company—not from the executive suite viewpoint, but from the grease-stained floor perspective. Instead of giving Edsel a seat next to his in the business office, Henry had Edsel put on a pair of coveralls and go directly into the factory. "I told Edsel that he ought to know everything that was going on in the place," recalled Henry. "I told him that he would do better to spend two or three years just wandering around, seeing how everything was done and getting acquainted with the men, always keeping in mind that, some day, he would be called upon to administer it all."[12]

That is precisely what Edsel did. "He was all over the plant, studying it," explained a factory worker who witnessed Edsel's development. "He would go from department to department—you'd see him almost anyplace." Edsel did almost everything—from engineering and manufacturing to sales and finance. Eventually, Edsel had worked in every area or department in the company. He worked on the assembly line, repaired cars in the "up-fit" area, and designed components in engineering.[13] He worked long hours and never received special treatment because he was the boss' son. In fact, Henry put more pressure on his son than he did on anybody else. Instead of merely allowing Edsel to learn the various functions of each department or area, Henry would lope through the factory, deliberately tangle routines, create problems, and leave them for Edsel to solve. "Father put me through the mill," Edsel confessed later.[14]

***

By 1908, the Ford Motor Company was nearly five years old and an overwhelming success. Since its humble beginnings, the company had generated more than $40 million in profits, and a significant portion of that sum went to Henry Ford. The years of hard work and frugality had paid off. To celebrate their new-found fortune, the Fords built a beautiful mansion in the ritzy Boston Boulevard district in the northern part of Detroit. Built to Henry and Clara's specifications, their new mansion was an expansive red brick and stone structure that cost nearly $300,000. It was their first new home since leaving Dearborn 17 years earlier. It had a full basement, large living and dining rooms, a large den, three spacious bedrooms on the second floor, and servants' quarters. On the third floor was a cavernous room where the Fords held dancing parties and Edsel played in winter. One of Henry's favorite features was the tunnel that ran from the main house to the coal-fired boiler and to the four-car garage. Clara's favorite feature was her magnificent rose garden, which stretched outward from the west side of the house all the way to the street and back alley. Clara set up her player piano in the family room, and when she was not playing the piano or Edsel was not practicing on it, the family would crank up the piano and listen to the latest tunes.[15] Located between the main building and the garage was a turn-around for automobiles, in the center of which stood a beautiful fountain.

Edsel's bedroom, located on the second floor above the downstairs hall, projected over the front porch overlooking Edison Avenue. In the front left corner of the bedroom was an alcove where Edsel read and studied, and to the side was his own bathroom, complete with tub and shower.[16] Hanging on his bedroom walls were drawings of automobiles that he had designed in school, and a frieze of automobiles ran around the ceiling of the room.[17] Although very young then, Edsel's cousin, Catherine Ruddiman, recalled seeing pictures of "cars that Edsel had designed in mechanical drawing classes at the Detroit University School...hung on the wall of his room."[18]

Edsel invited both of his old friends, Helen Gore and Faye Beebe, to his new home many times. "One day we were having lunch," explained Helen Gore. "For dessert, we had strawberry and vanilla ice cream. Edsel reached down and pulled some of the blue nap off the rug, put some around his plate, and said, 'Now we're patriotic.' "[19]

***

In spite of their growing wealth, the Fords were a "normal family," explained Henry's sister. Henry and Clara provided Edsel with "a wonderful home life," and they both "loved Edsel dearly."[20] For example, Henry would do his best to be home every night for supper, during which time the family would bring everyone up to date on their activities. Henry would explain what he was doing at work, Clara would share her day, and Edsel would tell his parents what he did at school.[21]

Henry, Clara, and Edsel did almost everything together. They went to church, visited friends, and went on holiday whenever they could. At a time when few Americans traveled far from their place of birth, the Fords had visited many scenic splendors in America by the time Edsel left high school: Niagara Falls, Pike's Peak, the Grand Canyon, and the giant redwoods.[22] They vacationed in Altadena, California, a small community near Los Angeles, nearly every winter during Edsel's formative years. They also went to Europe several times, taking their first voyage in 1912.

At home, the Fords' favorite winter pastime was ice skating, which they did on a pond on the Ford farm. Usually, Henry and Edsel did all of the skating, while Clara participated by cooking a big pot of oyster stew, the same kind that she and Henry used to enjoy at the Martindale House to keep them warm when they were courting.[23] Starting in the afternoon, they would skate well into dusk. In the old days, when Henry was courting Clara, he would build a large bonfire so that they could continue skating after dusk. However, Henry was now an automobile magnate. Thus, when it became too dark to see, he would have the pond surrounded by Ford cars that had their lights turned on, which illuminated the skating area as though it were day again.[24]

***

While Henry was busy building automobiles, Clara was busy rearing Edsel. That was her primary responsibility for the first 14 years of Edsel's life, which was a rather uneventful process. By all accounts, Edsel was an "awfully good boy," almost angelic.[25] He had his bouts of mischievousness, of course, such as the time he wanted a new sleigh rather than having his old one repaired, and the times he and his Aunt Margaret would playfully heckle the maids and "cause them a great deal of trouble."[26] However, that was as unruly as Edsel would ever become—the typical stubbornness of an eight-year-old boy and an occasional bout of mischievousness egged on by his aunt. That was simply Edsel's nature. He was quiet and shy, and Clara reinforced her son's intrinsic pleasant demeanor by teaching him "the way of simplicity and common sense."[27]

For example, when Edsel was still a boy in school, the vogue of the day was gloves with long gauntlets rising up over the arm. Of course, Edsel wanted a pair.

"Here's a swell pair, Mother, for twelve dollars," Edsel said as he and Clara walked through the Taylor-Woolfenden department store.

"Yes," replied Clara, "they are very nice, but too good for you. Your companions in school cannot afford twelve-dollar gloves. You must buy the same kind that they are wearing." Thus, Edsel had to choose a pair of gloves that cost only two dollars.[28]

Edsel was "very dear and very close to his mother, whom he affectionately called 'Callie,' " recalled Edsel's cousin.[29] Even after Edsel had left home, he would visit his mother frequently. His parents' Fair Lane estate was only minutes away from his office at the Ford Administration Building, so he would drop by the estate occasionally after lunch. All of the guards and servants knew Edsel, of course, so he would drive through the security gate, pull up to the front door of the limestone mansion, and walk right in, shouting to his mother, "Hi! It's me!"[30]

***

Around the time the Fords moved into their magnificent home on Edison Avenue, Henry took over the rearing of Edsel. Clara had done all that she could to instill proper manners, civility, and comportment in her son. Because of Clara's early direction, Edsel did not become a leisurely, self-indulgent playboy or a stuffy highbrow.[31] Now, his father began to set in place the tools for Edsel's "preparation"—to teach his son the skills to survive in a man's world—the world Henry himself was "overhauling." These were the tools Edsel used to gain the firsthand knowledge he needed to fit himself for the important duties his father would have waiting for him when he grew up.

In the large space above the garage of their new home, Henry outfitted a workshop for Edsel.[32] This was not an ordinary garage workshop with a bench, vise, and a few hand tools. It was an impressive machine shop replete with precision metal lathe, metal plane, metal shear, drill press, and numerous woodworking and hand tools.[33] Here, Edsel learned firsthand from his father the various manufacturing methods employed in the Ford factory. Henry and Edsel spent countless hours in the shop as Henry explained in detail how to operate each machine and handle each tool and how those lessons could be used in understanding the manufacturing methods employed in the Ford factory.

Henry and Edsel often spent Sunday afternoons in the workshop, building one thing or another. One of the first things Edsel built by himself in the shop was a wooden box in which to put crickets that he and his little cousin Catherine caught for his high-school science project.[34] He built all sorts of model boats and cars there, too.[35]

Not all projects went so smoothly, however. One evening after school, Edsel was trying to size a piece of metal in the metal shear when his hand slipped in front of the blade, slicing off the tip of his middle finger.[36] Biographers of Henry Ford have used this incident to prove that Edsel was no more mechanically inclined than his mother, which greatly disappointed Henry, who wanted Edsel to be just like him. A more accurate evaluation is that Edsel was still very young then and learning about the various machines and how they were operated. Furthermore, Henry had faced many close calls himself when he was a boy learning about machines. During Henry's first experiment with a steam-powered engine, for example, the boiler exploded, sending pieces of torn metal in all directions and severely cutting one of Henry's friends.[37] It was a miracle that Henry walked away unscathed.

Edsel readily admitted that he was not as mechanically adept as his father, but he understood machines. When it came to the theory of their operation, there were instances where Edsel knew more about them than his father knew. Edsel became a member of the Society of Automotive Engineers, read its technical material on motorcar design, and thereby kept abreast of the latest in automotive technology.[38] Henry did not, and his reluctance to keep up with the developments in the industry eventually would place his company in jeopardy.

Before all this came to pass, the garage workshop was Henry and Edsel's sanctuary. They went there not only to work but also as a place to talk privately.[39] Edsel learned the practicalities of life and how to solve problems using common sense, and Henry related to his son in the only way he knew: through machines.[40] Henry Ford lived by, through, and for machines; therefore, it is no surprise that the best way he related to people—including his son—was through machines. A sobering example of this characteristic became evident after Edsel's death. Having a difficult time coping with his loss, Henry had an exact replica of Edsel's Edison Avenue machine shop reconstructed, and he equipped it with all of Edsel's old tools. Although Henry could not bring back his son, he could at least walk among the machines and touch the hand tools with which he and his son had worked for so many years.

***

While Henry reveled in Edsel's mechanical interest, Clara saw another side of Edsel's nature. Edsel showed exceptional talent for painting, which Clara praised and encouraged. She took him to enjoy the works of the masters at the Detroit Institute of Arts.[41]

From rudimentary sketches of automobiles he made in elementary school, to oil paintings he did as an adult, Edsel's love of art started at an early age and continued throughout his life. In later years, he became a great connoisseur of fine art and held significant positions with the Detroit Institute of Arts. He also contributed greatly to the success of the Institute. Likewise, Edsel was musically inclined and tried the piano, violin, and drums, although without much success.[42] One time, Henry intimated to Faye Beebe's father that Edsel was very interested in music, especially the violin. "I'd like to have him take lessons," said Henry. Mr. Beebe replied, "Well, I'd be glad to make a violin for you to give to Edsel to use." We discovered later, however, from Henry himself, that his son never learned to play the instrument.[43]

***

There is a marvelous photograph showing Henry and Edsel posing in a 1905 Model F Ford, the fourth model in the so-called alphabet series of automobiles leading up to the Model T. Eleven-year-old Edsel is outfitted in a suit, with his hands on his knees and sitting proudly next to his father. Henry is at the wheel and also is handsomely dressed in a conservative three-piece business suit and derby. Where they were going or where they had been is inconsequential. What is important is that wherever Henry went, Edsel went with him. For example, on Sunday afternoons, Henry and Edsel would get into a new Ford car and meet Charles Sorensen and his son, Cliff, who were also in a new Ford car. Together, they would drive to Grosse Pointe, up Woodward Avenue, or to the Ford farm, ostensibly to enjoy each other's company. However, the real reason for the trip was to test the workmanship and performance of the vehicles.[44] Henry always seemed to combine business with pleasure.

In 1909, shortly after the Model T had been introduced, mining magnate Robert Guggenheim promoted an ocean-to-ocean race to celebrate the Alaska-Yukon-Pacific Exposition being held in Seattle, Washington. It would be a grueling race from New York City to Seattle, a distance of more than 4,000 miles. There had been other cross-country races in the past, but none terminating in the far northwest. Henry, who was well accustomed to racing and the publicity it afforded, wanted to test the endurance of his new Ford model and quickly signed up for the race.

Initially, 35 manufacturers entered the great race, but this number dwindled rapidly to only 14. However, at the time of the race, only five cars showed up: two Fords, an Acme, a Shawmut, and an Itala, the only foreigner in the group. To lighten their weight, the two Model T's were stripped of their bodies, leaving only the two front bucket seats. Frank Kulick, test driver for Henry Ford, and H.B. Harper, the editor of the *Ford Times* (the company's in-house publication), drove one car. Bert Scott and Jimmy Smith drove the other.[45]

The Acme was forced out of the race early, and the remaining competitors all made it to St. Louis at approximately the same time. From that point, the two Fords and the Shawmut alternated the lead position, the Itala breaking down in Cheyenne, Wyoming. The two Fords and the Shawmut met again in western Wyoming, but the Scott/Smith Ford took a commanding lead in Idaho and never relented.

Ford dealers along the route kept Henry Ford informed of the location of both Ford racecars.[46] Realizing that one of his new Model T's might win the great race, Henry, with Edsel beside him, took the train to Denver. There they met with the two managers of the Denver and Seattle Ford assembly plants, and all four of them motored up to Seattle in brand new Model T's—Henry and Edsel in one, and the two branch managers in the other.[47]

Henry and Edsel were at the finish line when the Scott/Smith Ford crossed it.[48] Edsel wrote a postcard to his mother, telling her of the experience[49]:

> *Dear Mother—*
>
> *The weather is fair. We are having a nice time, but wish you were here. Burt arrived last Wednesday and was cheered by 180,000 people. The streets were packed. Mr. Guggenheim is a fine fellow and the cup is great, but I'm afraid that we won't get it. The Shawmut has protested Scott for changing axles and Burt has protested Shawmut for leaving Toledo and Chicago before scheduled time. So there is a scrap on.*
>
> *Edsel*

Edsel's prediction came true, but not for the reason he stated. Judges disqualified the Scott/Smith Ford after learning that they had changed the engine in their car during part of their run.[50] Nevertheless, both Fords finished the arduous race, and the company "made hay" in its advertising because of it.

Two years later, Henry, Edsel, Sorensen, and a couple other Ford employees drove to Indianapolis in Model T's to watch the first 500-mile race. Thus began the tradition of the Fords attending the Indianapolis 500 nearly every year hence—with Edsel taking his sons there in later years.

Henry, Edsel, and Sorensen rode in one Model T, while three other engineers took another. They left the Fords' home at sunrise and took turns driving until they arrived in Indianapolis.

During one of Edsel's turns at the wheel, he came upon John and Gari Stroh, two friends from high school, who were driving a new Benz phaeton. (Their family owned the Stroh Brewing Company.) The roads were dusty and rough, and the Benz was not faring well—the Strohs had to keep their speed down for fear that a rut or bump might bend a wheel or break an axle. Tired of drowning in the Benz' dust, Edsel pulled down the throttle, advanced the spark, and passed the German car as if it were standing still.

The Stroh brothers did not know who had passed them until they caught up with the Ford troupe after they had stopped for lunch. They had not enjoyed being left in the dust, but they were embarrassed that their expensive import automobile could not do as well as the inexpensive Ford on American roads at the time.[51]

* * *

As we have seen, Henry pampered Edsel when he was a child by keeping him well supplied with toys.[53] When Edsel grew older, nothing changed except that his toys became larger and more expensive. It was obvious early on that Henry was not as strict as Clara had been. In fact, he allowed Edsel to do "about as I pleased," nearly reversing Clara's sensible and responsible approach.[54] What pleased Edsel was to buy every new kind of automobile that was introduced.[55] Although that was expensive, the Fords had plenty of money by this time to indulge Edsel's extravagance. Henry gave Edsel his first car, a 1903 Model A Ford Runabout, when he was only ten years old.[56] With the exception of not being strong enough to crank the engine (that was done by hand in those days), Edsel drove it with every competence.[57]

Edsel received his second car, the more refined Model N Ford, five years later. By this time, he was old enough and mature enough to operate a motor car on his own, but his father continued to watch him closely. The Model N was compact and simple to operate, but it was powerful enough to break a man's arm if cranked improperly. Concerned that Edsel might hurt himself starting his car, Henry ordered one of the servants to start the car every morning before Edsel left for school. He also instructed Edsel on what to do if he were not at home: "If you ever go downtown and you can't get it [the Model N] started, call your mother and she'll get in touch with Lou [their servant]."[58] These were the days before driver's licenses, stop signs, and traffic cops. The only vehicles on the boulevards of Detroit were streetcars and horse-drawn wagons and buggies, and the only restriction the city placed on driving motorized vehicles was the person's ability to buy one of the new contraptions.

The Model N was followed by a succession of other motorcars. "By this time," explained Edsel, "we began to have more money. After the velocipede [a three-wheeled German motorcycle that Henry had bought Edsel[59]] and my first automobile, there came another motor car, if you could call them that in the early 1900s. As I was able to handle them and knew everything about them in terms of performance, my father supplied me with one after another, of competing makes."[60] From records and photographs, we know that Edsel owned a Rolls-Royce, a Hispano-Suiza, a Locomobile, a Marmon, a Mercer, a Daimler, a Packard, numerous custom-bodied Lincolns, and a Bugatti during these early years.[61] At one time, he owned a total of 42 cars and commercial vehicles worth $50,000.[62] How many of these expensive and exotic automobiles Edsel owned in total is not known, but it was probably considerable. Soon, Edsel was more comfortable driving in one of these luxurious motorcars than in one of his father's Model T's. He certainly appreciated the styling of these magnificent cars, which was to spark Edsel's lifelong love of automobile design. "Naturally, I was interested in motorcars," explained Edsel. "How could I help it?"[63]

***

Although Edsel did not want for anything material, Henry and Clara were concerned because he was an only child. Coming from large families themselves, they understood the pleasantry that brothers and sisters provided. They felt that Edsel was lonesome and missing something because he had no brothers or sisters. To compensate for that, they regularly took Edsel to visit his cousins in Dearborn or had the cousins come to stay with him in Detroit. These visits lasted from a day or two to weeks, if Henry was out of town on business. In the fall of 1909, Edsel had an extended visit from his six-year old cousin, Mary Catherine Ruddiman, daughter of Henry's sister Margaret and her husband James Ruddiman, brother of Edsel's namesake. That summer, Catie's father died and Henry insisted that she and Margaret stay with him until they recuperated from their loss.

Catie was much too young to be the kind of companion that Edsel's parents wanted for him, but at least she was constant company. It was a little awkward for both of them for the first week, but eventually Catie became Edsel's "little sister."[64] (His experience in playing with Faye Beebe and Helen Gore enabled him to become acclimated to his new role of big brother quite readily.) Actually, Edsel enjoyed having his cousin in the house. For her birthday that winter, Catie received a pair of roller skates, and on many evenings after dinner, Edsel would take her to the roomy basement of the Ford mansion and teach her how to skate. "He was very patient with me," Mary Catherine recalled. "By spring, I tried the sidewalk and driveway. Edsel had a Kodak, and he took my picture under a tree in the yard. I remember one Sunday afternoon when I helped Edsel catch some crickets. We kept them in a glass jar, then went to his workshop above the garage. Then Edsel built a wooden box for the crickets. What would they eat? We asked Edsel's mother, and she said she thought they liked red flannel. She found a piece, we put it in the box. However, the crickets did not like red flannel or living in a wooden box. They died in a few days."[65]

What Mary Catherine remembers most about those days were the rides that Edsel gave to her in his retinue of automobiles. "That was great fun," she said.[66] Eventually, Mary Catherine and her mother moved into a new house that Henry had built for them.

In 1912, Clara invited her youngest brother, 26-year-old Roy Bryant, to live with them.[67] She may not have explained to Edsel at the time why she invited her brother to stay with them, but she confessed later that she knew it was difficult for a child to grow up alone.[68] "We thought it would be good for Edsel to have an older brother, one with whom he could have a close association, and I think it resulted as we hoped," recalled Clara.[69]

"Edsel and I weren't the same age, but we were close enough to be pretty good company for one another," recalled Roy. "We spent a lot of time together."[70]

Looking back, Roy obviously benefited much more than Edsel. "Clarie" thought she was doing her son some good, when in fact, it was her brother who needed the companionship. "Being the youngest in the Bryant family of 10, I was always kind of alone," he explained.[71] In fact, he was overjoyed at finally getting some privacy and some attention. If Edsel ever needed companionship, it certainly was not now. He had plenty of friends at school and was involved in all sorts of activities there. He also was working at the Ford factory in the afternoons and had been seeing a cute girl on a regular basis. "Call E.C.—Wednesday," he wrote in his calendar.[72] "E.C." was Eleanor Clay.

Nevertheless, the arrangement worked well for both Edsel and Roy. They went "nutting in the woods together," recalled Catherine, and they gave each other company when the Fords took Roy to England and Ireland with them in the summer of 1912.[73] There they visited Killarney, Dublin, and Cork, the area from where William Ford had emigrated in the preceding century.[74]

That December, the Fords took Roy with them to their winter home in Altadena, California.[75] In a letter that Edsel wrote to Joe Galamb, the chief body engineer at Ford, we get a rare glimpse into this time of Edsel's life and the interaction between him and his father[76]:

> *Well we are surely having a good time out here. The weather is very warm, about 85 degrees in the middle of the day. We can step out of our house and pick any amount of roses and oranges. I don't suppose you can do that in Detroit just at present.*
>
> *Today I was racing a Winton [in my Model T] and the speedster registered 54. Father and I have stripped another car and we are going to try out those Holley carburetors on the Mount Wilson trail, when they arrive.*
>
> *I think Los Angeles is the greatest automobile city in the country. It has more machines than Detroit or even Chicago. An awful bunch of Pierce-Arrows, Packards, Fiats, Peerless and, in fact, some Fords.*

While Edsel and his father were building speedsters, Roy was making the social rounds, and he met a charming girl at one of the local resorts. Shortly after he and the Fords returned to Detroit, Roy moved out of the Ford home, returned to Altadena on his own, and married the girl.

***

Whether he was lonely or not, nobody could have been luckier in the "gene lottery" than Edsel Ford. An early account predicted how fortunate his life would be, especially in terms of financial security. When he was ten years old, Edsel went to a friend's birthday party. The hostess made a caramel cake and baked into it seven pennies and a dime. "Wouldn't you know it that Edsel would get the cake with the ten cents in it?" recalled one of the guests many years later.[77]

From toys to motorcars, Edsel never wanted for material things, and Clara had instilled in him a sense of propriety and common sense. However, nothing could have prepared Edsel for what Henry gave to him for his twenty-first birthday.

That morning, as the banks opened, Henry took Edsel to a downtown bank and asked the manager to show Edsel the birthday present that he had prepared for him. The manager motioned Edsel toward the vault, opened it, and pointed to $1 million in gold stacked in the corner. "It's all yours, son," Henry said.

Being the sensitive person that he was, Edsel was more abashed by the gift than overjoyed. When he returned to work, he made more errors the remainder of the day than he ever did before or after that day.[78]

"He was a boy clean in mind and habits, and of exceptional strength of character," explained a close friend. "Few young men have been as little influenced by prospective millions as Edsel."[79]

Although Edsel respected wealth and appreciated all of the worldly goods that he had, few could have treated the lot into which he was born with such nonchalance. It almost seemed as though he were embarrassed by his wealth.

Chapter 5

College Is a Waste of Time

"Brainy Edsel."[1]
Notation in high school yearbook

In the fall of 1899, Edsel began his formal education when Clara enrolled him in Harriet Lodge's kindergarten for $2.50 per month.[2] At the time, the Fords were living on Second Avenue, and Miss Lodge's classroom was about eight blocks away. By the time Edsel began first grade at Farrand elementary school, the Fords had moved three more times. However, all of the moves were within the jurisdiction of Farrand, so Edsel was able to continue at that school through fifth grade. Edsel had skipped the second half of second grade and went directly into third, thereby finishing fifth grade a half-year early.[3]

Obviously, Edsel was a bright and diligent student. His intelligence was inherent, getting most of it from his father, whom Charles Lindbergh called a genius.[4] However, the credit for channeling and developing his intellect must go to Clara. Throughout Edsel's years in elementary school, Clara made Edsel do his homework while seated at table, so that she could keep a close watch on his progress and answer his questions as she made supper.[5] "Edsel had a hard time making the letter *E* made it too much like *L*," Clara wrote in her diary.[6] In later years, when the Fords moved into their stately mansion on Edison Avenue and Clara had servants to make the family meals, she allowed Edsel to study alone, either in his upstairs bedroom or in his father's den. However, she always reviewed Edsel's work after he finished. For studying hard and attending school steadily, Clara rewarded her son with little gifts of appreciation—a fountain pen, a small book, a writing tablet.[7]

Edsel must have received many rewards, for he developed strong study habits from the outset and was always an exemplary student. His grades throughout elementary and high school were always well above average. Frederick Searle, Edsel's eleventh-grade science teacher, recalled that Edsel was very good in physics, always had his homework completely done, and never failed to hand in his homework on time.[8] In addition to the sciences, Edsel also studied history, geometry, German, and French.[9] However, he was not a stellar student in all classes and, similar to most students, favored some courses more than others. Edsel did well in English and had an interest in the physics of the universe, but German was "hard," he said.[10]

***

Unlike Henry, who barely made it through the eighth grade and whose last book was one of the *McGuffey Readers*, Clara completed her mandatory schooling and loved to read. After Henry's business became successful and Edsel was in school, Clara had plenty of free time and went to the library almost every day and studied English literature.[11]

Edsel followed a routine throughout his school years. After school each day, he would go to the factory for a few hours, come home and eat supper, study, take a walk with his father, and then complete his studying or read.[12]

Edsel developed the same love for the written word as his mother had. He was a voracious reader all of his life, and his writing skills were clear and concise. Although precious little is available today to illustrate Edsel's intelligence, one surviving document clearly demonstrates how articulate he was, even at a young age. It is an essay that Edsel wrote in eighth grade about his father's new Piquette Avenue plant, and it illustrates his growing interest and knowledge of the automobile business[13]:

An Automobile Factory

> *This factory is in Detroit on the Michigan Central Railroad. It is 464 feet long and 75 feet wide and has 3 floors. The front half of the first floor is occupied by offices and the rear contains a shipping room and storage room. On the second floor is the experimental room and the assembly room for the touring cars, and a large machine shop. The third floor contains a wood and metal pattern shop, an assembly room for the runabouts, a stock room containing parts for earlier models, and a drafting room. This building is made of brick and has an automatic sprinkler system throughout. Adjoining the factory is a five-acre field, with a cinder track around the outside to test the cars. The railroad company has provided a siding so that the finished cars can be run out of the rear door and into the boxcars. Inside of the track is a one story building 200 × 250 feet which contains a stock room and paint shop. Another large building is the testing house for testing cars in stormy weather. The testers run the cars upon some large pulleys, which rise above the ground 2 inches, fasten the cars there and start the engine. The cars run upon big wheels all day and by night they are as free in the bearings and run as easy as though they had been tested on the track.*

Reading was probably Edsel's favorite pastime. Photographs of Edsel are precious few, but of those that remain, an inordinate number of them show him reading. The first one shows him at age nine, sitting cross-legged in a chair, with a book in his lap. In another, he is much older, perhaps fifteen or sixteen; however, again he is sitting in a chair and reading what appears to be a novel. Another photograph captures Edsel reading intently on the beach of his Florida home, and in the another, he is reading aboard a luxury liner during one of his many European trips. In one of his jot books, Edsel noted: "Read all afternoon."[14]

Edsel never lacked something to read. He subscribed to dozens of magazines (his personal secretary at the company kept track of all the subscriptions) and bought books on a regular basis.[15] In a single month, Edsel ordered five books. Among them were Sinclair Lewis' highly acclaimed *Main Street*, Booker T. Washington's autobiography, *Up From Slavery*, and Winston Churchill's novel about the American Revolution, *Richard Carvel*. These titles show the breadth of Edsel's interest, "which was amazing," recalls Ernest Kanzler, Edsel's brother-in-law and best friend. "When we played categories or twenty questions," Kanzler said, "Edsel always won, no matter what great brains happened to be in the game."[16] Whether the subject was bullfighting or sculpture or finance, Edsel seemed to have a knack for immersing himself in it and enjoying it to the fullest extent.[17]

***

After Edsel completed fifth grade, Clara withdrew him from the public school system and enrolled him in the Detroit University School—an exclusive, private, boys' school located near the Detroit River on Elmwood between Larned and Congress streets. Detroit University School (DUS) was a small institution—on the average, about only 200 students attended all twelve grades—but it was the most elite and prestigious school in town. All of Detroit's rich—the Books, the Oldses, the Strohs, and now the Fords—sent their children to DUS. The school offered a rigorous academic curriculum centering on college preparatory classes. However, it also offered classes in industrial arts (referred to as "manual training" then) and had a competitive athletic program as well.

Going from Farrand to DUS was not an easy transition for Edsel. All of his friends were at Farrand, and he had assumed that he would be going to Detroit Central High with them. Having been transferred to DUS, Edsel had to make new friends and become acclimated to the new environment. Edsel's housemother at DUS observed that the other boys stayed away from him when he first arrived, assuming he would be a spoiled, untenable type. After all, on Edsel's first day at DUS, he drove up in his own Model N Ford runabout.[18] As the housemother later confessed to Clara over a cup of tea, "The other boys soon found out that Edsel was a friendly, every-day sort like themselves."[19] Edsel knew he had advantages that the other boys did not have—even if they also came from wealthy families—but he did not try to win them over because of that. He was simply his usual quiet and pleasant self. In no time at all, Edsel was so well liked that he became a leader.[20]

***

During these years, Detroit was crisscrossed with trolleys, and that is how most DUS students got to school. However, for years, Edsel was the only DUS student to drive his own automobile to school. While his school friends crammed into the back of a trolley, Edsel sped past them, first in his Model N and then in one of his Model T speedsters. One of the favorite pranks of the boys who were forced to ride the trolleys was for all of them to crowd onto the back of the trolley and bounce up and down in an effort to derail the trolley. Sometimes their bouncing worked; sometimes it did not. However, when it did, Edsel was there to pick them up and take them home in his Ford, much to the chagrin of the trolley driver.[21] Edsel sometimes rode the trolley, too. However, on those occasions, he would drive his car to school in the morning, park it in a garage located there, and then ride the trolley back and forth until he decided to take his car home.[22]

Edsel met many of his lifelong friends while attending Detroit University School. As assistant editor of the school yearbook, *Helicon*, Edsel worked with assistant editor Charles Hodges, business manager Thomas Whitehead, and assistant business manager Horace Caulkins. They all worked together with their senior class president, Herbert "Bookie" Book. Leave it to Edsel to convince the other editors of the *Helicon* to change its format—from a format reporting only on fraternity activities to one dedicated to the activities of the entire school. Short stories and poems filled the book, making it a literary publication emphasizing creative writing and chronicling school life.[23] (Perhaps Edsel's father's company taking out full-page ads contributed to his clout.) In the back of the book were pages listing the nicknames of the students and funny quotations made by them throughout the school year. Edsel's nickname was "Brainy," and his joke was fittingly cerebral: "What's the number of the 24th experiment?"[24] When one of the boys gave Edsel a hard time, he would simply reach into his suit coat and hand them one of his cards: GO TO HELL. I HAVE PROBLEMS OF MY OWN.[25]

During his years at DUS, Edsel also met future art commissioner Robert Hudson Tannahill and Herbert Book's brother Frank.

During his last few years at DUS, Edsel was a member of the Gamma Delta Psi fraternity and a member of the athletic association, where he was a sprinter on the school's six-man track team. It was a small but formidable team during these years, especially in the 300-yard dash. In 1912, Edsel's last year at DUS, the track team took all places in that event, beating arch rivals Wyandotte High and Birmingham High. This enabled the DUS track team to attend the interscholastic meets at Adrian College and Cornell University.[26] Henry attended many of his son's track meets and was there when Edsel took third place in the 300-yard dash.[27]

***

The scholastic reputation of DUS enabled its alumni to enter any university in the country simply by presenting their DUS diplomas to the registrar, and many of the 25 classmates in Edsel's class did just that. Some went to Harvard and Cornell; others stayed locally and attended The University of Michigan. But not Edsel. After years of preparation, he decided not to attend college. Instead, he entered his father's business.

Why did "brainy" Edsel forego this next logical step of his education? He certainly had the intelligence to attend any university of his choice, and obviously finances were not a concern. He would have started his freshman year in 1912, nine years after his father founded Ford Motor Company and the year in which Henry and his fellow stockholders shared more than $5 million in dividends. Furthermore, it was not because Edsel capitulated under pressure from his father (as has been speculated), or because he was too afraid to ask his father if he could attend college. One story suggested that Henry was so opposed to Edsel going to college that Edsel asked one of his DUS friends (probably Thomas Whitehead) to plead with Henry to permit Edsel to attend Cornell with him.[28]

Although Henry never encouraged Edsel to attend college (Clara most certainly did), he did not stand in his son's way. The ultimate choice was Edsel's, and Edsel's alone, and he decided to enter the company rather than go to college. By this time, he was so anxious to go to work that he never even graduated from high school, opting to go to work immediately.[29]

Edsel's decision is easy to understand and is best described by an early Ford biographer: "When Henry measured his father's acres, all he could see was a mildly prosperous farm entailing a lot of hard work for a modest return. When Edsel glared out the window, however, it was on an incomparable colossus, and he had respect both as a son and an academic observer for what he saw and for the man who brought it to fruition."[30] That is why Edsel decided to enter the Ford factory instead of pursuing higher education. The flurry and marvel of his father's manufacturing company appeared more exciting to him than any university, regardless of its location or stature. Henry felt the same way, of course. He believed that college was fine for young men who wanted to become doctors or lawyers. However, for those who wanted to go into business—which is what Edsel wanted—there was no better place to learn about that than in Ford Motor Company.[31]

Thus, Edsel began working for Ford Motor Company in the fall of 1912, after returning from a European trip that he had taken with his parents and cousin Roy Bryant. However, Edsel later confessed that his decision to forego college was a mistake. "I wanted to go into business and not, as it seemed to me then, to waste time at college. I had my own way, and I think it was a mistake, for whatever else college may or may not do... it helps one to get more enjoyment out of life."[32]

***

Edsel never forgot his alma mater. He sent all three of his sons to DUS, and he was on its board of trustees for many years.[33] When times became tough for DUS during the Depression, Edsel issued a $50,000 guarantee to the school to carry it through until times became better.[34] Later, he donated $250,000 to establish a new Detroit University School in Grosse Pointe.[35]

Edsel's education eventually separated him dramatically from his father, and it probably was the first underlying reason why they never fully understood one another. Edsel's handwriting and spelling put his father's farmhand scrawl to shame, and the essay that Edsel had written when he was only 13 years old was written better than any of the notes or letters that Henry had written as a grown man.[36] Edsel's capacity to absorb and analyze almost any topic far exceeded that of his father, and Henry resented Edsel for it.

Chapter 6

San Francisco Sojourn

At age 16, Edsel ventured out on his own for the first time when he and his friend, Philip Worchester, drove through 600 miles of dirt and grime from Dearborn to Chicago and back.[1] Edsel had a special Model T roadster built for the trip—it was a boat-tail design and was painted yellow.[2] Today, that trip can be made in a matter of hours by automobile; however, back then, it took Edsel and his companion two full days, with the stretch between Dearborn and South Bend, Indiana, alone requiring 16 hours. Much of the delay going to Chicago was caused in fording a swollen creek where a bridge had been washed away. Instead of traveling many miles out of the way to find another crossing, Edsel and Philip inched their Ford into the swollen stream, hoping their trusty vehicle would make it to the other side. When they were halfway across the creek, the water rose above the wheels and stalled the engine. Stranded in midstream, the two boys had to get out and push the Ford onto dry land and wait for it to dry before it would start again.[3]

They made better time on their return trip, covering the same distance from South Bend to Dearborn in only 12 hours. Apparently, the travelers threw caution to the wind to hurry home. "As no word has yet been received from county sheriffs in regard to speeding," ran a newspaper account of the trip, "the boys are commencing to rest easier, in hopes that their spurt from South Bend went unnoticed by the watchful guardians of the village law."[4]

***

Five years later, and one year before he got married, Edsel decided to take a more arduous trip. He said he simply wanted to attend the great Panama-Pacific Exposition in San Francisco, celebrating the opening of the Panama Canal, and to inspect the Ford exhibit there, which featured a working replica of the recently innovated moving assembly line at the Ford Highland Park plant.[5] Edsel could have traveled in luxury aboard a transcontinental train if that is all he wanted to do. More likely, he wanted one last fling with his friends before getting married and therefore asked six of them to take the more adventurous avenue and drive the 2,400 miles to California.

Daring as it was, Edsel's friends agreed. After a couple of weeks of planning the route, getting the cars ready, and collecting supplies, the troupe left the Ford farm at precisely 7:55 on the morning of June 17, 1915, with Henry and Clara sending them off.[6]

Herbert Book and Robert Grey drove a Cadillac "8." Frank Book, Herbert's brother, and William Russell had left a couple of days earlier to pick up Russell's new Stutz touring car in Indianapolis, where the car was made. The other three—Thomas Whitehead, Horace Caulkins, and, of course, Edsel—drove off in a new Model T touring car replete with sporty wire wheels, multiple spare tires, a large spot light, and a rifle strapped to the left side of the car. For added security and protection, all of the boys carried six-shooters.[7]

Edsel's Model T and the Cadillac made good time on the first day, arriving in Indianapolis that evening, where they hooked up with Book and Russell. From that point, the three cars tried to stay together but split up time and again due to mechanical repairs on one car or another. Their itinerary was well planned, however. When any of them did become separated from the other two, they simply rendezvoused at the next checkpoint.

The trip took the young men through nine states and practically followed the line that Route 66 would eventually take. At the time, however, there were few roads, and those that existed were nothing more than dust-covered trails that quickly turned to mud at the slightest rainfall. It seems as though the troupe spent at least as much time digging their cars out of mud holes as they did driving. "Spent entire morning cleaning mud out of the inside of car and off the tools and equipment," noted Edsel in his logbook that he kept throughout the journey. "Left town at noon and ran into more mud holes."[8] This constant exercise of digging their cars out of muddy roads forced them to drive along the sides of railroad tracks and sometimes on top of them, simply to make better time.[9]

Their progress was also impeded by repetitious mechanical breakdowns and pernicious flat tires. That is why Edsel strapped so many spare tires to the side of his Model T, as did the others. While entering Colombia, Missouri, the left rear tire of Edsel's Model T blew out for the third time that day. Wearing a flannel shirt, golf trousers, gold stockings, and yellow oxford shoes (Edsel called these his "touring tweeds"), and the skin peeling off his nose from sunburn, Edsel climbed out of his car, took off his duster, and made the repair himself. As he was working, a crowd of curious locals had gathered to meet him and his mud-spattered companions. "Roads are very muddy east of here," Edsel said to one of the locals. "How are they west of here?" he asked as he tinkered with his machine.[10]

"We left St. Louis yesterday afternoon in the company of two other machines," he continued. "Owing to the floods, the roads are almost impassable, but the little old Ford rambled right along and we got through with very little problem."[11]

Somewhere along the route, Edsel hit a large rock and buckled one of the wire wheels on his Model T. He had to buy a new wheel at a local Ford agency, explained the dealer to a reporter.[12]. "He is now riding with three wire wheels and one of our regular [wooden] stock wheels. He told me that he thought before he got to San Francisco he would have all of our regular wheels on the car." Edsel explained to the dealer that except for the numerous flat tires and the problem with the accessory wire wheels, his little Ford had given him exemplary service. "The Stutz, however, has been limping in every night, and I think before we get to San Francisco, we will lose it entirely," Edsel said.[13]

Everyone at the Ford agency was excited to have Henry Ford's son visit their little dealership. The owner was particularly smitten. "Edsel was with us only a few minutes, but he made a friend of everyone he saw. He is certainly a wide-awake youth with a good head full of good, hard common sense," recalled the dealer.[14]

At Arrow Rock, Missouri, a small town about 40 miles farther west, the Ford party stopped at "Old Tavern," a roadside eatery, to have a late lunch that same day. They stayed only long enough to eat, but the "Eastern gentlemen" apparently made quite an impression on the local clientele, nonetheless. As soon as they had finished and walked out the door, their African-American waitress asked the owner, "Who was dem fine mans, Miss Kate? Dey certainly an de finest lot of gentlemans ever eten here at dis ol' tavernacle. I's guin to se who dey is." She went at once to the register and returned in a few moments. "Law bless your heart woman," she exclaimed, "no wonder, one of dem gentlemans is Mr. Ford hisself."[15]

***

Edsel wrote in his diary:

La Junta, Colorado, Sunday, June 27:

Got first sight of Rocky Mountains at this point.

The troupe spent five days in Colorado, four in Denver visiting a Ford family friend, and the better part of a day in Colorado Springs where they went to the top of Pike's Peak on the cog railway.

While in Denver, Edsel sent his father this ambiguous telegram:

Henry Ford, Detroit, Michigan.

Have lost books in Colorado.

Edsel Ford

Upon receiving the telegram at his Highland Park office, Henry immediately took action. "Edsel lost his guide books to the roads," he told his secretary Ernest Liebold. "Rush some more out to him."

Liebold said he would do that but then hesitated, explaining that guidebooks could be obtained in Denver as easily as in Detroit, "so why would Edsel ask you to send him more?" Before Henry had a chance to respond, Liebold came up with the answer. "Er—what's the name of the boys with Edsel?" questioned Liebold.

Henry gazed at his secretary, blinked, then smiled. "Well, I'll be hanged," he gasped. "Never mind the books."[16]

***

Any cross-country automobile journey in 1915 was expected to be arduous, but the Ford troupe took advantage of amenities at each opportunity. They ate and stayed at Harvey Houses—a chain of restaurant-hotels along the Santa Fe railroad—whenever they could, attended movies at major cities along the route, and telephoned their families regularly. At those places along the route that did not offer modern conveniences, they were forced to camp out; however, they were prepared. When they set up camp, they parked two cars parallel to each other and about 40 feet apart. Between the cars, they set up two tents, the backsides of each tent tethered to one of the cars, and the front flaps facing each other, with about eight feet between them. A rope was then stretched between the tops of the two tents and a kerosene lamp hung from it for light, halfway between the two tents. The boys sat between the two tents under the kerosene lamp and talked about the events of the day and plans for the next day, while listening to Edsel's phonograph.[17]

***

From the lush mountains of Colorado, the travelers proceeded due south to Las Vegas and on to Albuquerque. The 120-mile stretch between these two high desert towns was among the most desolate of the entire trip. Except for a few barking prairie dogs and an occasional coyote singing in the distance, the party went without seeing a house, man, animal, or anything except sagebrush and sand.

At their campsite near Socorro, New Mexico, Edsel made the following notation in his logbook:

> *Pitched tent among the tarantulas. All scared to "death." Caulkins decides to sleep in back of the Cadillac.*

With campsites such as this, it is little wonder that the young men rejoiced upon reaching Holbrook, Arizona, for there they caught up with the Santa Fe railroad and occasional sights of civilization. As they drove along the tracks of the Sante Fe, they waved at passing trains as the crew and passengers waved back at them.

Upon their arrival in Flagstaff, Arizona, on July 10, the Ford troupe visited the local Ford agency and then drove to the Grand Canyon. They had such difficulty reaching the canyon that Edsel made a note of it:

> *Fixed a puncture; started at 12 noon, fixed another puncture 6 miles out [from Flagstaff]. Cadillac also had puncture. Found Stutz crew at hotel [at Grand Canyon]. Left Stutz in Albuquerque to be shipped and arrived on afternoon train. Got rooms at 7:30 PM, retired early.*

Edsel had made a cursory visit to the Grand Canyon with his parents, but this time he wanted to see it in more detail and arranged for a tour guide to take him and his friends on horseback down to the bottom of the canyon. They stayed in the area for four days.

When they departed, they encountered more breakdowns, as noted in Edsel's logbook:

> *At 4-1/2 miles out...Ford broke rear axle shafts. Sent for new one by hotel chauffeur who happened to come along. When it arrived we found we had no wheel puller, so could not put new shaft in. All tossed coin, odd man to walk back to hotel for wheel puller. Tom Whitehead was odd, and started. Only food all day was can of beans and box of crackers for the 3 of us; other cars having gone on to Williams, Arizona.*

***

By now, the troupe was anxious to reach Los Angeles, but they encountered their biggest delay of the trip as they entered California. They were forced to wait at Needles Peak on the eastern edge of the desert. When they arrived at Needles, they learned that highwaymen had attempted to hold up two cars that were crossing the desert the previous day. Instead of taking any chances, Edsel and the others decided to wait in the "oppressive heat" until a group of travelers could cross the open spaces together. They kept themselves busy while they waited. "All had various things to do to cars," noted Edsel. "Got drive shaft from express office, bought a tire, repaired several tubes, bought some clothing."

Two days later, enough travelers had amassed to affect a crossing, and at 6:15 in the evening, Edsel's three-car party joined the ranks of five other vehicles—a Jeffery, two Chalmers, two Stutz, another Cadillac, and two more Model T's—and continued their trek westward. At the heart of the desert, the caravan stopped to make minor repairs and have a midnight lunch. To keep any intruders at bay, Edsel played his phonograph loudly.

Late the next morning, everyone arrived unharmed at Barstow. "No trouble in the desert," Edsel noted in relief. As a reward to themselves, Edsel and his friends sped the remaining 60 miles to Los Angles and spent the next 10 days relaxing on the beach and under the palm trees.

***

Rejuvenated, the troupe turned their motorcars northbound and arrived in the cool and pleasant surroundings of San Francisco, five weeks after leaving the heat and humidity of Dearborn. However, their tribulations were not over yet. When they arrived in town, they went directly to the Commodore Hotel, one of the finest hotels in the city, to rent rooms. After parking their cars in front of the hotel, all seven—dusty and dirty from the three-day drive from Los Angeles—walked inside to register. Edsel went first. "We would like three rooms," he said. Looking over the dirty troupe, the *maitre de* thought they were a bunch of bums who were making a joke, so he tried to deny them access as easily as possible. "With the Exposition going on," he said, "all we have left is the Presidential Suite."

"We'll take it!" Edsel quickly retorted.[18]

***

During their five-day stay, the young men visited the Exposition, which was even more exciting than the brochures described. The Ford exhibit, housed at one end of the Palace of Transportation, was particularly exciting. Men outfitted in white coveralls were building Model T's at the exhibit at the rate of one every ten minutes on a factory-like assembly line. The exhibit was one of the most popular at the Exposition, and throngs of spectators were separated from the assembly line by a waist-high, nickel-plated railing. When Edsel visited the exhibit, he was merely one of the crowd.

Frank Vivian, manager of the exhibit, was walking up and down the assembly line, watching his men and answering questions from the crowd along the rail. He noticed a young fellow dressed in a khaki shirt, khaki pants, a pair of leggings, and an old hat, waving at him to get his attention. Vivian did not pay any attention to him at first, but after "waving at me...two or three times in the course of a half-hour...I ducked under the rail and got through the crowd and I went over to him."

"Edsel Ford," said Edsel, as Vivian shook his hand.

Embarrassed, Vivian quickly responded, "I knew you all the time."[19]

***

Visiting the exhibits at the Exposition was not the only thing that Edsel and his friends had in mind while in San Francisco. They also wanted to see the real San Francisco, so Edsel persuaded Vivian, who was a native of the area, to show them around town. Vivian agreed, and he took the boys on an extensive tour of the city, including the Barbary Coast and Chinatown. As they were enjoying a show at one of the Coast's "hot spots," a fight broke out, involving more and more of the patrons by the second. Vivian's immediate reaction was to get Edsel safely out before they were all engulfed in the brawl. "I was tall, about six-foot," recalls Vivian, "Edsel was small, so I was able to cover him up." He then rushed Edsel out the door, with the others following, as club-wielding police came crashing through the other direction.[20]

Edsel and the others thought that was funny, and because the night was still young, they convinced Vivian to take them to Chinatown. Edsel wanted to see a Chinaman smoke opium, and he was willing to pay him for the privilege.[21] With such incentive, it did not take long for them to find a local to direct them to an opium den.

In the heart of Chinatown, down a dimly lighted alley, they entered a low door and immediately were standing in a den of paradise. It was a small area, perhaps ten feet by ten feet, and barely high enough for a tall man to stand in it. Along three walls were bunks where smokers would lie with head and feet together. In the middle of the den, a single lamp barely lighted the room.

A big Chinaman, lying next to the lamp, had his smoking apparatus next to him: a square tray, a glass lamp, and a tiny horn filled with opium paste. In his mouth was a long bamboo stem. "Have a smoke?" he said.

Edsel politely declined, and Vivian escorted them out of the den, realizing that Edsel had seen enough. "We nearly got all doped up with the smell of the stuff," recalled Vivian.[22]

Nearly exhausted from the journey and the San Francisco nightlife, Edsel and his friends boarded the Overland Limited on July 30 and headed back to Dearborn, having made arrangements for shipping the Ford home.[23]

* * *

Later that year, Edsel returned to California with his parents, cousin Roy Bryant, and Roy's new wife, Katherine.[24] Instead of driving, they traveled in luxury aboard Henry's private railroad coach, *Fair Lane*.

The first leg of their trip was to the Pan-Pacific Exposition, where Edsel proudly showed his family the exhibits. During their first night, they met with Harvey Firestone and Mr. and Mrs. Thomas Edison, and they watched a fantastic fireworks display.[25]

After spending a couple of days at the Exposition, Henry took the entire troupe to visit Luther Burbank, the famous horticulturist, in Santa Rosa, a small community about 60 miles north of San Francisco. Edison wanted to see how Burbank's experiments might be used in his own, and Henry was interested in the health benefits of Burbank's hybrid plants.[26]

During their visit, Frank Vivian had driven to Santa Rosa in a new Model T pickup. On a whim, Henry asked Vivian to take him and Edsel for a ride. Henry sat in the passenger seat, while Edsel and one of Vivian's helpers climbed into the bed of the truck and sat on a pile of hay.

Vivian was driving lickety-split, turned a corner too quickly, and Edsel and the other man went flying off the truck!

"My God, you've killed my son! You've killed my son!" Henry yelled.

Vivian immediately stopped the truck, and the two men jumped out and ran back to the boys. They found them sitting in the bushes and laughing.[27]

***

Before leaving Burbank's residence, the entire group posed for a photograph. Edison is the center of attention, and Burbank, Henry, and Clara are prominently posed. Only Edsel's head can be seen behind the crowd, and he would not have been seen at all had he not been standing on his tiptoes.

Chapter 7

Courtship and Marriage

Located on Boston Boulevard, only a few blocks north and on the opposite side of Woodward Avenue from the Ford family home, lived department store magnate Joseph L. Hudson.[1] In addition to his own family, Mr. Hudson's sister, Eliza Clay, and her two daughters, Eleanor and Josephine, lived there. Eliza and her family moved in with her brother after her husband, William Clay, died in 1908.

Similar to Henry Ford, Hudson was a self-made millionaire. He came to the United States from England in the middle of the nineteenth century and, after taking up shopkeeping, developed a department store that became one of the premier shopping emporiums in Detroit. And when the automobile business started to take root, Hudson started the Hudson Motor Car Company in 1906. When Eliza and her daughters came to live with him, Hudson was already a wealthy man. When he died only four years later, the two girls and their mother continued to live in the patriarch's mansion.

Although the Hudsons and Fords lived in the same neighborhood, Edsel did not meet Eliza's daughter until three years after she and her mother had moved in with her uncle. Eleanor was less than three years younger than Edsel, but that was enough to keep them from running into each other. Besides, Edsel spent much of his teenage years with his father, working either in his workshop or at the factory. He did not have much time to spend with girls. Because Detroit University School was an all-boy school, the only place Edsel had occasion to meet girls was at the Annie Ward-Foster School of social graces.

***

Children of rich Detroiters learned many of the social protocols at Miss Foster's school. Her classes were held in the basement ballroom of the Grosse Pointe home of Mrs. J.S. Newberry, and later at the Grosse Pointe Yacht Club.[4] Young ladies, accompanied by female chaperones, were required to wear frilly dresses and pink dancing shoes to class, and the young gentlemen wore suits, ties, and white gloves.[5] The dancing classes held by the school were veritable adolescent rites of passage. Miss Foster's students did not learn the Charleston or the rag—she taught only the waltz and the fox trot.[6]

Miss Foster's school was not part of either the all-girl Liggett School—the school that Eleanor attended—or the Detroit University School. However, nearly everybody from these schools attended Miss Foster's classes, including Edsel and Eleanor. There, in the fall of 1910 when Edsel was16, he met his future wife, Eleanor Clay. She had just turned 14 years old and was described as slender, tiny, and a little wistful, with blue eyes, a light complexion, and hair the color of flax.[7] When Eleanor was a little girl, one of her favorite pastimes was putting pennies on the streetcar tracks and then watching them get squashed. Similar to Edsel, she was an avid reader, "preferring," she said, "biographies to silly stuff."[8]

According to one of Eleanor's friends, "Edsel did not notice Eleanor Clay until after Miss Foster introduced them."[9]

***

Eleanor was a star pupil and an involved student at Liggett School. She was a member of the drama club, she was captain of the basketball team (she played forward), and she played hockey. In 1915, she was business manager of *Rivista*, the school yearbook.[10]

Miss Ella Liggett, the school's headmistress and daughter of the founder of the school, graduated from the first graduating class at Vassar in 1869. She encouraged many of her students to attend her alma mater, especially Eleanor Clay. Every winter, Miss Ella shepherded a number of Liggett School seniors onto a train to Poughkeepsie and marched them into the office of the Dean of Women at Vassar. "Well, here are my girls for this year," she announced.[11] Eleanor was among this group in 1916.

After their initial visit, Eleanor and some other seniors, including her close friend, Eleanor Everhard, decided not to further their education at Vassar or any other university. Eleanor and Miss Everhard, at least, concluded that everybody who went to college left home congenially but returned haughty and conceited. Thus, neither of them continued their education past high school.[12]

The Liggett School did not require uniforms, but it did have a strict dress code with which all the girls complied in their Peter Thom suits. Peter Thoms consisted of a middy-type sailor blouse over a slightly above-the-ankle-length pleated skirt. The outfit came in either white or blue with contrasting scarf. Eleanor and her friends purchased theirs in New York.[13]

***

Edsel and Eleanor's first meeting was an occasion Edsel would not soon forget; it was a turning point in his life. Until then, Edsel had spent most of his time with his parents, with his high-school friends, or with his cousin Roy Bryant. For Eleanor, it was simply another introduction to another wealthy Detroiter, something to which she was well accustomed. "I don't exactly remember when I first met Edsel," she recalled in an interview. "I guess it was some minor social affair."[14]

There is no evidence that Edsel dated anyone other than Eleanor, which is one reason he became so enamored by her. However, even if he had, he would have been much too young to become serious with any of the others. When he first eyed Eleanor, however, he was both old enough and vulnerable enough, and apparently lonely enough, to succumb to her innocence and charm. Apparently, Edsel was looking for more than what Roy Bryant's companionship could offer. "For every boy who's lonely, there's a girl who's lonely, too" ran the lyrics of a song Edsel jotted down in one of his notebooks.[15]

After their initial meeting, Edsel began seeing Eleanor on a regular basis. He reminded himself to call her by making a note in his appointment book. He took her to all of the dances at the Detroit University School and escorted her to parties at each other's homes.[16] One weekend, they would have a party at Eleanor's home. On another weekend, they would meet at Edsel's home, and on another, at the Books'. On many occasions, they would get together for a *fete* at the designated Tau Beta House (Eleanor's social sorority). They also spent time at Belle Isle on the Detroit River, cruising on boats in the canal, or driving their speedsters up and down cobblestoned Woodward Avenue—sometimes driving all the way to Birmingham.[17] Their courtship lasted for six years, until March 1916, when they reached "an understanding."[18]

Edsel and Eleanor decided to get married as soon as she graduated from Liggett School.[19] On June 15, while the handsome couple was attending another Tau Beta party, Edsel announced that he and the lovely Eleanor Clay would be married on November 1. Eleanor, wearing a white outing suit with a black hat, was, of course, the center of attention.[20]

During the ensuing five months, the couple's social calendar was booked solid. Supper at the Detroit Club. Dinner dance at the Detroit Athletic Club. Supper at the Bloomfield Hills Country Club. And on and on.[21]

***

A few months before Edsel was married, he and his fiancée and his future sister-in-law and brother-in-law went to New York City to do some shopping, attend the theater, and go dancing. "We all had a slick time in New York with Eleanor and Josephine. We went shopping with them two afternoons. Were in Lucille's saw some beautiful gowns. Eleanor is buying trousseau—fancy that," wrote Edsel.[2]

It is interesting to learn that Edsel knew how to enjoy himself. However, what is even more enlightening is what he mentioned in the last part of this letter: "The funny part of it all was that with all the late hours and strenuousness I feel great. Haven't had a pain or ache since I left Detroit."[3]

Aches and pains at age 23! Why would a young man, at the prime of his life and deeply in love, mention his health, unless it was something that he and his family monitored closely?

***

Henry Ford was not happy. It was not so much that his son was getting married; it was Edsel's young age. Henry thought Edsel was too young for marriage and hoped he would wait a little longer before taking this step. Until Edsel's announcement, Henry and Clara thought he would be with them for at least a few more years, and Edsel gave them no reason to believe otherwise. Edsel even helped them in the design and construction of their new mansion, as though he planned to live there, too—which he did, but for only a short time.

In February 1914, work began on the Fords' new home in the middle of the Dearborn woods, only a few miles away from Henry's birthplace. Built at a cost of $3 million, the gray limestone mansion consisted of nearly 32,000 square feet divided into 56 rooms. It had a 12-car garage and its own power plant, run by a dam that ran across the Rouge River, which passed only 50 yards from the rear of the residence.

For Edsel's entertainment, Henry included an enclosed swimming pool, a bowling alley, and a par-three golf course in the mansion plans. Edsel himself specified his own suite of rooms located on the second floor, down the hall from his parents' quarters. One of the more interesting features of his suite was its bathroom, which had a unique shower that sprayed Edsel from all directions with the turn of a single control.[22]

Before construction began, Edsel convinced his father to retain the services of Frank Lloyd Wright to design their home.[23] Henry agreed, but Wright had recently departed for Europe, leaving his office in the hands of one of his young apprentices, Herman Von Holst. Henry did not like the design developed by Holst and gave the project to a Pittsburgh interior decorator, William Van Tine.[24]

While Henry watched the construction, Edsel checked the accounting to ensure that the "financial side of it" was managed. He also assisted his mother in choosing furnishings for the expansive structure.[25]

Henry Ford's new mansion, which he called *Fair Lane* after the area in Ireland from where his ancestors came, was completed in November 1915. In January of the following year, Henry, Clara, and Edsel moved into the mansion. Edsel lived there only until the following November, when he and Eleanor were wed.

* * *

During Edsel's only summer at Fair Lane, the Fords put on an extravagant fireworks display to celebrate the Fourth of July and their new home. Many of Henry and Clara's friends were there, as well as Edsel and his new fiancée. Everyone sat on the patio as fireworks were shot one by one over Edsel's golf course and into the hazy Michigan sky. As the festivities were getting underway, a Roman candle prematurely exploded, sending fiery pieces into the crowd. One red-hot piece struck Eleanor on the cheek.

Edsel rushed her to the Henry Ford Hospital in Detroit and demanded that the doctor give her a tetanus shot. However, according to Dr. Smith, the attending physician, the injury was "nothing of consequence." It was only a superficial burn, causing some redness. Regardless, Edsel was still concerned and pressed the doctor about it further. "I suppose that you might kiss it and make it well," quipped the doctor.

"I've already done that," replied Edsel.[26]

* * *

Uniting two of the most prominent families in Detroit should have been the wedding of the decade, but both Edsel and Eleanor insisted on simplicity and privacy. A reporter of the social page asked Eleanor, "What will be the decorations of your wedding?"

"Oh, I cannot tell you that," she cooed. "If the guests knew what the decorations were to be and what we will offer them, they would lose all interest and there wouldn't be any surprise."[27]

The guests were surprised—not by the decorations Eleanor chose but by the lack of them. The wedding was void of all pomp and circumstance. From the beginning of the ceremony to its end, "there was an utter absence of the ostentation, show, glitter, and wealth," wrote one reporter in attendance.[28] Another reporter confirmed that. "I don't think I saw $1,000 worth of jewels in the entire crowd," he wrote.[29]

That is precisely what Edsel and Eleanor wanted. They did not want their wedding to be another social gathering for the elite of Detroit. They wanted to have a simple family affair. Most of the guests were immediate family and close friends. Older folks were there as well, but they kept to themselves in the background, at Edsel's insistence.[30] It was Edsel's wedding, and he was "tired of listening to all of his father's old fogies," although among them were such luminaries as Thomas Edison, Harvey Firestone, and John and Horace Dodge.[31] However, Edsel did not care. He was used to having such luminaries around him, and the last thing he wanted was to have any one of them steal his wedding day from himself and Eleanor. Even Henry kept in the background—one of the few times that he did—allowing Edsel to glow individually in the limelight. Newspaper accounts described Henry as standing with his hands behind his back and "the wraith of a smile hovering over his lean face."[32] Characteristically, the reporters made no mention of Clara, although she was definitely there.

Ernest Kanzler, Edsel's brother-in-law, Horace Caulkins, one of the adventurers who accompanied Edsel on his cross-country trip the previous year, Charles Hodges, Jr., and Robert Tannahill, Eleanor's bachelor cousin, were Edsel's ushers. Thomas Whitehead, another friend who drove to San Francisco with Edsel, was his best man. Eleanor's bridesmaids were her three closest friends: Ruth Woodruff, Agnes Taylor, and Esther Cutler. Eleanor's sister, Josephine, was her maid of honor.[33]

First to enter the parlor of the Hudson mansion, where the wedding ceremony was held, was the Reverend H. Lester Smith of the Central Methodist Episcopal Church, followed by Kanzler. Edsel, dressed in a conservative black tuxedo, was next, followed by his best man. Eleanor's mother came after him, with the three bridesmaids behind her and, finally, Eleanor.[34] The bride was dressed most fittingly in a girlish, ankle-length frock of silver cloth, the front of which was embroidered with pearls. Its bodice and sleeves were of tulle and silver cloth, also embroidered with pearls. She wore a Russian diadem made of silver and pearls. She carried no flowers and wore no gloves.[35]

The entire ceremony, which started promptly at 8:30 in the evening, was finished in a few minutes, ending with a simple, heartfelt prayer given by Reverend Smith for the lasting happiness of the young couple.[36] A reception and buffet dinner was held in the solarium after the ceremony, and there Edsel announced that he and his new bride would spend their two-month honeymoon in Honolulu.[37] For a wedding gift, Eleanor gave her new husband a handsome wallet—black in color with his initials and trim in silver. Inside, she placed a note: "A wedding gift to my Ned. ECF."[38]

***

The newlyweds did not go directly to Hawaii, but first took the train from Detroit to San Diego, stopping at Flagstaff, Arizona, along the way to visit the Grand Canyon. They next traveled to San Diego and spent a few days there before motoring up the coast to San Francisco.[39]

The couple did not need to return to Detroit anytime soon, so they allowed plenty of time for sightseeing, lounging, and writing correspondence. A letter Edsel wrote to his parents while at the Grand Canyon provides a rare glimpse into his new-found life[40]:

Dear Mother and Father,

A week ago today we were all getting ready for the big event and now it is all over and it seems like yesterday that we were saying "Good bye" and on the other hand it seems as though we have been married for years.

We are having a great time. Walking and breathing this great air. It is a lot colder [here] than at home, but not too cold. Yesterday we motored over to Grand View Point and on the way back were caught in a blinding snowstorm.

We felt quite as important as you and father when we arrived. There was a big Pierce touring car down to meet us and about six porters and bellhops.

I saw some real cowboy life here yesterday. They were all over here for election and a few drinks. They got into a fight in the barbershop and one was knocked cold but soon "came to."

Eleanor and I send lots of love and will be glad to hear from you at Pasadena.

Yours,

Edsel

From southern California, the young Fords drove Edsel's Mercer touring car, which he had shipped from Dearborn, up the coastal highway. "You wouldn't know us, we are so healthy," Edsel wrote his parents. "We are both very much sun burned and I have gained about six pounds and my face is peeling from the hard water."[41]

As a courtesy to his father, Edsel and "Ellie," as Edsel called his new wife, visited Billy Hughson, Ford Motor Company's first Ford dealer, when they arrived in San Francisco.[42] The amicable Hughson entertained the young pair as best he could for a couple on their honeymoon.[43] However, Frank Vivian, the man who escorted Edsel and his friends on their first trip to San Francisco a year earlier, was happy to show the newlyweds around the city.[44]

On November 27, Edsel and Eleanor set sail for Hawaii on the U.S.S. Great Northern, arriving on December 5 but wishing they had arrived more quickly. "We arrived here yesterday after an unpleasant voyage," Edsel wrote in another letter to his parents. "The sea was not very rough, but the boat rolled something frightful, due to her speed and light ballast. Neither of us could sleep at all on the whole trip. Perhaps three or four hours a night. So we are very tired, but feeling fine...."[45]

After they recuperated from their voyage, Edsel and Eleanor had a wonderful time in Hawaii, especially Edsel. "I am getting very good on the surfboard," he wrote home. "Am able to stand up and ride in on the wave, a feat which sometimes takes people a year to learn. It is so very exciting. I am more tanned than at any time last summer."[46]

***

What Edsel did not find exciting was the reporters who followed his every move. "To describe him is not an easy thing to do," wrote one reporter. "If there is any such characteristic, it is the trait of modesty. If anything, he is different and possesses the rare faculty of being at home in all situations and at the same time unobtrusive and, if the phrase be not a paradox, conspicuously inconspicuous."[47]

Chapter 8

Live Very Simply

Now that Edsel was married, Henry wanted Edsel and his wife Eleanor to build a home next to Henry's Fair Lane estate when they returned from their honeymoon. However, similar to their early marriage, Edsel and Eleanor had other plans. Eleanor said before her wedding day that she and Edsel were going to "live very simply."[1] She must have meant "simply live," for their lifestyle, while conservative and reserved, was neither simple nor inelegant.

Unlike the Henry Fords, who moved constantly during their early married life, the Edsel Fords had but three primary residences their entire married life—each one far from the scrutiny of Henry Ford.

Their first home—the one into which they moved upon their return from their Hawaiian honeymoon—was located at 439 Iroquois Avenue in the posh Indian Village area of Detroit, east of downtown and about 12 miles from Fair Lane. They paid $60,000 for the substantial but conservatively designed mansion, and they spent an equal amount to have its interior redecorated.[2]

The young Fords lived there for about five years before moving to a much larger and more elegant structure. Their new home was still in the Indian Village area but was a few blocks south of their first residence and nestled on a large lot between Jefferson Avenue, one of Detroit's major east-west thoroughfares, and the Detroit River. Costing a half-million dollars, their new home was more suited to their stature and economic clout. The house was made of plain gray limestone and was conservatively elegant, similar to their first. On one side was a *porte cochere*, where a motorcar could be driven to unload passengers.[3] In the back were tennis courts and a large swimming pool, illuminated at Edsel's request for night swimming, with specially designed blue lights to minimize glare.[4] All of the landscaping was done by Danish landscape architect, Jens Jensen. Edsel's favorite part of the property was its enclosed boat slip, which enabled him to take his motorboats directly from his house onto the Detroit River.[5]

***

In 1925, Edsel bought 55 acres of choice lakefront property in the affluent Detroit suburb of Grosse Pointe from, of all people, his father. Henry had purchased the parcel located on the western shore of Lake St. Clair, with the thought of building his own mansion there.

However, after deciding to build his dream home close to where he grew up, Henry sold the property to Edsel.

If there were any beautifully situated pieces of property in the whole of Detroit, this was surely the one. It was located on a small peninsula known as Gaukler Pointe, which juts out into the lake and provides beautiful views of the lake and extended shoreline. Edsel kept adding to the tract of land until he had acquired 65 acres and nearly a full mile of lakeshore frontage.[6]

The undeveloped property was in need of some improvements, however. Along this particular piece of shoreline, trees were scarce; therefore, Edsel had Ford employees transplant trees from the Ford farm in Dearborn to the Gaukler Pointe estate. Leaving at one o'clock in the morning to avoid both traffic and causing congestion, the laborers transported full-grown trees on logging wagons pulled by four-horse teams.[7] Edsel also had a lagoon dredged along the eastern side of the frontage so that his friends could moor their boats beside his 125-foot sailboat, *Acadia*.

The most important decision, however, did not pertain to landscaping, but to what type of building to construct on the exquisite site. During the numerous times that Edsel and Eleanor traveled to Europe, they became enthralled by the architecture of the Cotswold area of England. Located about 100 miles northwest of London, there is an area of grassy limestone hills, deep ravines, and barren plateaus known as *wolds*—Old English for "God's high open land."[8] With hills rising 1,000 feet, the area provides stunning vistas. Ancient villages with names such as Stow-on-the-Wold, Wotton-under-edge, and Moreton-in-Marsh dot this pastoral region. Since the sixteenth century, people of the Cotswolds made their living by raising sheep. From the income derived from the sale of wool, they built quaint homes and magnificent churches with gray Cotswold limestone. Early Cotswolds had thatched roofs, but that was soon replaced with split stone shingles, which required massive buttressing as a means of supporting the weight of the roof.

Edsel sent long-time associate, Albert Kahn, the architect who had created many of the Ford manufacturing plants, to the Cotswold district to see if he could adapt the Cotswold design into a home for Edsel and Eleanor. Edsel's only directive to Kahn's design would be that Edsel wanted it to be a modest, picturesque home rather than a palace.

***

The design that Kahn developed was nothing short of spectacular. It was huge—it had 60 rooms—but it retained all the style and charm of a tiny three-room Cotswold cottage. However, Edsel had to pay dearly in both time and money to obtain that charm in such a large structure. Construction of the home began in 1926 and took three years to complete—one year for the basic structure and two more years to custom-fit to the inside the various antique architectural elements such as the wood paneling and carved fireplaces that Edsel and Eleanor had acquired in England. The labor of 200 craftsmen and artisans was needed to transform Kahn's design into reality.[9] Not settling for simple cut stone, stone masons handworked each rectangular piece of Briar Hill sandstone until each had a uniform finish. The roof was made of split stone—not slate, which normally would have been used—because genuine Cotswold buildings used stone. Edsel hired a man from the Cotswold district to lay the stone.[10] Finally, to give the building a long-established appearance and feel, moss was grown on the roof and English grape ivy planted about the walls.

***

Construction of the home was temporarily halted when a silver bracelet and cross, and the lower jaw bone that was believed to have been that of an Indian princess, were uncovered during excavation. Legend had it that 150 years earlier, great numbers of Chippewa and Huron tribes stopped at Gaukler Pointe to camp and make merry, and the entire area could have been their ceremonial resting place. Edsel called in a team of archaeologists from the University of Michigan to investigate the artifacts, but they found the remains to be an isolated burial.[11]

Edsel also had to lay to rest the fantastic rumor that he was going to build a deep moat around the whole estate, with access to the property over only a single drawbridge. The rumor started because Edsel wanted a canal dug around the main building for aesthetic reasons. "That was never our purpose in the construction of the canal about the house," Edsel explained. "The architect suggested the canal as part of the landscape plans because it fitted in with the general scheme. I agreed, but there was never any talk of a drawbridge. It's nothing but a rumor."[12]

***

The interior of Edsel's mansion was no less painstakingly crafted. Each major room of the mansion was meticulously sketched and detailed by Edsel and Eleanor before work began.[13] The main staircase in the entrance hall was brought from Lyveden Hall in Northampton, England, and the morning room had a carved pine mantel from Spitalfields. The dining room was paneled with eighteenth-century panels that were dismantled from the Clock House in Upminster and shipped from England in sections.[14]

Edsel's favorite room and personal retreat was the study. It was not a large room compared to others in the house (about 300 square feet), but Edsel tastefully appointed it himself. The walls were paneled in original sixteenth-century Elizabethan oak taken from Heronden Hall in Kent, England. They concealed two full-size doors—one leading to a private washroom, and the other to a photographic darkroom.

Edsel's desk was a small but exquisite piece of nineteenth-century Georgian furniture made of burl-walnut. On top of it, he had a few books, a lamp, and a picture of his lovely daughter, Josephine.

The study was full of mementos of Edsel's life. A model of *Acadia* sat atop a fifteenth-century English cabinet, and one wall was filled with photographs of family and friends, including Admiral Richard E. Byrd, the Prince of Wales, Douglas Fairbanks, Mary Pickford, and Franklin D. Roosevelt.

In the end, the Cotswold mansion cost Edsel $3 million to build ($100 million in today's dollars), and when it was completed, it required a staff of 40 people to maintain it.[15] Edsel and Eleanor personally hired each of the servants, their strictest qualification being confidence. In return, the servants were well cared for and well treated. Rarely did a servant quit, and many retired from the residence. One of the custodians worked for the Edsel Fords for 60 years.[16]

***

Shortly before starting construction of the Gaukler Pointe estate, the Edsel Fords went on a vacation-home buying spree. They acquired a winter residence in Hobe Sound, Florida; a summer retreat in Seal Harbor, Maine; and a year-round residence—a sort of Boy Scout camp/farm combination—known as Haven Hill—located in a remote area northwest of Detroit. None of these properties was inexpensive—Edsel paid more than $6 million for the three of them.

Edsel was particularly proud of the property he had purchased at Hobe Sound, Florida, a small community on the Atlantic shore about 60 miles north of West Palm Beach. "We have bought a piece of property...and are very excited about it," he wrote to his mother. "It runs through from river to ocean and enables us to have our own dock and tennis beach on the ocean." There was no house on this property. Instead, Edsel bought a 125-foot houseboat (at a cost of $307,000) and docked it there.[17] Christened *Onika* (meaning "kiss me quick" in Hawaiian), Eleanor furnished the living room in Chippendale, the dining room in Hepplewhite, and the staterooms in French Provincial.[18] Edsel and Eleanor would usually spend a month to six weeks there every winter, starting after the first of the year. To get there, they took the *Detroiter* to New York—reserving drawing room A and connecting compartment B—then one of the Florida trains south.[19]

Edsel was enticed to purchase the Maine property—which he fittingly named *Skylands* for its breathtaking view of the Atlantic ocean—by Eleanor, who vacationed there with her family when she was a child, and by Nelson Rockefeller, a close personal friend of the young Fords, who also owned property in the area. Rockefeller's property was down on the beach, whereas Edsel's rested high on the shear rock cliff.

It was much cooler and less humid in the Maine countryside in summer than in Detroit, and Edsel would take his family to *Skylands* to rest and play during the hottest part of the year. They would be railed up in comfort in Henry Ford's private railroad car, *Fair Lane*, and they would take all of the conveniences of their Detroit home with them, including several servants to do the domestic chores, and a retinue of horses from the Ford farm.[20]

Edsel loved to prowl along the rocky Maine coast with Eleanor to find a cozy corner in the lee of a boulder and read to her in his soft shy voice. In the evenings, he and the family often sat around and played hearts, rummy, or backgammon. When they tired of that, they watched the latest movies in their private movie theater.[21] "I am so crazy to have you come here that I can hardly stand it," she wrote to Edsel. "You would just love the forest walks revealing mountain tops where you are unaware of anything human."[22]

"We have seen eight movies since we have been here," wrote 12-year-old William Clay to his grandmother, Clara. "We have tennis lessons in the afternoon and we go swimming in the morning. We ride horseback."[23]

***

One time, Edsel became ill en route to Detroit from his Seal Harbor home. So he got off the train near Bangor and hired a taxi to take him the 85 miles back to Seal Harbor.

"Sorry I haven't any car but a Ford," explained the taxi driver, who did not recognize his passenger. "If you don't feel well, you'd better try something else."

"A Ford's good enough for me," replied Edsel.

"Then all I can say about it is, 'I guess you don't know much about cars, mister.' "

"I guess I do," explained Edsel, "I make enough of them." He then revealed his identity.

Surprised by his celebrity passenger, the taxi driver rushed Edsel to the Bar Harbor hospital as fast as his little Ford could take them. After thoroughly examining Edsel, the doctors diagnosed his illness as simply indigestion.[24]

***

The Maine and Florida residences were too far away for the young Fords to frequent often, but they regularly went to Haven Hill, both in summer and winter.[25] It was a sprawling piece of property, covering nearly 3,000 acres. On it was a huge cabin, a lodge actually, built of logs hauled 500 miles from Ford Motor Company-owned forest land located at Iron Mountain in Michigan's Upper Peninsula. Pine needles were bagged and also brought down from Iron Mountain and spread on the paths around the cabin to supply a mountain aroma.[26]

Haven Hill was operated as a farm, as well as a retreat, and produced chickens, eggs, and milk. At one point, it had as many as 1,500 sheep, cows, and horses.[27] (Charles Sorensen said that Edsel had a fine horse stable and was a good rider.[28]) However, its main purpose was to provide a retreat for Edsel and his family. Edsel called it his "nerve retreat."[29] It was one of the few places where Edsel could not be bothered.[30] Here, the family rode horses on the rolling hills, went swimming in the kidney-shaped pool, or played tennis on the clay floor tennis court—the clay for which was imported from France.[31] In winter, the children had fun flying down the quarter-mile toboggan run.[32]

Eleanor's chauffeur would drive Eleanor and the children to Haven Hill on Friday afternoons, and Edsel would drive out himself later after finishing a full day of work. By early evening, Edsel and Eleanor would be dressed in hiking boots and would take long walks around the countryside.[33] Chain link fencing surrounded the several acres around the main cabin. At night, two guard dogs would be released onto the property to provide protection.[34]

***

All six properties owned by Edsel still exist today. A famous television personality now owns Skylands. The Hobe Sound and the Indian Village residences are still occupied. Haven Hill is owned by the State of Michigan and is used as an environmental and nature center. The Jefferson Avenue mansion is used by the United Automobile Workers union as a health facility.[35] The grand and stately Gaukler Pointe mansion is preserved in perfect condition, thanks to a $15 million endownment made by Eleanor at the time of her death in 1976. "Where once there were many extraordinary residences," she wrote in her will, "mine is the last to remain, for the changes in our manner of living... and in our attitudes have resulted in the demolition of all the others. I wanted it to be used for the benefit of the public."[36]

***

Edsel and Eleanor had four children: Henry Ford II, born in 1917; Benson, born in 1919; Josephine, born in 1923; and William Clay, born in 1925. The first three children took after the Clay family, with fleshy features, full lips, and a propensity for being overweight. William, on the other hand, took after Edsel—slight of build, thin, and handsome. He was the apple of Edsel's eye. "Billy was also Clara's favorite," recalled a Ford relative, "because he reminded her of Edsel as a child—there was that feeling of watching her own child grow up again."[37]

Clara also liked Eleanor. She told her niece, Catherine, that "if she and Mr. Ford had looked the world over, they couldn't have found anyone better suited for Edsel than Eleanor."[38]

That sentiment from Clara must have been expressed during a moment of weakness. If the truth were known, both Clara and Henry resented Eleanor for taking their son away from them. In particular, Henry had an intense hatred and contempt for the rich socialites of the Grosse Pointe set to which Eleanor had taken his son, and Clara tended to side with Henry on this regard. "It was even more galling to Henry because it was his money that made the young couple leaders in the social hierarchy he hated."[39]

Although Edsel was now a grown man and running a mammoth organization, Henry and Clara "wanted to completely tuck their child under their wings and guide him as they saw fit. They could not understand how Edsel could wander away on his own."[40]

It is not known if Edsel realized his parents' attitude toward Eleanor, but it probably would not have mattered. Edsel loved Eleanor dearly, and he preferred hobnobbing with the rich of Detroit, rather than enduring the dull agrarian lifestyle that his parents wanted for him. This caused the first breach between father and son, one of only many that would develop over the years.[41]

Chapter 9

Accusations of Evasion

"His country needs Edsel.
Where is he?"[1]

Edsel entered the Ford business right from high school. Four years later, in 1915, he was elected secretary of the company and in 1917 was appointed a seat on the board of directors. At age 24, he was assuming the responsibilities of a man at least twice his age.[2] However, these duties would soon pale against the challenges that confronted him at the outbreak of war in Europe.

The United States entered the Great War in the spring of 1917, despite the attempts of President Woodrow Wilson to keep America out of the conflict. Even Henry Ford did his part, sending a boatload of pacifists to Europe on a Peace Ship, hoping they could convince the belligerents to end the conflict. However, all efforts failed, and Wilson implemented the first draft since the Civil War, sending American troops to Europe. Soon, Ford factories were manufacturing both war materiel and Model T's.

***

In addition to his administrative duties, Edsel was assigned the responsibility of looking after the manufacturing of war products made by Ford. Taking advantage of Ford's mass production expertise, the company began making helmets, bullets, caissons, tanks, and submarine chasers known as Eagle boats.[3] In particular, the French were grateful for this materiel, but they wished the vehicles had come with men to drive the equipment as well.[4]

As an additional burden, on June 5, 1917, Edsel received notification from the Selective Service District Board of Eastern Michigan, calling him for military service and instructing him to "hold yourself in readiness to report for military duty...at a time to be specified." Fortunately, the "specified" time varied and could be far off because Edsel drew number 476. With such a high number, there was a chance he might not be called at all.[5] That would depend on how prolonged the war was, how many men would eventually be called, and how many others obtained exemptions.

***

The original civil service code governing the draft of 1916 provided several avenues for young men to receive exemptions. Those with poor health or physical maladies were automatically exempted, as were those seeking conscientious objector status, providing, of course, that they met the necessary criteria. However, the law also allowed men with extenuating circumstances, such as those who "were actually engaged in an industrial enterprise necessary to the maintenance of the Military Establishment"—Section 44, it was called—to apply.[6] Edsel clearly fell into the latter category. Although Edsel had a high draft number, he applied for an exemption—as did half of the other draftees—on the basis of "industrial necessity" to the local draft board, shortly after receiving his notice.[7]

One can only imagine Edsel's surprise when he received the board's denial. Everyone on the local board knew who he was. It was not as if some Washington bureaucrat were passing judgment on one of a thousand unrecognized applications that crossed his desk. This was Edsel Ford, the only child of Michigan's favorite son. Surely, he should not have to explain why he was applying for an exemption on industrial grounds. Even the local newspapers found the board's denial surprising. "All throughout the United States," editorialized one newspaper, "hundreds of men of less industrial weight than Edsel Ford are being exempted...because they are employed in the manufacture of products necessary to the successful prosecution of the war. Edsel Ford is worth at least a hundred [of those] men..."[8]

Fortunately, the denial by the local board was not the final disposition on Edsel's claim, and he was permitted to request an exemption from the district board. Edsel did so with the full support of his father and close confidant, Charles Sorensen.

At this point, Edsel was prepared to continue his philosophical fight against what he considered the built-in hypocrisy of the Selective Service system. However, both Henry Ford and Charles Sorensen convinced Edsel to submit a formal request for exemption under Section II-A: Married Man with Children, and Section III-L: Necessary Assistant or Associate Manager of a Necessary Industrial Enterprise of the Selective Service Code. The district board denied that claim as well and returned to Edsel his new status: Class I-A.

***

Now, Edsel was concerned. Had he lost his one chance of officially avoiding the draft? Was the board denying him his request out of hand, as a form of repayment for his well-known stance on its policies? He did not want to take the chance and asked the company's attorney, Alfred Lucking, to intercede on his behalf. From Lucking's assessment, Edsel had only two remaining options: (1) reapply for the exemption again, or (2) appeal directly to the President of the United States. Actually, Edsel had only one option, that of the appeal. The likelihood of the board reversing its previous decision was remote, but Lucking decided to reapply for the exemption nonetheless.

Instead of merely resubmitting Edsel's stance in a letter, Lucking decided to pursue a more rigorous approach. To help support Edsel's claim that he was a "necessary assistant" to the Ford Motor Company, Lucking obtained sworn statements—some 37 pages worth—from other key Ford executives, affirming that the company needed Edsel to produce war materiel.[9] Lucking compiled statements from a cross section of Ford managers involved in all aspects of the company, affirming that many departments and functions depended on Edsel's direction and guidance.[10]

The first executive to testify was Charles Brownwell, manager of the Ford advertising department:

> *Edsel Ford takes an active part in policies of dealer connections, the management of branches, questions of advertising, quantity, methods, copy, expense, in determining the lines thought to be expressed in our various publications, in formulating and executing the policy of factory papers and productions...*
>
> *[I have] found Edsel Ford to be consistently in his office, easily accessible and almost a daily necessity in getting prompt decisions upon matters of importance.*

Frank Klingensmith, assistant comptroller, was asked next:

> *Edsel Ford has become an important factor in the management of the company, having taken over important detailed work formerly referred to Henry Ford's office... That it is physically impossible for Mr. Henry Ford to determine all such important matters, and removal of Edsel Ford at this time would seriously impair organization of the company.*

Then, Peter Martin, engineering manager:

> *At the present time, [Edsel] is in constant consultation with the company's engineers, superintendents and factory managers with reference to their problems. These matters of such importance are so numerous as to make it impractical for Mr. Henry Ford to give them his personal attention and this practice is now fixed among company officials to consult with Edsel Ford only.*

Lucking also asked an outside supplier to support Edsel's claim, and John L. Allmand, factory manager of Fisher Body Corporation, was happy to respond:

> *Edsel B. Ford has always taken great interest in the construction and design of bodies to be produced for the Ford Motor Company by the Fisher Body Company, and [I have] been in conference with him relative to same many times; that some two or three weeks ago, Edsel Ford and the head of the Experimental Designing Department...called upon [me] with reference to certain changes, refinements and improvements to be made in the construction of closed bodies for the winter season of 1917–1918; [I] found Edsel Ford to be vested with full authority...to negotiate and authorize such changes and improvements and to plan and direct the production of such bodies for the coming year; [and he] showed a thorough practical knowledge of the construction of such bodies...*

These were all stellar remarks from authoritative men, but their comments could hardly be considered totally unbiased. Did the day-to-day activities that occupied Edsel's time really benefit the company, as these men proffered? According to Edwin Pipp, editor-in-chief of the *Detroit News*, the answer was "yes." "I got an illustration of Edsel Ford's intimate knowledge of the business in talking with him while the war was on," he said. "It had to do with the matter of cylinders for the Liberty motors (used to power Eagle boats). The government offered $19.50 each, and the Ford [Motor] Company turned them out for $8.25 each, saving hundreds of thousands of dollars for the government. Edsel Ford told me at the time exactly how it was done. He went into the detail of the mechanical operation; the experiments gone through before success was achieved, and told of the men, calling them familiarly by name, who had much or little to do with it, taking no credit to himself but showing at the same time a very intimate knowledge of and part in everything connected with it."[11]

None of these arguments, as persuasive as they were, satisfied the board. Five days after Lucking submitted Edsel's "Claim for Discharge," the district board sent its determination. "The Board has directed...that after considering the...affidavits presented in your behalf, they can find no reason to alter their previous opinions. The time for filing an appeal to the President will therefore begin today, and such appeal must be received in this office within seven days."[12]

Wasting no time, Lucking sent a telegram to the president immediately. However, in a shrewd move that probably saved Edsel from the front line, Lucking anticipated a delay and personally met with the board and persuaded them not to draft Edsel until the appeal had been decided.[13] All Edsel could do now was wait.

***

Meanwhile, Edsel's application to the president had become public knowledge and with it the stimulation of public debate over its probity. "Concerned citizens" began offering their advice as to how Edsel could use other, less noble ways of avoiding the draft than writing to the president. Men who supposedly understood the "spirit" of the law explained to him that he should voluntarily join the service and have the appropriate officials assign him to some cushy stateside assignment, probably in Dearborn.[14] It was done, he was told, in "hundreds and hundreds of cases."[15] One informed citizen even wrote to Edsel to provide him with the exact procedures[16]:

> *Dear Sir:*
>
> *In the event of your being refused exemption by the officials of the exemption board and as a precautionary measure against being taken into the army, the thought has occurred to me that you may be desirous of following the plan of action that one or two of the boys have here in Milwaukee.*
>
> *Two of the boys...who were desirous of beating the draft and had the necessary finances, purchased submarine patrol boats. The owners of these boats have donated them to the government and in consideration been granted commissions as second mates. They have been permitted to choose their crew to a certain extent, the government putting all these men through the general course of naval reserve training, consisting of three months at one of the training camps. These boats are then put in commission in the command of a competent captain to do patrol duty along the Atlantic Coast.*
>
> *This, as you can readily see, provides a delightful form of service and minimizes the risk to the greater possible degree.*

Such an arrangement would have been extremely easy for Edsel to accomplish, for the submarine patrol boats purchased by the two Minnesota boys were the same ones being manufactured by Ford Motor Company under Edsel's direction. If anybody could have bought his way out of military service, it would have been Edsel Ford.

However, that was not an acceptable alternative to Edsel. In his mind, he would either receive an exemption based on industrial grounds as he submitted, or he would willingly allow himself be drafted and suffer the consequences, the same as any other draftee. "I honestly believe that I can render my most valuable service here working at Highland Park, administering war work," he explained. "But if not here in connection with business, then in France. I want no stay-at-home appointment. I will accept none."[17]

In his response to the man who had made the suggestion, Edsel restated this position[18]:

> *Replying to your letter of recent date, wish to advise that your suggestion does not meet with my approval. The plan outlined by you is to my mind a good way of actually evading active service...*
>
> *My action in claiming exemption on industrial grounds is for the sole reason that I believe I will be of far greater service to the country right here in the plant that I am most interested in, which plant is now occupied with very important government work, than I will be in any other branch of the service. I fully realize that it is the duty of every citizen to do all he can for the government at this time, and if the Officials see fit to deny my exemption, I am perfectly willing to be drafted with the rest of the young men of this country.*
>
> *I sincerely thank you for your interest in the matter, but do not feel disposed to take up the suggestion you have outlined.*

Many believed Edsel's sincerity in these statements, but others thought he was using his position to obtain an exemption and that his claim based on industrial necessity was simply a cover-up for his not wanting to go. "I would be foolish to say that Edsel Ford is indispensable to the Ford industries," wrote one Ford adversary. "The local and district draft boards in Detroit, after an investigation, decided that he was not indispensable. The great Ford industries were built up without Edsel Ford. They most certainly can be maintained without him."[19]

* * *

While many disapproved of Edsel's stance, an equal number supported him, and as many letters were written in his behalf as were written against him. "The draft must be selective, not conscriptive," wrote one who tried to rationalize the entire sordid mess. "Industrial experts, skilled machinists, ammunitions makers, and industrial leaders must not be placed on the firing line at the cost of abandoning their tasks," wrote the *Detroit Free Press*.[20] One of the more supportive letters was from a Detroit mother who wrote a poignant and heart-wrenching letter to President Wilson[21]:

Dear Mr. President:

Henry Ford's only son, indeed his only child, has been refused exemption by the draft board and now it is up to you. The local Board were afraid to follow the dictates of their hearts because they thought the mothers of other only sons would come to them and say "Just because he is rich is no excuse, Henry Ford's son is no dearer to him than my son is to me, and you took my son." Curious how men go on forever, never able to understand the working of a woman's heart.

They say they can't exempt Edsel because the Ford Company can get along without him—probably it can. Henry Ford can get along without his son, but the industrial world cannot get along without the spirit of Henry Ford. The world has reason to expect that the spirit of the father will live in the son and in the son's son. Henry Ford has not spoiled his son—yet—as so many rich men do. He has put him into the business and is trying to train him to carry on his plans and purposes after Henry Ford and you and all the mothers whose sons are not coming back from "over the top" are dead. If our sons come back, they can't do for the working man what Edsel Ford can do because they will not have at their command the vast fortune and the big organization of the Ford Motor Company to do it with.

Edsel Ford will only count for one "over there." At the most, he will only stop one bullet. At home—if he's the man the world has the right to expect Henry Ford's son to be—he will count for a million men in the possibilities for helping humanity which he will hold in his hand.

Any Mother,

Mrs. Pound, Detroit

If all the board wanted was a body, someone to take up the spot of a chance lottery number—to "stop one bullet"—there were many volunteers who offered to serve in Edsel's stead. In an open letter to the editor, one such man volunteered[22]:

In regard to Edsel Ford being drafted into the United States service and seeking exemption, I wish to state that I will offer my services to Uncle Sam in his stead.

I am doing this purely as a matter of helping humanity and my country. I think Edsel Ford could serve his country best at his present position—helping carry on the great work his father established.

In offering my services in his stead I am not seeking notoriety, but a chance to do my bit. If I am allowed to serve in his stead and never return, the world will be better off that a life is spared to carry on a good work—that is, giving the working class a chance to exist properly. I do not know either Henry or Edsel Ford. I have never worked for them, nor have I applied for a position in their plants.

I am 42 years old and as sound as a dollar and could serve in some branch if given the chance on the offer I have made.

The emotional debate around Edsel's application was all but forgotten when he finally received his exemption on March 11, 1918, seven months after applying to the president. Except for a small blurb in the local paper, nothing again was ever said about it. That is, not until the senatorial campaign of 1918 went into full swing and Henry Ford entered the race.

***

At the insistence of President Wilson, who needed an ally in the Senate to help pass his peace initiatives, Henry Ford, Wilson's close friend, agreed to run as a Democratic nominee. This presented a real challenge to the Republican candidate, Truman Newberry. Both Ford and Newberry were multi-millionaires, but Ford's name was better known and his pockets were much deeper than those of Newberry.

Realizing that he had to come up with a campaign issue around which to rally support to offset Ford's popularity, Newberry seized on the rhetoric started by Republican Senator Lawrence Sherman of Illinois. Sherman, a megalomaniac and confessed Ford-hater—"I consider the Ford automobile an international pest," he once said—was the first to accuse Edsel Ford of obtaining his exemption through favoritism.[23] On the floor of the Senate (for which he was highly criticized later by his fellow senators), Sherman accused Edsel of being a "presidential pet" and that he had been saved from serving in the army by an abuse of executive power.[24] "It is generally believed," ranted Senator Sherman, "that Henry Ford made direct application to the president [on Edsel's behalf]...and as a result General Crowder was...required by the president to form the new regulations so as to create a class into which Edsel Ford could fall and be exempted without arousing too much attention and comment."[25]

Such accusations were absurd at the outset, and coming from a man who said he "knew everything from birth" made them more so. But it was too late; the damage had been done. In the heat of the senatorial race, the debate over Edsel's exemption became hot political fodder for Newberry and the press. "Of the seven persons in the world who were certain to go through the war unscathed," reported *The New York Times*, "six were sons of the Kaiser and the seventh was Edsel Ford."[26] The local newspapers, as we have seen, were more sympathetic, but that did not diminish the reaction.

Public reaction to these allegations was immediate. Letters-to-the-editor from enraged citizens appeared in newspapers throughout the country, calling Edsel a "coward" and a "slacker," and the process of appeal the "Edsel class."[27] Caricaturizing "swivel chair" commissions—the kind the two boys from Wisconsin received—a political cartoon appeared in one of the Eastern newspapers, showing Edsel sitting in an office chair in a bunker made of sacks of money, while fighting soldiers clashed in the background.[28]

***

The controversy over Senator Sherman and Newberry's accusations over Edsel's appeal brought such fervor that Attorney General Thomas Gregory asked President Wilson's secretary, Joseph Trummulty, for an explanation. "If the allegations about Edsel's application are untrue," wrote Gregory, "then someone should be in a position to deny them."[29]

Trummulty went directly to Provost Marshall General Enoch H. Crowder, who developed and implemented the regulations covering the draft, and asked him for a complete explanation of the circumstances that led to Edsel's Section 44. What he found was that all of the accusations about Edsel receiving preferential treatment were unequivocally false. First, General Crowder explained that President Wilson never acted on any of the 11,000 applications sent to him, and that included Edsel's application. Technically, of course, the president should have acted on each of these requests; however, in practice, he delegated that responsibility to General Crowder.[30] Too busy administering other aspects of the draft, the general did nothing with the applications except to pile them on an empty table. Second, even if Henry Ford had spoken to President Wilson about exempting Edsel (and there is no proof that he did), the draft laws were designed from the beginning to allow qualified men to gain an exemption if they held necessary positions in companies key to the war industry of the nation.

General Crowder then explained that he implemented all of the draft regulations after the first wave of inductees were sent to camp. The unabridged regulations divided the eligible draftees into several different categories. "All single men without dependents will be called first, and married men and industrial and agricultural experts will be selected in such a way as to supply the Army with at least 2,000,000 men with a minimum of disturbance to industry and home."[31] The general would have proceeded in this manner with the first wave of draftees, but there was not enough time. He was obliged to show the Allies some instant action by getting them to camp with all due speed.[32]

After the initial wave of men were inducted, General Crowder returned all of the presidential appeals to their respective local boards without action and instructed all registrants who had not yet been inducted to complete a questionnaire that would determine their new status under the full regulations. In other words, all inductees had to be reclassified "all over again."[33]

Edsel completed the new questionnaire, sent it to the local draft board, and was immediately reclassified in Class II: Married Man with a Child, and Class III: Necessary Assistant of an Industrial Enterprise.[34]

What Trummulty discovered and reported to the attorney general was that all of the clamor and controversy over Edsel's exemption was nothing more than political hot air. President Wilson never intervened on Edsel's behalf, and Edsel "was never made an exception."[35]

***

Unfortunately, Edsel carried the stigma of evading the draft for the remainder of his life. Although he had taken the steps that were well within the law and had never applied for an exemption based on home ties or other personal considerations, the ordeal left him uncertain about what he had done. "I was never sure that Edsel was sure he did what was right," recalled Charles Sorensen decades later.[36] However, Sorensen also said that he never knew anyone who felt that Edsel's decision was wrong.[37] One thing is certain: When World War II broke out, Edsel was glad to see his two oldest sons go into the service. Henry Ford II was drafted but, not wanting to go into the Army, joined the Navy in the spring of 1941. Benson, still too young to be drafted, joined the Air Force. No one could accuse Edsel's sons of being cowards or slackers, Edsel thought.

Ironically, the government quickly discharged Edsel's eldest son, Henry Ford II, when Edsel died so that young Henry could assist the company in its new war work. However, that was not the final irony in this tragic part of Edsel's life. According to Charles Sorensen, Edsel could have avoided the controversy and embarrassment altogether had he simply allowed himself to be drafted. Without saying a word about his predisposition, he probably would never have gone any further than the induction physical. Sorensen said that Edsel "could never have passed the Army's rigid physical examination anyway."[38]

Chapter 10

Buy-Out

Edsel Ford became president of Ford Motor Company too soon. He had barely turned 25 when his father appointed him to the position. Although Edsel was better educated than his father and had literally grown up with the company, he was being pushed into running a vast, worldwide empire. And what an empire the Ford Motor Company had become. First, there was the enormous 1,100-acre Rouge plant, where raw materials were brought in one end and new cars came out the other. There were 27 factory branches, of which 15 were assembly plants. There were hydroelectric plants, a fleet of ocean and Great Lakes vessels, coal and iron mines, and thousands of acres of timberlands. Add to this a vast rubber plantation in Brazil—a little world all its own—and 31 foreign plants.[1] Plunging Edsel into such enormous responsibilities at his young age was a disservice to Edsel and was irresponsible of Henry. Edsel's $75,000 a year salary would not be enough to compensate him for all of the grief he would soon encounter.[2]

All top officials of the company realized that Edsel would become head of the giant manufacturing company someday. It was his birthright. However, none of them expected the appointment to happen so soon, particularly Charles Sorensen, the production genius whom Henry admired almost more than his own son. Sorensen looked at Edsel "as a sort of youthful upstart, while he was doing the work."[3]

Henry knew Edsel was too young. He, himself, was a venerable 42 years of age when he became president of the infant company. So how could Henry reasonably expect his son, barely half that age, to be mature enough to step in and run a company that was 400 times larger? Henry made his son president not because he felt Edsel was ready or because, as he announced, "to devote my time to building up other organizations." He made Edsel president because of his selfish determination to gain complete, unimpaired control of the company.

***

With the resignation of Alexander Y. Malcomson, the local Detroit coal merchant who financed Henry's infant Ford Motor Company, in the spring of 1906 Henry Ford became the majority stockholder in the firm. When Malcomson left, he sold his 255 shares of Ford Motor Company stock to Henry Ford, edging Ford's ownership in the company to exactly 51 percent, a controlling interest. The 10 remaining original investors owned the balance of outstanding shares.

Sometime earlier, Albert Strelow, the last of the original 12 stockholders in the fledgling company, was the first to sell out, selling his 50 shares to James Couzens. Before 1907 ended, three additional stockholders divested themselves of the company, with Henry acquiring some of their shares and Couzens acquiring the remainder. When all shares were turned in and all checks cashed, Henry owned nearly 60 percent of the company.

***

From that point, Henry Ford did as he pleased with the company, much to the chagrin of the remaining stockholders, particularly John and Horace Dodge and James Couzens. All of them had become rich with the dividends earned on their Ford stock, but now that Henry had controlling interest, he began plowing more and more of the company profits back into the business for expansion rather than paying dividends to stockholders. This especially angered the Dodges, who counted on the steady flow of Ford dividends to finance their own automobile venture, which they started in 1914. In an effort to reopen the cash gates, the Dodge brothers brought suit against Ford a day after Edsel's wedding.

In their suit, the Dodges demanded that Henry Ford pay out at least 75 percent of the cash reserves of the company—some $39 million—immediately. However, Henry refused. "The stockholders have already received a great deal more than they had put into the company," Henry pointedly explained to John Dodge during Edsel's reception, and he did not propose paying any more to them.

John Dodge's response was equally direct. He told Henry that if he really wanted to be able to do whatever he liked with the money of the company, then he should pay for the privilege. In other words, Henry should buy him out, as well as the other five remaining stockholders. John and his brother Horace, at least, were ready to negotiate.

Henry, however, was not. Now that he had an overwhelming controlling interest in the company, he saw no reason to pay out large sums of money simply to pacify annoying minority stockholders, and he met the Dodges face to face on the courtroom floor.

In a legal battle that lasted two years, the Dodge brothers prevailed in their demands. Ford and his attorneys lost the first battle and appealed, only to be thwarted again. Finally, at the end of February 1919, a Michigan superior court ruled in favor of the plaintiffs again, and Ford was ordered to pay out more than $19 million in a "special dividend" to the company stockholders, to which Henry reluctantly complied.

Although John and Horace Dodge received a healthy dividend check for more than $2 million, it was a bittersweet victory. In addition to the court costs that they had amassed during the trial, Henry Ford kept the cash gates closed all the while. The court order benefited Henry much more than it did the Dodges or the others. Because Ford was the majority stockholder, he too was eligible for the court-directed dividend and, in essence, made out a check to himself for a cool $12 million.

***

Prior to his protracted ordeal, Henry Ford never imagined that minority stockholders could be so recalcitrant. Now, two years later, he realized that owning 60 percent of the company was not enough. He realized that to direct the company as he saw fit, he would need 100 percent control, and he spent the remainder of 1918 contemplating ways of obtaining it. By the end of the year, Henry had developed a scheme.

No stockholder then with the company would give up his lucrative shares lightly or cheaply. Henry knew that. However, he also knew that a defunct Ford Motor Company meant worthless shares, and worthless shares meant no stockholders. Therefore, Henry's plan, which came to light on March 15, 1919, as a headline in the *Los Angeles Examiner*, was to force the minority stockholders from the company. "HENRY FORD ORGANIZING HUGE NEW COMPANY TO BUILD A BETTER, CHEAPER CAR" ran the announcement.[4]

Here was Henry's blueprint, clearly spelled out in black and white for all the country and the stockholders to see. Ford was willing to buy out his minority stockholders, but he was not willing to pay them full value, and he threatened to start a new company to make them realize that their stock was only as valuable as Henry Ford said it was. To emphasize this fact, Henry resigned as president of the company and installed Edsel in his place on December 31, 1918. That done, he left for a protracted holiday at his winter retreat in Altadena, California.[5] From that point, the Dodge brothers, James Couzens, and the other stockholders would have to deal directly with Edsel.

***

All of the outstanding Ford stockholders were suspicious that Henry's announcement was only a ploy to negotiate better prices for their stock. Henry was notorious for "pulling fast ones," and Edsel never let on otherwise. In fact, Edsel carried out the coup in such a professional and businesslike manner that they could not help but believe that Henry meant what he had said.

Edsel was with his father when Henry decided to announce the formation of a new company, but he was not in favor of the strategy. For one thing, Edsel was concerned that his father would actually follow through with his threat by abandoning the company and starting a new one. His concern increased as Henry kept feeding one sensational story after another to the press, saying that his new automobile company would employ four or five times the number of workers currently at Ford Motor Company, and that the new car it would build would have none of the same parts as the current Model T.

During long discussions with his father throughout the ensuing days, Edsel argued that starting a new company would probably force out the minority stockholders but might also start a chain of events that ultimately could destroy the Ford Motor Company—and prevent a replacement as well. A better approach, argued Edsel, would be to simply buy out the stockholders. A buy-out would be much easier to achieve and cheaper to implement. The only drawback was that Henry might pay more for the stock than was reasonable, but at least he was assured that he would own a company intact in the end.

Henry reluctantly agreed and handed the task to Edsel—probably one of the more astute decisions Henry Ford ever made. Had he tried implementing the buy-out himself, it is possible that it would have spelled the end of Ford Motor Company. The hardheadedness of Henry, Couzens, and the Dodge brothers would surely have led to deadlock, if not more lawsuits, resulting in the complete dissolution of the company.

***

Edsel accepted the responsibility without hesitation, and he directed the task with the deftness of a manager with years of experience. However, he did not accept the challenge blindly. And he was well aware that many financial experts believed that the stock could not be purchased.[6]

Instead of jumping into negotiations with the stockholders himself, Edsel decided to assemble a team of experts to hammer out a strategy. First, he solicited the professional help of longtime Ford employee, friend, and Couzens' one-time right-hand man, Frank Klingensmith. "Kling," as Edsel called him, assisted Couzens in the early years in teaching Edsel the financial intricacies of the business. He was an astute bookkeeper and worked his way upward within Couzens' domain—which says something in and of itself—eventually becoming secretary of the company in 1913.

Edsel also recruited a relative newcomer, William B. Mayo.[7] Mayo was an engineer for the company and had directed the construction of the power plant for the mammoth Highland Park Ford plant. Edsel respected Mayo's analytical mind and trusted his friendship, not to mention Mayo's connections with the Old Colony Trust Company of Boston. It turned out that Mayo's brother-in-law, Fred Holmes, was vice president of both Old Colony and its banking affiliate, the Everett Trust Company, another Massachusetts concern.[8]

Early in their planning, Edsel, Klingensmith, and Mayo settled on an approach. All agreed to an outright buy-out of the minority stockholders; however, in an effort to keep prices under control, they agreed to keep the Ford name strictly out of any negotiations. Because Henry had already started the stockholders wondering about the viability of the company, Edsel and his associates saw no reason to erase their doubt.

Mayo then met secretly with Holmes to set up Old Colony as Henry Ford's agent. Holmes agreed to enter negotiations with the stockholders on a confidential basis and make it appear that Old Colony was simply another prospective purchaser of Ford Motor Company stock. The Dodge brothers, Couzens, and a couple of other Ford stockholders had each talked to potential buyers during the previous two years. A few weeks after seeing Henry's sensational headlines, the Dodges granted an option for their shares to an investment banker for $12,000 a share.

Thus, the timing was right. Another offer from another concern would not raise suspicions. However, Edsel was still worried that word would leak out, so he instructed Holmes to hire someone else—an unknown—to keep his plan concealed. To comply with Edsel's directive, Holmes appointed his most astute negotiator, Stuart Webb, to lead the campaign. Holmes said nothing to Webb except that Old Colony was prepared to buy all outstanding Ford Motor Company stock, and that if one minority stockholder refused to sell, then the whole deal was off.

***

Webb went to work immediately. He contacted each stockholder and immediately set the tone of negotiations by explaining to each of them his terms for purchase. All were interested.

Webb started negotiations at $7,500 per share, but the price soon reached $11,000, when it was evident that no one was interested in his initial offer. The stockholders were receptive to selling, but shortly after negotiations began, the parsimonious James Couzens raised the question of taxes, which made all the others reconsider their asking price for their stock.

Similar to all the other stockholders, Couzens also was willing to sell, but he had two stipulations. First, regardless of the price on which the others settled, Couzens wanted more. He had worked hard for the company, almost ruining his health in the process. For that effort, he felt he was entitled to a higher price than the others, and he would hold out until he received it.[9] Second, he wanted enough extra to pay for any income taxes that he would be assessed on the sale of his stock.[10]

To determine exactly how much extra that amount would be, Couzens needed to know the value of Ford Motor Company stock as of March 1, 1913, the date on which the federal income tax went into effect. He asked Webb to obtain an official determination.

Undaunted by Couzens' request, Webb, through a prominent Boston attorney, was able to persuade the Commissioner of Internal Revenue to sanction a 1913 value. The Commissioner sent a team of auditors to examine Ford Motor Company books, and they returned with a 1913 share value of just short of $9,500 a share.[11]

With that as a base line, Webb returned to the stockholders and achieved quick dispensation. The Dodge brothers' stock was obtained first, but only after Edsel paid nearly $700,000 to the investment banker for his option, plus the Dodges' asking price of $12,500 for each of their 2,000 shares. After that, however, all others followed suit, agreeing to the Dodge price as well—all except Couzens. At the last minute, he made one more demand.

Although he now knew what the government would assess him in taxes, Couzens told Webb that he would not sell until he knew who was behind the stock purchase. Was it, as was purported in the newspapers, General Motors? Or was it, as he suspected all along, Henry Ford?

Because that was the only obstacle between Edsel and complete ownership of Ford Motor Company, and because all the others had already capitulated, Edsel saw no harm in revealing his plan to Couzens. That would have been a risky move had Edsel not been so close to Couzens, for Couzens could have ruined the whole deal by refusing to sell at the last minute. However, that was a remote possibility, for the two men had been friends for years. Couzens used to send Edsel postcards while he was away on business or on holiday, and, as far as Couzens was concerned, Edsel was the only Ford worth talking to.[12]

With options from all the other stockholders securely in hand, Edsel took over the negotiation with Couzens and personally drove out to his country estate located on the outskirts of Detroit to "put all of his cards on the table." They had a good, long talk, and Edsel explained in detail the whole story. By the end of the evening, Couzens' curiosity was satisfied, and he and Edsel settled on a price of $13,000 per share.

Couzens did not fare badly. From the $2,400 he initially invested in Ford stock 16 years earlier, he received a check from Edsel for slightly less than $30 million.

In total, Edsel paid approximately $75 million to buy out the remaining Ford stockholders, a large sum indeed. However, that amount was significantly less than what it would have cost to start a new automobile company.

***

Henry was satisfied. In fact, he thought Edsel had done an exemplary job. When he was shown a news article explaining what Edsel had done, Henry simply laughed and said, "Well, if Edsel has bought it, I can't help it."[13]

Edsel did not mind, either. Now that his father owned the Ford Motor Company in its entirety, Henry redistributed its shares among himself and his family, giving Edsel 42 percent and Clara 3 percent. Henry, of course, held the controlling remainder.[14]

What about Edsel's presidency? Wasn't he installed as an expedient? Shouldn't Henry have resumed that role as the appropriate head, now that Edsel had accomplished his objective? Henry should have resumed his position as head, and despite the fact that Edsel had affected the stock buy-out almost perfectly, he was still too young to take over the reigns of the mighty Ford Motor Company. Likewise, Henry was not ready to surrender them.

Chapter 11

The Administrator

"Edsel was the administrator,
Henry was the boss."[1]
Charles Thomas

As president of Ford Motor Company, Edsel occupied an unpretentious suite of offices across the hall from his father's office in the Administration Building at Highland Park.[2] For its size, the suite was sparsely furnished with antiques that Edsel himself chose.[3] A few paintings and photographs adorned the walls, and in the corner was the crude gasoline engine that Henry had made so many years earlier. It was a constant reminder to Edsel of the family's humble beginnings.[4]

In the entryway was a portrait of Edsel's father, and two Gilbert Stuart paintings hung in the interior of the office. One painting was of General Henry Dearborn, the namesake of Henry Ford's hometown, for which Edsel paid $50,000. The other painting was of John Quincy Adams. There was also a John Trumbull portrait of Alexander Hamilton. On the wall close to Edsel's desk were photographs of his wife and children. However, what is more interesting than what Edsel had displayed on his office walls was what he did not have there—a picture of his mother.[5]

When the Ford offices were transferred to the new Administration Building on Schaefer Road near the new Rouge plant in 1928, Edsel moved to the new headquarters. (Henry did not. He set up his office in the Engineering Building.) Edsel's new office was more spacious than the one at Highland Park, and attached to it was a lounge where he would go when he did not want to be disturbed. "He was not too strong a man…and tired easily," explained Arthur Backus, a secretary on Edsel's staff, "and when the pressures of business got too great, he would go into the lounge, close the door, lie down on the leather couch and rest or have a relaxing cigarette."[6]

Edsel did his work on a substantial old English mahogany desk, which he always kept neat and well organized. Except for *The Wall Street Journal*, *The New York Times*, and *New York Herald Tribune* stacked on one corner, Edsel's desk hardly ever had more than a single, very thin sheaf of papers lying on it, awaiting his reply.[7]

The austerity of his office and the tidiness of his desk were indications of Edsel's style—no frills, no pretension, always serious and to the point. He was never one to waste time and was critical of those who did. This is why he was chauffeured to work nearly every morning.[8] (Occasionally in the summer, Edsel would take his speedboat *Greyhound* to work. He would motor from his Grosse Pointe home down the Detroit River to the Rouge plant, where he would be picked up by his chauffeur and driven to his office.[9]) Climbing into his Lincoln town car at precisely eight o'clock in the morning, Edsel made good use of the hour drive from his Grosse Pointe mansion to his office in the Rouge Administration Building.[10] During the ride, he would skim all of the Detroit newspapers—the *News*, the *Times,* and the *Free Press*.[11] Then he would respond to memos and answer letters.[12] By the time his Lincoln pulled into the executive garage, Edsel had done much of the daily administrative work, leaving the remainder of the day for more important tasks.

Edsel typically spent only the morning in his office, dedicating that time primarily to making phone calls, reading new memos and reports that had arrived since the previous day, and responding to correspondence. He rarely met his executives there, preferring to meet them in his father's office-turned-conference room across the hall or on their own ground—at the assembly line, in the Lincoln plant, or at the design studio. In all the years that Bob Gregorie, the chief designer at Ford, worked for Edsel, Gregorie could recall going to Edsel's office only once. Ralph Roberts, the chief designer for Briggs Manufacturing, who visited regularly during the1930s, never once met Edsel there. Roberts recalled that they always met at the Ford Design Department or at the Briggs design studio near downtown Detroit.[13] Edsel said he could keep his finger on everything easier through personal visits than through memos or reports—"that a manager should not make much use of a desk—he is supposed to be out seeing things."[14]

Because he spent so little time in the office, "When he *was* there, he was all business," said his secretary Albert Lepine. Except for the usual quick pleasantries of "Good morning" or "How are you?," Edsel did not engage in small talk of any kind. He proceeded straight to his desk and began the work of the day. In the many years that Lepine worked for Edsel, not once did he see him lean back on his chair and put his feet on his desk.[15] Edsel ordered the office security guards to "turn certain people away and let others in," so that he could minimize interruptions and adhere to his regimen.[16] However, Edsel always saw the "big executives," such as Charles Sorensen, Pete Martin, and Bert Craig, the company treasurer, whenever they asked. He always welcomed Jack Crawford, his engineering liaison, and Lepine, of course. Unless anyone else had an appointment with him, Edsel would refuse to give an audience.[17]

Even in the Design Department, his favorite domain, Edsel was purposeful. He would go to see his chief designer, Bob Gregorie, after lunch and talk about the designs on which the design team was working. If necessary, Edsel would approve some aspect of a design—maybe a trim piece, hood ornament, or an instrument cluster—but he rarely visited Gregorie to chat. "Edsel had no time for idle conversation—absolutely none," recalled Gregorie. "He made the decisions that needed to be made that day, and then left."[18] Edsel's industriousness and strict daily schedule are what enabled him to accomplish as much by noon as most executives achieve all day.

* * *

Next to Edsel's office was his secretary's office. Lepine assumed his position quite by happenstance. He was one of many secretaries who worked in the administrative side of the company, answering to both Klingensmith, the assistant secretary of the company, and Bert Craig. When Edsel became secretary in 1915, he too began using Lepine's services and eventually recruited him as his personal secretary.[19]

Lepine was a superb secretary, and he served Edsel well for 25 years. During that time, he and Edsel developed a rapport and a routine that freed up Edsel's time and greatly assisted him in his administrative duties. As memos, mail, and reports came into the president's office, Lepine sorted through all of it—both personal and business. (Edsel had some of his personal mail sent directly to his office, including income tax forms that Lepine completed.) Lepine then passed on to Edsel only those documents that needed his immediate attention. On items that were not urgent or needed clarification, Lepine would either set them aside for a later date or have one of his assistants make sense of them before forwarding them to Edsel. Even then, Lepine would not meet with Edsel until he had a number of topics or questions to review. Lepine was fully aware of Edsel's busy schedule and consequently would never bother him unnecessarily.

In the days before electric typewriters, word processors, and copy machines, Lepine toiled over a typewriter and pounded out letter after letter, always retaining for the files a carbon copy of whatever he did. He personally answered all inconsequential letters such as requests for money or for charity—of which there were many—and requests for autographs and appearances, which Edsel always refused. Lepine knew how Edsel would respond to such inquiries. Therefore, instead of taking the time to type a new letter each time one of these requests arrived, he simply made several form letters tailored to answer the typical requests. He then sent them out under the signature of "Office of the President."[20]

On letters that required Edsel's personal response, Lepine would write the responses for him. However, Edsel had his own unique way of signing those letters, depending on their genre. If it were business correspondence, he would sign his name, *Edsel B. Ford.* If the letter were more personal or if he knew the sender well, he would sign it simply, *Edsel Ford.*[21] His calling cards were printed in a similar format—the long version of his name was printed in block letters on business cards, and the short version on his personal cards. Cartier in New York made these simple but elegant cards.[22]

As Edsel read the countless reports, memos, and letters that crossed his desk, he made notes in strong, forceful handwriting in the margins with a No. 2 lead pencil. As he read, he wrote his decision, next steps, or questions that needed to be addressed before a decision could be made. On many occasions, Edsel would write the names of people who should be consulted. If his response required more than a quick note, he called Lepine into his office and dictated a reply. It was not uncommon for Edsel to rattle off a dozen letters a day.[23] This is when Edsel's true administrative prowess became apparent. His memory "was both accurate and retentive," recalled a close friend.[24] This "keen mentality," as Lepine called it, enabled Edsel to rattle off dictations that were precise and intelligible.[25] With his in-depth knowledge of the company, he was never at a loss for names, departments, or data.[26] "I have been amazed at the large amount of detail Edsel Ford can carry in his head as to the production and distribution of the Ford car and the men connected with it, as well as the broader grasp of Ford affairs," wrote a *Detroit News* reporter who witnessed Edsel's acumen firsthand.[27] "He knew everything that went on in the company," recalled a close associate. "It was Edsel who followed up everything in detail, on any promotion, experimental work, or anything..."[28]

His father was no less amazed at the grasp his son had on the organization. "Edsel knows everything that's going on everywhere," he said. "I don't know how he does it, but he does."[29]

The amount of work coming into Edsel's office became so voluminous that Lepine soon hired E.J. Houlihan and Arthur Backus as assistants to help carry the load. But even they were not enough. Eventually, Edsel had five secretaries working for him, with Lepine serving as his personal secretary and office manager.[30]

"Nobody should ever get the idea that Edsel was merely a figurehead," explains Ernest Liebold, "because if anybody every attempted to get into a job and run it, it was Edsel. And don't get the impression that he lived the life of ease. He didn't. He worked. He knew every phase of the business well. And he never avoided any responsibilities—he assumed them as fast as they came his way."[31]

***

During his quarter century as president of Ford Motor Company, Edsel was responsible for all aspects of the company except mechanical design and manufacturing—his father controlled those areas. However, everything else was Edsel's responsibility: finance, sales, advertising, purchasing, and styling—and for a number of years, labor relations. Especially between 1920 and 1926, when Henry Ford was getting involved in extracurricular activities—none of which pertained to the well-being of the company—Edsel shouldered most of the daily administrative responsibilities. Edsel, not Henry, was the real directing head of Ford Motor Company during these years, becoming one of the busiest men in American business.[32] However, Edsel performed his duties so quietly and efficiently that the public scarcely knew anything about him, an anonymity he both enjoyed and fostered.[33]

The area of the company that Edsel first grasped was finance. It was an area for which Henry never really cared and with which he rarely became involved, even when the company was in its infancy. Therefore, when Edsel became president, finance was a responsibility Henry gladly relinquished and an area Edsel quickly commanded.

Edsel took over the fiduciary duties of the company from James Couzens, who was as responsible for the success of Ford Motor Company as was Henry Ford. During the first few months of the existence of the company, Henry wanted to hold up a shipment of cars because he felt "they were not good enough." However, Couzens, more concerned with cash flow than with perfect machines, felt the cars were good enough and demanded that they be shipped so that the money could be received. The cars, he said, could be fixed later. The indomitable Couzens prevailed, and with him, the company.

Couzens taught Edsel the fine points of financial management. When Edsel assumed these duties, he watched the finances of the company as scrupulously as Couzens had done. Over the years, Edsel mastered the art of finance to the point where only he had the necessary understanding and background to make financial decisions with conviction and celerity. Although as treasurer, Bert Craig was responsible for monitoring and controlling the cash of the company, "Edsel dominated him completely," said Charles Sorensen. Craig could not pay out a single dime of Ford funds without Edsel's approval.[34] During the recession of 1920–1921, for example,

Edsel signed hundreds of notes obtained by Craig from many banks throughout the country to tide the company over until sales and cash started flowing again. As an observer recalled, Edsel signed "until he got tired."[35]

Edsel's other administrative duties would begin right after Henry and his cronies settled on some new idea in the manufacturing end. Regardless if it were a new part or a completely new automobile, Edsel would ensure that all departments throughout the company worked together to implement the change. First, he would meet with the purchasing manager, Al Wibel, and alert him about the changes being developed. Then he would start working with his sales and advertising managers to determine the best ways to promote and sell the new changes or new product.

There was never any love lost between manufacturing and the remainder of the company because Henry Ford ruled that manufacturing would dominate the company as long as he was alive. "What the manufacturing men said was final," says Henry Doss, Ford sales manager. "It was the law."[36] Thus, it is a testament to Edsel's diplomatic and administrative skills that he was able to get manufacturing to acquiesce into allowing his sales department to help set production schedules. He instructed his zone managers to survey their territories every 10 days and submit reports outlining what sales had been for the previous 10 days and for the previous month, what dealers' orders were, and what were the dealers' inventory levels. With this information, Edsel and his sales staff, led for many years by Jack Davis, were able to project what the sales volume of the company would be for the next 60 to 90 days.[37] Even while on vacation, Edsel instructed Lepine to forward weekly sales reports to him so that he could send his recommendations to Sorensen.[38] As their projections became more and more reliable, Edsel was successful in getting manufacturing to build not only the sales department's suggested volume, but to build that volume in the types and models of cars and trucks that were in demand.

Next to styling, Edsel concerned himself with sales more than any other aspect of the company, simply because it was imperative. Henry Ford and his manufacturing men could produce thousands of Fords a day, but none of them seemed to worry about selling the vehicles. Edsel did. "Your wire regarding May production would indicate that daily production is too low," he informed Crawford in the spring of 1925 from Miami Beach. "We must maintain dealer stock of at least 100,000 cars and trucks at this period of year to lose no business."[39]

In addition to monitoring sales, Edsel closely watched the company's advertising. He was astute in the advertising game, having studied automobile ads since he was a boy. The input and guidance he gave to Fred Black, the company advertising manager, and to the advertising agencies of the company, was always beneficial. "I remember one time," recalls Black, "when I was talking to [Edsel] about some copy, and I remarked that that copy would appeal to dealers."

Perceptively, Edsel recanted without hesitation, "Are we advertising to dealers or to the public?"[40]

Despite his stinging remark, Edsel was very supportive of the Ford dealers. Almost since the inception of the company, it had been advertising on a national basis. To bring the Ford story down to the local level, Edsel started a dealer advertising fund to entice dealers to advertise in their specific territories. The fund was not free—an additional $10 to $20 was added to the cost of each Ford shipped to the dealers. However, the dealers were ecstatic about having the professionally prepared advertising that the fund provided—at least for the moment.

Edsel approved every piece of advertising, right down to the copy. When the ads and campaign slogans began arriving, the dealers soon realized the extent of Edsel's conservative bent.[41] The ads were well done, equal to what General Motors or the other auto companies produced, but the ads were too honest. "We had to be very careful with Mr. Ford on using superlatives," recalled Doss. "Edsel was very meticulous about being factual and not bragging too much…."[42] Although Doss and the dealers thought Edsel's ads were not lively enough, they did believe his 1928 financing plan was a godsend. The best financing plan the company had until that time was the Ford Weekly Purchase Program, a pay-as-you-go arrangement instituted by Edsel five years earlier. This was not a true time-payment plan as the name implied, but merely a savings account that the customer could open at a Ford dealer. To buy a Model T under the plan, a customer would have to first register with a Ford dealer and then deposit at least five dollars in a bank and then continue making weekly installment payments until he had accumulated the price of the car. Then, and only then, could the customer return to the dealer and drive his new Ford home.

Edsel wanted to institute a *bona fide* installment plan through a Ford-owned finance company, similar to what General Motors was offering through its subsidiary, General Motors Acceptance Corporation. However, Henry was adamantly against such highfalutin financing schemes. Edsel kept pressing his father, and Henry finally capitulated by the time the new Model A was introduced in the fall of 1927. The new Ford, although peppy and stylish (thanks to Edsel), did not compete well with the increasingly dominant models of General Motors and other competitors. Edsel finally convinced his father to establish a finance arm when the Ford Motor Company was faced with the prospect of lower sales and production.

In the spring of 1928, Universal Credit Corporation (UCC) was organized under Edsel's direction. Of the 15,000 shares issued to form UCC, Ford Motor Company held 7,501, the Guardian Detroit Bank held 5,000 (of which Edsel was a major stockholder), and Edsel's brother-in-law, Ernest Kanzler, held 2,499 shares.[43]

When Edsel announced to the dealers two months later that the UCC was now ready to assist them in financing Ford products, they were exuberant. Not only would the new finance arm of Ford finance customers' new cars, it would also finance the dealers' inventory. Unlike the previous pay-as-you-go plan, customers could now drive off in a new Model A for one-third down and financing the remainder over 12 months.

As Edsel predicted, UCC also benefited the company at a time when the company needed it most. In a letter Edsel received from Ernest Kanzler three years later, Kanzler mentioned how UCC had supported Ford sales even as the country slid into the Depression. "Where UCC is operating," wrote Kanzler, "the decline in Ford sales in 1931 from 1930 was 43%, and where UCC was not operating, the decline is 65%."

In the area of worker welfare, Edsel was not as successful. It was not because Edsel failed to put forth the effort. When the company grew too big for Henry to recognize each worker by name and give him a slap on the back for a job well done, he passed the responsibility of labor relations first to Edsel and then to a pugilist named Harry Bennett. Until Bennett took over, however, Edsel did a commendable job.

Edsel was as paternalistic in the large corporate environment as his father had been in the small company. One year, Edsel and his staff had worked out a bonus that would distribute about $10 million among the workers, when one of the staff notified him that due to a coal shortage and the resultant loss of electrical power, the gigantic Ford factory would have to be shut down for an indefinite period. His advice to Edsel was to defer or even cancel the proposed cash outlay. However, Edsel did not agree. "All the more reason why we should go on with it," he said, "for if there is a prospect of closing down the plant, the men will only need the bonus all the more."[44]

Gestures of gratitude aside, Edsel could do little about the pernicious problems of absenteeism and turnover within the company. During a discussion with Bob Gregorie one autumn afternoon in the 1930s, Edsel digressed from discussing design to discussing labor relations. He pulled a piece of paper out of his pocket, which had a production curve drawing on it.

"What do you think of this?" Edsel asked rhetorically, showing Gregorie how the graph sloped downward, indicating lower production. "Everybody from the factory goes pheasant hunting this time of year, and look what happens to our production." Then he shoved the paper back into his pocket as quickly as he had pulled it out. "Well, that's the way it is!" he said. Then he and Gregorie returned to discussing design.[45]

Edsel achieved better success with the white-collar work force. He knew how to treat his managers and how to make them loyal workers.[46] He motivated them through praise, support, and honesty—techniques unheard of in the corporate world at that time. He was quick to praise and slow to criticize. It was not uncommon, for example, for him to take time from his busy schedule and send a letter or telegram to one of his men, congratulating him on a doing a good job. In a typical telegram, Edsel wrote: "We take great pleasure at this time of year in expressing to you...our appreciation of your services which you have extended over a period of years, having been particularly brought to our attention since your promotion. It is such conscientious and untiring effort as you have shown which has placed our company in the position we occupy today and which position we will continue to maintain through the concentrated and cooperative effort of our important [managers]."[47]

During Christmas, Edsel sent cards to a long list of Ford employees, and small gifts of appreciation to others. If he were in Florida for the holidays, he often sent boxes of freshly picked oranges to Charles Sorensen and other top executives. He presented cash bonuses to all of his executives. For those who were close by, he invited each to his office and personally handed them bonus checks, which usually exceeded $1,000.[48] These were tokens of appreciation, to be sure; however, they were sincere gestures, and Edsel's managers could not praise him enough for his generosity and sincerity. "I had never been so impressed with a man in all my life," expounded one manager.[49] "In my mind, Edsel was tops!"[50]

Edsel's employees liked the way he treated them and the way he managed them. Most of all, they liked the way he tried to move the company into a new era, one becoming more and more dominated by General Motors and other manufacturers run by boards of directors. Ford employees all hoped Edsel would eventually take over permanently and run the company the way he wanted it run—democratically and professionally. Then, the employees felt, Ford Motor Company would be a fine place to work, indeed. "We employees were all Edsel Ford rooters," cheered one employee.[51] "We stuck around because of him."[52]

In an industry run by callous autocrats who were indifferent toward the men they managed, Edsel was aware of and listened to people's feelings. He was far ahead of his time, practicing many managerial techniques common in today's business world. Edsel's was the reposed voice in a boisterous enterprise. He was considered by many as *the* gentleman of the industry.

He took an interest in all of his executives and managers. He learned of their expertise, their capabilities, and their limitations, and he willingly delegated tasks to them. When one of the men returned with a recommendation, he did not have to worry about stating the facts exactly as he had found them. Edsel's objective was to find solutions rather than fault. He "never questioned the authority of a person in whom it had been properly vested," said Fred Black. Edsel was "fair-minded and sincere," said Ernest Liebold, Henry Ford's personal secretary. Any questions he asked were always well thought out and pointed. He listened intently to recommendations, never interrupted anyone before hearing them out, and never ridiculed or responded with a snappy or sharp answer. Rather, he encouraged open dialog between himself and his employees. "I would tell Mr. Ford whatever I wanted," said a long-time Ford employee. "If he didn't think I was right, he would say so, and if I didn't think he was right, I would say so."[53] The Ford employees liked Edsel more for his ability to listen without pretension or arrogance. "Talking to Edsel," lauded one employee, "was just like sitting across the table from your best friend—that's just the way he treated you."[54] "You always knew where you stood...when you were dealing with Edsel."[55]

***

One of the more enjoyable aspects of Edsel's job was having cars made specially for friends. On a yearly basis, he would receive hundreds of requests to expedite production of their cars, and he took care of these requests himself. He even had a form developed to assist this process.[56]

When the new Model A Ford was introduced at the end of 1927, Edsel ensured that the one that had the same serial number as the 1903 Model A that James Couzens had owned was given to Couzens. In 1932, Edsel had one of his new Ford V8 sedans made especially for Will Rogers, who had visited the factory and had lunch with Edsel and Henry and six other Ford officials. To commemorate the event, Edsel had a gold plate attached to the dash of the car, inscribed with the signatures of all eight executives. "Well, I got the car and I got the nameplate, so the car is no good to me, for I cannot use it," Rogers jokingly wrote. "I am always standing still, having somebody look at the nameplate. So kindly let me know where to ship the nameplate so I can use the car. The car has done everything I asked it to. I have had it in two polo games, and it's won both of them. It's better than any horse I ever played on..."[57] Three years later, Edsel had a Ford cabriolet specially made for his friend Nelson Rockefeller, and in 1936 he gave a Ford phaeton to Franklin Roosevelt.[58]

***

Lunch was a central part of Edsel's daily routine. He met his father and other Ford executives nearly every afternoon in a small dining room located off the northeast corner of the Engineering Laboratory, a few miles from the Rouge Administration Building and the heart of Ford research. Piping-hot meals were served on a large circular table, and soon these executive get-togethers became known as the "Round Table" meetings.

Every day, Chef Jan (Jan Willemse), the Round Table's resident cook whom Edsel recruited from the posh Nautilus Hotel in Miami Beach in 1929, prepared a varied menu for the executives. They had their choice of split pea, vegetable, or soybean soup; two kinds of meat dishes; a fish dish; salad; and vegetables. Dessert was apple pie made with whole wheat and soy bean crust or soybean cookies made into the shape of a Model T hubcap. Edsel usually ate a full meal of soup, salad, and medium-rare steak. Henry, still adhering to his mother's dietary regimen, ate all of the soybean servings and "rabbit food."[59]

Henry could have held these meetings at the stately dining room atop the Rouge Administration Building that Edsel preferred, but that was too formal and businesslike for Henry. Instead, He preferred eating among the experimental chassis and engine parts in the Engineering Laboratory, so he had a small kitchen and dining room set up there.[60]

Edsel would be driven to the Engineering Laboratory and enter the little lunchroom through the front door, always arriving at precisely 12:15.[61] By that time, all of the other executives—Sorensen, Martin, and Wibel—were there. Henry, always the last to arrive, entered through the back door.[62] Before sitting down, Henry circled the table counter-clockwise and shook hands with everyone, ending with Edsel, who sat to his father's right. Sorensen sat to Henry's left, and any guests who attended sat next to the person who brought them.[63]

Henry started the Round Table protocol in lieu of formalized committee meetings, which he abhorred, and they were considered a replacement to those types of meetings, even though business was rarely discussed. "Sometimes they would talk about business," recalled a Ford manager. Most of the time, however, it was largely a social group with only everyday chatter on the subjects of the day, which one would hear at any informal luncheon table.[64] If business were discussed, Sorensen usually started the discussion. Even then, it was superficial. The "profound secrets" of the company, explained Ford branch manager Henry Doss, were not discussed there.[65]

However, if the profound secrets of the company were not discussed at the Round Table meetings or in boardroom meetings or executive offices, where were they made? Quite simply, and quite astonishingly, the weighty decisions of the Ford Motor Company were made in private, impromptu meetings between Henry and Edsel wherever and whenever they happened to run into each other.[66] Many employees, including Emil Zoerlein and Laurence Sheldrick, two Ford engineers who worked closely with both Fords for years, recalled seeing Henry and Edsel take up *ad hoc* conversations around a workbench in the Engineering Laboratory, in the design studio next to a drafting board, or while sitting on a bench in Greenfield Village. Usually, Henry would sit on one corner of a table or desk, with his foot up and his hands around his knee. Edsel, taking the opposite corner, would sit on his hands, feet dangling, and gaze directly into his father's eyes or out a window. There, they would talk, hardly moving from their positions, sometimes for hours at a time.[67]

Initially, Edsel never felt inhibited about approaching his father about any idea or subject. "They used to discuss a great many things," says Liebold, and Edsel would "argue with his father about things."[68] Of course, Edsel "knew his father was closed on some sides to some things," recalled a close associate, "but Edsel had a way of approaching him that, if anyone could, would plant a seed in his father's mind, hoping that someday it would germinate."[69]

***

Employees who were around Henry and Edsel frequently began noticing "something was going on" between them—something troubling.[70] They could not pinpoint exactly what the problem was, but they knew it was there by the way the two men acted toward one another. To these employees, the relationship between Henry and Edsel was not typical of a father and son.

It lacked warmth, compassion, and understanding. To them, Henry and Edsel treated each other more as diplomats than as father and son. (Every time the two men met, for example, they would shake hands.[71]) They were always too polite and too cordial to each other.

Throughout most of Edsel's adult life, episodes of true affection between himself and his father were rare. The last time anybody saw them display any kind of affection was in 1924, and nobody could remember the last time before that. Ray Dietrich, a prominent custom-body designer who had designed numerous custom bodies for Edsel, recalled the incident: "We had planned to meet at 10:00 A.M. in the Dearborn laboratories. Edsel was so punctual, I had made it a habit to be early. I was told by the receptionist that Mr. Ford [Edsel] was in a conference with his father. By 2:30, I had become disturbed and the thought crossed my mind that perhaps my program had been canceled. At three o'clock, a door opened at the rear of the lobby, and out walked Edsel and Henry, arm around shoulders, absorbed in deep conversation. When Edsel spotted me cooling my heels, he apologized profusely and grinned. 'Ray, such important business was being transacted, I forgot all about you. Dad and I have been designing an exciting new model, and we want your advice.' Then, they took me to a small shop. In the middle of the floor stood an unfinished baby carriage, which they were building for Edsel's youngest child. We laughed heartily and began our meeting."[72]

After that, the tenuous relationship between Henry and Edsel—boss and president, father and son—began to fracture. No longer did they joke with each other or tease each other. Rather, as they sat in some corner, discussing the destiny of the great company, no energy radiated and no animation emanated. As Zoerlein and Sheldrick and the others watched Henry and Edsel's discussions, they could not hear what the two men were saying to each other. However, they could see their facial expressions plainly. "Edsel always had a worried expression...a painful expression," recalled Zoerlein. "It seems that Mr. Ford was saying, 'Now, Edsel, this is the way it's going to be.' "[73]

Chapter 12

Differences in Direction

The business relationship between Edsel Ford and his father was not always strained. For 14 years—from the time Edsel entered the company right from high school until he began pressing his father for a car to replace the Model T—the two got along amicably. During construction of the Highland Park plant, for example, Irving Bacon found Henry and 16-year-old Edsel standing near the site, watching the workers and machines "with the avidity of a couple of boys."[1] "If that kid can only continue the way he is going," confided Henry to his office worker, George Brown, "I've got a boy that I can be proud of."[2] For the next 9 years, Edsel was the student, Henry was the instructor, and the topic was empire building. Henry was already anticipating the day when his son would take over the company.

When Henry appointed Edsel president of the Ford Motor Company, it was probably the biggest news event in the automobile industry since Henry had introduced mass production. However, Edsel assumed his new position without fanfare. When asked to comment on his new appointment, he replied humbly, using a Model T metaphor. "The hard part is finished," Edsel said. "It is easier to run something that has been cranked up and started and is running in high gear."[3]

Away from the limelight, however, Edsel was enthusiastic about his new position and entered it full of gusto and with an air of confidence that had never been seen in him. "The Ford interests have always been a one-man proposition and always will be," Edsel affirmed in the summer of 1924. "That's why father stepped out and let me steer. I'm playing his game—doing things myself when I want them done right."[4]

***

It appeared as though Henry fully supported Edsel, allowing him to take over more and more responsibility. Instead of making an immediate decision when confronted by one of his subordinates, Henry redirected them to his son. "I want you to take everything up with Edsel," Henry told them. "Things that you have been talking over with me, I want you to take up with him. Edsel is the boss."[5]

Maybe Edsel believed that, and maybe Henry wanted to believe that, but all of the men close to Henry Ford were dubious. "The smart ones," explained Fred Black, "also took up matters with the old man, usually saying, 'Well, I discussed this with Edsel, and he thought it was a good idea, but I wanted to be sure that you understood it and also thought it was a good idea.' "[6]

They did this because they had seen good men, talented men, come into the company and establish a foothold, only to be summarily dismissed—not because those men were doing a poor job, but because Henry thought they were doing too good a job and gaining too much power. The executives who had survived had no reason to believe that a similar fate would not befall Edsel, Henry's son or not. "It was generally understood," recalled a Ford executive, "that it was too dangerous to get too close to Edsel."[7]

Despite their behavior, all of the employees sincerely wanted Edsel to succeed. To them, he was the savior who would move the company into the next stage of its development. The manufacturing and production techniques so greatly influenced by Henry Ford were rapidly being replaced with administrative techniques, design, and sales. The employees believed that Edsel, with his education and experience, and the fact that he was part of a new generation that was receptive to these new developments, would move Ford Motor Company in this new direction. Edsel, too, felt that his father was losing touch with the business, "that he was more or less out of date."[8]

***

The year 1926 would have been the perfect time for Henry to relinquish control of the company to Edsel. Henry was now 63 years old and deserving of a less grueling lifestyle. Edsel, now 33, was ready to assume control. Thanks to Henry, there was probably no other man in the company more qualified than Edsel. Even Charles Sorensen could not match Edsel's skills in finance, administration, sales, and styling. Sorensen was an expert production man, but that was all he knew. A Ford executive who witnessed the transition said that business administrative techniques were beginning to crystallize at this time, and that Edsel had more of a concept for setting up these finer techniques—to analyze what had happened and use that information to make decisions about what to do in the future.[9] Another close associate remarked of this time: "I think the Ford Motor Company would have been better off if Henry Ford had said, 'Now, Edsel, I'm 60 years old and I can't last indefinitely. You go ahead for the next five years and let's see what you can do,' and let him do it."[10]

Unfortunately for Edsel and for the company, Henry never completed the baton pass to Edsel. The old-timers were right. Henry would never give up control of his company—not completely. During Edsel's formative years within the company, 1919 to 1926, Henry closely watched how Edsel managed and made decisions, and he was disappointed in what he saw. What Henry wanted to see in Edsel was someone similar to himself—someone ruthless, cunning, and dictatorial. What Henry saw instead was a considerate, unpretentious, and democratic person—someone whom he considered "too soft" and "too easily influenced."[11] As long as Henry felt that way about his son, he would never relinquish full control of the company to Edsel.

***

Because of Edsel's self-imposed obscurity, he has been portrayed as being "a chip off the old block"; in effect, he said as much. "I have not worked out a separate business philosophy for myself. It has not been necessary, because on all material points, I absolutely agree with my father's philosophy. I do not merely *accept* his beliefs, I feel as strongly about them as he does."[12] Edsel made this statement not to delineate his business philosophy or to confirm the general belief that he was like his father. He made the statement as a ruse to pacify the nosy public and publicists. It was the most direct way to divert attention from himself back to his father. Why would anybody want to talk to Edsel if all he would do was parrot his father?

The truth was, except for their similarities in appearance and a mutual interest in automobiles, the two Fords were complete opposites. They were as different as a Model T and a Lincoln. Edsel's cousin, Catherine Ruddiman, confirmed this. Henry once confided to her that "he and Edsel had not always understood each other, and at times, could not see eye to eye."[13] The way Miss Ruddiman described it was that Henry and Edsel "did not speak the same language."[14]

Henry and Edsel's differences were much deeper than that. In the same way that the homely Model T could go places the luxurious Lincoln could not, while the Lincoln was more suited for a night on the town than was the Model T, the differences between Henry and Edsel should have augmented each other, making each one stronger and more resourceful. Instead, Henry allowed their management philosophies, how they wanted to organize the company, and how they conducted themselves to create an environment of conflict—driving each other apart and causing Edsel untold frustration, humiliation, and sickness in the process.

Unlike Henry, who was autocratic and often dictatorial in running the company, Edsel used a democratic style of managing and delegated responsibility whenever he could. He rarely gave orders, preferring instead to solicit input, suggestions, and recommendations from his managers—allowing them to make their own decisions. For example, instead of giving a direct order, Edsel would say, "It would be nice if we could do this," or "It would be nice if we could change this to that."[15] He approached problems analytically and impartially, and he relied on consultation and discussion with his subordinates to reach decisions. "He would not compromise between what he thought was right or wrong," explained one executive, "but would seek adjusted agreement between extremes."[16]

Through his enlightened managerial methods, Edsel had as much control of the company as did his father. He also had the respect of everyone, and, consequently, he got things done through them. "The whole company would go to blazes for him," said one close associate."[17]

On the other hand, Henry was guided by hunches and intuition, and he did not seem to care how anyone else felt.[18] He intuitively felt that he knew all the answers, and that unless someone agreed with his position in its entirety, that person did not belong in the company.[19] "If I take my hunch," Henry explained to one of Edsel's schoolteachers, "I'm much more apt to be right than if I try to reason it out."[20] Obviously, this was the case on many of Henry's propitious decisions, but he used it as an excuse to prevent others from making contributions. One reporter who had known Henry for years commented that Henry was selfish, and another associate said that Henry was a "stubborn, hard-headed man."[21]

"Henry Ford always was apt to jump to a conclusion on the basis of too little information," recalled Ford advertising manager Fred Black. "We used to say around the office that if Henry Ford saw three blackbirds in the morning, all birds were black that day."[22]

Edsel's management style befit his docile personality. "He had a reasoning mind and liked to sit down and talk things over" before forming a conclusion.[23] More than that, though, Edsel's was a style that snubbed the years of taking orders from his autocratic father.

***

Edsel learned the mechanical side of the business from his father, and the business side of the company from James Couzens and Frank Klingensmith, but he learned his democratic management style from his one-time Detroit University School manual arts instructor, Clarence Avery. Edsel was so impressed with Avery's mechanical "horse sense" that he persuaded Avery to come to work for the company in 1912.[24] Avery had a knack for solving complex problems. (He was instrumental in helping Henry Ford develop the moving assembly line.) Equally important, however, Avery became Edsel's tutor in the art of managing.[25] He taught Edsel that he could get the most from his men by leading them rather than driving them—a concept not only new to this cutthroat business, but one that was all but anathema to Henry Ford.

Unfortunately for Edsel and the company, Henry did not understand this approach. He mistook it as a sign of weakness, and anybody who used it as spineless. From Henry's perspective, anyone who adopted Avery's democratic management approach could not effectively control or direct any company, let alone the mammoth Ford industries.

Henry eventually forced Avery out of the company, as he would so many other Edsel advocates, and he began criticizing and ridiculing Edsel for accepting his old schoolteacher's methods. Although Henry needed others' help and assistance, he alone always made all of the final decisions. Being the boss while also asking for opinions from subordinates was contradictory, according to Henry, so he tried to teach Edsel that same reasoning. Henry felt that Edsel was listening too much to others' opinions, and he told Edsel many times, "Don't be swayed by what somebody tells you."[26] Henry became obsessed with this point. He wanted Edsel to change and began taking steps to force Edsel to "be harsher."[27] Henry thought his son needed some "toughening up" to become "a little more shrewd, a little more crafty, and... a little more tricky," similar to Henry.[28] How Henry attempted to do that was the most tragic part of Edsel's life.

***

Many incidents over the years caused Henry to feel that Edsel was too easily influenced, but none was more poignant than Edsel's decisions to add more office space to the Highland Park Administration Building and to add more capacity to the steel-making operations of the company. In both instances, Henry abruptly and ruthlessly countermanded Edsel's decisions, hoping that this shock therapy would make Edsel tougher.

During the halcyon days of the Model T, William Knudsen and Frank Klingensmith convinced Edsel of the need for additional office space. They recommended that a wing be added to the existing Highland Park Administration Building. In his typical fashion, Edsel listened to their proposal, concurred, and ordered the construction to begin as soon as practical.

Bulldozers arrived in short order and began digging a hole for the foundation, piling the earth all around the hole. The contractor was making good headway until Henry came into Edsel's office.

"What's the construction for?" Henry asked his son.

"We need more office space," replied Edsel.

"What makes you say that?" asked Henry.

"Well, let's take a look, and I'll show you," replied Edsel, confident of his position.

As the two Fords walked through the Administration Building, Edsel showed his father the over-crowded offices and the additional desks that had to be lined up in the aisles, similar to cots in a hospital. After they had finished their tour, it was evident that more office space was needed. However, Henry outlined his position in short order, and it did not match his son's. "Edsel," he said, "we don't need that extra building."

"Well, all right, father," acquiesced Edsel. "We'll close up the hole."

"No, don't do that," replied Henry. "Just leave it that way for a while."

As a penance to teach his son not to listen to other people, Henry ordered that the hole remain open to remind Edsel of his apparent blunder. What a penance it was. Henry made sure the hole remained open not for a couple of days or a few weeks, but for several months. The frequent Michigan rains quickly filled the hole with water, and mud from the piled dirt washed down onto the pristine lawn and across the sidewalk. Everybody who entered the Administration Building had to walk by the mess every day.[29]

As for the congested Administration Building, Henry also took care of that situation in his own demented way. To make room in the overcrowded offices, he purged the administration offices of half of its personnel. Employees found out they still had a job after returning to work the next day and seeing if their desks were still there. "Are you still at Ford?" asked one employee to another. "Well, I was when I left this afternoon!"[30]

After the purge, Henry took Edsel on another walk through the Administration Building.

"See that?" Henry asked rhetorically. "We don't need any more room. We've got plenty of room, haven't we?"

"Yes, father," Edsel replied sheepishly.[31]

On another occasion, the affront to Edsel was equally cutting, and the expense was equally high. William Mayo, chief engineer at the time, convinced Edsel that the company needed additional coke ovens to keep up with the demand for steel requirements. Edsel told Mayo to proceed with building the ovens. When Henry heard about the new project, he could not see the need for the ovens but did nothing to stop the work, even though he should have stopped the project immediately. Instead, in what was to become one of Henry's most contemptuous examples of ruthlessness and insensitivity toward his son, he waited until the ovens were completely built before making his opinion known. Henry even told his comrade, Harry Bennett, of his intentions. "Harry," he said, "as soon as Edsel gets those ovens built, I'm going to tear them down."[32]

Whether that statement was made in confidence to Bennett is not known, but Bennett, in one of his more benevolent gestures—of which there were precious few for a man in his position—decided to warn Edsel of Henry's intentions. Edsel found Bennett's story too fantastic, perhaps with good cause. Edsel knew Bennett was out only for Bennett, and this story seemed to be another example of Bennett trying to undermine Edsel. By this time, whom was Edsel to trust?

"Bill knows more about coke ovens than Father," Edsel explained to Bennett. "I don't think he'll do anything of the kind." Nothing more was said, and Bennett waited for the explosion.

As Bennett had predicted, Henry followed through with his threat as soon as the coke ovens were completed. First, Henry fired Mayo. Then, as a climatic insult to Edsel, Henry ordered Mayo's son, William N. Mayo, who was also a Ford employee, to tear down the ovens.[33]

When Edsel heard about what happened, he shook his head in frustration. "I don't know what kick father gets out of humiliating me this way," he said.[34]

It can be argued whether or not the company actually needed more office space or more coke ovens, but there is no doubt that Edsel did the analysis and found that it did. In any event, these two decisions prove Edsel's decisiveness more than his gullibility. Edsel may have been reserved and soft-spoken, but he was not a pushover.[35] He was "inclined to make a detailed study and gather all the facts," instead of making the snap decisions for which his father was notorious.[36]

***

This difference in approach to managing and decision-making was the underlying reason for the derision that eventually developed between Edsel and Henry. Neither Edsel's approach nor Henry's approach was inherently right or wrong. However, because Henry had made a number of uncanny business decisions in his career—the low-priced car, mass production, and increasing the wages of the laborer—all without making a detailed study of any of them, Henry believe that his method was superior to Edsel's. He seemed to have forgotten that he had also made a number of out-of-hand decisions that were as devastating to the company as his propitious decisions were beneficial. These ill-advised decisions included not replacing the Model T sooner (Edsel's idea), not changing from a four-cylinder engine to a six-cylinder engine (also Edsel's idea), and not incorporating hydraulic brakes sooner (another one of Edsel's ideas). Joe Galamb, one of Henry's closest engineers, confessed years later that Henry made him work on many projects that were totally unfeasible from the outset.[37] Another long-time Ford engineer concurred with Galamb. "Henry Ford," he said, "was quite an ignorant fellow."[38]

Henry's worst error in judgment, however, was accusing his son of not having "enough boldness."[39] Henry told one employee that he was "very disappointed with Edsel" and confided to Harry Bennett that he thought "Edsel was a weakling."[40]

Contrary to his father's beliefs, Edsel was not easily swayed—not by his associates, not by the company's suppliers, and not by his father. Although he sought input from all corners, Edsel "was not one to be pushed into accepting someone else's ideas," explained Bob Gregorie.[41] Edsel always made the final decision based on input and his knowledge of the business. He delegated responsibility and listened to advice from many sources. He had definite preferences of his own, and when he liked something, he would say so. When he did not like something, he also would say so.[42] If Edsel wanted to know the status of a project or had to make a decision quickly, he went directly to the person in charge of the project, took him to one side, listened to all available facts, and then made a decision right there.

For example, during development of the Model A in the late 1920s, both Henry and Charles Sorensen witnessed Edsel's decision-making abilities. Sometime earlier, Edsel had approved the use of artificial leather as a new upholstery material for the new car, and he ordered the Purchasing Department to order a large quantity of the new material. When a sample of the material arrived, "Doc" Hueston, a clerk in the department, approached Sorensen to approve it. "I won't okay it," remonstrated Sorensen. "I don't know anything about it."

Hueston still needed approval before he could put the material into production, so he looked to Edsel, whom he tracked down a few days later at the Engineering Building. Henry, Sorensen, and Edsel were having a meeting, and as Hueston entered the room, Sorensen asked Doc if he wanted to see him.

"No," replied Doc. "I want to see Mr. Edsel."

"Well, Doc, what is it?" replied Edsel. Hueston unrolled the sample leatherette.

"Here's a red pencil," he told Edsel. "Would you please approve this material by signing your name here?"

Sorensen immediately interrupted, saying, "What's the idea? Why do you have to have Mr. Ford okay that?"

"Wait a minute, Charlie," Hueston recanted. "There isn't a man in this company who would put a yard of this new material into production—and you're one of them, because I came to you first."

"Mr. Ford," continued Hueston, "would you put your initials on it?"

Edsel took the pencil and scribed, "OK by Mr. Ford." Then, hesitating a moment, he scratched out "Mr. Ford" and wrote "EBF" instead, wanting everyone to know which Mr. Ford had given the approval.[43]

***

In addition to a different method of managing, Edsel wanted to restructure the company so that it operated like other large corporations. During its formative years, the company did not need a rigid organizational structure. After it became a formal gargantuan structure, with distinct lines of authority and specific positions, Edsel's management style made sense. He confidentially told Fred Black that he thought the company would be much better off with a half-million stockholders and a board of directors—that it would be better if "run by groups than by the opinions of one individual."[44]

That statement, of course, was a direct affront to Henry's philosophy and even to his authority. Edsel was treading on thin ice by even mentioning it. Henry loathed the formal organization and unabashedly said so in his first book, *My Life and Work*. In that collaborative work, Henry goes into excruciating detail of his anathema to the formal organization. "To my mind," he wrote, "there is no bend of mind more dangerous than that which is sometimes described as the 'genius for organization.' This usually results in the birth of a great big chart showing, after the fashion of a family tree, how authority ramifies. The tree is heavy with nice round berries, each of which bears the name of a man or of an office. Every man has a title and certain duties which are strictly limited to the circumference of his berry."

Knowing how committed Edsel was to the formal, organized method of running a company, and to show how vindictive Henry could be, even toward his own son, Henry decided to test the system that Edsel was trying desperately to implement. During one of their rare but mandatory board meetings, Henry asked Edsel whether the books of the company were all in balance.

Receiving a nod from Bert Craig, the company treasurer, Edsel said that of course they were. However, he should have suspected something fishy was in the offing as soon as his father asked the question, because Henry rarely exhibited interest in that part of the business. Unknown to Edsel, Henry had been renovating his yacht and supposedly had spent a substantial sum of company money in doing so.

"Oh, they are?" sneered Henry. "You don't know anything that's going on, and you're surrounded by crooks!"

"What do you mean?" stammered Edsel.

"There has just been a half-million-dollar expenditure on building a yacht—somebody around here is a crook!"

It was one of the few times Edsel talked back to his father in public. "Well, in that case," replied Edsel, almost surprised at his own daring, "the crook would have to be you!"[45]

If Henry had been using company money to rebuild his yacht, he would have to have been covering up the expenditure some way, because Edsel and his finance men would never have allowed that much company money to be spent without somebody knowing about it. "Edsel kept a finger on everything," explained his secretary. "He had a hundred thumbs stuck in as many holes in the boat," is the way one executive described it.[46] A half-million dollar leak was simply too big to go unnoticed.

Moreover, Henry did not have the patience to monitor such an intricate scheme, which obviously would have taken months to complete. Unlike Edsel, he detested the minutiae of the day-to-day activities of the company, preferring instead to look only at the big picture, such as what type of power plant his cars should have, or producing as many cars as possible at the lowest price. Henry could care less about the thousands of details required by as many people to accomplish his grand schemes. Recalled one Ford executive, "It was almost impossible to keep Henry Ford in a meeting for more than fifteen minutes."[47]

Albert Lepine, Edsel's long-time secretary, glimpsed Henry's distaste for detail one morning while Edsel was meeting with some of his managers. Peeping through Edsel's office door, Henry asked Lepine what was being discussed at the meeting.

"They're having a conference on taxes, Mr. Ford," said the secretary. "Would you like to join them?"

"Heavens, no," replied Henry. "That would bore me to death."[48]

Meetings never bored Edsel; they were central to his democratic style of managing. Indeed, his methodical approach to running the company required that he solicit the cooperation and recommendations of his subordinates. "Edsel had good judgment," says a close associate, "but he sought guidance from others." Edsel was particularly receptive to the ideas he received from dealers, customers, and staff about improving Ford products.[49]

Edsel's father was too obstinate for meetings or any other means of formal discourse, and the phenomenal success of the Model T reinforced his stance. Now, Henry believed he could do no wrong—that he possessed some sort of sixth sense about the wants and needs of the car-buying public. He became jaded by the tremendous demand for Ford products and began to ignore customers' opinions and recommendations altogether, although they began asking for legitimate features such as more power, better luxury, and enhanced styling. Henry was determined to give his customers what *he* termed was best for them.[50]

From the beginning of Edsel's reign as president, it was apparent that he would do things differently than his father had done. Unlike his father, Edsel surrounded himself with a well-educated, sophisticated, and resourceful group of men. They were the antithesis of his father's cronies, who were truculent, aggressive, and uncouth. For example, Henry's men settled disputes among themselves and with others by verbal assaults, fist fights, and guns. Edsel's managers approached problems openly and analytically. So dichotomous were the two groups that they became two separate factions within the company and were known as Henry Ford-men and Edsel Ford-men. The two most disparate men, Henry and Edsel Ford, led each group.

***

Even if Edsel had overcome his father's distaste for the formal organization, his obstinacy, and his unpredictability, he still would not have had the opportunity to take control of the company. Henry was too mercurial and too mutable to allow that to happen. The instant Henry realized that Edsel had finally grasped his stratagem and understood his nuances, he would switch again, baffling Edsel and all others. During a casual conversation with Irving Bacon, the resident artist of the company, Henry revealed his true *modus operandi*. "A big Ford family discussion was going on," Henry explained. "Mrs. Ford wanted something done her way, and Edsel wanted it done another way. I let them fight it out for a while, and when they all got through and decided on what they wanted, I stepped in and had it done my way."[51]

Therein lies the fundamental difference between father and son, a difference that would haunt Edsel for most of his adult life. "Henry Ford never wanted to have peace around him," explained Charles Sorensen. "A good battle, a real problem, is what pleased him."[52] There was nothing that Edsel or anybody else could have done to change that, although to Edsel's credit, he tried. Edsel understood that problems were a part of business, but to deliberately create them was, in his mind, counterproductive and created unnecessary chaos. Thus, Edsel desperately tried to rule through meetings and interaction with his managers and to implement a formal organization, to ensure that real problems could be dealt with effectively and efficiently. All Henry did was to assign two or more people to work on the same project—not to see which person could come up with the best solution, but to determine the weaker of the two and then to get rid of the slacker.[53]

What kick did Henry get from tormenting his son this way? None. He did not enjoy it, but in his Darwinian form of management, it was the only way he knew to teach Edsel to become tougher, more shrewd, and more independent—to become, in effect, identical to himself.[54] The obstacles Henry created for Edsel as a young man learning the business were understandable—they were better than any case study Edsel would surely have covered in college. However, to continue the affronts, now that Edsel was well experienced and had developed his own method of managing, crossed the line from education to persecution.[55]

"I think Henry Ford was trying to educate Edsel the hard way and didn't know when to stop," explained Emil Zoerlein.[56] Theodore Gehle, Sorensen's secretary, was more emphatic. "Henry Ford was so unmerciful in embarrassing Edsel's well-thought plans, it was unbelievable," he said. "If I didn't know it was true, if somebody told me a story of the way Henry Ford treated his son, I would be inclined to say it was impossible, that it must be wrong—it was that bad. If I were Edsel, I would have told Henry to stick the job."[57]

Ironically, Henry blamed himself for Edsel's behavior. Since the time when Edsel was a little boy, Henry overprotected his son, and he thought that was why Edsel turned out the way he did.[58] That was the easy answer. Subconsciously, however, both Henry and Clara never wanted their only child to grow up. "They wanted to keep him close to themselves and to guide his every thought," recalled Sorensen. "Henry never got away from the feeling that his little boy had to be protected and looked after."[59]

The only way Henry believed that his son would change would be for him to put Edsel through the ringer until he did change. He actually believed that this process would do Edsel some good. "It was his [Henry's] idea of the best way to educate Edsel and get some sense into him. I don't think Henry had any idea the pressure this was putting on Edsel or that this would suppress him more," explained Charles Voorhess, a long-time Ford employee.[60] It did, of course. Every time Edsel tried to make his own decision, his father would rescind it. This caused Edsel to become more introverted rather than more open. The more shy that Edsel became, the more pressure Henry put on him to change. It was the classic reward-and-punishment motivational theory, but without the reward.

***

The reprimands, insults, and constant humiliation eventually overwhelmed Edsel. Between 1919 and 1926, he held his own, made many decisions on his own, and brushed off his father's attacks as learning experiences. Edsel once related to a reporter, "Father seems to know in advance how things are going to turn out, but he just waits and does not say a word, hoping that the man will discover his own mistakes. If the man does not discover that he has made a mistake, father steps in and sets things right. I only hope that I can make all of my mistakes while he is still here to show them to me."[61]

In the years after 1926, however, "Edsel developed a cynical bitterness and resignation toward everything in relation to the company," explains Emil Zoerlein. Edsel took the stand that he would do what he could and get along the best he could until the time came when he could change things.[62] "If that's the way father wants it," Edsel told his associates in confidence, "that's the way it's going to be."[63] Unfortunately, Edsel never got his chance.

Henry usually had two reasons for doing or not doing something—the reason he told people, and the real reason. He once revealed this duality to Harry Bennett. "Harry," he said, "never try to out-guess me."

"You mean, I should never try to understand you?" questioned Bennett.

"Well, that's close enough," replied the auto magnate.[64]

When another confidant of Henry's told him that he was inconsistent, Henry became angry. "If you mean I change my mind," the old man replied, "I reserve the right to change my mind at any time I want to!"

Humiliating Edsel was not the end of Henry's insufferable treatment of his son. Not satisfied with merely pointing out Edsel's errors, Henry taunted Edsel by reminding him about them whenever the opportunity presented itself.[65] Leaving the foundation hole open after Edsel attempted to build additional offices in the Administration Building was only one example.

After visiting the Lincoln plant to see the new cars come off the line, Henry and Edsel invited their favorite dealer, Billy Hughson, to take a ride in one of the new models. It was in the middle of winter, and the roads were rutted with frozen snow and ice. Edsel was driving, and they were traveling along lickety-split when suddenly a large ice truck loaded with blocks of ice turned onto the road in front of Edsel's speeding car. Edsel slammed on the brakes, but to no avail. The heavy Lincoln simply skidded over the frozen ground. A collision seemed inevitable.

Fortunately, Edsel managed to veer the massive automobile out of harm's way onto the other side of the road, but directly into the path of another oncoming vehicle. "Well, here we go!" shouted Hughson. The oncoming car was equipped with tire chains and was able to maneuver far enough to the right, allowing Edsel to squeeze the Lincoln between the ice truck and the car.

Aside from the right front fender being somewhat smashed and upsetting the passengers, no great harm was done. Edsel brought the Lincoln to a halt to regain his composure. As soon as he did, Hughson blurted, "Has anybody got any toilet paper?" Everyone looked at each other and laughed, even Edsel—although deep down inside, he was lamenting.

That evening, Edsel invited Hughson to his home for dinner. As they were sipping cocktails in the den, Edsel said, "You know, I wouldn't have had that happen for anything in the world."

"Why, Edsel?" consoled Hughson, seeing that Edsel was still upset about the incident. "It could have happened to anybody."

"I know," replied Edsel, "but father will never forget it. He will always throw it up to me."[66]

***

For all the struggles he endured and all the challenges he faced every day in running the Ford Motor Company, Henry Ford always appeared calm and collected. With the exception of an occasional complaint about his stomach, it seemed that he never worried. He had no reason to worry. He was the boss, completely and totally, and nobody would dare to threaten his authority. (Those who did, of course, were quickly booted from the company). As Charles Sorensen aptly said, "Only a fool would have ambitions to dethrone Henry Ford."[67] "As long as Henry Ford lives," proclaimed Sorensen, "no one else but Henry Ford will run the Ford empire."[68]

Unlike his father, who never seemed to have a care in the world, no matter how tough business became, Edsel worried constantly about business. It must have come with the territory, because his predecessor, James Couzens, who was dogmatic and callous, also worried about the business and eventually succumbed to the pressure. At the time of his resignation, Couzens said he quit because he no longer agreed with Henry Ford's pacifist views on the war in Europe. However, the real reason was Couzens' failing health. Debilitating migraine headaches brought on by the pressures of business got the best of him, and those headaches limited his capacity for carrying on, not Henry's views on world events.[69]

Edsel was never robust, but he held on almost twice as long as Couzens did. Eventually, though, the pressures of business, augmented by his father's constant humiliation, took their toll on Edsel as well. Where Edsel was once a patient and tolerant man, he eventually became a nervous wreck. "When Edsel is talking to you," wrote a *Fortune* reporter who was doing a report on the company, "he tears sheets of paper into little bits, rolls them up, and pops them into a waste basket, one by one."[70] Even when Edsel was engrossed in a meeting, his head would be the first to pop up the instant somebody walked into the room.[71]

In the beginning, Edsel had a high degree of tolerance and a steady hand, and he never fidgeted or showed undue excitement.[72] He could become annoyed and aggravated, of course, but that seldom occurred. When it did, he always handled it as a gentleman.[73] He never stayed angry for long. In fact, the only way one could tell Edsel was angry was that he would become silent, his neck muscles would tighten, and he would purse his lips.[74] Toward the end of his tenure, however, the constant humiliation and frustration caused by his father, combined with daily business pressures, eroded Edsel's fortitude. At that point, when Edsel became angry, his neck would tighten, his stomach would cramp, and he would vomit.[75]

Chapter 13

From Model T to Model A

"Edsel's greatest victory."[1]
Charles Sorensen

The real rift between Edsel and Henry occurred in 1926, but it had been developing since 1922. Since Edsel had become president, he had been pressing his father to change the Model T. Edsel had been studying automobile design since he was a boy, and he knew even by this early date that the Model T was becoming outdated.

In the fall of 1922, Henry allowed Edsel to make some stylistic changes to the Model T. He told Edsel that he could make any cosmetic changes he wanted, but he was to leave the mechanics of the car alone—a philosophy Henry maintained until his death. Edsel was not that fond of mechanics, and because he did not obtain the approval to make changes until late in 1922, he would not have time to make any substantive mechanical changes anyway. Edsel barely had time to make changes in the sheet metal of the Model T.

Nevertheless, with the help of Joe Galamb, Ford's chief body engineer, Edsel managed to dress up the "Tin Lizzie" by incorporating more curved surfaces and smoother lines in the body and fenders, as he had always wanted. He also added height to the radiator, blending its decorative apron at the bottom into both fenders, and he raised the hood lines so that the hood flowed more graciously into the cowl. These were simple changes, but they gave the car a more solid appearance and the impression that it was designed with an overall objective in mind.

Although these changes were by no means everything that Edsel might have wanted to do to the stodgy Ford, they were nonetheless beneficial. The changes in 1923 were minor, but Model T sales reached their zenith that year and exceeded two million vehicles, exemplifying the impact that styling was already having in the industry.

***

If such subtle changes caused such a boom in sales, reasoned Edsel, then making more substantial modifications should produce even greater rewards. While his father was visiting his foreign factories in Europe, Edsel decided to press forward on a design prototype the Ford employees called the "Australian Job." Starting with a standard Model T touring car, Edsel instructed his engineers to cut, chop, section, and weld the body and frame of the car until they managed to make a car that was lower and longer. By the time they were done, the engineers had removed no less than 4.5 inches from the ground-to-roof height of the vehicle and added 12 inches to its overall length, dramatically softening the top-heavy and stubby appearance of the one-time Model T. The final touches included a one-piece windshield and a coat of bright red lacquer.

"It was a beauty," exclaimed George Brown, the purchasing agent who had bought many of the special parts for the car. "You put that in a showroom window, and you wouldn't need salesmen!" Edsel thought he had a winner also and placed the prototype in the officers' garage in the Highland Park plant to await his father's return. He wanted to beat his father to the office so that he could be there to explain the purpose behind the new model. However, to everyone's surprise, Henry decided to go directly to the plant the evening he returned to Detroit, instead of waiting until morning, as everyone had anticipated. Henry's chauffeur dropped him off at the Woodward Avenue entrance, and Henry proceeded directly upstairs to his office. From there, he went downstairs to the garage, where he ran into Brown, who had been working late.

"I glanced up," recalled Brown, "and oh, my God, my heart stopped! There was Mr. Ford at the top of the stairs, looking down at the car."

Brown hesitantly walked over and met Henry at the foot of the stairs. They stood there and made small talk for a few minutes. All the while, Henry was eyeing the bright red mock-up.

"What's over there?" Henry finally asked suspiciously.

Practically overcome with anxiety and barely able to speak, Brown muttered, "Well, Mr. Ford, that's the new car."

Henry stared at it for a few moments, and then questioned, "Ford car?"

"Yes, sir."

"It is?" replied Henry in bewilderment.

Henry kept staring at the car as he walked closer to it. "He'd tip his head this way, then he'd tip his head the other way," explained Brown. "He walked around the car three or four times, looking at it very closely. Finally, he gets to the left side of the car, and he gets hold of the door, and BANG! one jerk, and he had it off the hinges! He ripped the door right off! God, how the man done it, I don't know! He jumped in, and BANG! goes the other door! BANG! goes the windshield! He jumps over the back seat and starts pounding on the top. He wrecked the car as much as he could."

Brown ran up to the second floor where some of the engineers who had helped develop the car were working. "Look out, boys!" he cried. "Mr. Ford's kicking the car all to pieces downstairs!"

They all jumped up and ran down to the garage to witness the commotion. When they arrived in the garage, Henry was still at it. "His hands were going, his feet were going, and talk about cussing! It was the first time I had ever heard Mr. Ford cuss."[2]

This was a major setback for Edsel and was indicative of what he had to endure from his father. Edsel thought he had finally persuaded the old man to give at least as much credence to design as he did to engineering, and he wanted to prove what could be done to the stodgy Model T, given enough time and resources. Unfortunately, whatever styling propensities Henry accepted with the changes Edsel made in 1922 were wiped out with the Australian Job. To Edsel's credit, however, he did not give up.

Ironically, if Henry had agreed to produce Edsel's Australian Job, perhaps he could have stemmed the precipitous decline in Model T sales that began shortly after the garage scene. A number of executives explained years later that had Henry Ford introduced an entirely new automobile at this time, he could have remained the Number One producer in the industry.[3]

Unlike his father, Edsel kept a keen eye on the market and thoroughly understood that the Model T was becoming outdated. Although it remained the most inexpensive and durable vehicle available, other manufacturers were making significant inroads into Ford's market by offering a lot more luxury for only a little more cost. Edsel realized that it would be only a matter of time—and not much time at that—before the Model T would become customers' *second* choice.

Even with the changes incorporated the preceding year, Model T sales dropped in 1924 and again in 1925. By 1926, they were plummeting. In an effort to stem the regression, Edsel pressed his father for more styling changes. As far as Edsel was concerned, the only way to regain sales would be through additional aesthetic changes—or, better yet, by introducing an entirely new car.

***

Faced with ever-decreasing sales, Henry finally capitulated in August 1925, giving Edsel permission to make the most pronounced changes in the Model T since its introduction almost two decades earlier. Taking cues from what he had done previously, Edsel enhanced the curvature of the fenders and lowered the height of the Model T. Its chassis was lowered 1.5 inches, the radiator was raised, and a redesigned cowl merged with longer body lines. In all models except the four-door model, the steering wheel was dropped 3 inches, and the seats were lowered and moved backward for greater legroom. In an effort to emulate the Chevrolet, Edsel introduced new body colors on closed models, such as Channel Green, Windsor Maroon, Fawn Gray, and Gunmetal Blue. Wire wheels, made optional, gave the car a sportier appearance and were available in Casino Red, English vermilion, yellow, and black, offering contrast to the new body colors. It was exciting for Edsel to finally see a Model T outfitted with a green body and contrasting vermilion wheels, after the millions of Tin Lizzies that had been produced in Henry's ubiquitous black. Charles Sorensen, who had overseen the production of every one of the more than 15 million Model T's produced, agreed. "I was sick of looking at them," he said.[4]

Unfortunately, Edsel's changes were too few and too late. Although the new Model T appeared more stylish, it was a Model T. Except for its outward appearance, everything under the skin—engine, transmission, suspension, and steering—was 1908 technology. Wrote one Ford dealer about the so-called improvements: "Yes, you can paint a barn, but it will still be a barn."[5] In fact, with Edsel's new and heavier bodies, the Model T barely had enough power to get out of its own way.

Edsel knew that a completely new automobile was needed—one that would satisfy the public's growing demand for more refined cars. "Father," insisted Edsel, "we've got to dress up the Ford. The Model T will have to be scrapped so that we can do something about the competition Chevrolet is giving us!"[6] But Henry was not ready.

Everybody in the company, except Henry, was on Edsel's side. However, nobody except Edsel would bring up the subject with the old man. "The sales department hardly had the temerity to tell Henry Ford that they could not sell his product," recalled Sorensen. "They told

Edsel, Edsel told his father, and, for the first time, a real break between father and son was apparent."[7] However, Edsel could not persuade his father to introduce a new car. "Every suggestion he made was ignored," recalled Sorensen.[8] "Edsel suffered more than I can tell. He could not get anywhere with his father."[9]

***

While all of the other executives were afraid to bring up the subject with Henry, Edsel had an ally in his brother-in-law. Ernest Kanzler was a graduate of the University of Michigan, earned his law degree at Harvard, and married Eleanor's sister, Josephine, in 1917. Kanzler was working for the plaintiffs in the Dodge brothers' stockholders suit when he first met Henry Ford. As a friend and relative of Edsel's, the two couples would visit Henry and Clara quite often, and Henry would openly discuss business issues. "Please, Mr. Ford, don't talk about your affairs when I am here," pleaded Kanzler.

Henry simply laughed. "You ought to be on my side," he retorted. "Why do you want to be a lawyer anyway? They're just parasites. Come and work for me."[10]

Kanzler joined Ford Motor Company in a roundabout way. When Kanzler told Henry that he was interested in coming to work for him, Henry put him to work—not in the automotive company but in his newly formed tractor company called Henry Ford & Son. Henry took Kanzler to the tractor plant and gave him a seat on the opposite side of Sorensen's desk. "Charlie, show this young fellow the business."[11] This was in the summer of 1917.

It was not long before Kanzler started showing Sorensen a thing or two. "Mr. Kanzler had an analytic mind," explained a co-worker.[12] "Dog-gone it," said another, "Kanzler's the smartest individual that I've ever ran into. He's ahead of you all the time. He asks you a question and you start explaining it and you can see him immediately light up. He'd be way ahead of your answers... He could ask more intelligent questions than anybody who ever came into this company."[13]

While at the tractor plant, Kanzler was not content with merely doing the odd tasks that Sorensen assigned to him. He became fascinated with the relationship between production and shipment and realized that failure to synchronize the two built up inventory—and that tied up money. He began analyzing the entire flow of parts from procurement to shipment, and he developed several routines that practically eliminated the oversupply or undersupply of inventory. So exact were Kanzler's schedules that supplies arrived practically as needed. Trainloads of wheels, radiators, castings, and other parts were used in production a few hours after their arrival.[14]

Henry was so impressed with Kanzler's system that he made him business manager of Henry Ford & Son, giving him complete power over purchasing, sales, and office operations.[15]

Edsel was also impressed with Kanzler's business acumen. Over the years, he and Kanzler had talked about business aplenty. The Kanzlers were frequent guests of the Edsel Fords. As relatives, they vacationed together, played golf together, and drank together.[16] The Kanzlers even built a home next to Edsel's home in Hobe Sound.[17]

With his father busy building his new manufacturing plant along the Rouge River, Edsel presided over all other aspects of the company. "The truth is," wrote one reporter, "that Henry Ford has delegated to Edsel so much of the responsibility, particularly on the side of business management, that the younger Ford is one of the busiest men in the realm of American business."[18] Clearly, he needed help.

In 1924, Edsel transferred Kanzler from the tractor plant, gave Kanzler an office next to his in the Highland Park Administration Building, and promoted him to second vice president of Ford Motor Company.[19] Organizationally, Kanzler was third in command between Pete Martin and Charles Sorensen, but Henry Ford never ran the company according to organizational lines anyway. Thus, although Kanzler had an official title, Henry continued to call the shots. "Kanzler threw his chest out and said a lot," explained a high executive, "but Mr. Ford didn't pay much attention to him. He ignored him."[20] Sorensen was also aware of Kanzler's braggadocio and rarely listened to him.[21]

Nevertheless, Kanzler assumed responsibilities as quickly as Edsel could delegate them, and Edsel delegated them willingly. "Kanzler would go out and dig up information and bring it back to Edsel, data which would otherwise have taken a crew of 15 or 20 men to dig up," said a Ford manager.[22] "He carried out many things that Edsel wanted."[23] The two men thought alike in terms of how a business should be run: clear lines of authority, study problems to develop reasonable solutions, and, above all, keep abreast of the market and make changes accordingly. In a way, Kanzler was Edsel's business expert in much the same way that James Couzens had been Henry's 20 years earlier. In the same way as Couzens had done for Henry, Kanzler shielded Edsel. "Everyone wanting to see Edsel had to get by Kanzler," explained Sorensen. "The only way he could see me or his father would be when he came to Dearborn for lunch."[24]

***

Meanwhile, Edsel and Kanzler were becoming more and more concerned about the competition. Model T sales continued to drop as the 1920s were coming to an end. The car that Henry Ford had developed while his son was only a teenager, the one that had made the Ford family fortune, was sorely out of date by 1925. While other manufacturers were continuously making improvements in their products to gain customer attention, Henry had concentrated on simplicity, rigidity of design, and low cost to lure his customers. "Anybody can add on to a car, but it takes a damned smart man to take one nut off, to reduce the price of the car," Henry once said. "That was his aim in life—to have a cheap car for the public," recalled one of Henry's engineers.[25] However, according to a Ford executive, Ford's penchant for simplicity was both his strength and his weakness. "Mr. Ford wanted everything simple, to the point where he stifled progress."[26]

Henry's formula had worked well for many years, but by the mid-1920s, the competition had been able to keep down the price of their products by mimicking Ford's mass production techniques. They had also improved styling, power, and comfort.

Henry was not totally against the thought of developing a new car. However, in his mind, if he were going to come out with an automobile to replace the Model T, he wanted something that would astonish the marketplace in the same way that his Model T had done 20 years earlier. "He knew so much was at stake that he wanted to generate those thoughts himself," explained Sorensen. "I lived on top of Henry Ford trying to catch his idea. The same thing we did when we went into production on the Model T."[27]

Sorensen had to wait a long time. Sometime during the mid-1920s, Henry decided that he would not introduce a new car until after 15 million Model T's had been produced. Nobody knows why Henry chose that number because he had confided the information only to his secretary, Ernest Liebold.[28] Regardless, Henry knew the Model T would have to be replaced at some point, and he began working on some ideas as early as 1920.[29] He gave thought to several ideas and combinations of ideas. For engines he considered four-cylinder, five-cylinder, and V-6 configurations. For the chassis, Henry contemplated a rear engine with rear drive, a rear engine with front drive, and a front engine with front drive. In addition, he considered conventional steering, four-wheel steering, and a composition body made of fiber-cork.[30] At that moment, though, Henry put the chassis and body configurations on the back burner and pursued an X-8 engine design—an engine shaped like two X's hooked together back to back, so that there were four cylinders facing upward and four cylinders facing downward. Attached to this radical engine design would be a hydraulically operated, three-speed planetary transmission—in effect, an automatic transmission.[31]

***

All the while, Model T sales were falling. From a high of more than two-million units in 1923, sales dropped in 1924 and again in 1925. By 1926, sales were 30 percent less than they had been only two years earlier. While Model T sales remained more than three times the sales of Chevrolet, Ford's sales were dropping while those of Chevrolet were increasing. Between 1924 and 1925, for example, Ford sales dropped 10 percent, while Chevrolet sales increased a staggering 168 percent! Henry did not believe the numbers. He said that General Motors was influencing R.L. Polk & Company, the independent firm that tracked sales.[32]

Edsel was doing the most he could, within the constraints of the relationship with his father, but his concern fell on deaf ears. On the other hand, Kanzler wanted Edsel to force the issue. "God damn it, Edsel!" he exclaimed, "we are behind the times! Let's get on the ball! Why are we manufacturing an out-of-date car? This is terrible! We are losing sales left and right!"[33] Edsel could not have agreed more with Kanzler and did what he could, but at the time, Henry was too engrossed with the X-8 project. "Henry Ford didn't like mechanical ideas to come from too many sources," explained Ford publicist William Cameron. "He liked to work in his own way. For instance, if he was working on a new model, the hardest thing in the world was to get him to drop that model so they could get it into production. He would keep it for another week or another month and do a little of this and do a little of that. He would never let it go. He objected to being pushed."[34] But pushed Henry was, especially by Edsel and Kanzler. Sorensen relates this period to the time when Henry was trying to buy out the minority stockholders. "When Edsel and those he could gather around him started screaming at Henry, much like his partners used to do, I just could not join in. I remained neutral, confident that he probably had another new idea and we'd better let it develop."[35]

"Edsel set out to get under way on a new model, and he started lining up his idea of a car," continued Sorensen. "When he asked me to go along with him, I asked him if he was sure he could do this without getting him and me in trouble with his father. I knew Edsel had a designer working alone on a new car. I wanted his father to know that. This is not what Edsel wanted. He must avoid his father. It was this struggle between father and son that delayed bringing out the Model A."[36]

Henry's new idea never came. After four years of development, Henry was still working on the X-8 engine. Trial after trial begot more and more problems. Nobody except Henry believed that it could work. Even if it proved to be an engineering success, Edsel and Kanzler were sure the public would dismiss it out of hand.[37] Too much was riding on such a wild concept. "Why doesn't Mr. Ford take his God damn fiddle and go somewhere and play, and let us run this plant!" exclaimed Kanzler.[38]

That was not the first time Kanzler had criticized Henry Ford, and Kanzler did not like the way Henry was treating Edsel. Kanzler "started blasting Henry Ford, telling people how unfair Henry was to Edsel."[39] He thought all employees felt the same way, particularly the other executives. While visiting Sorensen one Sunday afternoon, Kanzler and his wife "could not resist attacking Henry for the way he held Edsel down. They both went so far as to say that Henry Ford was crazy."[40]

Whether or not Sorensen agreed with Kanzler, he was not foolish enough to admit it openly. He kept Henry informed of Kanzler's and Edsel's plans and how they were "messing things up in general."[41]

When Henry heard about these comments, his support for Kanzler vanished and his displeasure with Edsel increased. He had hoped Kanzler would assist Edsel, but it seemed that Kanzler was trying to turn Edsel against Henry. Henry felt Kanzler was getting a little too big for his own good.[42] "I remember one time Kanzler walked in and everybody had to get up and report to him," explained Gene Farkas, a Ford engineer. "He was really a big wheel. He was trying to tell Mr. Ford about a lot of things that he should've changed."[43]

With Kanzler's support, Edsel began challenging his father even more. In the process, Henry became resentful. He simply could not believe that Edsel would act this way unless someone was goading him on. And that someone was Kanzler. "Mr. Ford thought that Kanzler was going through Edsel to push his ideas, that he was making balls for Edsel to throw," recalled a Ford executive.[44] However, what disappointed Henry most was that Edsel was listening to a man who did not know a thing about motorcars. "Here is a man who is advising my son that I don't know anything about a modern motorcar!" Henry confided to another executive. "Gee, the old man hit the ceiling!"[45]

In addition to a new car, both Edsel and Kanzler wanted to formalize the company, to set up a corporate structure that would have a central administrative system similar to those of other corporations. Specifically, they "wanted control to be in the office of the president."[46] With such an organizational structure, based on logical lines of authority, they could force the development of a new car "and bring out a new car with style, appearance, and what the public wanted."[47]

Of course, Henry objected to this thinking because, as Liebold explained it, "he was very jealous of the manner and methods of administration."[48] Henry wanted Edsel to be his own man and to act on his own volition, but he also wanted Edsel to be independent of everyone except Henry.[49] Henry said that people such as Kanzler were "piddling in Edsel's ear."[50]

All of the Ford executives realized the discord that was developing between father and son over the debate of introducing a new car, but Eleanor also was seeing the changes. While she hesitated to talk to Henry personally, she went to Sorensen on many occasions, crying as she asked him to intervene on Edsel's behalf.[51]

While Sorensen did not become too involved at this point, he did mention to Edsel several times that he thought Kanzler was coming between Edsel and Henry, and that Henry wanted to get rid of Kanzler, not necessarily by firing. Henry once told Harry Bennett, "Harry, I wouldn't care if a bad accident happened to Kanzler." Bennett rebuked Henry for making such a gruesome comment and told him that he should forget any such ideas. But Henry did not. A short time later, Henry ran into Chester LaMare, a notorious Detroit underworld figure who worked for Bennett, and told him the same thing.

Bennett was furious with Henry when he learned of Henry's conversation with LaMare, and Henry tried to wriggle out if it when Bennett confronted him. "Now, Harry, I didn't ask LaMare to arrange any accidents. I just told him what I'd said to you."

"You knew very well how LaMare would take it!" Harry fired back. Bennett averted any problems by meeting with LaMare, warning him that if anything happened to Kanzler, he would contact the authorities himself, leaving LaMare holding the bag.[52]

***

In January 1926, Edsel and Kanzler decided to take a more adamant stand. Instead of talking to Henry face to face, Kanzler instead decided to write a memo to Henry[53]:

This memorandum is given to you so that I can feel that I have dealt honestly and squarely with the responsibility you have given me. It hurts me to write it because I am afraid it may change your feelings for me, and that you may think me unsympathetic and lacking confidence in your future plans.

Please, Mr. Ford, understand that I realize fully that you have built up this whole business, that it has been your battle and your creation, and that all of the Company's successes day after day regardless by whom personally conducted are nevertheless a direct result of your conception and will really be your personal accomplishment for many years even after your lifetime.

Any powers I may have are mostly due to the opportunities you have given me and have not created in me any exaggerated ideas about myself. You have allowed me to play with the throttle of your engine. That's all.

From things you have said, you above all others recognize the need for:

(A) An intermediate car

(B) Greater power, smoothness, refinement of the Model T type of car

(C) Something for European requirements

If we face the facts, we must know that there is little chance for the production of a tried and thoroughly tested X type car within 18 months which would not be before the summer of 1927.

Those of us who have been privileged to follow the X development look into the future and hope for great things, BUT, and this is what worries me, I feel that there should also be other development in process on a power unit along conventional lines so that we would have if necessary a power unit to maintain our position in the automotive field until the X motor is perfected—something which will serve until you will have been given a fair chance to produce the X motors to their final stages of development so that when once adopted they will lead all others for another 20 years like the Model T.

Such a motor can be installed in the intermediate chassis and hold this market for us against competition until such time as we would sweep all before us with your revolutionary X power plant substituted when its perfection has been achieved.

While there is every logical engineering argument in favor of the X type, yet we have all our eggs in one basket. It might take much longer than expected to get it perfected and until it should perform greatly better than the present conventional sixes the public will hardly welcome the change.

I know of no one who feels more deeply than you the obligation to 180 odd thousand employees who have started to make their life work with this company.

I think there will always be a field for 4,000 to 5,000 Model T's per day, but I do not think the Model T will continue to be a satisfactory product to maintain our position in the automobile field until the X models shall have been developed.

We have made over 100 million dollars the last two years each and will probably make 100 million next year, BUT, we have not gone ahead in the last few years, have barely held our own, whereas competition has made great strides. You have always said you either go forward or backwards, you can't stand still.

In the past 12 months, the other manufacturers have gained tremendously. Our production and sales in 1925 were less than in 1924.

Our Ford customers, particularly the Two and Four door customers, are going to other manufacturers, and our best dealers are low in morale and not making the money they used to.

In spite of higher prices, the public are choosing six-cylinder cars. In 1924, 14 makers sold 30 percent as many sixes as we sold Fords. In 1925, these same makers sold 54 percent as many six-cylinder cars as we sold Fords—and each one of these 14 increased his sales, some 400 percent.

There may be theoretical engineering objections to a six, but in every one I have ever driven there has been a most satisfactory smoothness and power range entirely different from and far superior to four-cylinder performance and almost as good as the Lincoln. This is not only my view, but also that of the public as demonstrated by the way they have opened their pocketbooks to buy sixes.

Practically every man in your organization to whom you have entrusted the greatest responsibility holds this same opinion.

This same opinion has been cautiously expressed at different times by certain Ford executives whose opinion merely as automobile purchasers has great value.

These men are all in favor of the X type when completed, but feel a great danger if it is not almost immediately available.

We spend millions in advertising, millions in extra dividends on investment certificates, millions in additions, therefore, could we not spend 1/10 of one percent of last year's earnings to help insure our dominance in the field and that we can keep all of our employees busy, allowing the X motors the time necessary for thorough development.

Could we not carry out your ideas that the product must be made right as expressed in your page of the January 16, 1926 Dearborn Independent. *Won't you permit the organization to develop a refined six-cylinder motor without imposing in any way on the time of those working on the X motors. Such a power plant would never be used unless its performance satisfies you that it had real merit.*

Even if such experimental motor would cost $100,000 to design and work out, it would show you what kind of an engineering outfit the company could produce. If nothing else, let us try it out on the English market.

Moreover, this idea would quickly develop a fine spirit of competition between the workmen on the straight sixes and X-8 parts and designs which would bring out the best in everybody to the undoubted best interests of the company.

It is fair to say that with all the difficulties of the new body designs, the dual high transmission, the Fordson truck, the Fordson bus, all Lincoln engineering, and the new front axle, steering gear, rear axle, spring suspension, body design, and transmission of the intermediate car that there is much more work ahead than the organization as at present constituted in Dearborn can handle satisfactorily.

The advantage of this memorandum is that I can write certain things that I find it difficult to say to you. It is one of the handicaps of the power of your personality which you perhaps least of all realize, but most people when with you hesitate to say what they think.

It is unique in the commercial history of the world that one man should run away with the field as you have done in the motor industry. We have had a wonderful head start because your first designs of a car were 20 years ahead of the world, as well as your methods of production and marketing.

But we are losing our position because the world has learned from you and with its combined efforts each learning from the other it has now developed a product that is alarming absorbing the public's purchasing power.

The best evidence that conditions are not right is in the fact that with most of the bigger men in the organization there is a growing uneasiness because things are not right—they feel our position weakening and our grip slipping. We are no longer sure that when we plan increased facilities that they will be used. The buoyant spirit of confidence expansion is lacking. And we know we have been defeated and licked in England. And we are being caught up in the United States. With every additional car our competitors sell they get stronger and we get weaker.

Even on the basis of equal design value we could still out-distance all competition because of our "from the mine to finished car" ability to produce, and unified ownership. But with our competitors' volume increasing, they are rapidly approaching our formerly unique powers of producing at lowest cost. Inwardly we are alarmed to see our advantage ebbing away, knowing that the counter-measures to prevent it are not immediately at hand. We all realize that an epoch making motor such as we expect the X lines to be cannot be the product of an immediate future.

This feeling exists not outwardly, but I will stake my reputation it exists in every important man in the Company. I, personally, have helped to stamp this feeling down wherever it has tried to break through.

The writing of this has not been pleasing, Mr. Ford, but I have always tried to tell you what I see and feel.

These thoughts have been uppermost in my mind the last year, and I cannot keep from expressing them any longer.

This memo was Kanzler's death-knell. For years, Henry had been looking for an excuse to get rid of Kanzler. Since Kanzler had moved next to Edsel's office, Henry thought Kanzler was influencing Edsel too much. Henry honestly believed that Edsel did not really want a new car, and that it was Kanzler who "was going around through Edsel to push his ideas."[54] Kanzler's memo proved that to Henry.

During conversations at the Round Table, Henry Ford all but ignored Kanzler. Occasionally, he would cut him off in mid-sentence. At other times, he would ignore him completely. "This was very embarrassing to Edsel," explained an executive in attendance. "Edsel would side with Kanzler, but not *too* far."[55]

The row that was developing between Henry and Kanzler "got so damned hot," explained one Ford executive, "that they packed up Edsel and sent him to Europe. He was over there for about six months."[56]

A few days after Edsel left for Europe, Henry told Sorensen, "Well, by this time, I think Edsel is several miles out on the ocean, so tomorrow we get rid of Kanzler."[57] On July 26, 1926, Kanzler and Ford Motor Company parted ways.

Edsel had learned about Kanzler's dismissal from Kanzler himself. He wrote to Edsel, explaining the episode that occurred shortly after Edsel had left New York. There was nothing Edsel could do at the moment, and by the time Edsel returned from Europe, the storm had passed but not Edsel's resentment toward Henry for what he had done.[58] "Things were never the same between them," explained a close associate. "After that, Edsel seldom took his father's word for anything. When Edsel wasn't deeply embarrassed by the things his father did (which was often), his attitude was one of cynicism."[59]

***

During his exile, Edsel had plenty of time to think about the dire position in which his father had placed the company. He felt that if a new car were not started soon, there might not be a company over which to fight.

By the time he returned to Detroit, Edsel "was determined to abandon the Model T," recalled Charles Sorensen. Edsel went to the Engineering Building to see his father and met with him for quite a long time. Nobody knows what Edsel said to his father or how he finally convinced Henry to stop playing with his X-8 project. However, by the time Edsel walked out of his father's office, the old man had agreed to give up Model T.[60]

This would be the first time that father and son worked so closely on a project. After all, the Model T was designed when Edsel was a young boy. All the changes that Edsel had made to the Model T until this time were merely superficial, and he did not need much direction from his father in carrying them out. Now, father and son were faced with designing an entirely new automobile in an industry that had been making significant advancements in all aspects of the motorcar, from engine design to body styling. Many wondered if they could pull it off.

True to form, when Henry decided to make a new car, he faced the challenge head on. If the public wanted a new Ford, Henry said, a new Ford they would get. Except for the size of the tires and the Ford logo script on the radiator shell, nothing on the new Ford that he and Edsel engineered and designed would either fit or resemble anything on the old Model T.

However, instead of working on the new car while still producing the Model T, as all other manufacturers would have done, Henry decided to shut down the entire Highland Park plant, design a new car, and then restart production when it was ready. That was a major mistake. Designing and building a new car was an expensive proposition—it cost Ford more than $250 million to change to the Model A (as they called the new car)—but Henry also lost untold millions in lost sales. Henry Doss, Ford sales manager at the time, recalled later that the company had no trouble getting rid of the remaining production of Model Ts and could have gotten rid of any subsequent production had Henry kept making them during the transition. "The cooperative dealers took what the others didn't want," said Doss, "and those that didn't take them lived to see the day they wished they had before they got any more cars to sell."[61]

"Edsel and I settled on the wheelbase and size right away," said Henry. "After that, it was a matter of working things out on the drawing board until we got them right."[62] The overall length of the chassis was to be slightly more than 113 inches, and the wheelbase was set at 103.5 inches. The other matters were not so easy. Edsel wanted a six-cylinder engine, sliding-gear transmission, and hydraulic brakes. Henry felt a four-cylinder engine was sufficient, that he could perfect the automatic transmission on which he had been working in conjunction with the X-8 project, and four-wheel mechanical brakes would be adequate.

Henry took off working on the mechanical aspects of the new car, and Edsel on the aesthetic parts—the body design, the interior design, and color schemes.

While Henry had a complete engineering staff to assist him, Edsel had only Joe Galamb. For additional design expertise, Edsel turned to the Murray Corporation and the artistry of Amos Northup. In fact, Edsel had contacted Murray in April 1926, months before his father had given the go-ahead to build an entirely new automobile. Edsel sent Murray a 1926 Model T chassis, complete with radiator shell, hood, front and rear fenders, running boards, and splash aprons, and asked Murray to design and build a new body that was more "beautiful in appearance, [with] European lines, wider doors, and low effect." By November, Murray had a complete prototype ready for Edsel's inspection. "The job...was constructed not to alter in any way the interchangeability feature of your chassis...which you laid down...as a fundamental," wrote a Murray executive. "We have, however, put on smaller wire wheels of 27 × 4.40 size. We [also] used crimped springs...which brought us possibly one inch closer to the ground."[63]

The design that Murray developed was a handsome car (even though it rested atop a Model T chassis), but the design went no further than the one prototype. By the time the prototype was completed, Henry had given the okay to pursue a completely new automobile, and the initial Murray proposal was dropped.

After Henry had given the go-ahead for development of a new car, Edsel recruited Galamb and Northup again, and the three men began work on an entirely new body design based on the new chassis. The mechanical aspects of the new car were done at the Ford Engineering Laboratory, whereas the body was designed at the Highland Park plant. "Sample bodies of the Model A were built up on the sixth floor of the 'Y' building," recalled one of the engineers working on the project. "They were always under lock and key."[64]

Unlike today where automobiles are designed using computers and full-size clay models, at Ford in the late 1920s, automobiles were designed primarily on large backboards. "They never made clay models then or anything of that sort," recalled Bob Gregorie. "You couldn't tell very much how the car would look until a prototype was made. It was all very crude and unprofessional."[65] Actually, the "blackboard" was a large piece of black canvas-type material that could be rolled up and filed away. "We colored these drawings, too," recalled Sheldrick. "We would fill in with colored pastel crayons certain parts so that we could distinguish the cylinder block from a bearing cap, or a piston from a connecting rod. Different parts carried different colors so you could refer to the red piston or pink cylinder block, and so on."[66]

With Joe Galamb, Edsel scrutinized every aspect of the new model as Northup developed it. When different ideas were presented, Edsel and Galamb would debate them at length. No detail was too minute or feature too trivial to warrant their scrutiny. They debated the profile, the ride, what materials the seat cushions should be made of, and even what windshield to use to give good vision. "He [Edsel] knew what he wanted," explained Galamb, "and insisted that he get it."[67] One of the details to which Edsel paid particular attention was the front end of the car. He knew from years of experience that the front end of the new Ford would receive the most critical attention, so he insisted that it have an appropriately designed radiator shell. Eugene Farkas, a long-time Ford engineer, was given the assignment.

"Initially, we didn't have any radiator shell on the front, just the radiator itself," Farkas explains. "Edsel wanted a shell, but he didn't want it perfectly flat like the Model T's. He wanted to have a little body to the shapes, so I bumped it out a little bit so that it had a curved, warped surface. In its final form, the radiator shell had a little reverse peak in it—just for style."[68]

As Edsel had predicted, the front end of the new Ford received the most attention—but it was acclaim rather than criticism. Rounded at the top and square at the bottom, the radiator shell was perfectly proportioned. With Farkas' reverse peak feature, it became the focal point of the entire car and enabled the remainder of its design to have low contours.

To complement the striking appearance of the Model A, Edsel devised various color schemes for the new car. Unlike the last versions of the Model T, for which he added only a handful of colors, Edsel provided a complete palette of body colors for the new Model A.

Henry Ford's favorite color, black, was a given, but Niagara Blue, Dawn Gray (light or dark), Arabian Sand, Copra Drab, and Andalusite Blue were among the other colors on which Edsel insisted. "The joke on me," lamented Harley Earl, General Motors' chief designer at the time, "is the painting of [the new Ford] is almost exactly like my LaSalle, which I had painted as a special job. Hereafter, I shall have to go to Highland Park for my coach work!"[69]

Now that Edsel had the body design under control, he turned his attention to the work his father was doing. Visiting the Engineering Laboratory regularly—looking over blueprints and examining parts—Edsel found that his father was making woeful progress.[70] As the plant lay silent, Henry dragged his feet on almost every decision. The engineers had to wait for Henry's approval on every part. Sometimes he would make a fast decision; sometimes he dallied for weeks.

To make matters worse, the components that should have been easy for Ford to design and make were difficult. Parts that should have been difficult were easy. The sliding gear transmission, for example, the part that Edsel finally convinced his father to use in the new car, came together without incident, although the company had never made one. The four-cylinder engine should not have presented a problem; after all, the company had made millions of them for the Model T. The problem lay in its power output. The engineers wanted a 40 horsepower engine that could propel the car at 60 miles per hour, but their redesigned engine was developing only 34 horsepower. Sorensen became so frustrated that he reassigned one of the engine designers of the Ford Tri-motor airplane to the project. "If you can get us 40 horsepower in three weeks," Sorensen told him, "I'll make you the biggest man in the Ford Motor Company!"[71]

Edsel was present when Sorensen made these brash remarks, but he could understand his frustration—everybody was aggravated. However, as Edsel analyzed the blueprints, he noticed an obvious flaw in the design of the engine. To Edsel, the valves appeared too small, so he suggested to the engineers that the valves be made larger. "Edsel knew that the right way to get power out of an engine was to get fuel and air into the cylinders and explode it," explained one of the engineers working on the engine. "You couldn't have small valves and get power out... It was Edsel's idea alone that we should open up and make the valves bigger."[72] After the engineers resized the valves, the horsepower of the engine increased to 40, as they had wanted.

Although Henry was having his tribulations over the engineering of the new car, he did not hesitate throwing his "two bits" into the body design. He tried forcing Edsel and Galamb into making the new car as high as the old Model T. Henry sat in one of the prototypes one day and said that there was not enough head clearance. When he sat in the car, he had about 5.5 inches of head clearance. When Sorensen got in, he only had 4 inches of clearance. That was not enough for Henry, so he ordered Galamb to redo the prototype and add an additional 1.5 inches to the roof height. "I'll have to clear that with Mr. Edsel," explained Galamb.

"Fine. Then do it," ordered Henry.

When Galamb told Edsel what his father had said, Edsel told Galamb to leave it alone.

Six weeks later, when there was a lull in his engineering activities, Henry visited Edsel's prototype again. "Did you add the inch and a half?" he asked Galamb.

"No, we didn't," replied Galamb.

"Why not? I told you to."

"I know you did, Mr. Ford, but I talked it over with Mr. Edsel, and he said to leave it alone."

"I don't care what Edsel said. Make it an inch and a half higher!"

So Galamb ordered new body panels, at great expense, with the higher profile that Henry wanted. After they arrived, he asked Henry to look at them. "Here's that higher body," said Galamb.

"Who asked for that?" asked Henry.

"You did, Mr. Ford," replied Galamb, now really confused.

"Scrap it," ordered Henry. "It looks terrible!"[73]

That is what Edsel had to work with—poorly educated engineers being managed by an obstinate Henry Ford. When another engineer had calculated the proper size of the crankshaft based on the projected horsepower and rpm range of the engine, Henry balked. He instructed the engineer to reduce the size of the dimension by a quarter of an inch "so that the crankshaft will be limber enough to follow the bearings..."[74] That was absurd. If anything, the larger the crankshaft, the better, and under no circumstances would a crankshaft "follow" the bearings.

Then there were the brakes. Edsel wanted hydraulic brakes on the new car. They were the growing trend in the industry because they offered better braking efficiency and less effort to apply—something that would be particularly appealing to women buyers. The expensive Duesenberg was first to offer hydraulic brakes starting in 1920, but then Chrysler began offering them in 1924, and Edsel learned that Chevrolet was going to be offering them. In contrast, the Model T did not have brakes at the wheels—it had but one brake, a band around a drum in the transmission. It was not very effective, but by the late 1920s, after the Model T had gained a considerable amount of weight, the transmission brake was all but worthless. On the new car, Henry was receptive to adding four-wheel brakes, but he wanted mechanically operated brakes, with steel rods hooked up to steel levers. ("Safety of steel from pedal to wheel," was Henry's mantra.) He did not want Edsel's newfangled hydraulic brake system. "The fluid might leak out," Henry said.[75]

Nevertheless, Edsel kept pushing. During lunch one afternoon, Edsel brought up the topic of hydraulic brakes again with his father. William Cowling, Edsel's sales manager, and Jack Crawford, his engineering liaison, had supplied Edsel with enough engineering and sales data for Edsel to make a strong argument for hydraulic brakes. "Father," Edsel started, with all of the executives around, "I believe the time has arrived for us to give serious consideration to the hydraulic brake system."

Henry looked at Edsel as though Edsel had just shot somebody. "Edsel!" he yelled, getting up from his chair, "you shut up!" And he walked out.[76]

Sorensen followed after Mr. Ford, and as he passed Edsel, "he turned to Edsel and just raised his lip to him."[77]

A few minutes later, Henry Ford asked Frank Campsall, his personal secretary, to get Sorensen. "Henry Ford told me that he wanted me to tell Edsel to take a trip to California. 'Make it a long stay,' he said. 'I will send for him when I want to see him again.' "[78]

One of the executives who had witnessed this unnerving scene could not believe what happened and felt terrible for Edsel. He went home and told his wife what had happened. "If I were Edsel Ford today," he told her, "I would go out and get me a couple of big guns and go to Africa and shoot lions and elephants for a while."[79]

In one of the rare instances where Sorensen did not follow Henry's directive to the letter, he never told Edsel to go to California. Instead, Sorensen waited for a few days to let things cool down and then told Edsel how upset his father was with him—not only about the hydraulic brake question but about the entire way Edsel was pressuring his father to complete the new car. "It's useless to press him," Sorensen explained. "He would do the job his way, and assume full responsibility for it."[80]

***

In the end, Edsel's persistence and demands paid off, and the new Model A was introduced in the fall of 1927—nine months after Henry had shut down the Highland Park plant. Although not the car that Edsel had wanted—one with a six-cylinder motor and four-wheel hydraulic brakes—the new Model A was good enough. It was well designed, well built, and spunky. With its 40 horsepower, four-cylinder engine, and three-speed, sliding gear transmission, the lightweight car could hold its own against larger cars. For the first few months of production, Edsel spent a lot of time "around the hole" (the end of the production line), and walked around the plant with Pete Martin to keep his eye on production and quality control.[81] In this way, Edsel could react to production problems quickly, which were many with the new startup. "It seems impossible for me to get down to New York at this time," Edsel wrote to a friend, "as things are coming up in connection with the new car every day."[82]

While Edsel had men such as Pete Martin around him, he did not have to worry much about quality and production. However, Edsel was burdened with all other aspects of the company. Kanzler, who had been taking up the slack in many of these ancillary activities, was gone, and shortly before production of the Model A began, Edsel lost another close associate.

Clarence Avery, Edsel's one-time schoolteacher and confidant, finally had had enough. From helping Henry Ford with his first moving assembly line, to working night and day on the development of the Model A, and constant frustration due to Henry Ford's idiosyncrasies, it all finally got the best of Avery. Similar to Edsel, Avery's stomach was giving him trouble—he was eating a whole bottle of aspirin a day—and he decided to leave before his health failed him completely. His resignation letter is both poignant and telling[83]:

> *For the past year or more, I have felt my executive authority being gradually removed. I have been unhappy in my work to the extent that it has seriously threatened my health.*
>
> *In view of these facts, and in consideration of my health, peace of mind, and self-respect, I feel that I must ask you to accept my resignation at your earliest convenience.*
>
> *I have enjoyed the greater part of my 15 years with the company, especially the contact I have had with you. I want you to know that I can never respect any man more than I do you, and I sincerely hope that our friendship may continue.*

Edsel released Avery immediately, probably wishing he could go with him.

***

While the Fords were busy working on their new car, the public at large could hardly stand the wait. Many people who had owned the reliable Model T over the years wanted the new Ford and refused to buy another vehicle until they saw the new creation. The press, kept at bay for months, was now welcomed by Edsel to view the new car. On a cool, rainy day in November 1927, Edsel invited the press to the Engineering Laboratory and allowed them to inspect a pre-production prototype. For once, Henry was not there to steal the limelight, and Edsel, in a rare moment of exuberance, proudly answered questions. He remained with the press for an hour, and despite the rain, he drove the roadster bareheaded out of the laboratory to a side lawn so that reporters could take pictures of him and the new Ford car.[84]

While the press made an extravaganza out of the introduction of the first new Ford model in 20 years, there was only a subdued celebration at the assembly line. Henry and Edsel were there, as were Galamb, Sorensen, and Martin. They all gathered around the car to watch Henry stamp "1" on the engine block, and then they all stepped back to watch him get behind the wheel.[85] "Here you go, Mr. Ford," said William Klann, one of the plant engineers who was in attendance in case something went wrong. "Drive it off." Henry balked. "No, you drive it off," he said.

Surprised, Klann turned to Edsel and asked him if he would do the honors of driving off the first Model A. In deference to his father, Edsel also declined, as did Sorensen and Martin. Therefore, Klann jumped into the new car and drove it off the line himself.

As the group of executives broke up, Henry Ford walked over to Klann. "I wanted to drive the car off the line," he confessed, "but I don't know how to drive a shifting transmission!"[86]

***

The Model A Ford was a big success. People who had been waiting for the new car thronged Ford dealerships and were willing to pay to see the new car before anyone else.[87] The Cartier brothers wrote to Edsel, applauding him on the design of the new car. "Perhaps you could let us mount a few Model A's up as broaches and bracelets," wrote Pierre. "I could personally sell 50 on Monday."[88] Whether they did that or not is unknown, but each brother did buy one of the new models. "The car is a sensation, and my chauffeur is losing his voice, answering questions."[89]

Edsel gave his friend, Franklin Roosevelt, a Model A Cabriolet, and the only thing Edsel wanted in return was an autographed picture of the governor in his new car. "I am delighted to autograph the picture for you," wrote Roosevelt. "The little Ford Cabriolet is a real joy. If you and Mrs. Ford will come to Hyde Park sometime soon, I will take you for a nice drive in it."[90]

***

The new Ford showed the public and the competition that Ford Motor Company was still a major player in the industry and that Henry Ford was finally willing to give as much credence to design as he had to engineering. "The mere fact that Edsel was able to take a country-boy vehicle," explained one reporter, "and extract from it something beautiful, tractable, and urbane, testifies to his inner strength of purpose."[91]

Henry Ford preferred the useful Tin Lizzie to its more graceful and comfortable successor. Realistically, however, he understood that the success of the Model A was due primarily to Edsel's design mastery. During a conversation with one of his old acquaintances, Henry credited his son with giving the new Ford its "style and much besides."[92] From that point, noted Sorensen, Henry would not let anything leave the Ford factories "until Edsel had passed on it."[93]

"It was Edsel's greatest victory," recalled Sorensen.[94] Said another engineer working on the project, "Edsel Ford should get all the credit for getting Henry Ford to switch from the Model T to the Model A."[95] However, it also signaled the end of the cordial relationship between father and son. "With the coming of the Model A," said Sorensen, "Edsel became more of a factor in matters of policy. He was willing to step out on his own."[96] This new-found independence, however, was limited. "Officially, Edsel was it," explained Liebold. "Unofficially, he wasn't it."[97]

Edsel Ford

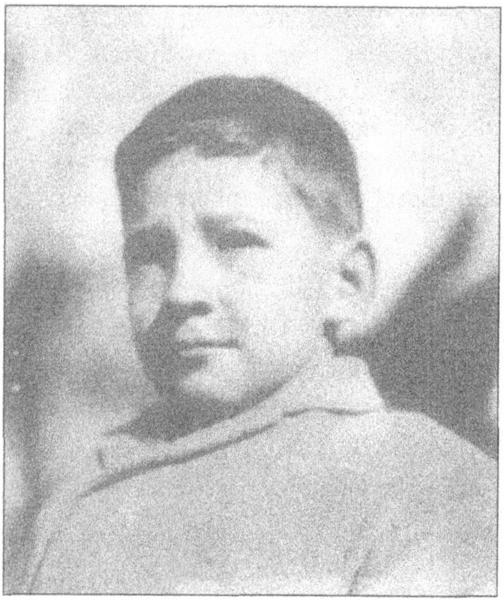

Age 10.

Age 20.

Age 35.

Age 47.

Except where noted, all images are from the Collections of Henry Ford Museum & Greenfield Village.

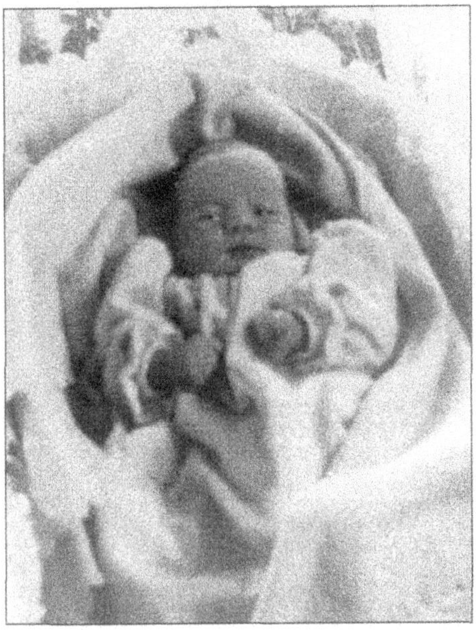

Earliest picture of Edsel Ford. Only a few weeks old here, he is resting in the bassinet in which he was sleeping when his father first started his experimental gasoline engine in December 1893.

Four-year-old "Eddie" Ford, fishing at Lake Orion in summer 1898.

Clara and Edsel on the Ford farm, January 1, 1901.

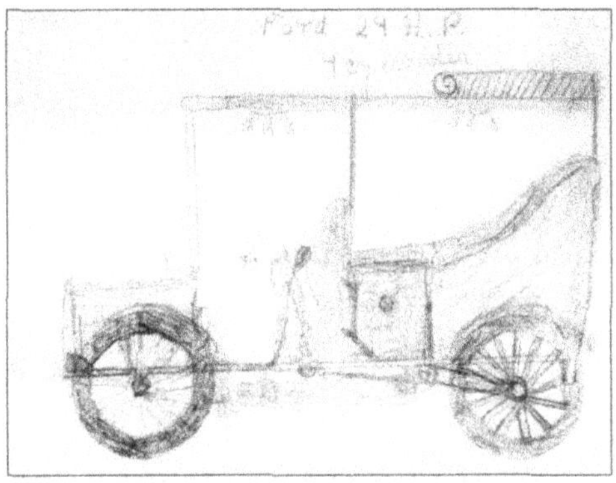

Automobile drawing done by Edsel when he was about ten years old. Although the sketch is rudimentary, the design is well proportioned and detailed.

Sketch of a speedster, done by Edsel, circa 1911.

Except where noted, all images are from the Collections of Henry Ford Museum & Greenfield Village.

Ten-year-old Edsel was the hit of the neighborhood in winter 1906. Here he pulls friends on sleighs behind his new Model N Ford.

Father and son—Henry and Edsel—in a 1905 Model F Ford. These were amiable times, when Henry was beginning to teach his son the ways of the world.

Detroit University School's track team, circa 1911. Edsel is at the far right.
(Photo courtesy of Edsel & Eleanor Ford House)

Edsel poses with high-school friends, circa 1912. Left to right: Thomas Whitehead, Herbert Book, William Russell, Edsel Ford, Robert Grey, Frank Book, and Horace Caulkins. (Photo courtesy of Edsel & Eleanor Ford House)

Except where noted, all images are from the Collections of Henry Ford Museum & Greenfield Village.

Edsel and Eleanor's wedding party: The men, left to right: Ernest Kanzler, Robert Tannahill, Charles Hodges, Thomas Whitehead, Edsel Ford, and Horace Caulkins. The ladies, left to right: Ruth Woodruff, Josephine Clay Kanzler, Eleanor Ford, Agnes Taylor, and Esther Cutler. (Photo courtesy of Edsel & Eleanor Ford House)

Edsel (standing) and Horace Caulkins take a well-deserved break during their trip to San Francisco in 1915. Now 21 years old, Edsel is "a spitting image" of his father. (Photo courtesy Edsel & Eleanor Ford House)

Edsel and Eleanor on their honeymoon in Honolulu, Hawaii, December 1916. (Photo courtesy of Edsel & Eleanor Ford House)

One of Edsel's favorite pastimes was fishing. Here, he enjoys a day of deep-sea fishing off the Florida coast, circa 1930.

Edsel and Eleanor at their Grosse Pointe mansion, circa 1937.

Edsel's Gaukler Pointe estate. Edsel lived here for only 14 years, Eleanor for nearly 50.

A boyish-looking Edsel with his first child, Henry Ford II, 1917.

Edsel with his two oldest children, Henry Ford II (left) and Benson (right), circa 1921.

The young Edsel Ford family, left to right: Benson, William, Edsel, Josephine, Eleanor, and Henry II, circa 1929. (Photo courtesy of Edsel & Eleanor Ford House)

Relaxing in the sun in Miami, Florida, circa 1930. Edsel is in the foreground. Eleanor is under the parasol in the center of the picture, and her sister, Josephine, is at the right.

Heirs apparent: Henry Ford II (right) and his brother, Benson, with Edsel and Eleanor, circa 1936.

The Edsel Fords on their way to Europe in 1936. Left to right: Eleanor, Edsel, Josephine Kanzler (Eleanor's sister), Benson, and Henry II. (Photo courtesy of Edsel & Eleanor Ford House)

The mature Edsel Ford family, circa 1940. Left to right: William, Josephine, Edsel, Eleanor, and Benson. (Photo courtesy of Edsel & Eleanor Ford House)

Edsel and Eleanor with their first grandchild, Charlotte, in 1942.

Ernest and Josephine Kanzler (right) were Edsel and Eleanor's closest friends, circa 1929. (Photo courtesy of Edsel & Eleanor Ford House)

Edsel and Eleanor Ford, circa 1935.

Old-fashioned dance at Lovett Hall, Greenfield Village, circa 1930. Edsel is shown bowing in the right foreground. His mother, in the bobbed dark hair, is directly behind him. Henry, with his back to the camera, is at the bottom right.

Except where noted, all images are from the Collections of Henry Ford Museum & Greenfield Village.

Edsel and Eleanor relax on their sailboat, circa 1930.

Edsel at the wheel of his yacht, Onika, *circa 1922.*

Edsel behind the wheel of one of his race boats, circa 1922.

Albert Lepine, Edsel's secretary for 25 years. (Photo courtesy of Monica Lepine Starrett)

Henry (right) poses with Edsel, shortly after appointing him president of Ford Motor Company. Edsel had just turned 24 years old.

Leaders of the automobile industry, circa 1934. Left to right: Edsel, Alfred Sloan of General Motors, and Walter Chrysler.

Ford Motor Company executives, 1933. Left to right: Bert Craig, Pete Martin, Charles Sorensen, Henry Ford, Archie Johnson, Edsel Ford, and Albert Wibel.

Even at his Florida retreat, Edsel read constantly, circa 1938.

Edsel overlooking the mammoth River Rouge manufacturing plant in Dearborn, Michigan, circa 1940.

Edsel (foreground) and Henry, bird watching, circa 1915.

Left to right: Unknown, Edsel Ford, Henry Ford, and the Prince of Wales on a tour of Ford plants in Detroit in 1924. (Photo courtesy of Edsel & Eleanor Ford House)

Henry (left) and Edsel, in 1942, with Henry's first automobile. When he was a little boy, Edsel used to sit next to the left front wheel and hold on to the tiller when his father took him for rides in this quadricycle.

In this revealing image taken in Greenfield Village in 1940, Henry had pulled Edsel (left) aside to lecture him about something. Edsel, the dutiful son, always listened attentively. (Photo courtesy of The Detroit Public Library)

Edsel, at the wheel, gets ready to drive the fifteen-millionth Model T off the assembly line. It was a solemn occasion for Henry (left), but to Edsel, it signaled a new era for the company and for him.

At right is a prototype design developed by the Murray Corporation at the behest of Edsel to replace the aging Model T. This prototype was completed several months before Henry gave the go-ahead to place a new car into production.

Ford executives during the development of the Model A engine. Left to right: Charles Sorensen, Edsel Ford, Henry Ford, Pete Martin, Charles Hartner, and Fred Rockelman.

Edsel Ford drives a new Model A roadster out of the Ford Engineering Laboratory in Dearborn to show a throng of automotive journalists, October 1927.

Henry (left) and Edsel at the time of the introduction of the Model A in the fall of 1927.

Edsel being sworn in by Senator Bankine at the Senate Committee on Currency and Banking, 1934. Senator James Couzens is seated at the left.

Chapter 14

Bank Holiday

Part of the success of the Model A was not Henry's engineering or Edsel's styling, but the customer's ability to finally buy a Ford on a payment plan. For years, Henry Ford refused to allow his dealers to sell Fords on credit. The closest he came to a credit program was his so-called $5 Per Month Plan, whereby a customer could make monthly payments in advance of obtaining a new Ford. All the customer had to do was to deposit $5 per month at the local Ford dealer. When enough money had accumulated, the customer could drive home in a new Model T. Obviously, that was not what the customer wanted, nor is that what the competition was offering. In 1919, General Motors established its financial arm, the General Motors Acceptance Corporation. Shortly thereafter, Chrysler established Chrysler Credit Corporation. With both companies, a customer could place a small down payment on a new General Motors or Chrysler product, and then make monthly payments until the car was paid for, all the while enjoying his or her new car.

***

Earlier, Kanzler and two friends, Carlton Higbie and George Fink, established a securities company called Kanzler, Fink, & Higbie to take advantage of the astronomical stock market of the 1920s. "Edsel used K.F.H. as a channel for his own investments, and he made money," recalled a close associate.[1]

As their investments increased, Edsel and Kanzler discussed ways of expanding their financial expertise, and Edsel suggested opening a bank. He was not new to the banking business. He had been a board member of the National Bank of Commerce, the Dearborn State Bank, and the Highland Park State Bank for years. Because Ford Motor Company was a major depositor in many banks of Detroit, it made sense to channel some of those deposits into a bank owned by Ford interests.[2] Henry felt that was a good idea also, and he encouraged Edsel to become more involved in the finance side of the business.[3]

Together, Edsel and Kanzler organized the Guardian Detroit Bank in the summer of 1927, and they enticed their friends to join them in their new venture. Other stockholders were Roy Chapin of Hudson Motor Company; Walter O. Briggs, owner of Briggs Manufacturing, a major body supplier to Ford; Howard Bonbright, Edsel's friend and an associate of Briggs'; Alvan

Macauley, president of Packard Motor Car Company; and, of course, Kanzler's associates, George Fink and Carlton Higbie.[4]

Initially, the Guardian Detroit Bank was a small player in the banking world of Detroit. However, with the power of Ford Motor Company—not to mention that of Packard and Hudson and Edsel's persuasion—the small bank soon became a powerhouse. At the insistence of Kanzler and Macauley (and probably against Edsel's better judgment), Edsel began sending solicitation letters to Ford's major suppliers, asking for their lucrative investments and associated deposits. "Your influence with them would be considerable," wrote Macauley.[5] In short order, Edsel had written letters to Bethlehem Steel, Firestone Tire & Rubber Company, Goodyear Tire & Rubber Company, and others[6]:

> *I am writing an expression of my interest in the Guardian Detroit Bank. The bank is, as you know, a new one but very firmly established and with strong backing here in Detroit. The policies are sound, and with the young organization in command, the future looks very bright. I am sure that anything you could do for the bank in the way of a deposit would be very highly appreciated.*

With Edsel's mild disposition, he must have felt uncomfortable sending this letter to his suppliers; however, once he agreed, the floodgates opened. A few of the major corporations were not enough for the new managers—they wanted them all. "The bank should have accounts from the A.O. Smith Corporation and the Briggs & Stratton Company," wrote Macauley. "I believe you are buying from these people."[7] A short time later, Macauley again wrote to Edsel: "I have...been working on Mr. R.H. Wurlitzer of the Rudolph Wurlitzer Company. He is unquestionably impressed very favorably with our setup and possibilities. Apparently, all he needs is a little urging. If you could merely drop him a note stating your confidence in the Guardian and its future, I feel sure you would turn the trick."[8]

In all instances, Edsel sent notes of encouragement, and, in most cases, his bank received sizeable deposits in return. Through his name alone, Edsel was directly responsible for bringing millions of dollars into the coffers of the Guardian.

***

With the Guardian doing well, Edsel again began talking to his father about establishing a finance arm of Ford Motor Company, and he was well prepared with his argument. For quite some time, Edsel had been pressured by his dealers to provide financing, and he had already met with representatives of the Agriculture Bond & Credit Corporation to learn about their operation. As an independent financial institution, ABC was already providing financing to Ford dealers for their tractor sales.[9] Additionally, Edsel and Kanzler had many serious discussions, over dinners at Edsel's home, concerning the establishment of a Ford finance company.

Unlike Edsel's previous attempts, Henry acquiesced this time. Similar to automobile styling, Henry finally realized that the time had come to offer financing, and he thought it would be a good idea for Edsel to gain experience in that area as well. After all, the company had such large deposits in these banks that somebody in the family should know more about their operation.[10] Henry gave his approval but did not want the new finance company wholly owned by Ford Motor Company.

Therefore, in March 1928, Edsel and Kanzler established the Universal Credit Corporation to offer credit to both Ford dealers and Ford customers. With an initial capitalization of $11.5 million, 15,000 shares of stock were issued: 7,501 were owned by Ford Motor Company, 5,000 shares were owned by Edsel's new bank, the Guardian Detroit Company, and 2,499 shares were owned by Ernest Kanzler.[11] With Kanzler's substantial interest in Universal Credit Corporation, Edsel made him chairman of the new finance company.

When Kanzler was fired from Ford Motor Company, Edsel felt obligated to help him find a new career. Edsel knew that he could never get Kanzler back into the company directly, so he did it indirectly—through this new financial arm. "Edsel insisted later on that something ought to be done for Kanzler," explained Liebold, "and Mr. Ford, in order to acquiesce in that, agreed to organize the Universal Credit Corporation, so that Kanzler would have something to do."[12] What Henry had wrought two years earlier, Edsel had finally undone.

Thanks to the boom times of the late 1920s and the success of the Model A, Universal Credit Corporation was an instant success. In the first year of business, UCC arranged loans for more than 400,000 Model A buyers.[13] The company pushed the Model A with the slogan, "$25 a month will buy a Model A."[14] By 1933, the financial arm of Ford had financed nearly $900 million.[15]

***

Edsel never intended to become more involved in high finance than through these two financial institutions, but soon he was more embroiled than he ever could have imagined. During the time that he and Kanzler established the Guardian Detroit Bank and the Universal Credit Corporation, a trend in banking began to develop that overtook the industry like a prairie fire. It was called group banking—a method by which one or more large banks took over a number of smaller banks to gain power in a region. The first of these conglomerates in Michigan was the merger of the National Bank of Commerce and the Griswold First State Bank.[16] Shortly thereafter, these banks again were consolidated into the Union Trust Company, and a new company—a holding company—called The Union Investment Trust was established.[17]

While Kanzler was vacationing in Europe in the spring of 1929, Edsel and the other board members decided to join the conglomerate fray.[18] They combined the Guardian Detroit Bank with the Guardian Detroit Trust Company and formed a holding company called the Guardian Detroit Group. With Edsel's backing and the tremendous deposits of Ford Motor Company, the Guardian Detroit Group began gobbling up banks throughout Michigan. Before long, it had 25 banking companies within its umbrella, from Saginaw to the north, Grand Rapids to the west, Port Huron to the east, and Jackson to the south.[19] Before 1929 ended, Edsel's new holding company had acquired seven additional banks, including the Highland Park State Bank, the Highland Park Trust Company, the Bank of Dearborn, and the Bank of Detroit, and merged again, this time with the Union Commerce Corporation.[20] They called their new mega-conglomerate the Guardian Detroit Union Group.

Edsel himself converted 30,000 shares of stock from a myriad of banks—Bank of Detroit, Fordson State Bank, and National Bank of Commerce, among others—into Guardian Group shares, and then bought additional shares. His father bought 1,188 shares and eventually gave them to Edsel as a gift.[21] All told, Edsel owned 55,000 shares in the Guardian Group, which had a face value of more than $1 million, and the Ford Motor Company invested nearly $11 million in the new conglomerate.[22]

To counterbalance the tremendous power of the Guardian Detroit Union Group, 40 other banks backed largely by General Motors and Chrysler money merged into the Detroit Bankers Company in January 1930. Between the Guardian Detroit Union Group and the Detroit Bankers Company, they controlled nearly 90 percent of banking in Detroit and held more than 50 percent of all loans and investments in the state.[23] Almost every family in Michigan had its savings in banks owned by one of these two banking groups.[24]

Within two short years, Edsel's bank was second only to the General Motors-Chrysler conglomerate. Robert Lord, then president of the bank, wrote to Edsel in January 1930, expounding on their success: "I think you will be interested to note that Guardian Detroit Bank shows the greatest percentage gained in deposits of all Detroit banks during the past 12 months."[25] Kanzler also was quick to point out to Edsel how well the bank had done over this same period.[26] In the annual report of 1932 for the Guardian Group, Kanzler mentioned that "in three cities, our units have 100 percent of the deposits in the city; in another three cities, our units have more than 75 percent of all deposits; in two cities, our units have more than 50 percent; in five cities, our units have more than 25 percent; and in three cities, we have less than 25 percent."[27]

***

To symbolize its new stature in the community, the Guardian Group directors approved the construction of an office building on the corner of Griswold and Congress streets in the heart of Detroit. Rising more than 500 feet above the street, the Guardian Building was the epitome of glitz and glamour of the Roaring Twenties. A local architect was commissioned to design the building and decided on an Aztec theme. It was constructed of expensive metals, elaborate tile work, and exotic stone from the far ends of the earth. The papers rightfully dubbed the monolith "The Cathedral of Finance."

At a cost of $12 million, no expense was spared. For the exterior, a specially formulated "Guardian brick" of a red-tan color was used. "At the sidewalk level is a two-foot sill of Somes Sound granite," explained a reporter. "Above this is a six-foot band of Montrose granite, then another band of the Somes Sound, above which is Mankato stone to the sixth story. Somes Sound granite comes from Maine, Montrose granite is quarried in northern Wisconsin, and Mankato stone comes from Minnesota."[28]

Customers entered the building through massive doors made of Monel metal, a nickel alloy, and were welcomed by an expansive main lobby. A mural designed by Ezra Winter made of thousands of pieces of mosaic glass decorated the lobby. The walls were made of Belgian black marble and a rare Numidian marble, chosen for its blood-red color. The architect went to Africa to choose the pieces of marble himself. To keep the noise level inside the lobby to a minimum, the high vaulted ceiling was made acoustical; a three-quarter-inch mat of horsehair was covered with plaster, and over this was a layer of perforated canvas, which was painted. Light entered the lobby through stained-glass windows. The conference rooms on the sixth floor were done in knotty pine and furnished with Cape Cod chairs and tables. There were offices of indescribable luxury for the president, and these were flanked by retiring rooms, a massaging room, a barbershop, and baths. The rugs in the offices "were thick as a cat's tail."[29]

However, the executives rarely used these lavish offices, recalled a Ford executive. "They were just ornate reception rooms."[30]

***

Things were going well for Edsel's banking empire—mortgages were being made, stocks were being sold, and, more importantly, the stock of the Guardian Group was appreciating nicely. Then disaster hit. The Wall Street stock market crash of October 1929 struck the United States hard—especially Detroit. Detroit was a one-product town, and when men started losing jobs throughout the country, they stopped buying automobiles. When they stopped buying automobiles, the factories of Detroit came to a halt. In one government report, the employment rate in Detroit one year after the stock market crash was half of what it had been immediately before the Depression.[31] At the nadir of the Depression, the unemployment rate in the country was 25 percent. In Detroit, it was twice that.[32]

Before the Crash, when the automobile factories were humming, men flocked to Detroit from all over the country and found jobs. In 10 short years, the population of Detroit grew by 600,000. This fantastic growth spawned widespread speculation in real estate and, with it, the price of homes went through the roof. However, the men were working and earning good wages, and Edsel's bank gladly gave them mortgages to buy their homes. When the stock market crashed and the automobile companies began laying off workers, formerly industrious men soon began defaulting on their mortgages. The banks could do little to recover their loans—with no buyers, liquidation was all but impossible. The Guardian Detroit Union Group was heavily leveraged on these inflated real estate properties. Therefore, when customers began defaulting on their loans, the deposit ratio of the unit banks began to drop dramatically. Edsel did not know it, but his banking empire was on the brink of collapse. He was too busy running the Ford Motor Company, monitoring the production of the Model A, and working with Briggs Manufacturing in designing another new car that soon would replace the Model A. Competition was stronger than ever in the automobile business, and Edsel needed to devote his time and expertise to product development, not to high finance. He left that to Kanzler. When Edsel was asked how many bank board meetings he had attended, he replied, "I do not recall."[33]

***

The problem with the group system was that only the stock of the holding company was offered on the stock exchange. Because the holding company itself was unable to generate income, it depended on its units to make the money—from banking, securities, and mortgages—and then have the units pay dividends to the holding company. The income from these dividend streams supported the stock price of the Guardian Group. When the economy was good—when people were working and buying homes and paying their mortgages—all of the units of the holding company made money and were able to pay handsome dividends to the holding company. The first dividend declared in 1930, for example, was for 20 percent per annum.[34] These high dividends, in turn, drove up the price of the stock of the holding company, making the stockholders rich. When the Guardian Group was organized, its stock was valued at $20 per share. However, with the wild speculation in real estate and in the stock market, its stock price rose as high as $350 per share. Everybody expected it to rise to $1,000 a share.[35] Edsel hoped his $1 million investment would soar to $55 million.

Additionally, with an amalgam of companies and subsidiaries under the umbrella of the Guardian Group, it became almost impossible to keep track of transactions among them all. When the securities company of the Guardian Group issued bonds, for example, the banking units of the Group would loan its customers money to buy those bonds using the customers' stock in the holding company as collateral for the loan.[36] To make matters worse, the Guardian Union Trust Company gladly made loans to the officers of its sister banks on little or no collateral. What little assets they had, reasoned the bank managers, would certainly rise in value in this booming economy. One of the larger unit banks had loaned nearly $3.5 million to officers and directors of the other unit banks. Another unit bank loaned approximately $1 million to a number of its officers and managers, on collateral of only $15,000. These vast sums were being offered at a time when a sizeable home in Detroit cost approximately $8,000! However, that was only the beginning. The holding company borrowed $14 million from its unit banks so that they could put up a good front, and by end of 1932, it held nearly 150,000 shares of its own stock as collateral for these loans.[37]

When the economy collapsed, people began defaulting on their loans and mortgages. Consequently, the units were having difficulty in remaining solvent, let alone paying dividends to the mother company. If dividends ceased to flow to the coffers of the holding company, the stock of the Guardian Group would drop, and with it, the fortunes of its stockholders.

To prevent this, Kanzler and the other board members began forcing the unit banks to pay dividends, even though the units did not have the funds to do so. When that did not work, Kanzler resorted to issuing dubious financial reports that made the Guardian Group appear more solvent than it was.

John Stalker, president of the Union Guardian Trust, wrote a letter to Robert Lord, questioning this policy:

> *We have your letter of the 4th instant with respect to the 5 percent quarterly dividend which you suggested that we pay this month. I presume that a dividend of this amount is necessary to the fulfillment of your plan... However, as you are aware, a dividend of this amount has not been earned. In addition to that, the trust company is setting up no reserves, and we feel that is not as it should be. There is no doubt in my mind that the company will suffer some losses.*

A few years later at a Senate hearing, when asked about these dubious business activities, Kanzler could not recall this letter or any discussions regarding the units being unable to pay dividends. However, he did confirm that the policy of the Group was to have a fixed dividend to maintain the price of the stock. This connection between the price of Guardian Group stock and the dividends paid by its unit banks prompted the president of one of the unit banks to make a suggestion in December 1931[38]:

> *My own opinion is that the most constructive move you could possibly make at this time would be to make arrangements whereby the Guardian stock would be immediately taken off the market. Only last week, one of our depositors withdrew $5,000 from the bank in cash and put it in a safe-deposit box, plainly stating that while he had every faith in our bank as a separate institution, he was worried about the declining market on Guardian stock—and would keep his money in cash until things got straightened around.*

Although that was a good idea, the directors felt that "it would be too dangerous to take it off the market...."[39]

***

As the dividend stream dried up, the price of Guardian Group stock plummeted. From a high of $350 per share in August 1929, it was only $75 a year later.[40] To keep the price of the stock of the holding company from collapsing, the directors agreed to form a stock pool—Kanzler called it a "purchase group"—and purchase 60,000 shares (a value of nearly $5 million) and hold them for one year. A trustee committee, which included Kanzler and Roy Chapin, was formed to oversee the purchase of these stocks and to ensure that the holders did not sell it prematurely.[41]

While Edsel was part of the pooling group, he did more than his part to protect his investment and shore up the bank. In December 1930, he loaned the Group $1 million in cash and $5 million in municipal bonds. A year later, he endorsed a $2.5 million loan obtained by the Guardian Group from the Continental Commercial Bank in Chicago. That loan could be increased at any time—at the discretion of the Guardian Group—to $15 million.[42] Additionally, at the end of 1932, Ford Motor Company made a $3.5 million loan to the Guardian Group. By the beginning of 1933, Edsel and Ford Motor Company had loaned the Group more than $9.5 million in cash and securities, and Edsel was liable for another $12.5 million if the Group decided to expand the loan request it had with Continental Bank. "That was not a gift," said Kanzler, "that was a way to provide liquidity."[43]

As Edsel explained later, he supported the Guardian Group because "their assets were being appraised and depreciated all the time, and that is why the different units required assistance of this kind—to satisfy the depreciation and the criticism of the securities against the loans in these banks."[44] Edsel had willingly made the loans because at the time he did not think the situation was desperate.[45] He believed—the same as everybody else believed—that these dire financial times would soon end.

When they did not end, Edsel was asked to lend additional support and perhaps unwittingly contributed to the thickening malaise. Year after year, he instructed Bert Craig, Ford Motor Company assistant treasurer and secretary, to make substantial deposits in the Guardian Group immediately before year end, and then withdraw those deposits a few days later. This enabled the bank to show better deposit ratios than would otherwise be possible.

During a Senate hearing investigating the Michigan banking crisis a year later, Senator Ferdinand Pecora asked Edsel about this dubious tactic. In answer to Senator Pecora's question, Edsel replied: "Those requests would undoubtedly go through Mr. Craig. I do not recall any request ever having been made directly to me, but there might have been."[46]

"Do you know the purpose for the making of those temporary deposits?" asked Pecora.

"I suppose to show increased deposits in their statements," replied Edsel.

"Now, as a bank director, Mr. Ford, and as a businessman, do you think it was an ethical method of enhancing the appearance...of the bank's condition?"

"No, I don't suppose it was."[47]

When Charles Sorensen heard about these shenanigans, he was livid. "Edsel simply didn't know what he was in for," recalled Sorensen. "Bert Craig was asleep—he was pitiful.

If Craig had been a stronger treasurer of the Ford Motor Company, he would have kept Edsel out of trouble."[48]

***

These were trying times for Edsel. People who were close to him noticed that he was "very depressed."[49] At one point, while discussing the crisis with Ernest Liebold, Henry Ford's secretary, Edsel lost his composure and started crying. "Well, I'm cleaned," he sobbed. "All the Continental Bank has to do is to call me on my note and that will clean me up of all I've got."[50]

"Well, son, I see they took you to the cleaners," Henry said in his typical unsympathetic tone. However, that is about all Henry said to Edsel regarding his financial woes. "Henry just brushed it off."[51]

Meanwhile, Edsel began watching his expenses carefully and took expedient measures to conserve his capital. He cut his expenses at his various residences to the bone. "At Grosse Pointe, a great number of employees were laid off," recalled an associate. "Normally, there would be approximately 25 gardeners, now the number was reduced to 4 or 5. Watchmen, boatmen, engineers, and all the household staff were reduced to a minimum."[52] Edsel tried his best to keep his mind preoccupied with other matters, but it was useless. He could not even get a haircut without being reminded of his financial straits. During a break in his hectic schedule, he walked into a barbershop that he did not usually frequent and took a seat. The *Detroit Times* had just arrived, and blazoned across the top was EDSEL FORD OWNS 6,000 SHARES IN BANK OF DETROIT. The barber picked up the paper and handed it to Edsel, not knowing who he was. "Sir, would you like to look at the scandal sheet?"[53]

***

By no means were the banks in Detroit the only banks in the country suffering from the Depression. Hundreds of banks had already failed because of the business slowdown, the drop in real estate prices, and from unpaid mortgages. To maintain public confidence and stem the tide of these bank failures, the federal government established the Reconstruction Finance Corporation (RFC) at the end of 1931. Through this agency, the government made up to $300 million available to banks and other financial institutions to help them weather the storm. In its first official act, the RFC loaned $90 million to a banking consortium in Chicago run by Charles Dawes, who had been the first president of RFC. (He resigned that position to obtain the government loan.)

In Detroit, the situation had worsened day by day. However, Kanzler and the others were eternal optimists, thinking that the Depression would soon end. Therefore, they carried on with their grandiose schemes, only to be caught in a whirlpool from which they could not escape.

The breaking point occurred at the end of 1931. For two years, Kanzler and the others had been juggling books, shifting loans, and boosting stocks. By June 1932, the largest unit of the Guardian Group, the Union Guardian Trust, was on the brink of collapse. The oldest trust company in Michigan and the flagship of the Guardian Group, it had under its aegis the Guardian National Bank, the National Bank of Commerce, the Highland Park State Bank, and the Bank of Dearborn. Combined, these financial institutions served more than 50,000 customers, and their combined resources exceeded $500 million.[54] Ford Motor Company routinely had $30 million of deposits in the Trust company.[55] The general feeling of the board members of the Union Guardian Trust was that if the Trust failed, it would have a ripple effect throughout the units of the holding company, and they too would go under.

With RFC funds now available, Kanzler went to Washington to secure a loan to shore up the Trust, and he easily received $14 million.[56] Ironically, and most astoundingly, while Kanzler was in Washington securing the loan, the Trust was declaring a 5 percent dividend to the Group.[57]

But that was only the beginning. Two weeks later, Kanzler told Edsel that he would need to go to Washington again and ask for additional relief from the RFC. The $14 million loan was not enough. "The Union Trust Company cannot hold up much longer," Kanzler explained. "We have to get considerably more money or the whole Group is going to collapse."[58] According to his calculations, the Guardian Group needed an additional $21 million to keep its doors open. "I concurred with that, and steps were taken to proceed," Edsel said later.[59]

Fully aware of the desperate situation of the Guardian Group, and in an effort to maintain public confidence and the price of the stock, Kanzler issued false financial statements. He stayed up all night on January 23, 1933, compiling a report to present to the stockholders of the Guardian Group the next day. It was as carefully worded as his highly trained legal mind could make it. "The consolidated net operating earnings of the group companies, banks, trust companies, and all other affiliated companies amounted to $1,316,952," he reported, without blinking an eye.[60] He also reported that during 1932, $1.7 million in securities were sold at a net loss of only $42,000. That would have been a laudable accomplishment, had it been true. However, the truth was that those securities had been written down over the previous two years to that level. In an effort to indicate that things might not be so rosy in the future, Kanzler did mention the precarious condition of the banking industry in general, that there had been many bank failures, and that people in Port Huron were taking their money out of their bank and taking it to Canada.[61]

***

No matter how Kanzler presented the situation, the fact was that Edsel's and Kanzler's banking empire was on the edge of collapse. Instead of making a profit in 1930 of more than $1 million, as Kanzler had indicated at the stockholders meeting, the Guardian Group had actually lost more than $39,000. The deceit did not end there. In 1931, net earnings before charge-offs were publicly reported as $3,887,052, but the official report to the Michigan Securities Commission listed a loss of $288,930.[62] Actually, the Group had been losing money at a prodigious rate since the stockmarket crash. Between 1930 and 1932, the Group had suffered a deficit of more than $700,000. Kanzler had the audacity to tell the Senate hearings "that was not material, to my mind...it really was not a figure that the stockholders should have been or would have been interested in."[63] He stated that the balance sheet of the company would have helped explain the situation of the Group, but it was never published.

On January 25, 1933, Kanzler and Ford attorney, Clifford Longley, took the overnight train to Washington and met with RFC officials the next morning, in hopes of obtaining more money. Present were Charles Miller, president of the RFC; Jesse Jones, an RFC director; Ogden Mills, Secretary of the Treasury; Undersecretary Arthur Ballantine; and Ohio Senator Atlee Pomerene. Kanzler told the officials that $20.5 million would be required to liquidate the deposits of the Union Guardian Trust Company, but the Group could offer only $6 million in bona fide assets as security for such a loan. He stated that conditions had worsened considerably since his last request and that if a loan could not be obtained, he would have to close the Union Guardian Trust, which would result in the collapse of the entire Guardian Group. Because his units were much healthier than other banks, the collapse of his units would effect the collapse of all other banks throughout the State of Michigan and perhaps the entire country.[64]

This did not sit lightly with the RFC members. They had already advanced $14 million to Kanzler's Group, and now the Group wanted millions more, on very little security. Plus, this was Edsel Ford's bank, and they felt that he had enough money to rescue his own bank. "Why should we bail out Mr. Ford?"[65]

Nevertheless, federal bank auditors were sent to Detroit to appraise the assets of the Group. "The various officers of the banks came in with their collateral in pouches," recalled John McKee, RFC chief appraiser, "and we attempted, working day and night, to prepare a schedule."[66] In the end, the auditors determined that to save the Guardian Detroit Union Group, a loan much greater than what Kanzler had requested would be required. According to their calculations, the Guardian Group would need an additional $50 million—$1.7 billion in today's dollars. It is unclear why Kanzler did not request this amount in the first place—perhaps his managers were lying to him, or perhaps he misinterpreted the financial statements of the Group.

Unfortunately, the Group as a whole could generate only approximately $35 million in bona fide collateral, and that included the magnificent Guardian Bank building. By law, the RFC could loan only the amount of money supported by collateral. Therefore, the Guardian Group was $44 million short. Edsel said that "the assets which were first presented were not of sufficient value; that the officers who came down here were told to put in the slow assets; that the slow assets, provided they were all right, would be just as suitable for a loan of that kind; and that was done. Then, when these figures were made up, the collateral was insufficient, and they came back to Detroit and put in additional collateral."[67]

When President Hoover was told of the dire situation in Detroit, he called a meeting with Chairman Pomerene, Secretary Mills, and Senator James Couzens, Henry Ford's old business manager and now a Michigan senator, to discuss the matter. In a rare change of philosophy, Hoover openly suggested that in light of the situation, the RFC should make the loan to the Guardian Group in spite of the insufficient collateral. However, Senator Couzens objected strongly. "If you make that loan without sufficient collateral," he yelled, "I will shout against it from the roof tops and on the floor of the Senate!"[68] Shortly thereafter, Couzens withdrew $300,000 from one of the Guardian's units and sent it to Canada.[69]

Meanwhile, Kanzler and Longley were working day and night to devise a plan by which the RFC would make the loan. They might have been able to negotiate terms with the RFC had Couzens not become involved. However, with Couzens' intervention, they needed to secure the exact amount of collateral—good collateral.

Kanzler thought that perhaps the Fords could provide further assistance. Maybe he could convince them to subordinate $7 million of deposits that Ford Motor Company had with the Trust. Both Kanzler and Longley called Edsel.

Edsel talked to his father about the request, and they both agreed that if that was the only thing left to do, they would subordinate their deposits.[70]

It seemed that everything was now in place for the loan, but then the matter of facilitating the loan was mentioned. If each unit of the Group were required to apply to the RFC for an individual loan, it would take months for all of the transactions to take place, and time would run out. The situation in Detroit was becoming more aggravated each day, and there was concern that mass panic could occur at any time. During those few weeks, scattered runs on banks had already occurred throughout the country. To speed up the process, the suggestion was made that a separate company be created—a mortgage company—which could legally accept the assets from the unit banks and then pledge them back to the RFC. This would eliminate the "multiplicity of transactions" that would have complicated the loan. However, to accomplish this, another $5 million would be needed.

Again, Kanzler and Longley asked for the Fords' assistance. This time, Edsel and Henry refused. They would subordinate their deposits but would not offer additional capital. "We had put up what we thought was a great amount of money to help this situation through, and we came to the end of it," Edsel later explained. "We decided we could not do any more."[71]

President Miller and Secretary Mills met with Edsel and pleaded with him to offer additional assistance to the Detroit bankers. A few days later, the Undersecretary of the Treasury, Arthur Ballantine, and the Secretary of Commerce and Edsel's good friend, Roy Chapin, met with Edsel in the Guardian Building. They also urged him to supply additional capital so that the RFC loan could be made and a banking disaster averted. Edsel said he would approach his father again but made no commitments.

For weeks, Henry was convinced that the banking crisis was nothing more than hot air. However, when his confidant, Harry Bennett, confirmed the crisis, he became concerned and called a meeting with Longley and Liebold. "It was Mr. Ford's purpose to find out by indirection just how stable the Guardian Trust Company was," explained Bennett years later. "He didn't trust these men to give him direct information. So he told them that he intended to draw some millions of dollars out of the First National Bank of Detroit, where he also had money on deposit. These men told him that he could never get his money. That was when Mr. Ford knew the worst."[72]

When Edsel brought up the subject again, Henry became incensed. Henry told Edsel that he would neither subordinate his deposits in the Trust nor offer additional capital. In fact, he would withdraw the $25 million that he had in the First National Bank (an institution now owned by General Motors) in the morning. "That came about," explained Edsel, "because Ballantine said that the Trust should not be allowed to open in the morning because it was insolvent. Ford said, 'if the Trust could not open, then First National should not open as well.' He wanted both to be treated the same. Therefore, if First opened but not the Trust, he would withdraw his deposits."[73]

"The situation was serious," recalled chief national bank examiner Alfred Leyburn. It was apparent that no deal could be made between the RFC and the Guardian Group. Although there had not yet been an "out-and-out run" on the Trust, "it was about through." The only thing that the bank and government officials could think of doing to provide some "breathing room and maybe work out something," was to declare a bank holiday—closing the banks by government declaration so that depositors could not withdraw their funds. As Leyburn explained it, "...if you let one of the largest banks in Detroit close, you cannot keep the parent bank open across the street, and [the Guardian Group] had a lot of other weak units out through the State that could not stand up anyway. It would kill the whole thing off. Then we would have a run on the First National Bank, which was in a bad situation, too."[74]

Sunday, February 12, was Lincoln's birthday, making Monday a legal holiday. Therefore, the banks would be closed that day anyway. However, what about Tuesday? And the next day? And the next? That evening, Michigan Governor William Comstock joined Kanzler and the others in the ornate conference rooms of the Guardian Building. Henry's threat made it clear that closing the banks was the only way to prevent a statewide panic the next day. The bankers pleaded with the governor to declare a bank holiday. Thus, at 1:32 in the morning on Tuesday, February 14, 1933, the governor signed the proclamation declaring the bank holiday, instructing the banks to remain closed for a week.[75]

The Michigan bank holiday froze deposits of more than $1.5 billion in the state. A total of 436 banks and trust companies closed. To meet payrolls, industrialists such as Walter Chrysler went out of state and returned carrying sacks of money.[76] Henry Ford, anticipating a statewide panic, had ordered Liebold to store millions of dollars in cash on Ford Motor Company premises, "in case we ever happen to need it." Liebold bought four large safes and began filling them with cash, withdrawn from numerous banks throughout the country—$100,000 at a time.[77]

Nonetheless, things grew worse. Before the week ended, the governor extended the moratorium for another six days and then for another four days. Eventually, Michigan's statewide freeze affected banking systems throughout the country. During the next ten days, banks all over the country closed their doors. First those in Indiana closed, then banks in Maryland, Arkansas, and Ohio, followed by Alabama, Kentucky, Nevada, and Tennessee. On March 2, banks in Arizona, California, Louisiana, Mississippi, and Oregon closed.[78] On March 4, most other banks in the country closed. The next day, newly elected President Franklin Roosevelt, as his first official act, was forced to close every bank in the nation.[79]

***

Ten days after the bank holiday had been declared, and in Henry Ford's typical fashion, he had a change of heart. Instead of threatening to make massive withdrawals from the Detroit banks, he issued a statement to the press, saying that he and Edsel wanted to buy the Guardian Group as well as the Detroit Bankers Company. He said that the reason for their collapse in the first place was poor management, and he promised to run the two new banks as efficiently and profitably as he did his automobile plant. He and Edsel offered $8.5 million for the two banks.[80]

The public was elated. At last Detroit's banks would be run by an honest, hard-working man such as themselves, rather than by money-grubbing financiers. (They must have forgotten or had not been aware of Edsel's involvement.) Henry's fellow citizens in Dearborn marched through the streets, waving banners saying "BANK WITH HANK."[81]

However, Henry and Edsel's rock-bottom offer fell on deaf ears. General Motors had already stepped in and, working with the RFC, reorganized the Detroit Bankers Company and the Guardian Group into the National Bank of Detroit. That did not dissuade Henry, and the involvement of General Motors probably encouraged him all the more to open his own bank. In August 1933, Henry and Edsel used their own money to found the Manufacturers National Bank of Detroit, a bank that remains in existence today.

***

After the smoke had settled, James Couzens called for a Senate investigation by the Senate Committee on Banking and Currency into the banking fiasco. The hearings began in Washington on December 19, 1933, and lasted for eight weeks. All of the major players were called to testify: Ernest Kanzler, Alfred Leyburn, Clifford Longley, and, of course, Edsel Ford. Each witness was asked about his involvement in the Detroit banking system and what may have caused its collapse.

Kanzler's testimony was most enlightening. It described his deep involvement in the operations of the bank and in the cover-up. Ferdinand Pecora, council to the Senate Committee, and Senator Couzens drilled Kanzler, and at times the discussions became very heated. Their primary questioning centered around a stockholders meeting over which Kanzler had presided on January 24, 1933. In that meeting, Kanzler told the stockholders that the Guardian Detroit Union Group had a higher "degree of liquidity" in 1932 than "any group of banks in the entire State of Michigan." He also said that during this same period, the Guardian Group experienced "net operating earnings...amounting to $1,316,952."

This puzzled both Pecora and Cousins, for immediately after this meeting, Kanzler took off to Washington to obtain a loan from the Reconstruction Finance Corporation.

"Does this compromise substantially what you gave to the stockholders...?" asked Pecora.

"It does," replied Kanzler.

Then Pecora presented a financial statement from this same period, a document that Kanzler did not share with the stockholders on that January afternoon. "Will you look...at the...Consolidated Statement of Earnings and Expenses and tell us what the last row of the figures on that page represents?" asked Pecora.

"The figure shows $1,798,069," replied Kanzler.

"In the red or black?"

"There is an 'R' behind it—red, I presume."

"What does that indicate?" grilled Pecora.

"I would say that indicates a profit for the year 1932 in the red," replied Kanzler with a straight face.

"Did you make any mention at all in any statement made to the stockholders of that figure?"

"In the way our reports were being made, that was not material, to my mind," said Kanzler.

After several more hours of bantering back and forth about what Kanzler had and had not shared with the stockholders, Pecora finally got Kanzler to admit that he deliberately kept this information from the stockholders. "This is an operating statement," explained Kanzler, "and I do not think one usually burdens the stockholders with operating statements."[82]

The following week, Edsel was questioned. Pecora began the questioning by asking Edsel how he had become involved with the Guardian Group, and then he asked him directly what he knew of the finances of the Group.

"Did you know, Mr. Ford, that the Group, in the years 1930, 1931, and 1932, had a deficit at the end of each of those years?"

"I do not remember," replied Edsel.

"Would that not be a fact that would linger in your memory?"

"Not necessarily," replied Edsel. "There are losses in many other corporations."

Then Pecora directed his questions about the loans Edsel had made to the Group. "I loaned them $5 million in municipal bonds and $1 million in cash. And that receipt is dated December 11, 1930."

"What was the reason for the loan?" asked Pecora.

"Well, as I understood it, the Guardian Detroit Company had borrowings in various other banks, outside banks on which collateral was posted. This was at the time when collateral values had shrunk materially, and as I understood it, the collateral was insufficient for the loans that were outstanding at that time. One of the officers of the Group came to me and asked for this loan, which I made so that they could rectify the loans that had been made outside."

"And the fact that it needed money, in your opinion, would not that be an indication that the company was unsound?"

"I wouldn't say so necessarily," replied Edsel. "It might. The reason for making the loan was to help the company...and I did not press them to pay it back because I felt that when they could repay it, they would."

"Did you get those securities back?" asked Pecora.

"Well, actually I did get them back, yes. The Ford Motor Company paid off that loan and the securities were returned, and then I satisfied the Ford Motor Company later on. In other words, I repaid the Ford Motor Company."

Then Pecora and Couzens began asking Edsel detailed questions about the other loans he had made to the Group, and about specific operational details of the Group, to which Edsel had very few answers. That upset the Committee. "Have you a good memory, Mr. Ford? Up to the present time, you have not recalled most of the important things that I have asked you about concerning the business and the operation of the Group and unit banks."

"Well, I am very sorry, sir, but there are a great many things going on from day to day that I have to go through with, and it is not always possible to retain all kinds of facts and figures. Some people are more able to do that than others."

Eventually, Edsel recalled enough information to tell the Committee that the loans made to the Group by both himself and by the Ford Motor Company were never repaid.

At the end of the questioning, Couzens asked Edsel, "If you had to start all over again, would it be the individual [banking] system or the group system?"

"Well," replied Edsel, "past experience has been mad, but it may have been on account of economic conditions that caused their downfall. I would not blame the group for that."[83]

***

Before Edsel's high finance troubles were behind him, he had to overcome one more hurdle. As the Senate Hearings were concluding, the Justice Department brought an action against Ford, General Motors, and Chrysler, charging them with violating the Sherman Antitrust Act. The charge was that these three, through their separate finance companies, were trying to monopolize the financing of automobile sales. A federal grand jury returned an indictment.

Edsel was deeply upset and fearful. He sent Longley to Indiana where the legal action was taking place, to see if he could settle the case. Edsel did not want to be dragged through another legal case, as he had endured in Washington.

It was a serious matter, but Longley learned that the government would be willing to drop the charges if the company signed a consent decree, agreeing to stop the alleged practices.

Edsel was all for this, and so was Longley. However, Henry Ford had a completely different reaction. "He got mad and wanted to fight," recalled Liebold. "He said he wouldn't sign a consent decree, and wouldn't let anyone else do so."

"Edsel was white with fear," continued Liebold. " 'Sure, Father wants to fight,' Edsel said. 'But I'm the president of Ford Motor Company—and I'm the one who will go to jail!' And the tears ran down his cheeks."

"You don't take very good care of your son," Liebold told Henry, as he explained the situation to him. At first, Henry was not too concerned. However, when Liebold showed him evidence that General Motors hoped Ford Motor Company would fight the case, Henry "softened up right away."

"Go ahead. Fix it up!" ordered Henry.[84]

Liebold immediately sent word to the District Attorney's Office that the Ford Motor Company would sign the consent decree, and "that ended what was a tense and difficult situation between Mr. Ford and Edsel," recalled Liebold.[85]

***

Although the case was settled, it raised other concerns with Henry about the wisdom of keeping the Universal Credit Corporation. He did not want it to begin with, and perhaps this first lawsuit was only the beginning of further troubles later. The UCC still owed $85 million to various banks, and Henry was concerned that he would have to come to their assistance as well. Kanzler had assured Henry Ford that neither he nor the Ford Motor Company was liable for any of those debts, but Henry remained concerned. "I don't like it," he told Liebold. "I want to get out of it. Close it up."

Liebold visited Kanzler and told him that Henry Ford wanted to close down the UCC. Kanzler was not surprised in light of their recent financial woes. However, being a shrewd businessman, Kanzler asked, "Why not sell it?" Henry Ford agreed.

"What terms do you want?" asked Liebold.

"Cash," replied Henry.

That startled Kanzler. After all, all of the banks in the country were closed, and money was tight. Only a handful of companies in the country could pay cash at that time. Nevertheless, Kanzler canvassed the market and, surprisingly, found a willing buyer. The Commercial Investment Trust Corporation bought the Universal Credit Corporation in 1933 for $33 million.

"Of course, that was a terrible mistake," Kanzler later explained. "UCC had been making money when Henry Ford abandoned it, and it continued to make money afterwards."[86]

***

The entire banking fiasco was a drain on Edsel. There is no question that he anticipated making some money in the venture, but that was not his primary reason for becoming involved in high finance. He simply wanted to gain more experience in financial operations, but he got much more than that.

Although Edsel lost upwards of $12 million in the fiasco, he was never in a position of losing all of his wealth. His salary from the Ford Motor Company during the early 1930s exceeded $1 million per year, and he also received substantial dividends from stock he owned in Ford of Canada.[87] He owned nearly 375,000 Class A shares and more than 30,000 Class B shares in the Canadian subsidiary, which provided him with significant income.[88] To replenish his personal treasury to an amount with which he was more comfortable, Edsel sold nearly all of his

Class A shares to the Ford Motor Company at the end of 1933.[89] Edsel and his father used these funds to establish the Manufacturer's National Bank.[90]

Edsel's financial losses were great, but his loss of face was greater. Somewhere in the back of his mind, he knew he should not have solicited the funds of his fellow businessmen. He knew it was not a good idea for him to lend his name in the solicitation of stockholders to his new bank, and it was inappropriate for him to write letters to Ford suppliers, asking for deposits and business. They had trusted him and the Ford name. If only he would have kept his associations with them at arm's length, he would not have felt guilty about their plight.

Now, four years later, all of their investments were lost. Everyone who had deposits in the Guardian Group eventually got their money back, but those who had purchased stock in the grand scheme lost it all. Wrote one customer: "We of course felt that with the backing that your name and your effort could give to this bank, it would come through these times without any trouble whatever."[91] However, the investors were more forgiving. "At the last meeting of the Guardian Directors," wrote a board member, "a committee was appointed to draft appropriate resolutions for the purpose of expressing the appreciation of the Guardian directors for the wonderful support which you have rendered the Group organization from the beginning.

"I want to add a personal note to any such resolution, an expression of my personal appreciation to you...for the perfectly wonderful cooperation and assistance you have rendered the Group Company... Many of us are lamenting the fact that we have lost our money as stockholders, but you and your associates suffered not only as stockholders, but also as creditors and depositors."[92]

"While we have tried to be of as much assistance to the Guardian Group as possible since its organization," replied Edsel, "it is very discouraging to have the final outcome as disastrous as it has been."[93]

Chapter 15

Patron of the Arts

"Art really is part of Edsel's life."
Henry Ford[1]

Edsel purchased the two Gilbert Stuart paintings that he had hanging in his office not so much for their historical significance as their artistic appeal. Stuart was a noted portraitist during the American Revolution. Among his subjects were George Washington, John Adams, and Thomas Jefferson. Every elementary student is familiar with Stuart's portrait of Washington.

Stuart had painted three portraits of Washington, and Edsel hoped to own one someday. He put the word out of his desires but had difficulty finding one for sale.[2] Eventually, a resourceful art dealer located a Stuart Washington in the possession of a wealthy Eastern family who had purchased it from John Jacob Astor. The new owner wanted $150,000 for it, but Edsel had to decline the offer. "I might say that in view of the present economic conditions," wrote Edsel, "I cannot continue to consider the Stuart of Washington seriously. I had hoped at one time that this picture might become available, but feel at the present time I cannot make a large investment in anything in the way of fine arts."[3]

The economic condition of the country notwithstanding, Edsel Ford could afford to purchase any piece of artwork his heart desired. However, with his frugal upbringing, he purchased artwork only when he thought the price was reasonable, and he never competed for a piece at a public auction.[4] Therefore, he was content to have the two Stuart paintings of Dearborn and Hamilton.[5]

***

To relax after impossible days at the factory, Edsel took up painting as a hobby. For about two years, he spent several hours on Saturday mornings, taking painting lessons from John Carroll, a modern Detroit painter.[6] During the week, Edsel attended art classes at the Detroit Society of Arts & Crafts.[7] "He loved to paint," recalled Irving Bacon, the resident artist at the Ford Motor Company. "Several times he expressed a desire for me to come to his home to paint with him."[8] Bill Ford remembered his father painting and sketching everything from cars to landscapes at home, and he said that Edsel took his easel and paints with him on vacations.[9]

How many works of art Edsel created is not known, but a few pieces have survived—done when he was a young man—reposited at the Ford Archives. These include a red roadster with brass radiator shell that he did in watercolor, a perfect pencil drawing on paper of a roadster, a nicely done pencil sketch of the Washington Monument, a watercolor painting of a flower, a charcoal still life, a penciled bottle and vase, and a charcoal drawing of a boy sitting on a stool. There is even a wallpaper design that Edsel did in an automotive motif.[10]

None of these pieces is a masterwork, but all are well done and clearly indicate Edsel's artistic bent. "You have become quite an artist," wrote his Aunt Margaret.[11] One can only imagine the quality of work Edsel did when he was older and had completed his art lessons. And apparently, he was skilled on more than one medium. "Edsel Ford's talents as a sculptor, for example, were so considerable that many felt he might have become an outstanding artist," recalled George Pierrot, a member of the Detroit Institute of Arts. "He possessed unusual sensitivity, and a rare feeling for form, composition, and color. His taste and critical judgment were extraordinary."[12] Another acquaintance agreed with Pierrot's assessment. "Edsel Ford had a very keen sense of color harmony," he said, "and a remarkable color memory."[13] Even Henry recognized his son's artistic skills. "Shake hands with my son, Edsel," Henry said as he introduced Edsel to Irving Bacon. "He is the artist in our family."[14]

Although Edsel inherited his artistic talent from the Bryant side of the family, Henry was aware of Edsel's artistic bent early in Edsel's life.[15] However, he could not have guessed the importance it would have in the development of Ford Motor Company in the years ahead.

***

In 1885, 40 prominent Detroiters founded the Detroit Museum of Art, the predecessor of the Detroit Institute of Arts. It was located on Jefferson Avenue, not far from Edsel's second home. In a special election in 1918, the people of Detroit approved a new city charter, which, among other things, established for the first time in the growing metropolis a municipal Arts Commission. The four members were to be appointed by the mayor and charged with the responsibility of managing a city-supported art museum to be renamed the Detroit Institute of Arts.

It seemed like the right thing to do, for the city was in an economic boom due to the expanding automobile industry. James Couzens, Henry Ford's one-time business manager and major stockholder in the company, was the mayor of Detroit at the time. He appointed William Gray and Albert Kahn as trustees, both of whom were associated with Henry Ford. The first president of the Arts Commission was Ralph Booth, an art connoisseur and long-time supporter of the arts in Detroit.

One of the first objectives of the commission was to construct a new museum, to be built on a site located two miles north of city center on Woodward Avenue. The area would become known as the Cultural Center because the Detroit Public Library and other museums eventually would be located in the same area. In 1919, the Detroit Museum of Art, a private corporation, transferred the ownership of its collection and its building to the City of Detroit, and planning of the new museum was begun in earnest.

In 1921, Paul Philippe Cret, a professor of architecture at the University of Pennsylvania, was selected by the Arts Commission to design the new building. To assist the commission in giving direction to Cret, they hired William Valentiner.

Valentiner was the first professionally trained art historian employed by the Detroit Institute of Arts. He received his education at the Heidelberg and Leipzig universities and trained at The Hague. In 1908, he became a curator at the Metropolitan Museum of Art in New York. Subsequently, he became an expert in Dutch painting, Islamic art, and Oriental carpets, and he was published extensively about these specialties.

Valentiner's idea for Detroit's new museum was to display its artwork not in separate picture galleries or textile galleries or rooms full of sculpture, but in period galleries—rooms showing the historical context of all types of art. That is how Cret laid out the new Detroit Institute of Arts building. Designed in an inverted T shape, the entrance to the museum was at the apex of the horizontal and vertical leg of the T. Immediately behind the entrance was the main hall. To the left of the main hall was the American section, to the right was European art, and past the main hall was a garden court, around which were 10 smaller galleries. At the far end of the building, past the garden court, was a large hall for musical and dramatic performances.

With new-found public monies, cost was no object in creating the new museum. Originally estimated to cost $2 million when the plan was created, the new Detroit Institute of Arts cost twice that amount by the time it opened its doors to the public in 1927. However, nobody seemed to complain after seeing the magnificent structure—covering approximately 250,000 square feet—that Cret and Valentiner had created.

The museum received great acclaim and significant donations almost immediately. Sir Joseph Duveen, a prominent art dealer in New York, contributed to the new museum. Julius Haas, Edsel's friend and president of the People's State Bank, also made substantial contributions. Horace Rackham, one of the original stockholders in Ford Motor Company, bought Rodin's massive bronze *The Thinker* for the new museum.[16] For the grand opening, Edsel donated one of the great masterpieces, Nino Pisano's sculpture *Madonna and Child*.[17] It was only the first of many pieces that Edsel and Eleanor would donate.

Meanwhile, Valentiner went from being Professor Cret's design advisor to becoming the first art director of the Detroit Institute of Arts. The Arts Commission appointed him in October 1924, and the following year, at the behest of Mayor John Smith, Edsel became a member of the Arts Commission. "I have known for a long time of your deep interest in the work which the DIA performs in the community," the mayor wrote to Edsel. "I feel that the work can be greatly furthered if you will give your thought officially…to this splendid effort."[18] Six years later, Edsel was elected president of the Arts Commission, a position he held for the next 13 years.[19] With Edsel as president and Valentiner as director, the museum thrived.

***

As the years passed, Valentiner developed an uncommon relationship with Edsel, something that few people were able to do. However, because Valentiner did not work for the Ford Motor Company, Edsel began inviting him to his home, first at Jefferson Avenue and later at his new mansion at Gaukler Pointe, so that Valentiner could give Edsel and Eleanor lectures on art. "Edsel knew nothing about art history," recalled Clara, but through Valentiner's lectures, he hoped to learn.[20]

Dressed in formal wear, Valentiner would join the artistic troupe for dinner and then talk well into the night about all aspects of art and art history. Many times, Ernest and Josephine Kanzler attended these formal gatherings, as did Eleanor's bachelor cousin, Robert Tannahill. Occasionally, Henry and Clara would join the troupe for dinner but leave when the lectures started.

Valentiner was more than pleased to lecture this group of Detroit's upper crust. He said that Edsel enjoyed all periods of art—from Italian Renaissance to Post-Impressionist to abstract.[21]

Edsel asked many questions about the objects that Valentiner was acquiring for the museum and listened intently to his explanations. "When I once recommended that he purchase a still life by William Kalf," recalled Valentiner, "Edsel asked whether he was the best Dutch still-life painter, and I answered in the affirmative. Two years later, I proposed the acquisition of a still life by Abraham van Deyeren (who is just as good as Kalf, but in a different way), saying that he was the best Dutch still-life painter. Edsel laughed and said, 'Sometime ago, you said Kalf was the best!'"[22]

***

For some time, Edsel and Eleanor had planned their "forced exile" to Europe during the development of the Model A. They were in the midst of planning and constructing their fabulous mansion on the shores of Lake St. Clair and decided to take a trip to Europe to buy furnishings and ancient paneling for their new home.[23] When Valentiner learned of their plans, he told them that he was also planning to go to Europe and would be glad to consult with them on their purchases.

As was the usual practice, Edsel and Eleanor took the Detroiter to New York and boarded the *Acquitania* to Europe. To make their trip as comfortable as possible, Eleanor brought 26 pieces of luggage, and Edsel brought a custom-bodied Lincoln touring car. He also had one of his secretaries, Arthur Backus, accompany them as their bodyguard.[24]

One time during the voyage, Edsel asked Backus to get some white lilies for Eleanor. When Backus returned, he knocked on the door and thought he heard someone say, "Come in." "Upon entering the suite, I caught Mrs. Ford partially undressed," recalled Backus. "She hollered and ran into the adjacent room. The boss came out and just laughed like the deuce. He thought it was a good joke."[25]

Valentiner took his wife Caecilie and daughter Brigitta to Hamburg and Berlin to show them where he had grown up and to see the sights. They met with Edsel and Eleanor later in Genoa, Italy. Valentiner wrote in his diary that the young Fords had sent a basket of fruit to their hotel room, with a note to say that they had purchased Fra Angelico's *Annunciation* from the Duke of Hamilton collection. Valentiner saw this as a pleasant result of his art lectures, for it was the first important painting that the Edsel Fords acquired.[26]

Traveling in luxury in the Lincoln touring car, the Fords and the Valentiners toured Italy, drove to Paris, and ended up in Monte Carlo.[27] From there, the Valentiners returned to Detroit. Edsel and Eleanor left several weeks later, but not before acquiring two additional works of art: a terracotta relief by Donatello, and a rare sixteenth-century Isfahan carpet.[28] Valentiner made such a fuss over the two works that the Fords eventually donated the magnificent carpet to the Detroit Institute of Arts.[29]

Over the years, Edsel and Eleanor purchased many paintings and other artwork to decorate their lovely mansion at Gaukler Pointe. "He [Edsel] buys some of the canvases for himself and hangs them in his house because he likes them," explained a friend. "However, he does not intend to establish a private gallery."[30] Nevertheless, the Edsel Fords decorated their home as if it were a museum. Paintings, sculptures, and tapestries by Cézanne, Pisano, Van Gogh, Raphael, Degas, Matisse, Frans Hals, and Diego Rivera decorated their walls and furniture. On many occasions, they allowed curators from museums throughout the world to visit their home and view their masterpieces. "I am sorry that I must bother you once more," wrote Valentiner. "I had a letter from the director of the Victoria and Albert Museum in London... He is coming here on Wednesday...and asked if he could see your Renaissance sculpture. I wonder if you would be so kind as to let me go with him to your house and show him the objects. The worst of being the owner of great masterpieces is that it combines with pleasure, the disadvantage that scholars are always wanting to see them."[31]

Through these actions, Valentiner believed that the future of art in Detroit depended on Edsel's activity as an art collector and benefactor. To a large extent, Valentiner was correct. Hardly anyone in the great state of Michigan appreciated art as much as Edsel. If Edsel were not already the president of a major corporation, said a close associate, "he would be a superb art museum director."[32]

***

The combination of Edsel's interest in art and his commitment to the community at large compelled him to make numerous and significant contributions of art to his beloved Detroit Institute of Arts. At an early date, he established the Edsel B. Ford Fund, into which he made yearly contributions so that the museum could purchase art. Over a 20-year period, Edsel and Eleanor contributed nearly $350,000 toward the purchase of classic artworks by Titan, Raphael, Leonardo, Donatello, and others.[33] True to form, most of Edsel and Eleanor's contributions were made anonymously.[34]

On many occasions, Edsel acquired one of the paintings from the loan exhibitions that were held at the Detroit Institute of Arts. After the English Sixteenth Century Exhibition, for example, he acquired some of the English; after the Frans Hals Exhibition, he acquired some of the Frans Hals; later, a Titian, two Fra Angelicos, a Benozzo Gozzoli, and a Perugino.[35] During these days, Edsel also became most interested in Donatello and Desiderio and the other great Renaissance sculptors. This speaks for Edsel's unusual understanding of art in general. He acquired a large coat of arms by Donatello and an exquisite relief of a girl by Desiderio for himself. In the mid-1930s, he purchased the cream of the J.P. Morgan collection—a portrait of Giovanna Tournabuni by Ghirlando—for less than $50,000.[36]

Not all works of art purchased during the Depression were obtained at bargain prices, however. At the suggestion of Valentiner, Edsel bought a sculpture by Mino de Fiesole for $30,000. A year later, while at Gaukler Pointe for dinner, Valentiner asked Edsel if he could talk to him in private. A light snow was falling outside, and the two went outdoors to enjoy the ambiance. "The damned thing isn't right," Valentiner confessed to Edsel about the de Fiesole. "It's too pretty...it looks worse every time I look at it..."

"In other words, you think it's a fake," suggested Edsel placidly.

"I'm afraid it is."

"Well, such things happen. Don't worry."[37]

That was not the last time Edsel was notified that some object of art that he had purchased was found to be a fake. A few years later, Valentiner notified Edsel that a sculpture supposedly created by de Fiesole had really been done by one of the artist's students. "I have made a great mistake—the worst perhaps in my whole career," wrote Valentiner. "This is still more embarrassing since it was I who called your attention to the sculpture and induced you to give it to the museum."[38]

Again, Edsel was not critical of Valentiner. He was well aware of the pitfalls of dealing in ancient art. If an expert such as Valentiner could be fooled, then Edsel was prepared to be lenient when mistakes occurred.

***

Word of Edsel's passionate eye and his involvement with the Detroit Institute of Arts spread throughout the art world. In the summer of 1935, Abby Rockefeller, wife of John Rockefeller, Jr., wrote a complimentary letter to Paul Sachs, curator of the Museum of Modern Art in New York, perhaps the quintessential art museum in the country. She suggested to Sachs that Edsel be invited to become a member of the board of trustees. "I have only met Mr. Edsel Ford once, a number of years ago," she said, "but I then gained the distinct impression that he would like to be of real service in the field of art…"[39]

Sachs responded to Mrs. Rockefeller's suggestion by sending a letter directly to Edsel[40]:

> *We would particularly hope to have you as a trustee because of your wide knowledge and experience, not only with the Detroit Museum, but in industry and other fields.*

Abby Rockefeller wrote a letter to Edsel as well[41]:

> *I want you to know how eager we all are to have you join us and how valuable we feel the contribution would be that you could make.*
>
> *I am particularly eager to have you become a trustee from a rather selfish point of view, because I should be so happy to be associated with you myself in the Museum, and it would mean so much to Nelson to have the opportunity of knowing you.*

Although honored by the request, Edsel, in his typical fashion, had to deliberate the invitation before accepting it. He wrote[42]:

> *I feel highly honored by your suggestion that I join the Board of Trustees.*
>
> *If it is not asking too much, I would like to have a little time before deciding whether I can conscientiously accept your kind invitation. My visits to New York are most irregular, and it is questionable in my mind whether I could spare the time to make a special trip at the time of the board meetings.*
>
> *It would be the greatest pleasure I can think of to be on the Board of the Museum of Modern Art, both for the experience and also to be associated with you and Nelson in this most interesting enterprise.*

Edsel did not take long to make up his mind. He could hardly refuse a request from the Rockefellers, and he wanted to be involved in the activities of the Eastern aristocracy. Thus, he accepted their invitation a few weeks later. "If you and the other members of the Board are still of the opinion that I may be of some service, I will be glad to accept membership," Edsel wrote.[43]

"We are perfectly delighted that you are willing to be a trustee of the Museum of Modern Art," replied Abby. "There is not one in the whole country that we would rather have than you."[44]

Edsel remained on the board of the Museum of Modern Art for the remainder of his life and made substantial financial contributions to his second museum.

***

During the summer of 1928, tennis champion Helen Wills came to Detroit to play in a tennis tournament. One evening, she and her mother, a self-taught artist of some repute, visited the Detroit Institute of Arts. It was late and the museum was closed, but they were persistent. They saw that lights were still on in the building, so they knocked on the door until someone answered.

Fortunately, Valentiner was working late that night. Although he was perturbed at being bothered, his annoyance quickly dissipated when he opened the door and saw Miss Wills standing there.

"We're so sorry to bother you," she told Valentiner, "but our train leaves for California in the morning, and we so much would like to see your collections. We have read so much about them."

Valentiner was smitten by Wills' svelte figure and vivacious personality. "In her, nature has produced perfection," he later wrote.[45] He gladly obliged their request and took them for a tour of the museum collection. From that point onward, Valentiner and Wills developed a "rare friendship."[46]

A year later, Valentiner and Wills met again in Berlin. She was in the German capital for a tennis tournament, and he was in Germany to buy works of art for the museum. The two spent a lot of time together, and Valentiner took Wills on tours of all of the important art museums in the German capitol.

Shortly thereafter, they met again in London. In 1930, they met in New York. Later that year, Valentiner returned to London and saw Wills win the world tennis championship at Wimbledon.

That December, Valentiner traveled to San Francisco. His excuse was that he had never been any farther west than Chicago and was anxious to see this part of the country. Although that was true, his real reason was to see his beloved Helen Wills.

Before leaving, Valentiner spent Christmas dinner with the Henry and Edsel Fords at Edsel's new home on Gaukler Pointe. Henry and Clara always spent Christmas dinner with their son, and this time Henry had brought to each of Edsel's children a miniature replica of his first automobile.[47] Edsel was delighted and proceeded to explain to Valentiner how he used to sit in the little contraption when his father took him for rides in it when he was a little boy.[48] Edsel and Eleanor gave Valentiner a gold travel clock as a Christmas present and informed him that a chauffeur-driven Lincoln would be waiting at his disposal upon his arrival in California.[49]

***

The day after Valentiner arrived in San Francisco, Miss Wills gave a tea party for Valentiner at her apartment. The tasteful artwork that decorated her apartment impressed Valentiner. During the party, Wills invited Valentiner to watch her play tennis the next day, and Valentiner quickly accepted. She said that she would pick him up at his hotel and that she would bring her friend, noted Mexican painter Diego Rivera. Rivera had been in San Francisco for quite some time and had recently finished painting murals in the California Stock Exchange Luncheon Club and the California School of Fine Arts.

"Her little green Cadillac was overflowing when it stopped for me," recalled Valentiner, "and jammed into the rumble seat I saw a strange-looking, heavy-set man, wearing a black serape and a large Mexican hat. And scrunched up next to him was his lovely and artistic wife, Frida Kahlo."[50]

Six feet tall and weighing more than 300 pounds, Rivera was indeed a big man. On the return trip from the tennis match, Valentiner sat in the back of the car with Rivera. "While we discussed balance and harmony in composition," continued Valentiner, "we found ourselves pitched backwards to a horizontal position and looking directly up at the sky, because Helen Wills took delight in driving us up the steepest streets in San Francisco. Indeed, Rivera's weight carried him so far backward that I was afraid he might tumble out, and I with him!"[51]

When they arrived at Wills' apartment, Valentiner met Lord and Lady Hastings, aristocrats from Europe, who at the time traveled everywhere with Rivera.

Rivera asked Valentiner many questions about industry in Detroit and told him that he would like to visit that city some day and create a painting depicting the interaction of men and machinery, for which America was famous.[52] That idea struck a chord with Valentiner. For some time, he and the directors of the museum had been discussing what to do with the barren walls of the garden court at the museum. At the time, one wall was decorated with a large tapestry, and the other wall carried a large mural by Augustus John. However, neither of these works of art was satisfactory because each was dwarfed by the immensity of the room.[53] Valentiner did not mention to Rivera the idea of him coming to Detroit to apply his talent to the walls of the museum. First, Valentiner wanted to be sure that he could convince the directors of such a plan.

***

Diego Maria Concepcion Juan Nepomuceno Estanislao de la Rivera y Barrinetos Acosta y Rodriguez was already a painter of great repute when Valentiner met him in California. Within the month, Rivera was to stage a one-man show at the prestigious Museum of Modern Art in New York. His medium was fresco—a process of painting directly onto wet plaster so that the paint was not only on the plaster but imbedded in it. It was a process quite popular in sixteenth-century Europe, and the art of fresco had all but died before Rivera revived it. Between 1921 and 1930, he had completed eight major murals in fresco in Mexico on historical subjects. One of his creations was called *The Billionaires' Banquet*, a large fresco on a wall of the Ministry of Education Building in Mexico City. Ironically, it portrayed John D. Rockefeller, J.P. Morgan, and Henry Ford, sitting around a table and eating a dinner of ticker tape, with Henry Ford "quaffing a glass of champagne."[54] This was a political statement, for Rivera was a professed Communist.

Rivera was much more than a painter; he was an intellectual. He could speak Russian, French, and Spanish, and he had read Huxley, Darwin, Kropotkin, Taine, Renan, Zola, the

Russian novelists, and Karl Marx.[55] He had an inherent interest in machines and industry, and he understood engineering principles.

When Valentiner returned to Detroit—his thoughts of Wills replaced by walls of frescoes—he recommended to the Arts Commission that Rivera be brought to Detroit. However, the commissioners were in no position to acquire any more artwork. The Depression had lain over the city of Detroit similar to a slow death, and there were no excess funds in the budget for new acquisitions. However, Edsel was entranced by Valentiner's proposal and offered to donate an additional $10,000 to the Edsel B. Ford Fund to have Rivera paint the two large walls in the garden court—50 square yards on each wall, $100 per square yard.[56] "Edsel set only one condition—that in representing the industry of Detroit, I should not limit myself to steel and automobiles, but take in chemicals and pharmaceuticals, which were also important in the economy of the city," recalled Rivera. "He wanted to have a full tableau of the industrial life of Detroit. He said good-humoredly that he wished to avoid any impression of partiality toward the industry served by his father and himself."[57]

***

Diego Rivera and his wife, Frida Kahlo, arrived in Detroit from New York City on April 22, 1932. They were met at the train station by Valentiner, the press, and a reception of approximately 20 Mexicans living in Detroit, led by the Mexican Consul.[58] Although Diego was impressive, the attention of the press was particularly drawn to his beautiful wife.[59]

"I have been ill," Diego told a reporter as he got off the train, "and am not really well now. But all that will be forgotten when I get to work. That will not be at once. First, I must see the city, the factories, the machinery. I will drive around the city and think about what I see before I start to work. It may take several days or even weeks."[60]

The first stop on Rivera's itinerary was to visit the Detroit Institute of Arts. Valentiner gave him and Frida a personal tour, and the two were impressed. Valentiner said that Rivera was knowledgeable in all aspects of art and was particularly fond of the El Greco, Titian, Rembrandt, Rubens, and Velasquez works on display.[61] At the end of the tour, Valentiner showed Rivera the great garden court, the area where he was to conduct his work. It was a large room—67 feet long, 50 feet wide with a floor-to-ceiling height of 43 feet. The ceiling was glass, capturing what little sunshine the eternal gray Michigan sky offered. As the focal point of the substantial museum building, Paul Cret had designed the room in an Italian baroque style. The walls were expansive, but their surfaces were broken into smaller rectangular shapes by small windows, ventilators, and cross moldings. At the base of each of the longer walls were large marble planters, and a massive fountain took up most of the floor space. "It was in such a frame that I was to represent the life of an age which had nothing to do with baroque refinements," recalled Rivera, "a new life which was characterized by masses, machines, and naked mechanical power. So I went to work consciously to overpower the ornamentation of the room."[62]

For three months Rivera wandered through the factories of Detroit. "I studied industrial scenes by night as well as by day," he wrote, "making literally thousands of sketches of towering blast furnaces, serpentine conveyer belts, impressive scientific laboratories, busy assembly rooms, and of the men who worked them all. I was afire with enthusiasm."[63] Rivera confessed years later that he would have liked to remain in Detroit eight or ten years to digest and understand the great workings of the city. "Unluckily," he said, "I had but one year in Detroit, and during that year I did as much as I could and lived what was perhaps the best and most fruitful period of my life."[64]

Edsel made all of the arrangements for Rivera to visit not only the Ford factories, but also the Chrysler and Hudson automobile factories, the Parke-Davis chemical company, the Michigan Alkali company, and Edison Electric. During Rivera's first visit to the Ford Rouge plant, Edsel picked up the Rivera entourage—Diego and Frida, Lord and Lady Hastings, and Lucienne Bloch, a budding artist hired by Rivera as an assistant—at the Wardell Hotel located across the street from the art museum. Edsel's tour lasted most of the morning, and then he took them to have lunch with his father at the Engineering Laboratory. After lunch, Edsel took his party to visit the Rouge foundry and explained that the men who worked there worked only four hours at a time because it was such grueling work.[65]

"I have found in Detroit the typical American city which I have sought, for so long," Rivera said. "Here is the machine and the man united in the production of one object—the motor car. Here is a chaotic unity that gives me the subject I most love to paint."[66]

During initial discussions among Edsel, Valentiner, and Rivera, all had decided that at least one of the large walls of the garden court be devoted to the automobile industry. Although no decision had been made as to what to put on the opposite wall, they had discussed everything from the Michigan landscape to famous industrialists and other manufacturing giants of the region—not only automobile manufacturing. However, Rivera had collected so much material from his tours of the Ford Rouge plant that he conceived the idea of utilizing both walls of the garden court to depict the Detroit automobile industry—specifically the operations at the Rouge plant. On one wall, he would depict the creation of the motor and on the opposite wall the assembly of the finished car.[67]

After Rivera decided on his approach, he created two large drawings of what he had in mind for the walls of the garden court, and he invited Valentiner and Edsel to his hotel suite to have dinner and to look at the finished studies. "After supper we inspected the two large, precisely executed studies (about two feet by three), for the two main walls in the museum court," recalled Valentiner. "It was an exciting experience. Edsel Ford was carried away by the accurate rendering of machinery in motion and by the clearness of the composition, which was not confused by the great number of workmen represented, each occupied with his assigned job. The function of the machinery was so well understood that when engineers looked at the finished murals, they found each part accurately represented."[68]

Although mesmerized by the project, Edsel, in the back of his mind, was somewhat concerned that the public would perceive Rivera's depiction of the Ford assembly line as some sort of Ford advertising scheme that Edsel had developed. "Nothing was further from the truth," recalled Rivera.[69]

The next day, Valentiner presented the studies to the full Arts Commission, and they approved the project after only a few minutes of discussion. After that, Valentiner walked across the street and told Rivera that he could start working as soon as he wanted.

***

For the next year, Rivera was consumed with his work and beleaguered by the illness that he was suffering when he first stepped off the train in Detroit. His doctor had instructed him to lose weight and put him on a strict diet of fruits, vegetable juice, and salads.

The diet was totally inadequate for a man of Rivera's size, especially considering the work he was doing. He did most of the work himself, first sketching the figures and scene and then applying the paint—his assistants merely applied the wet plaster to the walls for him. The work was continuous and arduous. He could apply paint only while the plaster was wet so that the paint would be fully absorbed. On many days, he worked 18 to 20 hours at a time in heat in the high parts of the court that reached 120 degrees. However, he kept steadily onward, early and late, right through Sundays and holidays.[70] By the time he had completed his work at the museum, Rivera had lost more than 100 pounds.[71]

As he was completing the second large wall, Rivera asked Valentiner if he might expand his project by covering the two remaining walls of the garden court. Rivera said that he had been pondering that possibility for some time, for the two large walls told only a partial story of the industry of Detroit. He wanted to use the other two walls to represent the elements from which the automobile factories and other industries obtained their resources.

This was a grand gesture on Rivera's part; however, for this to be accomplished, large moldings and decorative work would have to be removed from the corners of the ceiling, which would add to the cost. Then there was Rivera's salary.

Valentiner was ecstatic about Rivera's proposal, but where would he obtain the additional funds? The museum had already used the money that Edsel had donated earlier, and Valentiner knew there was no source for additional funds other than Edsel. He would have to work up the courage to ask Edsel for more money.

A few days later, Eleanor Ford invited Valentiner to her mother's birthday at the Kanzlers' home, and there Valentiner asked Edsel for additional funds for Rivera. "I was terribly nervous and began talking rather intensely about Rivera's work," recalled Valentiner. "I simply did not dare to broach the real subject. But with a woman's intuition, which I often admired in her, Eleanor Ford, without pressing me directly for an explanation, called across to Edsel, 'William is very nervous. Can't you take him into a corner and find out what he wants to tell you.' "[72]

"So I was able to explain the situation," recalled Valentiner, "and Edsel, amiable as usual, said that he would be glad to help. 'We must keep artists in a good humor if we want them to do good work,' he told me."[73]

For the additional work, Edsel paid another $15,000: $5,000 for the removal of the marble planters along the bottom edge of the longer walls and the end sections of the center fountain, and $10,000 for Rivera's additional work. "Rivera fell about my neck the following morning after I told him that Edsel would subscribe the additional work, and then hurried to begin the drawings for the new section," Valentiner said.[74]

According to one of the painters who worked with Rivera, Edsel visited the museum often to watch the great painter work. When Rivera noticed his patron standing down below him, he would climb down from the scaffolding, and the two would shake hands and make small talk.[75]

Unfortunately, with his busy schedule, Edsel could not visit the museum as often as he liked. Therefore, he sent one of his corporate photographers to the museum regularly to take both still photographs and motion pictures of Rivera as he worked. The artist was high up on a scaffold when one of the photographers first came into the room. As the photographer was setting up his camera, Rivera asked him what he was doing. "Mr. Ford instructed me to come here every day and take pictures of you in action," explained the Ford photographer. "I must have each day's shooting ready to be shown to him in his home by evening. He is eager to watch you paint, and since he can't spare the time during the day, he thought of this idea. It will make him feel as if he were actually here."[76]

How could Rivera reject Edsel's peculiarities? After all, it was through Edsel's benefactions that Rivera was able to create such a magnificent work of art, and Edsel's hospitality was more than generous. When it was found that sand brought in from Lake Saint Clare had too many impurities in it for Rivera to use, Edsel provided bags of white silica from the Rouge plant glass works. He also provided large panes of unpolished glass, brought from the Rouge plant, on which Rivera could mix his colors.[77] Rivera later confessed to Valentiner, "Edsel had none of the characteristics of an exploiting capitalist, that he had the simplicity and directness of a workman in his own factories and was like one of the best of them." [78]

***

Several months into the fresco project, Rivera decided to break from his arduous task and paint a portrait of his patron, Edsel Ford. Rivera had already planned to include both Edsel and Valentiner in the fresco project as a way of recognizing them for their contributions, a custom harking back to the Italian Renaissance.[79] He had made quick sketches of each of them earlier, but when he began the individual portrait of Edsel, he requested several photographs of him to aid in his work.[80] Edsel had his secretary, Albert Lepine, fill the request.

For three days, Edsel posed to allow Rivera to capture his likeness on canvas, and it took the painter several more days to complete the work.[81] "Rivera would never have missed the opportunity to capture Edsel's true character in a portrait of him," recalled Lucienne Bloch, "so it is highly likely that there was a lot of discussion between the two of them about Edsel. These discussions brought out Edsel's automobile styling philosophy and artistic talent which Rivera captured on canvas."[82] In a setting that we will see later was perfectly suited for Edsel, Rivera painted Edsel as though he were in an automobile design studio. Because Ford Motor Company did not have a design studio at that time, Edsel must have taken Rivera to the Briggs studio where Edsel had much of the design work on Ford products developed.

In the portrait, Rivera captures Edsel's inner strength. Edsel stands upright behind a drafting board and appears confident, which is the way he acted when designing automobiles. In the background are three large chalkboards, one with a full-size rendering of a body design proposal developed by one of Edsel's early designers in the company.[83] On the chalkboard to the left is an inscription written in Spanish, which reads: "True portrait of Mr. Edsel B. Ford industrial engineer and President of the Arts Commission of the City of Detroit, State of Michigan, U. S. of America. Painted by Diego Rivera the month of November of 1932."[84] This is enlightening, for no where is it mentioned that Edsel was president of Ford Motor Company. "Rivera deliberately chose this setting to draw attention to his [Edsel's] styling talent," explained Bloch.[85]

Shortly after the portrait was completed, Lepine asked for the return of the photographs that he had loaned Rivera. "Some time ago," wrote Lepine to Valentiner, "we sent you 10 pictures of Mr. Edsel B. Ford for Mr. Rivera's use. I believe I should ask for the return of these if they are available and have served their purpose, because we issue photographs here only when we obtain permission, and in a general way it is the purpose to avoid having photographs fall into strange hands."[86]

Over thc ycars, Edsel had a number of portraits painted of himself but never seemed to like any of them. Rivera's portrait was no exception. Although he admired Rivera's talent, Edsel confided to Fred Black, Ford advertising manager, that he did not think Rivera's portrait "was a very good likeness of him."[87] Nevertheless, Edsel hung Rivera's painting at the top of the majestic stairway at Gaukler Pointe, and there it remained for the next 50 years.

Rivera did not charge Edsel for the painting, but Edsel tried to show his gratitude by offering to Rivera a new Ford V-8. Rivera graciously refused.[88]

***

Rivera spent 228 days in completing his garden court masterpiece. The two large walls, as previously explained, showed the production sequences of automobile manufacturing. Above these two panels were figures representing the four races that shaped North American culture and made up its work force at the time. It also included the medical, pharmaceutical, and chemical industries of Detroit. On the east wall were two female nudes holding fruit and sheaves of wheat, representing the origins of human life. On the west wall, Rivera depicted images of airplanes and shipping. Scattered throughout the murals are images of Henry Ford, Charles Sorensen, Edsel Ford, Valentiner, and Rivera himself.

The frescoes were artistically magnificent and technically accurate. "You may wish to correct me by reminding me that Rivera is not an engineer by profession, but this fellow has fused together in a few feet, sequences of operations which are actually performed in a distance of at least two miles, and every inch of his work is technically correct. That's what is so amazing!" said a spokesman for a group of automotive engineers who came to admire the artist's work.[89] So realistic were Rivera's scenes that one could almost feel the heat of the molding rooms, hear the pound and clash of heavy machinery, and smell the acrid odors of the paint shops. "I have just come away from Rivera's murals, awed as perhaps the first spectators were who entered the Sistine Chapel after Michelangelo took down his scaffolds," is the way one visitor described his experience.[90]

The technical accuracy of the frescoes was not what drew all of the attention, however. It was Rivera's depiction of the medical industry that was to raise an ugly controversy, threatening the existence of his monumental work.

In a relatively small scene at the upper right corner of the north wall, Rivera painted his homage to the pharmaceutical industry of Michigan. To symbolize that industry, he painted an infant in the arms of a nurse, being vaccinated by a doctor. In the background were several pharmacists creating new vaccines, and in the foreground were a horse, a cow, and several swine—the sources of vaccine. The infant stands naked, except for a loincloth, and around the child's head Rivera had painted a halo.

***

Meanwhile, the Detroit Institute of Arts was receiving enormous pressure due to the lack of funding caused by the Depression. In the pre-crash days, the budget of the Detroit Institute of Arts was more than one-half million dollars. By 1932, however, the year in which Rivera was commissioned to do the frescoes, the budget was less than one-tenth of what it had been only two years earlier.[91] Mayor Jeffries was forced to take expedient measures. He ordered that all salaries be frozen, educational and curatorial staffs be eliminated, and all additions to the art collection be halted. There was talk that the museum would have to be closed altogether because of the lack of tax revenues.

Concerned with these cutbacks, Edsel wrote a letter to the mayor, complaining that such cuts would be detrimental to the museum[92]:

> *... in view of the fact that we have been so drastically cut in the past and are operating on the most meager of budget and staff, it seems very difficult for us to make the saving of $4,400 as requested by the Budget Director without reducing our personnel... It does not seem to me that we can function with less on our Curatorial Staff than the six we have, and we most certainly have great use for the two educational workers. It might be possible to make some saving on guards and charwomen, but this is only lowering the standards of our maintenance and endangering the valuable collection to a great extent.*

Edsel's plea received little support. Mayor Jeffries managed to squeeze a paltry $15,000 from the city council, but that was hardly enough to keep the museum open. Valentiner asked Edsel for his help, and Edsel responded by contributing funds to support the staff salaries. "It's gotten to the point where I'm paying the salary of the Director and the staff," he grumbled to Fred Black, "and if we are going to keep the place open, I'm going to have to pay the salaries of the guards and janitors!"[93]

Valentiner wrote in his diary about Edsel's discontent. "Edsel Ford is more depressed than I have ever seen him. He had deposited his private fortune in the banks, probably against the will of his father, who was always against banks. Every day there is more bad news, the museum has lost all its funds, and there is no money for the salaries of the staff, for me, or for Rivera. It is doubtful whether Ford, who has suffered such financial losses, will be willing to rescue us again."[94]

Nonetheless, Edsel did come to the rescue. In May, he contributed $30,000 to the operations of the Detroit Institute of Arts, and he continued to make regular contributions until the museum got back on its feet.[95] Edsel paid Rivera the money he was owed, and paid him again when Rivera lost his money when the banks closed.[96] Edsel also would have paid Valentiner's salary indefinitely, but Valentiner volunteered to take a leave of absence during this tumultuous period, saving himself the embarrassment of receiving a handout from Edsel and saving Edsel the expense. Valentiner returned to Europe where he said it was cheaper to live. Even then, he had to sell a number of Rembrandt drawings that he had treasured since college days, simply to get by.[97]

Additionally, Edsel continued to purchase works of art and donate them to the Detroit Institute of Arts. Throughout the Depression, he purchased *Judith with the Head of Holofernes*, by Titian, *St. John* by Andrea del Castagno, *Portrait of a Florentine Lady* by Verrocchio, *The Fruit Vendor* by Caravaggio, and *Interior of an Amsterdam Church During Service* by Emanuel de Witte. Edsel donated a wooden horse by Donatello, a bronze by Pollaiuolo, a marble, *Madonna and Child*, by Nino Pisano, and the bronze, *Niobid*, by Francesco d'St. Agata. He also presented specimens of Near Eastern and Far Eastern art, including Persian rugs and Chinese vases, and gave a collection of Colonial furniture to the American section of the institute.[98]

In total, and over a period of 50 years, Edsel and Eleanor Ford contributed nearly 300 pieces of artwork valued at more $600,000 to the Detroit Institute of Arts.[99] Because most of their contributions were made anonymously, the enormous support came to light only after their deaths. "I am quite sure that without Edsel Ford," said Valentiner, "the Detroit museum would never have developed in so many different directions, which makes it one of the outstanding museums in the country."[100]

***

While the public coffers were indeed low on funds, part of the reason that the city council did not appropriate more funds was the general consensus of the public that the Detroit Institute of Arts was a place that only the rich could visit and enjoy. "One trouble with the museum," Fred Black told Edsel, "is that it hasn't made any effort to interest the general public. The general feeling among people around Detroit, who don't feel they know a lot about art, is that it's something they don't understand, so they don't have any great interest in it. They don't know enough about it to protest if the place were to close up."[101] Because the city council had not heard any complaints from the general populace about funding to the Detroit Institute of Arts, they suggested that if the Grosse Pointers wanted the institute to remain open, "let them pay the bills!"[102] To that sentiment, the Grosse Pointers replied, "Well, Edsel Ford is Chairman of the Arts Commission. He has a lot of money. Let him pay the bill!"[103] Edsel understood this perfectly well, as "he had always envisioned the Institute as a whole lot more than just a big picture gallery," explained Fred Black. "He wanted it to be helpful to every segment of community life."[104]

With this in mind, Edsel asked Black to see if he could devise a way to build public support and entice the average man on the street to visit the museum. Black's idea was the Peoples' Museum Association, wherein the average Detroit worker could become a member of the museum by paying only $1.00 per year in dues. For that fee, he could visit the museum as much as he wanted and participate in special events. "We'd find some way to convince them that they were getting more than their dollar's worth in terms of service from the museum," explained Black. "And if we could get a large enough membership…we might impress City Hall that voters of that kind really were interested in the museum."[105]

One of the early activities of the association was to try a travel lecture by the well-known Swedish explorer, Sven Hedin. He had just returned from a round-the-world exploration, and he easily filled the auditorium of the museum.

The attendance and interest at Hedin's lecture was such a success that the association decided to undertake a large program of travel lectures.[106] Therefore, the public relations manager of the Detroit Institute of Arts, George Pierrot, started the *World Adventure Series*. Held on Sunday afternoons in the auditorium of the Detroit Institute of Arts, the programs brought famous lecturers to the stage, providing Detroiters a glimpse of faraway places. One of the first speakers was Lowell Thomas, who gave his illustrious presentation, *With Lawrence in Arabia and Allenby in Palestine*.[107]

World-famous Arctic explorer Admiral Richard Evelyn Byrd was booked with the then-new story of his second Antarctic expedition. To pay Byrd's high fee while keeping ticket prices low, the Masonic auditorium with its capacity of nearly 5,000 was engaged instead of the hall at the Detroit Institute of Arts. "We had a huge promotion job on our hands, and Edsel Ford shouldered most of it," recalled Pierrot. "He caused the *Josephine Ford*, the plane which Byrd flew over the North Pole, to be semi-dismantled and, during night hours, hauled through the streets from Dearborn. Ford mechanics set it up on Washington Boulevard, adjoining the Statler hotel, and its great wings straddled the street. The towering plane attracted big crowds from dawn to midnight, and as a result, the big Masonic auditorium was filled, both matinee and night."[108]

Although the activities of the association were warmly received, they did little to attract the middle class of Detroit to the exhibits of the art museum, and fewer came to see Rivera's work. Occasionally, one or two people would wander into the garden court and watch the Mexican perform his craft. From this came a serendipitous event that would put the Detroit Institute of Arts on the front page of newspapers throughout the world.

***

As Rivera was putting the finishing touches on his frescoes in the spring of 1933, an Episcopal clergyman walked into the garden court. He pondered the work of art for some time, becoming fixated on the small mural representing the Detroit pharmaceutical industry. "He made the remark that the picture looked like it might be a travesty on the Holy Family," recalled Fred Black. "He said that there was a child with golden hair, that looked something like a halo... There were animals in the picture...a doctor and a nurse... The mural could, by using a little imagination, be associated with a representation of the Holy Family with Christ the Child."

The minister walked over to Rivera and asked him why he had painted a nimbus around the child's head.

"Because," replied Rivera, "all children have nimbuses around their heads."[109]

Pierrot told Fred Black what the minister had said. "Here's a grand opportunity," Pierrot explained to Black. "I'm going to get a committee of preachers to come in here, and invite some of the press. Whatever they say will become news."[110]

Pierrot started by calling attention to the fact that a respected clergyman had criticized Rivera's work, saying that the painter had created a travesty of the Holy Family. "What do you think?" asked Pierrot. "How do you feel about it?"

Eager to voice their opinions, they had made enough remarks within a half-hour to create a terrific publicity story.

The controversy broke in the local newspapers the next morning, and within a few days, news of the "travesty" in Detroit was known all over the world. *The New York Times* made it front page news, and the story was printed in England a few days later. Pierrot and Black kept priming the media by conducting daily meetings in the music hall of the museum, with the rule that anybody could talk for two minutes about the Rivera murals. "We soon had the hall overflowing at every meeting," said Black. "There was always enough said at every meeting that newspapermen there would spot certain people and get little interviews. We kept that pot boiling for three or four weeks."[111]

All told, nearly 90,000 people visited the museum that March, most of them coming in the last two weeks, after the furor had started. It was the largest attendance since the museum had opened five years earlier.[112] Edgar Richardson, head of the education department of the museum, responded by giving lectures about the murals to the throngs of visitors, standing on the edge of the fountain because there was no more space on the floor.[113]

Edsel was in favor of Black's plan from the outset and was delighted with the results they were receiving from all of the publicity. Edsel "would read all these stories we put out as I would get them...and in most cases, he'd laugh," recalled Black. "He thought it was a great scheme. We'd accomplished the thing he'd wanted done."[114]

At the height of the controversy, Pierrot hosted a party at his residence to celebrate the coup. As everyone gathered around the piano, they all sang an ode to Rivera[115]:

Rivera was a jolly lad,
He lived in Mexico City,
He painted pictures on the walls
That weren't very pretty.
He painted here, he painted there,
With many a flip adroit,
Till Edsel offered him twenty grand
If he would paint Detroit.

Oh, jolly old Diego,
His nudes were fat and round-o,
That fresco-paintin', aggravatin',
Son-of-a-gun Diego.

This satirical sonnet continued for nine more stanzas, poking fun at all aspects of Rivera's murals.

* * *

Hardly a day passed on which comments were not made about the murals—both pro and con. It became a sword fight between those who appreciated art for art's sake, and those who simply wanted the garden court returned to its original, pristine quietude. Paul Cret, the designer of the museum, was especially upset that his garden court had been "destroyed."[116] "Señor Rivera has perpetrated a heartless hoax on his capitalist employer, Mr. Ford," exclaimed a college president.[117] The editor of the *Detroit Free Press* wrote a scathing editorial about Rivera's work, and a national art critic wrote in an Eastern paper, "A bull in a china shop could be little more incongruous than a mural of automobile machinery against a Renaissance background."[118] One local resident said that he could well understand Detroit's desire to have an example of Rivera's artwork, but "could not the *locus classicus* have been more appropriately selected—for example, the banking room of the Union Guardian Trust Company?"[119]

With the criticism came demands that the murals be painted over, and that precipitated the cry that they be completely obliterated. "We don't want the walls whitewashed," protested one woman. "Whitewash is too easily removed. We want them destroyed."[120] So great had the threats become that extra police protection was required to keep an eye on the garden court.[121] "If my Detroit frescoes are destroyed," said Rivera, "I shall be profoundly distressed, as I put into them a year of my life and the best of my talent."[122]

The raging controversy compelled Reverend H. Ralph Higgins, the clergyman who started it all—with a little help from George Pierrot and Fred Black—to write to Edsel Ford directly to clarify his position:

In view of the widespread furor precipitated by my recent criticism of the Diego Rivera frescoes... I should like to take this opportunity of presenting to you a few personal observations. I am prompted to do this because of the character of the controversy as it has been determined by the press.

It should be understood that when I made public what I felt to be a restrained personal protest, I had not the remotest idea that this would be the signal for the ignition of so widespread a public controversy. I mention this because of certain insinuations that have been made in some quarters. It was unfortunate that the newspapers seized upon the fact that I am a clergyman, to endeavor to characterize the controversy as a religious and sectarian one. For the purpose of clarifying the situation, I am enclosing herewith a statement of the chief objections which have been raised and which I believe fairly represent the point of view of the great majority of the critics.

I should like to say that I am not now and never have been in favor of the destruction of the murals. I believe that Mr. Rivera's work is worthy of preservation. But I feel strongly that it should not be in the central portion of our public Museum..."

In a separate sheet, the reverend spelled out his chief objections:

The Garden Court of the Museum, a long source of delight and inspiration to thousands, has been sacrificed.

The Murals are supposed to portray the spirit of this city and her people.

The frescoes over-emphasize the ugliness and brutality of the contemporary scene.

As for the symbolism employed, interpretations naturally differ widely. In any event, Mr. Rivera, outstanding realist that he is, could readily have depicted the spirit of Detroit in such fashion as to raise no reasonable doubt as to the political and religious impartiality of his work.

American artists should at least have been considered for the execution of this work.

The controversy involves an important question as to the proper relationship between the [Detroit] Institute of Arts and the people of Detroit.

Edsel had been out of town when Reverend Higgins wrote to him; therefore, he was delayed in replying to him. However, he did reply, distinctly and to the point[123]:

I greatly appreciate your letter of April 6th giving me the opportunity to have your views on the Rivera frescoes in the Detroit Institute of Arts.

The frescoes have been accepted by the Arts Commission and will have to survive on their own merit. I believe they are giving many thousands of people pleasure and education in spite of the criticism they have had from a religious and political point of view.

As a testament to the greater senses of Detroit, more people were in support of Rivera's frescoes than were against them. One Detroiter wrote, "I am sorry Detroit is laying itself open to be laughed at all over the world. My only objection to the murals is that they have been placed in a room too small for them."[124] The director of an Eastern art museum said, "I have never seen so strong a statement of the achievements as well as the failures of our civilization."[125] Another "man of the cloth" put it better than anybody. "Mr. Rivera came to Detroit under no false pretense," wrote Rabbi Leon Fram. "He came at a time when he could see the splendid machinery of production and its present paralysis. He might have painted soup kitchens, bank failures, riots, men sleeping in Grand Circus Park, and many other things."[126]

"Instead of criticism," said Albert Kahn, "there should be paeans of praise. These frescoes are a great credit to the painter, and to the Detroit industrialist, Edsel Ford, who believes in the poetry and beauty of industry. To reveal that poetry and beauty, Edsel Ford engaged the man most qualified in the world to bring out the poetry and beauty of industry. He has put Detroit on the map artistically."[127]

When Edsel was asked if he thought Diego Rivera's frescoes accurately interpreted industrial Detroit, he simply replied, "Well, they are Mr. Rivera's interpretation."[128]

Although somewhat concerned about the safety of his masterwork, Rivera was fascinated by the controversy and welcomed it. "This is not the first time my work has been criticized," he said. "If these Detroit frescoes did not bring criticism, then I would be worried. I would know that my work was not alive, that it made no impression on people who came to see it."[129]

Men, women, children, factory workers, and office workers—many of whom had not had a day's work in many weeks—turned out to view the much-disputed work of Señor Rivera. However, few comments were heard about the so-called travesty panel. The laid-off factory workers "pointed to the grim faces of Rivera's factory workers; the hands eloquent in toil; the lashing expressions in the faces of shop foremen; the significant absence of any evidence of pleasantness—of anything that smacked of a smile—in any of the artist's conception of industrial Detroit." Some workers said that Rivera's paintings did not depict the factory workers accurately, but a preponderance of them said that he had caught the plight of the worker and had painted him as he really was.

Rivera was pleased to hear the workers' comments. "The frescoes...were inspired by a belief in the worker," he explained. "I am a worker myself. I know the worker, and I have painted him as I know and see him. And I have tried to put into these frescoes, that struggle as I understand it."[130] "I have heard the objections raised that I have not painted any spiritual thing in these frescoes," said Rivera. "Well, I have never seen spirits. Every day I see automobiles and workers. And I paint only what I see."[131]

***

Valentiner returned as head of the Detroit Institute of Arts at the height of the controversy. As he entered his office, he was confronted with a massive amount of mail that had been sent to his attention. He read through it all and then called a meeting of the Arts Commission to see what should be done. "Dr. William Valentiner...opened the meeting by calling attention to two heaps of correspondence on his desk. One, a small pile of letters and petitions, representing the objectors to the frescoes. The other pile, almost covering his desk, containing petitions favoring the murals and letters from art critics all over the country, congratulating Detroit on its acquisition."[132]

Valentiner came right to the point. Should they keep the murals, or should they be destroyed?

The commissioners, in a unanimous and almost instantaneous decision, adopted a resolution accepting the frescoes, and they expressed their appreciation to Edsel Ford for having donated them to the museum.[133] Therefore, by edict of the Arts Commission, the Rivera frescoes would be saved for posterity.[134]

After the vote, Valentiner said, "I do not believe that anyone can question the artistic quality of Mr. Rivera's work. No one expects any work of art to please everyone equally… I can assure the public that no article is included in our collections because of either religious or anti-religious reasons; it must have an historical or an artistic significance."[135] Edsel Ford told reporters, "I admire Mr. Rivera's spirit. I really believe he was trying to express his idea of the spirit of Detroit."[136]

Regardless of how the Arts Commission voted that day, it is unlikely that the frescoes would have been whitewashed, let alone destroyed. Any decision to the contrary would have required Edsel's approval, and he would never grant it. "I paid for the murals," Edsel said defiantly, "and I'm going to keep them."[137]

Chapter 16

Aviation Pioneer

"Well, you can say I'm an aviation enthusiast."
Edsel Ford[1]

While Henry Ford was putting the final touches on his Model T in the summer of 1908, Edsel had his own audacious plans. In a large barn that his father had leased on Woodward Avenue, not far from the Piquette Plant, 14-year-old Edsel and his friends Charles Van Auken, William Theisen, and James Smith were working on an experimental airplane.

Airplanes fascinated Edsel ever since he had learned about the Wright brothers' successful flight off Kill Devil Hill in North Carolina five years earlier. With Edsel's propensity for bookwork, he and Van Auken searched libraries and bookstores, studying everything they could find about flying machines. Edsel, in particular, absorbed the theories of air flight. "He sat and talked such things as 'lift' and 'load' and 'stress' and 'resistance,' " recalled a reporter, "until it is plainly to be seen that he had made a thorough study of aeronautics."[2]

Work began in the summer of 1908. They used Van Auken's design, and Edsel utilized pattern makers and machinists at the Ford factory to build the parts. It took the two young men two years to complete the crude-looking airplane. It had one wing and tricycle landing gear—two features unique for the time. The power came from a souped-up Model T engine, which they had drilled full of holes to make it more lightweight.

For the maiden flight, the boys towed the little plane behind a Model N out to the Ford farm, where they had constructed a plank runway on a gentle slope of ground. Because Henry would not allow Edsel to fly, Van Auken had the dubious honor of piloting the craft.

"The thing did leave the ground," recalled Edsel many years later, "and probably it is just as well that it did not get too high, for it might have fallen and killed someone. It finally landed against a fence of the flying field, and that was the end of the experiment. But we all had a good time!"[3]

Although Edsel's first experiment with flight was marginally successful, it was only the beginning of his adventure into aviation.

***

In the autumn of 1921, Edsel was approached by Ralph Upson, an engineer formerly with the Goodyear Tire & Rubber Company, with an idea for a new type of lighter-than-air aircraft. Upson's idea was to make dirigibles—the workhorses of the air at that time—out of a new type of metal called Duralumin, instead of rubber and fabric. "It is perfectly feasible," Upson explained to Edsel, "to build an airship with an all-metal hull or envelope, eliminating the fabric envelope used by the British, Germans, [and] Italians...[4] For aircraft manufacturing, there were many benefits to using Duralumin. It was strong, lightweight, and non-porous. Because helium, the gas of choice for lighter-than-air aircraft, was so expensive, gas leakage was ordinarily a great cost factor. In Upson's design, gas leakage was almost nothing. In one test, one of his prototype balloons stood in a hanger for months without replenishment of gas.[5] However, the much cheaper and highly flammable hydrogen gas could also be used in Upson's design. "It is recommended that hydrogen gas be used for inflation on account of its availability, cheapness, and lifting qualities, and the fact that it can be used for reserve fuel," wrote one engineer. "It is believed that hydrogen carried in a metal hull should be no more dangerous than gasoline in a properly constructed metal tank."[6] (This, of course, was before the Hindenburg disaster in 1937.)

Edsel was immediately interested in Upson's proposal and in his credentials. Upson had started the Goodyear balloon division years earlier and had made it into a highly successful division, particularly with orders from the U.S. government during World War I. However, when Upson presented this new idea to Goodyear officials, they balked. Therefore, he left the company and started the Aircraft Development Corporation in 1922.

Edsel could foresee the day when aircraft such as Upson's design could transport large quantities of parts and materials from one Ford plant to another. From a strictly personal standpoint, he envisioned a three-day passenger service from Detroit to London using such an aircraft.[7] This, of course, was years before dirigibles were used to transport people across the Atlantic.

Without much coaxing, Edsel subscribed to the Aircraft Development Corporation. Initially, he bought 100 shares of its stock for $1,000. Later, after he had become a director of the company, he invested another $5,000 for an additional 1,500 shares. He also recruited Alex Dow, George Holley, and other prominent Detroit industrialists and friends to invest in Upson's fledgling enterprise. Before long, Thomas Edison, the famous inventor and friend of the Ford family, invested in the Aircraft Development Corporation at Edsel's suggestion, as did Charles Kettering, the creative engineer of General Motors, William Stout, an unconventional aircraft engineer, and William Mayo, Ford Motor Company's chief engineer.[8]

Thanks to Edsel's backing, money began flowing into the Aircraft Development Corporation, and Upson immediately went to work developing his new aircraft. Several months later, he presented to Edsel and the other stockholders a one-tenth-scale model of his metal-clad airship, with a full-size prototype of one of the typical structural members of the ship.[9] Upson's next step was to construct a full-size airship and enter it in the National Balloon Race to be held in Dearborn in 1923. "The Aircraft Development Corporation is assembling the world's first metal-clad airship," wrote a reporter for *The Aeroplane*. "If it succeeds, it will mark an immense step forward in air transport. The recent failures in transatlantic flights only emphasize the possibilities of airships for long non-stop journeys."[10]

However, the reporter also had words of caution. "Those who are working seriously on airship development deserve every credit. About 10 times as much money, perseverance, and patience are needed to tackle the lighter-than-air problems as are needed for heavier-than-air work." This statement was to be prophetic as far as the Aircraft Development Corporation was concerned.

Upson and his company completed the first prototype—a 90-foot dirigible powered by two 85-horsepower engines—in time for the Dearborn race. Although small, the craft had more carrying capacity and was faster than the fabric dirigibles twice its size.[11] In short order, the Aircraft Development Corporation received $300,000 from the U.S. Navy to build an experimental airship. Similar to Edsel's thoughts, a West Coast firm was interested in purchasing a one-million-cubic-foot displacement craft to fly passengers and cargo between California and Hawaii.[12] After that, the company built a much larger Duralumin-clad airship for the Navy. Playing off the name of the Ford *Tin Lizzie*, the new airship was dubbed the *Tin Bubble*.[13]

***

While Edsel was financially knee-deep in the Aircraft Development Corporation, one of the board members was working on a project of his own. William Bushnell Stout, an inventor and self-taught engineer from Illinois, was designing and building an all-metal airplane in a shop on Beaubien Street in Detroit, near the Ford Piquette Avenue plant. A college dropout, Stout studied engineering on his own. He had been a teacher and a journalist and, in 1912, became editor of his own aviation magazine called *Aerial Age*. He also wrote an aviation section for the *Chicago Tribune* and lectured on the development of the new technology.

As the outbreak of World War I became imminent, a number of American manufacturers began thinking of ways to produce goods for the war effort. Alvan Macauley, Edsel's friend and president of Packard Motor Car Company, recruited Stout to head a department that would monitor aircraft development. A few months later, war broke out in Europe, and Stout was asked to go to Washington to consult with the best aircraft engine designers in the country to design an engine that could be built for the government. Within months, Ford, Packard, Marmon, Cadillac, and the recently formed Lincoln Motor Car Company were building the twelve-cylinder Liberty engine that Stout had helped design.[14]

***

During this time, Stout had calculated that only about 25 percent of the horsepower in an aircraft engine was used to move the plane through the air. The remainder was consumed by drag of the struts, and wires holding the upper and lower wings of biplanes together.[15] Stout's solution was to place all of the support structure of the airplane inside one single wing.[16]

Stout drew up some plans of his idea and showed them to people in Washington and to several industrialists in Detroit. William Rands, president of Motor Products, a supplier of automotive parts to the industry, was highly interested in Stout's idea and allowed him to build a mock-up at his plant. Approximately 15 months later, in November 1918, the first successful flight of Stout's mono-wing plane occurred. The flight was not long, but it proved the feasibility of a mono-wing design (as did Edsel's rudimentary plane 10 years earlier).

By this time, Rands dropped his support, but Robert Stranahan, president of the Champion Spark Plug Company in Toledo, Ohio, picked up the slack. With $15,000 of new capital and several engineers, Stout moved his headquarters to the building on Beaubien Street and began building a new version of his monoplane design.

The revised plane had a 200-horsepower Packard engine, and, as far as aircraft development was concerned, it was a resounding success. The new plane was able to carry a 1,200-pound load to 5,000 feet.[17]

News of Stout's activities had been well publicized in all of the local Detroit newspapers, and several prominent Detroiters dropped in from time to time to check on his progress. Roy Chapin of Hudson Motor Car Company, Walter Chrysler of the Chrysler Corporation, and William Mayo, chief engineer at Ford, were among Stout's visitors.[18] Some of these men were stockholders in the Aircraft Development Corporation and therefore were somewhat familiar with Stout.

These first two planes were made of plywood over an aluminum structure. However, with his involvement in the Aircraft Development Corporation, Stout wondered whether or not the new Duralumin material could be used for the "skin" of the aircraft. With a contract from the U.S. Navy for a prototype of a lightweight, high-performance bomber, Stout went to work again. This time, he built a plane entirely of aluminum and equipped it with a 150-horsepower Hispano-Suiza engine. Two years later, the aircraft was finished and hauled to Selfridge Field at Mt. Clemens for trial runs.

In the early morning of May 21, 1922, Stout and his men drove to the airfield. William Mayo, Robert Stranahan, test pilot Eddie Stinson, Admiral William Moffet, in charge of U.S. Navy aviation, and other Navy personnel met them there. "Quite nonchalantly Stinson climbed into the plane," recalled Stout. "We started the engines, and away he flew. Things went off in excellent shape. He made his runs and landed."[19]

Unfortunately, after Stinson's successful fights, Admiral Moffet wanted one of his test pilots to try the plane. Against Stout's reservations, the Navy man took the plane up, circled the field once, came down over the trees, and crashed it while landing. Fortunately, the pilot was not injured, but Stout's two-year project lay in ruins and, with it, his dream of a lucrative contract with the Navy. Immediately after the crash, Admiral Moffet canceled the bomber contract with Stout.

Broke and with nowhere to turn, Stout came up with an ingenious idea for raising new capital.

* * *

That summer, Edsel Ford received a letter from Stout, who was soliciting funds to start his own aircraft company. Edsel was not the only one whom Stout had contacted—he had sent the letter to 100 other Detroit industrialists, among them Henry Ford, Roy Chapin, Walter Briggs, Phil Wrigley, and Willard Dow. In his letter, Stout wrote[20]:

> *If a man were to walk into your office with a design of a motorcar with the engine upside down and under the rear seat, you would quickly be able to answer him as to its value, and would have the authority to make a decision.*

But suppose some man should come to you with an airplane design with the engine upside down and under the rear seat, what would be your reply—and more important—what would be your authority for making the decision? On your decisions and those of other leading men of Detroit depends the future of Detroit in aviation.

I'm going to send out a weekly series of letters, purely educational, giving the fundamentals of aviation—what makes a plane fly, etc.—for the education of Detroit's management. Will you have your secretary call or drop me a line if you would like to receive this series?

Of the 100 letters sent, Stout received 65 replies, all asking for his newsletter. However, Stout's objective was not to merely spend his time writing a newsletter for some of the wealthiest men in America. He wanted money.

After Stout had sent his fourth issue of the newsletter, he began calling on the subscribers. He preyed on their inherent propensity for adventure, innovation, and pride in the ability of their community to manufacture almost anything. "If Detroit is going to take its place in aviation," Stout told them, "someone has to build some airplanes. This city has started the metal airplane, and this city knows more about it now than any other location. We're going to take our knowledge and build another airplane. It will cost money. It will not be profitable. We want from you $1,000. No more, no less. You will get for it one definite promise, and that is that you'll never see your money again!"[21]

"That's the damnedest solicitation I've ever heard," was the most common response to Stout's proposal. However, his audacity and unusual approach won him more support than he could ever have imagined. Nearly everyone he met gave him a check for $1,000, fully expecting to lose every cent of it. When he finally was able to see Henry Ford, Henry listened patiently.

"How much do you want?" asked Henry.

"One thousand dollars—no more, no less."

"Well, I can't quite do it that way, but I'll put down a thousand for me and a thousand for Edsel."[22]

The money, of course, was of no concern to Henry Ford. He invested in Stout's scheme because of Edsel's interest in airplane development. "While I don't think that Mr. Henry Ford was ever particularly keen about going into aviation," recalled Ernest Liebold, "he felt that as long as Edsel had an interest in it, he was willing to encourage him along that line just to see how far we could go in it."[23]

Edsel became engrossed in Stout's experiments, and Mayo spurred his enthusiasm. Mayo was also one of Stout's subscribers and had been keeping Edsel abreast of Stout's activities at Selfridge Field. Mayo was present for the maiden voyage of Stout's first all-metal airplane and soon was acting as intermediary between Stout and Edsel.[24]

Throughout 1923, Stout and his men worked on their new all-metal airplane called the Air Sedan. While Stout was enticing other executives to contribute to his new enterprise—he used Henry and Edsel's name quite liberally—he kept his current investors well informed and excited. "We want every subscriber to put at least one rivet in the new ship," wrote Stout. "We have saved a space to take care of our entire list, each rivet being labeled with a name."[25] Henry, Edsel, and Mayo all accepted his invitation.[26]

The Air Sedan was completed in early 1923 and took its maiden voyage on February 9. Test pilot Walter Lees took the plane up and flew around the field a few times. After this initial flight, he took off with one passenger. The next day, he took up three passengers.

All three flights were successful, but the plane's 90-horsepower engine was under-powered. Thus, they took the plane back to the shop, equipped it with a Swiss-designed, Spanish-made 150-horsepower V-12 Hispano-Suiza engine, and brought it back to Selfridge Field for more tests. Now, the plane took off without difficulty, even with a load of passengers. Over the next few weeks, the Air Sedan made hundreds of demonstration flights, taking up courageous visitors from Detroit. Among its first passengers was Eleanor Ford's friend, Mrs. Evelyn Book.[27] "We have covered nearly 3,500 miles with the plane so far and have carried nearly 600 passengers," Stout wrote to Edsel. "It is operating with all the precision of any other transportation..."[28]

The Air Sedan was a success, except for its cost. Equipped with the Hispano-Suiza engine, Stout knew that he could not build an affordable aircraft using that engine. Through Mayo, however, Stout had learned that the Wright Company (the aircraft company founded by the Wright brothers) was working on a radial engine design that promised high output and low weight, but there was no timetable for its availability. Until that engine became available, Stout decided to adapt his new plane design around the much-improved Liberty engine.

***

Edsel must have been impressed with Stout's developments. One day, Glenn Hoppin, Stout's brother-in-law and an engineer on Stout's staff, was at the Detroit Athletic Club for lunch and ran into Mayo. "I wish you would talk to Bill Stout and get him to tell me what Mr. Edsel Ford can do that will help aviation the most," asked Mayo.

"I know what Bill will tell you. He will tell you to give us an airfield where we can operate from."

"Talk it over with Bill, and have him write me a letter," replied Mayo.[29]

Stout wasted no time in sending a letter to Mayo. Because Mayo had instigated the request, Stout felt it was the perfect time to make his feelings plain. "We are in need of three things," Stout wrote. "First: the moral and financial support of the community and its big men. Second: A place to do our work well, with a flying field close at hand. Third: A motor development, connected with our work that will enable us to have a cheap engine developed and ready for production by a year from this summer." If that were not bold enough, Stout concluded the letter by asking three questions that he hoped would be rhetorical: "Can we count on yourself and Mr. Edsel Ford for a subscription of $5,000 each for our present step? Is there available near Detroit or Dearborn perhaps, vacant land owned by your company, which we could use for a period of time for our development work? If a building was available, it would save us the expense of erecting a hangar. Would Mr. Edsel Ford be interested in working with us on the development of the engine and in some way arranging mutual interests in the result?"[30]

Mayo forwarded Stout's letter to Edsel, and Edsel was enthusiastic about the proposal. In fact, about a year earlier, Edsel had considered the possibilities of establishing a manufacturing plant and airport in Dearborn in which to build and test the lighter-than-air machines designed by the Aircraft Development Corporation.[31] In a letter, Mayo encouraged Edsel to support Stout, reminding him that Wright's radial engine could prove the answer to a reliable and affordable power plant[32]:

> *I believe that Mr. Stout's efforts have been of a high enough character to warrant giving him some assistance, either in cash, a free leave for a certain length of time, a piece of property, a free lease of some building we may not be using, or any other combination that might suggest itself to you. Mr. Stout... has a plane of the size that I think would be readily adaptable to our radial motor when it has been fully developed, and if we had some connection with him through his development stage, it would, I think, keep us in more intimate touch with the flying program and the all-metal construction and this might be of considerable assistance to us should you and your father decide to start into the aerial game.*

Edsel then discussed the idea with his father. Henry was not enthusiastic about building airplanes—he was scared of the proposition.[33] "It wasn't something he grabbed at all," recalled Mayo. "I personally know that it took an awful lot of talk from both Edsel and my father to put this thing over," recalled Mayo's son.[34] Eventually, Edsel and Mayo talked Henry into pursuing the manufacture of airplanes, saying that it would give prominence to the Ford Motor Company in a new area of manufacturing. "This work in the air is not primarily my work," wrote Henry Ford. "It is my son Edsel who first became interested in aviation, and it took a long time to convince me that it had a commercial feature. The direction of the work depends on my son."[35]

One of Henry Ford's greatest strengths—and weaknesses—was that when he made up his mind to do something, it was done. A few days after he and Edsel had agreed to help Stout, Henry called Stout to Dearborn to choose some property on which to build an airport. Three sites were determined to be suitable, and Stout chose the one most expensive to prepare. "That doesn't make any difference," said Henry. "If it will make the best field, that's the one we will take."[36]

Within 24 hours, an army of men with Fordson tractors began clearing 240 acres of land covered with 100-foot-high trees. Sixty days later, the airport was completed. Built to Edsel's specifications, it had two concrete runways (the first of their kind in the world)—one 3,400 feet long running northeast to southwest, and the other one 3,700 feet long, running northwest and southeast. The name *FORD* was laid down in white crushed rock 200 feet high and could be seen from an altitude of 10,000 feet. "Mr. Henry Ford asked what kind of a building we would want for our experimental work," recalled Stout. "Awed at the expenditure of even small amounts on such a precarious thing as aviation, we designed the cheapest type of steel and brick construction that would fit the need. Very shortly the building was put up... (concrete floor and all) and we were told to move in on a basis that would be decided later."[37]

***

Ford Field officially opened on January 15, 1924. To celebrate, airplanes from all over the Midwest flew into the new field. All morning, Edsel, Henry, and Stout watched the planes landing. After that, they went to the Engineering Building and had lunch at the Round Table.[38] To cap off the day, Edsel bought 16,000 shares of stock in the Stout Metal Airplane Company, making him the largest stockholder in the company—next to Stout himself—and a member of the board of directors. Other board members included Walter Briggs, George Holley, Charles Kettering, and William Mayo.[39] "A letter cannot really express our feelings regarding the very wonderful field and plant which you have made available for us at Dearborn," Stout wrote Edsel. "I wonder if even you realize what a tremendous incentive this is going to be to the development of commercial aviation in America."[40]

The airport terminal, purported to be the first of its kind in the country, was patterned after an old-time railroad station. The two-story brick structure had offices upstairs, and a passenger waiting room and ticket window on the main floor. It was air-conditioned, receiving its cool, dry air through underground ducts from the power plant at the Ford Engineering Building. Eventually, buses ran from hotels in downtown Detroit to the air terminal. For those who wanted to stay near the airport, they were put up at the Dearborn Inn—a luxurious hotel that Edsel had constructed across the street from the terminal.[41]

***

The first all-metal airplane that Stout built in his new factory at the Ford Airport incorporated all of the elements of design that he had learned in his earlier planes. With its lightweight, all-metal construction and the improved Liberty engine, the *Maiden Detroit* could out-perform and carry more than any other airplane of the time. It made one of the first multi-passenger flights from Detroit to Chicago, and after that, flew six passengers to Dayton, Ohio, and back.[42]

With the success of the plane, Stout received his first order—the U.S. government wanted one for its airmail service. Stout finally was in business, and Edsel, who had been monitoring Stout's achievements with enthusiasm, decided to become more involved. "Edsel took a deep interest in the airport and the airplane activities," recalled Ernest Liebold. "He spent a great deal of time over there."[43]

As usual, Mayo had been keeping Edsel apprised of Stout's financial status. Although Stout now had an order from the U.S. government, he was "still in a mess financially."[44] Working on a shoestring budget, Edsel knew it would take years for Stout's company to make any significant progress. Therefore, Edsel and Mayo met with Henry and discussed the possibility of buying Stout's company. Eventually the two men sold the idea to Henry, and Mayo started the negotiations.[45]

Stout was more than interested in selling, but he did not want to be kicked out of the company either. In an agreement that he worked out with Mayo, Stout assigned all of his patents to Ford Motor Company, and Ford kept him on as a consultant at a handsome salary.[46] At the last moment, Stout asked if he could also start his own airline company. The deal was done, and Ford took over the company on August 1, 1925. Edsel assumed the post of president, Stout and Mayo became vice presidents, and Bert Craig was treasurer. Mayo was actually the operating head, and Henry Ford, because he controlled Ford Motor Company, held the final power.[47] "Just a note to thank you for the spirit which has surrounded your taking over the Stout Metal Airplane Company, and the purchase of the stock," wrote Stout to Edsel. "The way the proposition has been handled with us, by yourself and Mr. Mayo, has been exceptionally fair in every way."[48]

Stout also wrote publicly of his support of the sale. "Naturally we are all for aviation first and must be if the industry is to come. There are no resources in the world, governmental or otherwise, that can move as fast or as accurately as the Ford organization. For me to hold back this development in order to satisfy the personal angle of either pride or cash would be a very definite hindrance to aviation. I have no fears however but that the eventual is going to work out..."[49]

The new aircraft division of Ford Motor Company had airplanes under construction, but no market. However, Henry Ford solved that problem also when he bought six planes and started his own air freight line. Initially, the planes flew parts and supplies between Ford plants in

Dearborn and Chicago. Later, flights were expanded to Buffalo. Stout also purchased several planes and started the Stout Air Service, carrying passengers between Dearborn and Grand Rapids.

With the experience that Ford airplanes were gaining in both freight and passenger service, the United States Post Office awarded Ford the first commercial airmail contract to a private company. In the agreement, Ford planes would carry mail between Detroit and Cleveland and between Detroit and Chicago. "This letter is to wish you greetings by Air Mail," Edsel wrote to Ford's Branch Manager in Chicago. "I am posting the letter in Detroit at 8:15 A.M. this morning, and it will be transported to Dearborn by motor and flown to Chicago on the first privately operated Air Mail Service flight for the United States Post Office Department."[50]

***

During these early days of air transportation, there were few places for planes to land, let alone well-built airports. Ford Field was probably the best in the country. Air mail pilots were expected to fly in all sorts of weather, without radio communications and without clear indications of airport locations. "No airfield was lighted unless the ground crew built a bonfire when a plane was due in to let the pilot know where he was," recalled Stout.[51] Edsel used his extensive dealer network to assist in the navigation of Ford mail planes along mail routes. It was crude, but the system was better than anything else available at the time. "I had all of our dealers between Detroit and Chicago send in a telegram every morning at 8:00 as to the weather conditions, the humidity, barometric pressure, prevailing winds, the condition of the atmosphere, the visibility, and so on. From that, we made up a weather map. Upon the conditions that were revealed by the weather map, I would advise the Ford pilots whether or not they could fly."[52]

However, this was not enough, as far as Edsel was concerned. He wanted to do more to advance aviation. Thus, at the end of 1925, he wrote a letter to his nationwide network of Ford dealers, asking for their cooperation[53]:

> *I am now going to ask you as a Ford dealer to help commercial aviation. It seems that one of the very great difficulties of cross-country flying is in trying to distinguish over what town or city the pilot is traveling...*
>
> *Therefore, will you not paint on the roof of your garage or place of business the name of the city or town in which you are located? The letters of the name should be white, spaced well apart, not less than twelve inches in height, facing toward the sky, and whenever possible the name should extend due east and west, with the top of the letters to the north. Also an arrow, pointed due north, should be painted immediately following the name of the city or town.*
>
> *If every Ford dealer does this in the 10,000 villages, towns, and cities where you represent us, we shall have placed a network of sign boards throughout the United States for the use of air pilots, and we, as well as every air pilot, will appreciate this action on your part.*

Edsel received an immediate and enthusiastic response. "Many of our dealers are planning letters five and six feet high where size of roof permits," wrote one Ford roadman.[54] In California, flyers were complaining that they had difficulty in determining exactly when they had arrived at Pomona; therefore, the local dealer, following Edsel's request, painted "POMONA" in twelve-foot high letters on the top of his Ford agency. "The name could be easily read by aviators at an elevation of 10,000 feet, beyond which airplanes are seldom flown," wrote the dealer. "The very first day this sign was completed, a pilot landed here and immediately rendered us thanks for such assistance." A Ford dealer in Tennessee did the same thing, except that he painted directional arrows and the distance to the airport, in addition to the name of the town.[55]

***

Meanwhile, Edsel decided to pursue building a "plane for the masses," something similar to what his father had done with the Model T approximately 20 years earlier.[56] According to Harold Hicks, the engineer who was given the task of designing the plane, "Edsel explained to his father that he had worked up an idea for producing a flivver plane, a small mono-wing that would carry two to three passengers. His idea was to build five planes for various testing operations, settle on a design, and then produce 2,000 of these planes, putting them in various Ford agencies all over the country."[57] Surprisingly, Henry "was receptive" to the idea."[58]

The first of these flivver planes was completed in June 1926. With a wingspan of slightly less than 22 feet, and powered by a three-cylinder air-cooled engine, it had room for only the pilot. Films show its chief test pilot, 21-year-old Harry Brooks, flying the little plane with abandon. He would fly back and forth from Southfield to Dearborn and would land in front of the Engineering Building.[59] "It was quite a little plane," recalled Ernest Liebold.[60] Of course, such shenanigans did not take long to find their way into the newspapers. *Detroit Free Press* editor, Arthur Brisbane, "just had to tell the world the grand and glorious news that soon, everyone who had a $5,000 bill to spare could go to the nearest Ford dealer and fly away with the newest product of the automobile Croesus."[61]

In 1927, a second flivver plane was built. Similar to the first version, this plane was also a one-seater, and Harry Brooks tested its worthiness. On January 25, 1928, Brooks took off from Ford Field and headed for Miami in an attempt to break the world endurance record for small planes. He was forced down in North Carolina due to bad weather, but he had still managed to beat the American record. A month later, he tried again, and this time he made it all the way to Florida, breaking the world record. On his return trip, however, disaster struck. Brooks landed on the beach at Melbourne to get gas. When he took off again, "he got up in the air and made a circle, as they usually do to get their bearings," recalled Liebold. "Just as he was out over the lake, he sideslipped and went down in the water, and that was the last of Harry Brooks."[62]

"After Harry Brooks died," continues Liebold, "Henry Ford told us, 'Nobody around here can fly.' Even if we had to go out of town, we couldn't take an airplane. We had to take a train or car. He forbade everybody to fly."[63]

That did not affect Edsel. "Oddly enough," Edsel told a reporter, "I've never gone up yet. You see, I promised my wife not to go up until I had converted her to my own views on flying.' "[64]

Undaunted by the accident, Edsel was still determined to build an airplane for the masses. An article issued by the Department of Commerce may have goaded him onward. The article said that there was an abundance of licensed pilots and a severe shortage of personal-sized airplanes. It wanted to see an affordable airplane built that could carry two people.[65]

Edsel put some of his men to work on an all-new small plane. Called the Bat-wing, it was an updated flivver plane with two seats: one for the pilot, and one for a passenger. A Ford V-8 engine mounted behind the seats supplied power, and a driveshaft, which ran between the pilot and the passenger, connected the engine to the propeller.[66] Use of a standard Ford engine would allow an owner to get his plane serviced by any Ford dealer in the country.

Unfortunately, the little plane was too unstable in test flights, and it was dropped.

***

By this time, Edsel Ford had become well known for his support of the fledgling aircraft industry and was considered an expert in the field. Although he was not accustomed to granting interviews, he agreed to talk to Admiral Richard Evelyn Byrd, also a pioneer aviator.

Byrd asked Edsel about his small plane projects and the future of aviation[67]:

"Is the flivver plane practical today?" asked Byrd.

"Yes," replied Edsel.

"Then your huge plant will soon fill the air with planes. And since we have 26 million automobiles on the earth's surface today, the sky will be black with planes tomorrow."

Edsel quickly headed off Byrd's grotesque speculations. "Its range of usefulness is not very wide," Edsel said.

"How much do you think the flivver plane ought to cost when it comes out?" asked Byrd.

"Under proper conditions of quantity manufacture, its cost should not exceed $1,000. But, of course, this is dependent on quantity production. In quantity the cost decreases rapidly, and selling prices automatically decrease with it."

"What would it look like?"

"I think it would be comparable to a motorcycle," explained Edsel. "For, after all, a plane such as you have in mind would be a mere messenger plane for light duty."

"How many passengers do you think the flivver plane will carry?"

"The smallest flivver plane would of necessity be a one-passenger machine. But the popular plane would have to carry two or three passengers. Of course, when you discuss the small plane, it should be borne in mind that there is less chance for change and improvement of design in the small planes than in the larger ones. Small planes, I think, will remain practically what they are now."

"You see," continued Edsel, "people are inclined to forget an important point—a point that is obscured by exaggerated statements about the progress of flying. This point is that for some time to come, perhaps for a long time to come, the control of airplanes will be mostly in the hands of professional pilots. As a rule, the amateur who can afford to own a plane hasn't time enough to train, as he ought to. It is true that some have done so, but not a sufficient number to warrant our believing that the amateur pilot will submerge the professional pilot for a long time to come."

"Mr. Ford," asked Byrd, "exactly what is the business aspect of airplane manufacturing today?"

"Well, the business is here, and a great deal of it," explained Edsel. "Airplane manufacturers have plenty of orders on hand, we certainly have. And I know others that have. All of the aircraft plants are busy, and most of them are, at this moment, behind in their orders."

"How can American business be induced to use aviation to a greater degree than it does now?"

"We cannot expect American business to step in and develop aviation for us. It is the task of those of us who believe in flying to develop aviation for American business. That is an idea which should be clearly understood. Instead of depending on business to supply the opportunity for aviation, it is up to aviation to supply the air-opportunity for business."

"Mr. Ford," queried Byrd, "what is the next big step in airplane manufacture and in commercial aviation—is it in airplane design or in airplane piloting?"

"Neither," replied Edsel.

"Then what?"

"Motive power."

"Engines, you mean?"

"In a sense, yes," replied Edsel. "The designs of our present planes are all determined by the type of their engines. Change the type of engine power, and you change the type of plane. The design follows the power."

"But won't competition in airplane design have a great hand in determining the future of building and flying airplanes?"

"Yes, but it can lead us only one way, toward increased safety and efficiency. You may be perfectly sure of that."

Byrd ended his interview with Edsel by asking him what he thought of the recent transatlantic flights, in which many lives were lost before a successful crossing was attained.

"I am not sure that the term *stunt* is always deserved in speaking of such flights," replied Edsel. "Men and women who undertake to prove certain convictions of theirs at the risk of their lives are entitled to full credit for sobriety of purpose. Regrettable losses have occurred, but were they not necessary to the advancement of the art? Surely they were not wasted. Those who have failed deserve honor with those who succeeded."[68]

A year later, Edsel gave another interview, this time to a reporter with *Nation's Business*.[69]

> *There are sound economic reasons for believing that a great epoch of air transportation is being born. I believe that 1928 will go down in history as the year in which American business accepted the airplane.*
>
> *A new commercial and industrial era is beginning with the airplane just as surely as new eras began with steamships, railroads, and automobiles. Distrust and catastrophes marked the early days of those eras and befogged the true pictures of their futures. Likewise the heroic hazards of the war, the daring of pioneer fliers, and the foolish use of antiquated machines have diverted public attention from the true achievements in air transportation.*
>
> *The arrival of this new and tremendous force in our national life does not mean the obsolescence of the old means of transportation. It foretells only the extension of them and a great increase in man's power to cover great distances swiftly and unerringly. Aviation holds for us another safe and comfortable way to save time.*
>
> *As for the airplanes that will fly into these terminals, perhaps the greater number will be operated by transportation companies somewhat similar to the rail-*

roads. I believe we are about to see a repetition of the history of the automobile, but I do not anticipate an airplane in everyone's garage.

The time is near when commerce and industry may cease to depend on the thousands concentrated in cities. This swifter, more flexible means of transportation may mean the decentralization of some of the great metropolitan areas. A man may commute 100 miles to his business.

***

Although Edsel was heavily involved with the aircraft operations of the company, not to mention his work with the automobile side of Ford Motor Company, Edsel took time to do something that he rarely did—he wrote an article. With his penchant for anonymity, he seldom gave interviews, let alone wrote articles under his own name. However, he felt strongly enough about the development of the aircraft industry that he penned a well-written and incisive article in which he gave his thoughts on the need for aircraft development, as well as the future of the airplane. It appeared in the May 1 edition of *Nation's Business*, and it was titled, "Why We Have Faith in Aircraft"[70]:

The problem of aerial navigation is now a matter of educating the general public to its uses and possibilities.

Since the automobile motorized the world and turned every other man into a mechanician, there will probably be no need for the airplane to pass through a stage similar to the spectacular racing stage, which was necessary to keep the early automobiles in the public eye. That preliminary work had been done for all forms of motor transportation, and aircraft will naturally benefit from it.

It is conceivable that some day we [Ford Motor Company] shall be engaged in the quantity production, if not of air sedans and air transports, at least of their motors.

The fundamental of all commercial transportation is speed, and the field in which greater speed can be obtained is in the air. The center of speed and security in an airplane is the power plant, and the motor industry, of course, is legitimately concerned with that.

It is my opinion that both types have their distinctive places, like passenger automobiles and motor trucks. The airplane will be used for speed or any form of fast express work, while the dirigible will be used for carrying heavy loads. It is probable that dirigibles will traverse the longer main routes of air travel, while the planes act as feeders on the branch lines. The heavier-than-air machines will probably not be used for regular transcontinental or transatlantic journeys.

We have progressed far enough with the cabin-type (or, if your prefer, the sedan-type) of the all-metal airplane to be fairly well convinced that increased production would result in decreased cost that would put the plane within the means of a wide circle of users. At Dearborn, we have established an airport and will presently have a mooring mast for dirigibles.

Amongst the commonest experiences of the not distant future will be the daily use of airplanes for personal purposes in many parts of the country, the preparation of landing fields everywhere, and acceptance and use of aircraft as widespread, comparatively, as the radio. It is truly the Age of the Air.

***

Edsel did more than his part in promoting air transportation through the formation of the Detroit Air Olympics and the so-called Ford Reliability Tour, and by writing articles and giving interviews. Through these tests of endurance and competition, he hoped "to establish and prove the reliability of the airplane."[71] "This air tour is conducted to sell aviation to the United States," Edsel said. "It is to the aviation industry what the Glidden Tour of old was to the infant auto industry. When the demand for airplanes has been created, Detroit will supply it..."[72]

The first Airplane Reliability Tour was held on September 28, 1925. Edsel "was the moving spirit on these tours," having suggested the activity to the Detroit Aero Club of which he was a member. To stimulate interest, Edsel had his friends at Tiffany & Company design a four-foot-tall trophy that cost $5,000. (He contributed a similar amount for each year that the Tour lasted.[73]) On the splendid trophy was inscribed: "This trophy is offered to encourage the building-up of commercial aviation as a medium of transportation."[74] A separate committee from the Aero Club established rules, and the tours were conducted under the supervision of the National Aeronautics Association.

In addition to the Reliability Tour, the Gordon-Bennett International Balloon Race and the finals for the Boys' National Model Airplane Contest were also conducted at Ford Field. Balloonists from all over the world came to Dearborn to participate in the race. Eight countries were represented the first year, including the United States, Argentina, England, and Germany. Edsel gave permission to anchor a balloon over the Ford plant to display dates of the activities.[75]

For the first tour, airplanes were to take off from Ford Field and fly from Dearborn to Chicago, Des Moines, Omaha, Kansas City, St. Louis, Indianapolis, Cleveland, and back to Dearborn, a distance of approximately 1,800 miles. There were 12 entries in total, including a Stout-built Air Sedan. It had been raining for days prior to the start of the competition, but the rain stopped just prior to the planes taking off. Edsel, with flag in hand, and outfitted in raincoat and knee-high flying boots, circled the flag several times over his head, and then in a final swoosh downward, started the great race.[76]

Six days later, all but three of the planes returned safely to Dearborn, to a crowd of 35,000. The Stout plane was one of those that had successfully completed the long journey, but Tony Fokker, in his tri-motor, crossed the finish line first.

The six subsequent Ford Reliability Tours, as they were then known, were run in the same manner as the first, except that each year, the distance the planes traveled increased. In 1926, the route covered 2,600 miles and included Milwaukee. In 1927, the route was expanded eastward to encompass New York and Philadelphia. By 1928, the route completely circumnavigated the country, a distance of more than 6,000 miles.

The Air Tours were discontinued in 1931 because their purpose of proving the reliability of the airplane had been fulfilled. "In the five years after the tour began," wrote an aviation magazine, "commercial aviation grew from a dubious sport for daredevils, into an industry involving millions of dollars of capital. Pilots, instead of being objects of curiosity, became the envy of every small boy and the guardians of countless human lives as they ferried great air liners across hundreds of miles of scheduled daily flights. By the end of the Tours in 1931, the number of miles flown daily had grown to the amazing figure of 68,000."[77]

***

Shortly after receiving the airmail account from the government, Stout began hearing pilots' concerns about flying over mountainous terrain in Pennsylvania and New York. A one-engine plane, they said, left little margin for error.

This lesson had already been learned. About a year after the plane was built, Ross Kirkpatrick, a Ford pilot, crashed the *Maiden Dearborn* and was killed while attempting to land in Chicago. If it had been a two-engine plane, or even a three-engine plane, perhaps their good friend would not have been killed.[78]

Stout discussed this with Mayo, and Mayo brought the situation to Edsel's attention: "Colonel Henderson is very anxious for us to develop a safe plane for this [mountainous] work. He and all of his associates are thoroughly of the belief that a three-motored plane is the only safe job for this particular piece of...flying, and I think it is well that we develop this plane as quickly as possible. This fits in with the thought we have had on the most safe type of passenger plane, and I think I am safe in saying that at least 75 percent of all those who are giving the passenger business serious thought, are convinced that a three-motored plane is the best answer for safety."[79]

The immediate obstacle was what power plant to use. They had the powerful and reliable Liberty engine. However, as Edsel had mentioned earlier, the design of an airplane follows the power plant, and utilizing more than one Liberty engine would mean a much larger airplane than what was needed for air mail service.

Fortunately, the Wright Company had finally refined its radial engine and placed it into limited production. The *J4*, as it was called, was an engineering marvel: small in size, air cooled, and capable of producing 200 horsepower. One of these engines would not be sufficient to power the Ford plane, and two would only match the horsepower of the Liberty engine. Therefore, it was decided to incorporate three Wright engines into a new Ford plane—one engine on either wing, and one in the nose.[80] Three engines would give the plane the margin of safety that Ford was seeking.

Work progressed rapidly, and on November 25, 1925, Stout's first tri-motor airplane was rolled onto Ford Field for tests. Unlike his previous designs, however, Stout's new plane was a failure. With its stubby nose, the plane resembled a fat, ugly bird, and Henry Ford called it a monstrosity.[81] All three test pilots who flew the plane agreed that it had poor lift and inadequate reserve power. What was needed was a completely new design.

To say the least, Henry was not pleased with Stout's lack of progress. Henry had invested a lot of money in the project, and he wanted to cut his loses at that time rather than invest more in the newfangled venture. "I've bought a lemon," Henry said, "and I don't want the world to know about it."[82] Edsel quickly intervened. He suggested that Ford engineers be put in charge of the project, and he reassigned Harold Hicks from working on Henry's X-8 engine to begin working on a new airplane design. Edsel was aware of Hicks' talent because the two men had worked together on the Liberty engine during World War I. Thus, Edsel was confident that Hicks could come up with a suitable design. Henry agreed but told Edsel to "keep Stout out of the design room."[83] Henry had lost all faith in Stout, so much so, that he wanted all of Stout's work erased from his memory and from the prying eyes of the public.[84] Purportedly, Henry asked one of his henchmen to go to the airplane factory one night and remove all of Stout's engineering drawings and take them to the Ford Engineering Building.[85] The next night, the airplane factory went up in flames, destroying the entire factory, including the ill-fated tri-motor and three other Air Transports that were under construction. "I hurried over to the site the next morning and saw Mr. Henry Ford looking over the scene from his car," recalled Stout. "He called me over, noting my long face. 'Don't be discouraged,' he said. 'This is the best thing that could have happened. Now we can build the type of factory and hangar that we should have in the first place.' "[86]

The cause of the fire was never determined, but many thought it highly coincidental that an all-encompassing fire occurred so soon after Stout's failure. Although Henry kept his word about building a new airplane manufacturing plant, Stout's involvement was not in the picture.

During the next few months, the Fords built a 60,000-square-foot factory, and by June 1926, the new Hicks-designed Tri-motor was ready for testing. It was a resounding success, and the company soon was fulfilling orders from major companies and the government as well. Rodman Wanamaker, son of the department store magnate, purchased the first Ford-designed Tri-motor and placed it on sale in his department store. It quickly sold.[87] Soon afterward, Ford Tri-motors were sold to the U.S. Navy, U.S. Army, Standard Oil, Royal Typewriter, and Maddux Airlines in California.[88]

There was a lot of activity at Ford Field during this time, but it all came to a halt for a few hours in the summer of 1927. Three months after Charles Lindbergh made his successful solo flight over the Atlantic that May, he flew his *Spirit of St. Louis* to Ford Field as one of the stops on his cross-country celebration tour. Ford's entire executive entourage was present to meet the hero, and Lindbergh took several of them up for short flights around Dearborn. (Henry said that Lindbergh was the only man with whom he would fly.[89]) The little plane barely had enough room for one person, but Lindbergh managed to take up two passengers at a time. Edsel went up first, then Henry.[90] When they returned, Edsel had a new Ford Tri-motor all warmed up so that everyone could take a ride together. "We had Sorensen and Martin, Mrs. Dahlinger, and Henry and Edsel," recalled Ernest Liebold. "All the heads of the Ford Motor Company took a 20-minute ride over Detroit."[91]

***

By 1928, production of Ford Tri-motors was under full operation. Thirty-six planes were sold that year—a significant number for that time. It appeared that Edsel's dream was coming true. Things were going so well that Edsel authorized the purchase of some property and a manufacturing plant in Hammond, Indiana, to expand Tri-motor production.[92]

However, as Edsel's airplane division was taking hold, disaster struck on three fronts. Similar to the problems Edsel was facing with his banking interests, the Depression wreaked havoc on his aircraft division. Between 1925 when Edsel took over the Stout Metal Airplane Company, and the end of 1929 when the Depression hit, the division had sold more than 100 planes, making it one of the largest airplane manufacturers in the world. After the Depression, however, orders were canceled and sales fell precipitously.

At this same time, Mayo, who had kept a close watch on the development of the industry, told Edsel that William Boeing was making significant developments in a highly sophisticated monoplane of his own. Mayo had seen the plane with his own eyes and stated emphatically that if Ford did not pursue a new design to take the place of the Tri-motor, Boeing could put Ford Motor Company out of the airplane business. Edsel was in favor of starting a new project and again put Hicks in charge of designing a new plane.

Henry was losing patience with the step-by-step approach that Edsel and Mayo were taking. Everybody knew that airplane flight was possible, but in Henry's mind, the keys to *profitable* airplane manufacturing lie in passenger-carrying capability. Therefore, Henry ordered his engineers to make a radical jump and design a plane capable of carrying 100 people.

By February 1932, a colossus of a plane made its way onto Ford Field. Fashioned loosely on the well-designed Ford Tri-motor, Henry's monstrosity had a wingspan of 110 feet and an overall length of 80 feet. It had one 650-horsepower Hispano-Suiza engine mounted in each wing and a thunderous 1,000-horsepower Hispano-Suiza engine mounted, not on the nose of the craft, but on a pedestal on top of the center of the wing.[93] Even with its size and horsepower, the plane was capable of carrying only 40 passengers, not the 100 passengers for which Henry had hoped. "The plane was so poorly designed," recalled Mayo, "that it never left the ground."[94] Henry's experiment was scrapped a short time later, after hundreds of thousands of dollars had been invested. Ironically, Henry's direction of building a multi-passenger airplane was correct, but its implementation was executed hastily. This was not the first time that Henry spent a lot of time and resources on a project that ended at a dead end.

Another incident that placed a pall over the aircraft division was a fatal crash of a Ford Tri-motor, killing all aboard. A few months before Henry's monstrosity was wheeled onto the runway, a Tri-motor that had been converted for bomber use as a prototype for the U.S. Army crashed, killing its pilot LeRoy Manning and mechanic L.H. Garriot.[95] Henry Ford had never gotten over the tragic death of his good friend Harry Brooks and pilot Fitzpatrick. Now he had lost two other Ford employees. Those who were close to Henry knew that he felt their loss deeply.[96]

By that summer, things were beginning to look up again as Hicks began wind tunnel tests on his new design. It was almost a dead ringer for the DC-3 plane that Douglas ultimately produced. Had it gone into production, it might have kept Ford Motor Company in the forefront of airplane manufacturing. However, as the project was getting under way, Henry Ford halted it and ordered Sorensen to shut down the aircraft division entirely.[97] "Airplanes are no good," Henry said.[98]

***

There were many reasons that Henry abruptly ended Edsel's aircraft division. Ostensibly, it was because the division was not making money—after all, it had sold only 50 planes in the past three years. However, insiders knew the real reasons. The death of Brooks, Manning, and Garriot, to be sure, weighed heavily on Henry's mind, but the most important reason was that Henry simply wanted "to humiliate Edsel."[99] The Ford Tri-motor, although not selling well at the time, was a great success. If Henry had allowed Edsel and Hicks to develop their DC-3-type model into a full-size airplane, Ford Motor Company might have eclipsed Boeing and Douglas from taking over the manufacturing race. As it turned out, Boeing came out with the 247, the twin-engine monoplane that Mayo had seen earlier. Likewise, Douglas introduced the DC-2, the predecessor of the popular DC-3. Aerodynamically and acoustically, these planes were years ahead of the Tri-motor, and they quickly captured what remained of Ford's market.[100]

Henry was simply jealous of Edsel's success with the aircraft division, particular in light of his own failure with the 100-passenger plane. "I don't think the Depression played any part in the end of Ford's aviation industry," recalled Mayo years later. "I blame it mostly on Henry Ford."[101] Charles Lindbergh agreed. "Henry Ford never really wanted to go into the airplane business," he said.[102]

This was probably one of Henry's most contemptible and costly decisions. The company could have maintained the aircraft division through the Depression by reducing its workforce. Had it maintained its development work on Hicks' new design, perhaps a prototype could have been built during the slack time, and the division would have been ready for production with a truly competitive airplane by the time the Depression ended. The onslaught of World War II, which was not in the picture at the time, would have ensured that. Instead, Edsel's dream of making Detroit the airplane capitol of the world was thwarted by his father.

Nevertheless, Edsel made his mark in aviation history. "The greatest single thing I accomplished for aviation was getting Edsel Ford interested in it," recalled Stout in his autobiography. "His participation proved that flying had become something more than a harum-scarum, hair-brained venture. From that moment on, Wall Street and the country began to take aviation seriously, and it became a business, not a state fair exhibition. Edsel Ford's interest in aviation got America into the air industry."[103]

Chapter 17

The Byrd Expeditions

"So long as I live I shall always know that you have helped me more than anyone else."
Admiral Byrd to Edsel Ford[1]

Although Edsel Ford was an early proponent of the airplane, there were many others, of course. But there were few, particularly in the early 1920s, who thought that the airplane could be used in the Arctic regions of the world. One of those men was U.S. Navy Commander Richard Evelyn Byrd.

A graduate of Annapolis, Class of 1912, Byrd received an early retirement from the Navy in 1918 due to an injury. However, he was able to continue conducting Navy business. He met his calling one day in 1914 when he took his first airplane ride in one of the Curtiss flying boats that the Navy was testing. After that experience, Byrd was smitten by air travel. During the next three years, he learned how to fly and how to repair airplanes. He also studied meteorology and the theory of aeronautics (to the extent of what was known at the time). Byrd's greatest strength, however, lay in navigation. Over the years, he invented the bubble sextant, drift and wind indicator, and course and distance indicators to assist in airplane flight.

***

Ever since he was a boy, Byrd dreamed of adventure. He dreamed of being the first man to reach the North Pole and was disappointed to learn that Robert E. Peary had beat him to it. By the time Byrd became involved in airplanes, he quickly set his goal of being the first man to fly over the North Pole.

While perhaps not as romantic as being the first man to reach the North Pole on foot, it was nonetheless an ambitious goal. After all, aviation was in its infancy, and many thought that the severe conditions of the Arctic made it impossible for airplanes to function—airplanes were not doing well in mild climates. Byrd was determined to prove otherwise. He believed in aviation and realized early that for the industry to progress, it must be commercialized. For that to occur, air travel must be made safe and reliable, a point in sync with Edsel's views. In fact, Byrd was to become not so much an Arctic explorer as he was an aerial explorer—an aviation pioneer.

***

Ever vigilant of prospective contributors, Byrd had known about Edsel's interest in aviation for some time. He recalled seeing newspaper accounts of the Ford Air Races and of Edsel's contributions toward airplane safety. And, of course, there was Edsel's vast wealth. If Byrd could convince Edsel to contribute to his Arctic flight, it would be a boom to business. Therefore, Byrd headed to Dearborn to see Edsel.

Byrd did not go empty-handed, however. During his time in and around Washington, DC, Byrd had met many government dignitaries. Senator James Couzens was happy to give Byrd a letter of introduction.

"One morning, in the early spring of 1924, Byrd arrived at my office with a letter of introduction from a friend of mine in Washington," recalled Albert Lepine.[2]

"What's your business with Mr. Ford?" asked Edsel's secretary.

"The North Pole," replied Byrd.[3]

Byrd gambled on this direct, forceful approach. He had never met Edsel and obviously did not know his general distaste for boastful, pushy people. However, Lepine was impressed with Byrd's candor and his position in the Navy, and he suggested to Edsel that he meet Byrd. Soon Byrd was inside Edsel's spacious office at the Ford Administration Building, sitting at Edsel's desk and outlining his plan.[4]

"It was a big moment," Byrd wrote several years later. "Should Edsel Ford deny a subscription, there would be no 1926 expedition for me. There were then four or five North Pole expeditions underway. I laid my cards on the table."[5]

They met for 20 minutes—Edsel intently listening to Byrd's plan—and by the time their chance meeting ended, Edsel agreed to give Byrd $15,000 toward his North Pole expedition.[6] Edsel also mentioned to Byrd his involvement in the Stout Metal Airplane Company and that perhaps by the time Byrd was ready for his flight, he might use one of Stout's airplanes.

As Byrd was leaving, he confided to Edsel that if he crashed during the flight, it would mean sure bankruptcy, if not worse.

"What difference does that make?" replied Edsel, trying to alleviate Byrd's concern. "I'm simply betting on you, am I not?"[7]

With backing secured, Byrd immediately started to plan his expedition. However, he soon realized that the venture would cost him much more than he had anticipated. In December, he called on Edsel again. Edsel understood Byrd's plight and increased his support to $30,000, which was significant but still shy of the total amount that Byrd needed. With nothing to lose, Byrd asked Edsel if he could write a letter to his friend, John D. Rockefeller, Jr., to match his donation. "As the Expedition is going to be a very costly one," Byrd told Edsel, "it is vitally important to me that Mr. Rockefeller subscribe."[8]

Edsel was happy to comply and wrote to Rockefeller, asking him to support Byrd's expedition. "...it is a constructive venture inasmuch as anything that can be done to further aviation at the present time, and help formulate the basic idea in the minds of the public..." wrote Edsel. "In addition to the possibility of being the first nation to fly to the North Pole, Commander Byrd lays great stress on the point that there is a vast area to the westward of any of the former expeditions which could be taken in the name of the United States, and may turn out to be another large continent."[9]

Rockefeller quickly agreed, saying, "...if you fully approve it and would like to have me do so, I will be glad to join you in the enterprise and duplicate your contribution...."[10]

Subsequently, Rockefeller gained additional support for Byrd from Vincent Astor, and Edsel convinced the Fisher Brothers, wealthy automobile body manufacturers in Detroit, to also contribute to Byrd's expedition. Thus, Byrd obtained his money through Edsel's support and recommendations.[11]

"I do want you to know that I have not forgotten that I am looking to you as my Board of Control," Byrd wrote Edsel. "I have not made a single important decision without first putting it up to you. I should appreciate it very keenly if you would let me know if you do not approve of anything I am doing. I am doing what I think is best to bring success, but I owe what success I have had to you, and I want you to be pleased with the way I am playing the game."[12]

Edsel appreciated Byrd's humility and courage, and he offered him a new motorcar to show his admiration. "It would give me a great deal of pleasure if you would accept a Lincoln car from me as a token of my personal regard," he wrote. Byrd chose a seven-passenger sedan with channel cord upholstery and had the car painted maroon.[13]

***

With new funds in hand, Byrd began working on the Arctic expedition in earnest. By the end of 1925, he had all of his supplies and equipment secured, except for an airplane. He had visited various manufacturers and had inspected many used planes that could perform the task, but he waited until the last possible moment to make a decision, hoping to obtain a Ford plane. "If you think the plane you manufacture would be suitable for long flights in the Arctic, please let me know because I would, of course, much rather purchase it than any other," wrote Byrd. "I know you are on the right track in the development of your planes, but I am not sure that you have range enough yet for the project at hand."[14] While awaiting Edsel's reply, Byrd sent a telegram with more specifics[15]:

> *Would much prefer to use your three-motored plane than any other. Please notify me if you have one available to sell that will remain aloft with one motor dead. If so quote price. 1900 miles cruising radius advisable. Could plane be delivered in time? Leave states March 15th.*

Unfortunately, Byrd's request came at the time of the mysterious fire in the Ford aircraft plant. Edsel replied[16]:

> *Our experimental three-motored plane destroyed in recent airplane factory fire. Will not have another such ship ready for at least two months and it will be only experimental.*

Through no fault of his own, Byrd was forced to look elsewhere for a suitable plane for his expedition.

During this time of airplane development, few planes were being built that could meet Byrd's requirements of safety and reliability. His desire was to obtain a multi-engined plane, preferably one with three engines. Ford Motor Company was developing its tri-motor plane, and the German company, Fokker, had also developed a reliable tri-motor plane. In fact, during his search for suitable aircraft, Byrd came upon a used Fokker tri-motor that had been flown 15,000 miles—"long enough to get the kinks out," he recalled.[17] Because of Edsel's inability to provide a plane in time, Byrd was forced to obtain a tri-motored airplane from Edsel's fiercest competitor.

Indeed, Byrd found himself in an awkward position. Edsel was his single largest benefactor, but Byrd would have to use a Fokker airplane for his assault on the North Pole, rather than a Ford plane. However, Byrd had no choice. Naturally, Edsel understood Byrd's predicament and continued his support, nonetheless. "I believe your expedition will do a lot to increase popular interest in aviation in America," Edsel told Byrd.[18]

The least he could do, Byrd thought, was to name the expedition after his benefactor. However, on further consideration, Byrd felt that Edsel would not favor that. Thus, he asked Edsel to name the plane. "I want to offer you the opportunity of naming the aeroplane," wrote Byrd. "I shall be delighted to give it whatever name you may like it to have."[19] Edsel knew what Byrd was seeking, but he was not willing to allow *his* name to be put on the plane. Instead, Edsel suggested naming the plane after Admiral Robert Peary or Captain Robert Scott, two early Arctic explorers, or after Mrs. Calvin Coolidge, the president's wife.[20] Instead, Byrd decided to name the plane himself and appropriately named it *Josephine Ford* after Edsel's only daughter.[21] He sent instruction to the Fokker factory to have the name "Josephine Ford" emblazoned on both sides of the fuselage.[22] Anthony Fokker complied, but only after he had "written his name all over the plane."[23] "Fokker" was painted on both sides of the fuselage, in two places on the leading edge of the wing, and prominently on the underside of the wing, with "FOK" on the starboard side and "KER" on the port side.

***

With a ship borrowed from the Navy, Byrd set off from New York on April 5, 1926, bound for Spitzbergen, a city on the northern end of Greenland, only 750 miles from the North Pole. When the ship arrived three weeks later, Byrd found Roald Amundsen and his team well under way in their attempt to fly a dirigible to the North Pole. Therefore, Byrd could not waste time in preparing for his attempt.

Byrd's men retrieved the plane from the ship, assembled it on shore, and made a couple of test flights to ensure that everything was working properly. During the last flight, Byrd found that the engines consumed less fuel than expected, giving him confidence that he could fly to the North Pole and back in one non-stop flight.

On May 9, the U.S. Weather Bureau gave Byrd a satisfactory weather forecast. His crew filled the plane with fuel and food. Byrd and his mechanic, Floyd Bennett, climbed into the plane and took off, setting a course directly for the North Pole.

Eight and one-half hours later, Byrd and Bennett were flying over the northern apex of the world. They flew over the North Pole for 13 minutes and then headed back to Spitzbergen.[24] When they landed, Captain Amundsen was there to greet them.

Both Byrd and Bennett were exhausted from the trip. They had been awake for almost 36 hours straight. However, before Byrd went to bed, he sent a telegram to Edsel. "We report to you that wc have reached the pole..."[25]

***

Byrd was given a hero's welcome when he returned to New York, but his excitement was dampened when he learned that he was nearly $30,000 in debt.[26] Earlier, he had set his minimum financial requirements at $140,000, and he thought he had commitments for more than that amount. He was disheartened to learn that he still owed so much.

However, Byrd realized that he could never have made the trip without Edsel's support and never failed to give credit to Edsel for enabling him to carry out his expedition. As an act of appreciation, Byrd offered Edsel the propeller from the center engine of the *Josephine Ford*, a gesture that was very touching to Edsel.[27] However, Edsel had his eye on more than one of the propellers from the plane. He wanted the entire airplane.

Byrd had mentioned to Edsel earlier that he had been contacted by several museums throughout the country, asking that the historic plane be reposited in one of their institutions. Edsel decided to make a bid for the plane, too. "With regard to the finial disposition of the *Josephine Ford*," wrote Edsel, "I am thinking of the possibility of repurchasing this plane from the expedition and putting it on a permanent display in our proposed new Museum of Transportation. Give me your opinion as to what the plane would be worth in its present condition, and its value from a historical standpoint. I would like to suggest the figure of $30,000, which I believe is a generous price, not considering the historical value of the plane. From that standpoint it is invaluable and impossible for me to estimate in any way."[28] This offer was more than generous, considering that Byrd had paid only $15,000 for the plane a few months ago.

Byrd was more than happy to accept Edsel's offer. The sum Edsel was willing to pay for the plane was significantly more than he would have received from any government agency or private museum. "I am writing [to] the National Museum and others who have wanted the *Josephine Ford*," responded Byrd, "that I cannot place it with them." [29] With that, the historic plane became one of the first acquisitions placed in the Edison Institute—Henry Ford's personal museum of industrial artifacts—when it opened in 1929.

Where Edsel came up with the $30,000 figure is unknown. However, it may have been more than a coincidence that it was the exact amount needed by Byrd to eliminate his debt.

Before giving the *Josephine Ford* to Edsel, Byrd and Bennett took it on a cross-country tour—not to raise money but to promote aviation. They visited 50 cities, showing the plane and touting the advancements in air travel. At the end of each talk, Byrd showed pictures of his benefactors and publicly thanked them for their support. "You received considerable applause in Chicago when your picture was thrown on the screen," Byrd wrote to Edsel. "In fact, you do everywhere. People seem to appreciate what you did for the expedition."[30]

Because of his desire for anonymity, Edsel was probably not thrilled to learn that his picture was paraded in every major city of the nation. Nevertheless, he allowed Byrd this liberty because of Byrd's great accomplishment. "I am sure that the trip of the plane around the country has done a great amount of good for the commercial side of flying," replied Edsel, "and your never ceasing interest in this great new mode of transportation is very inspiring."[31]

***

After his successful flight over the North Pole, Byrd now set his sites on Antarctica and the South Pole. Similar to Byrd's northern expedition, the South Pole had already been conquered. Captain Roald Amundsen, a Norwegian explorer, was first to reach the South Pole in 1911. By 1928, nobody had attempted to reach the South Pole by aircraft. Byrd wanted to add this accomplishment to his personal list of "firsts," and the first person whom he contacted for funds was his good friend, Edsel Ford. "He was near my own age," recalled Byrd, "a big man with ideals...who thought in terms of America's tomorrow."[32] Byrd hoped to capitalize on those ideals.

Byrd spent most of one day in early August with both Edsel and Henry Ford, explaining his anticipated trip to the South Pole, and the Fords showed Byrd their latest activities in aircraft production.[33] With its new factory now in operation, the Ford Aircraft Division was now producing four Tri-motor airplanes a month.[34] Byrd was impressed with the operation.

Although the Fokker airplane served him well on his trip to the North Pole, Byrd desperately wanted a Ford airplane for his South Pole expedition. "I will use your three-motored plane or none," Byrd told Edsel.[35] Byrd was genuinely impressed with the Ford Tri-motor, but the probability of obtaining one gratis made it all the more appealing.

Within the month, Byrd and Edsel had reached an agreement as to how much Edsel would contribute to Byrd's second Polar expedition. While the expedition to the North Pole was expensive, it paled in comparison to the expense to conquer the South Pole. Regardless, Edsel came through with funds, machinery, and solicitations. In the end, he would donate a total of $90,000 in cash and equipment toward Byrd's anticipated expenditure of $750,000, and again Edsel was able to convince both Rockefeller and the Fisher brothers to contribute $50,000 each.[36] "You are a peach to write that letter to Mr. Rockefeller," said Byrd. "I believe there is absolutely no doubt now that he will subscribe. You certainly have stuck by me."[37] In terms of equipment, Edsel gave Byrd two Fordson tractors; a sturdy snowmobile, made from a new Model A chassis, which had skis on the front and a caterpillar track on the back; and a Ford Tri-motor aircraft, built especially for Byrd's special needs.[38] "Changes were made in the Ford Tri-motor to enable it to fly over the Pole," recalled Harold Hicks, the chief engineer of the plane. "It was a skinned-down fuselage. In other words, we had nothing in there but the bare necessities. Our big battle was to try and get weight out. For that reason, we put in such things as celluloid windows in the place of glass windows, and we had lighter gauge skin on the fuselage and on certain sections of the wing. We had trouble preparing the plane for cold weather. The oil tanks were particularly large, due to the extent of the flight. Special provisions had to be made for jacketing the oil tank so it wouldn't be chilled so much to starve the engines. Then the engines had to be completely jacketed for stoves in order to heat them up to enable them to get them started."[39]

It was a beautiful airplane, and Byrd christened it *Floyd Bennett*, in honor of his stalwart companion who flew with him to the North Pole, only to be felled by influenza in the spring of 1928.[40] Edsel thought it proper to name the plane after Byrd's fallen comrade and had *Floyd Bennett* painted in bold letters on both sides of the fuselage of the plane.[41] Byrd offered to pay for the plane, but Edsel gave it to him on the condition that he be allowed to purchase the plane when Byrd returned.[42]

The news hit the papers immediately: "BYRD'S POLAR FLIGHT BACKED BY EDSEL FORD," wrote the *Detroit Free Press*. "Edsel B. Ford will back Commander Richard E. Byrd in his flight to the South Pole 'just because he is a scientist and because that is a great service which he can do for aviation and the world.' "[43]

"Byrd is a great fellow," said Edsel. "He is a gentleman, a scientist and altogether likable. I enjoy being behind him in such enterprises. It is not for any personal interest that I am helping

Byrd, although I do like to live the adventures he has when we can get together and talk them over. It is principally because he is doing a service to aviation and that is all we are trying to do with our aviation enterprise at Dearborn."[44]

During the time that transpired between his North Pole trip and his Antarctic trip, Byrd wrote a book about his earlier exploits. *Skyward* was published in the summer of 1928, and Edsel was among the people to whom Byrd dedicated his book.

***

In mid-October 1928, Byrd and his crew of 100 men, two ships, two snowmobiles, a single-engined Fokker, one autogyro, and the *Floyd Bennett* set sail for Antarctica. This expedition was to be a much more significant and arduous task than Byrd's North Pole flight. To reach the South Pole, Byrd had to set up a base—which he called Little America—on the southernmost continent, and he awaited favorable weather to attempt a flight over the South Pole. It turned out to be a grueling 24-month endeavor. In comparison, Byrd's North Pole expedition took only three months to complete.

Much had to be done before Byrd could take the Ford plane over the South Pole. A base camp had to be built, many experiments had to be conducted, and the airplane had to be tested in the frigid environment. Finally, in November 1929, the *Floyd Bennett* was taken up for its first test flight. It was loaded with six tons of snow blocks (to simulate actual provisions) and taken up to 10,000 feet. The port engine began to cough, but it was only a minor problem, and the mechanics corrected it in short order. Soon, the Ford Tri-motor was making satisfactory flights at 12,000 feet, the altitude they needed to attain to safely clear the mountain range in the path to the South Pole.

Later that month, Byrd and his crew flew the plane over 400 miles south of Little America to set up a secondary base camp. Byrd had determined that it would be too dangerous to attempt a nonstop flight from Little America directly to the South Pole and back. Therefore, they established this provisional camp, where provisions and gasoline were stored.

On Thanksgiving Day, Byrd received word that the weather was as good as it would ever be, so he told his men to get ready. They loaded the plane with provisions, filled it with fuel, and took off into a 20-mile-an-hour headwind.

About 10 hours later, Byrd and his crew were flying in Edsel's airplane over the southern geographic center of the world. After making several passes in different directions to ensure that they had indeed passed over the southern axis, Byrd ordered the pilot to return to base camp.

With significantly less weight than when they had started, and with a welcomed tailwind to speed their return trip, they landed the Ford plane at the provisional camp five hours later. They rested for only a short while, then refueled the plane and took off again. By ten o'clock, Byrd and his crew had returned safely to Little America.

The next day, Byrd could not wait to notify Edsel of his accomplishment and tell him how well the Ford airplane performed. "It's a great plane," telegraphed Byrd.[45]

Edsel was ecstatic. "I have been worried over the possible failure of the plane," Edsel replied, "and am so glad to know that the flight has been made, and with great success."[46]

Edsel was unable to collect on Byrd's promise of returning the *Floyd Bennett* to him after the expedition. The weather in Antarctica was unpredictable at most times of the year, but during the Antarctic summer of 1929–1930, the ice pack engulfing the Bay of Whales, the entry point for Byrd's ships to enter and dock, refused to subside. Byrd had hoped to be out by February 1, 1930, because in previous years, his ships were able to enter the harbor in early January. However, this year, too much ice remained for his ships to risk entry. They waited at the end of the ice pack until a time arose where they could make entry. Based on this, Byrd made plans for a quick escape. He and his men divided all of the materials at Little America into A, B, and C classes, with A material certain to go home.[47] With this system, Byrd was forced to leave all three planes at Little America. His crew secured the planes atop a windswept ridge facing into the prevailing winds, with the hope that they would not be buried by snow for several years and with the promise that they would return for them in the not-so-distant future. Also left behind were the Ford snowmobiles and other equipment Edsel had given the expedition.[48]

Byrd was heartbroken that he was unable to bring back the *Floyd Bennett*, but he did bring Edsel a pair of sealskin gloves, a handful of Eider feathers, and a ptarmigan skin as souvenirs.[49] He had wanted to bring back a couple of penguins. "They are quite tame and interesting," Byrd telegraphed. "I believe you and the youngsters would get a kick out of the little fellows." Edsel politely declined.[50] His most significant gesture of gratitude was his naming a geological landmark, heretofore undiscovered, after his magnanimous benefactor. Byrd named a range of mountains running at least 250 miles north and south and approximately 300 miles due east of Little America as the Edsel Ford Range. "How much further it goes," Byrd said, "I don't know."[51]

***

While Byrd and his staff made many discoveries during their extended expedition to Antarctica, much remained to be done. He wanted to return to the frozen south, Byrd told Edsel, to conduct more scientific research, perhaps to discover precious natural resources, gain more power and prestige for the United States, and because exploring was in his blood.[52] This had become Byrd's new mission, and he immediately began making plans—and soliciting support—for a second Antarctic expedition. Unfortunately, Byrd left on his first South Pole trip almost a year to the day before the onslaught of the Great Depression and returned when the country was sliding precipitously into its quagmire. As Byrd was making plans for his second trip, things grew worse and worse. It was not the time to be asking for money or anything else, particularly for something that had already been accomplished.

With some reservation, Edsel again agreed to support his old friend on another mission, and again Edsel solicited the assistance of John D. Rockefeller, Jr. Similar to the previous two expeditions, which became more and more expensive, the second Antarctic expedition was to cost much more than $1 million. Edsel contributed $100,000 in cash and two new half-track Ford snowmobiles. With the cessation of airplane manufacturing at Ford, Edsel had no aircraft to give to Byrd, which greatly disappointed both men. "If I had the whole world to select from for a plane for the Antarctic," Byrd told Edsel, "I would take your stock three-engine Ford."[53] In the end, Byrd had to manage without a Ford-built aircraft, but he scrounged four other aircraft for the expedition: a large two-engined Curtiss Wright, a Kellett autogyro (an early rendition of a helicopter), a single-engine Fokker, and a small single-engine plane.[54]

Byrd still hoped he would be able to salvage the three planes that he had left in Antarctica and bring them back, at least the *Floyd Bennett*. Edsel was interested in retrieving the historic tri-motor for The Edison Institute, but he was pragmatic in the possibility of that happening, contrary to what Byrd had told him. "I hope you do not take too seriously your promise to bring back the Ford plane from Little America. It would be fine if you can do so, but I fully realize the limitations on your capacity and the lack of knowledge as to the condition you will find the plane in when you arrive down there. I hope that the ship will be in condition to fly, as that would be such a great victory for an all-metal ship. I feel that you would be taking altogether too many risks in not taking a plane down with you for such use as you plan around the South Pole. It seems to me an unnecessary hazard to depend on the Ford plane."[55]

Byrd needed more—more cash, more equipment, more supplies—but he was reluctant to ask Edsel for additional assistance after all Edsel had already done. In July 1933, however, Edsel received a letter from Victor Czegka, general manager of Byrd's Antarctic Expedition II. "Commander Byrd made it clear to me that he is undoubtedly in desperate need of funds, and I am taking the liberty of addressing these few lines to you."[56] As an enticement, Czegka explained, Byrd would name a ship that he would take on the expedition "after someone who would contribute a substantial sum towards his enterprise."[57] Unfortunately, Edsel was in the throes of his banking fiasco at this time and was in no mood to extend more support to Byrd's self-centered exploration. In the margin of Czegka's letter, he wrote: "Not interested—have done all I can."[58]

After learning what Czegka had done, Byrd wrote a letter to Edsel, apologizing for his employee's error in judgment. "First I want to give you my word of honor that I knew absolutely nothing about this, and I certainly would have disapproved it had I known. The sort of proposition he made you, as I understand it, is close to being an insult, for which I apologize for him and express my sincere regret that anyone representing me would do such a thing."[59]

Nothing more was said between the two men for quite some time. In fact, Byrd left on his expedition without asking for further support from Edsel. To make this trip, Byrd secured funds by lending his name to companies that supplied him with goods and services and by government backing.

However, why go in the first place? Byrd had already achieved fame by founding new territory for the United States and for being the first man to fly over the South Pole. Byrd said that there was much more exploring to do in this vast continent—perhaps there were untapped resources of precious minerals, gold, oil, or diamonds.[60]

Byrd left on his second Antarctic expedition in October 1933 and spent two years on the frozen continent—seven months of it in complete isolation in a shack in the far interior of Antarctica. The initial plan was to send three men to a base in the interior of the continent to monitor weather conditions and compare the data with meteorological information gathered at Little America. However, their arctic tractor broke down, cutting the party's hauling power in half, which prevented them from supplying the outback base with enough supplies for three men for half a year. Byrd felt it would be better that only one man go alone rather than two men—they might end up murdering one another.[61] Thus, Byrd manned the post alone for seven months—four months of it in total darkness.[62]

With his expedition complete, Byrd and his men loaded the ships and left Antarctica on February 7, 1935. This time, the weather was cooperative, and Byrd and his men had plenty of time to load the ships.

Surprisingly, Byrd was able to retrieve the *Floyd Bennett* and return it to Dearborn. The trusty Ford Tri-motor was covered with snow and ice when he found it, and it was not usable during his stay. However, the plane was intact and worthy of bringing home.

Edsel was thrilled to have the *Floyd Bennett*, but he was through with Byrd and through with becoming personally involved in philanthropic activities, no matter how exciting they seemed. "I tried to get you on the phone," Byrd wrote, "but your assistants would not tell me where you were."[63] A testimonial dinner was held on behalf of Byrd at the Waldorf-Astoria Hotel a few years later. Although Edsel was invited, he did not attend.[64]

Chapter 18

The Lincoln Motor Car Company

"Father made the most popular car in the world, I would like to make the best car in the world."
Edsel Ford[1]

Henry M. Leland was a master machinist. Born in Vermont in 1843, he worked as a toolmaker at the Colt firearms factory in Hartford, Connecticut, and at the Brown and Sharpe machine tool company in Providence, Rhode Island. In 1890, he moved to Detroit. With his son Wilfred and two investors, he opened his own machine shop, called Leland, Faulconer & Norton.[2] No job was too large or too small for the new company. They built everything from gears for bicycles to large machine tools. At the turn of the century, they began building parts for the budding automobile industry, initially supplying transmissions and engines for Ransom E. Olds.[3] Leland's greatest contribution to the automobile industry was that he proved that automobile parts could be made to exacting tolerances, paving the way for automobile mass production.

Around this time, the stockholders of the Henry Ford Company, frustrated over Henry Ford's lackadaisical progress in producing a saleable motorcar, contacted Leland to analyze the situation. This, as we have mentioned, caused Henry Ford to leave the company. With his departure, the name of the company was changed to the Cadillac Motor Car Company.

Eight years later, the man who had been building horse-drawn carriages in Flint at the time of Edsel's birth was in the throes of creating the world's largest automobile company by assembling a conglomerate of smaller companies. Thus, he came courting Leland's Cadillac company. Billy Durant would not take "No" for an answer, and his $4.5 million offer was too good to refuse. Therefore, the Lelands sold Cadillac to Durant.

Although the money was nice, the part of the deal that the Lelands liked best was that Durant allowed them to remain with the company and run it on his behalf. This allowed Durant the freedom to seek and buy other companies, while allowing the Lelands to continue working in the industry they loved. Under their aegis, Cadillac introduced the first electric starter, which eliminated the need to hand-crank the engine, and the first 90-degree V-8 engine. When war broke out in Europe in 1917, the Lelands offered to build Liberty engines for the government, but Durant refused to give them permission. Therefore, the Lelands quit General Motors and, on a shoestring, started the Lincoln Motor Car Company in August 1917 to build the Liberty engines.[4]

Throughout history, business founders frequently named their businesses after themselves. Thus, it would not have been a surprise if Leland had named his new company after himself. However, he did not. Leland's boyhood hero was Abraham Lincoln (he voted for Lincoln in the 1864 election), and thus he named his new company after the "Emancipation President."[5] Leland never used Abraham Lincoln's image on his products or on his letterhead, but he did have a large statue of Lincoln erected in front of the plant.

***

Although they had no plant nor employees, Henry and Wilfred Leland were able to gain an order for 6,000 Liberty engines from the U.S. government, based solely on their reputation for building quality products. To get the new company up and running quickly, they recruited many of their former Cadillac employees. With a $10 million advance from the government and the sale of $2 million in Lincoln stock, they built a sizeable plant on the corner of Warren and Livernois avenues in Detroit. Then they began manufacturing precision-built Liberty engines.

During the next two years, the Lelands built the plant and successfully fulfilled their contract with the government. After the war ended in 1918, the government cancelled the contract with Lincoln, and the Lelands were left with an idle plant and considerable debt.

At this point, the Lelands decided to return to the motorcar business. On January 26, 1920, they formed a new company of the same name and issued new stock that brought in more than $6 million. Their first Lincoln motorcar rolled off the assembly line in September.[6]

Business took off immediately, and they had 1,000 orders before the first car had been built. However, sales took a precipitous drop from there. In the summer of 1920, the country fell into a severe recession, and the sale of motorcars dropped. Henry Ford felt the pinch with his inexpensive Model T—the Lelands were in dire straits with their expensive Lincoln. By the early part of 1921, the Lincoln Motor Car Company was forced into receivership.[7]

***

The Lelands were not going to give up easily, however. They hoped to either obtain a loan with which they could buy back the company at the receiver's sale, or have someone else buy the company and allow them to manage it, similar to what they had done with Durant. In both instances, they were not being realistic.

With Henry and Edsel's command of the motorcar industry at this time, they must have been aware of the Lelands' plight long before it became public knowledge. Regardless, the Lelands approached the Fords first. Late in October 1921, the Lelands had their attorney, Harold Emmons, approach William Mayo to test the waters. Emmons explained the situation to Mayo and asked him to see if the Fords would be interested in making a loan to the Lelands.[8]

After Mayo informed Henry Ford of the Lelands' request, Henry Ford invited Henry and Wilfred Leland to Fair Lane to discuss the situation. Initially, Henry Ford did not have much interest in acquiring the company, but Edsel did. He saw this as an opportunity to acquire a company that was capable of producing quality automobiles, and he discussed the possibility of buying the company with his father "three or four times."[9]

The day of the meeting, Clara Ford had received a letter from Mrs. Wilfred Leland, asking for aid. Clara had overheard much of the meeting between the Lelands and her husband and son, and it became clear to her that her husband was not interested. Therefore, Clara prevailed upon Edsel. After the Lelands had left, Clara pulled Edsel aside and told him about the letter she had received. "Can't you do something to help them?" she pleaded, as only a mother could. "It's a shame that all Detroit should stand by and see that company wrecked."[10]

This was the break that Edsel was seeking. He wanted Lincoln badly, but his father was reluctant. Now, Edsel had his mother's support, and Henry rarely went against his wife's wishes. The following day, Henry invited Edsel, Mayo, and Sorensen to dinner to discuss the Lelands' proposal at length.[11] Purchasing the Lincoln Motor Car Company, probably at a rock-bottom price, was an "unusual opportunity," they all told Henry. No decision was made at this time, but shortly thereafter, Edsel met with his father, and this time, Henry told Edsel to purchase the company.

***

Edsel had several meetings with the Lelands at their office downtown in the Dime Bank Building to work out the details, and then all three met at the home of Judge Arthur Tulle, who had authority to act on behalf of the receivership. Using the skills he had honed during the buyout of Ford stockholders two years earlier, Edsel first gave them a cut-rate offer of $5 million.

"The judge felt that was not sufficient," Edsel recalled later. Nor was he patient. Judge Tuttle did not want the negotiations to drag out indefinitely, so he quickly made a counter-offer. "If you will pay $8 million for the plant and equipment, and promise to pay off all of the outstanding creditors of the company," the judge said, "I will call for a sale immediately."

"That will be satisfactory," replied Edsel.

"It is my understanding that you and your father will pay every Lincoln creditor 100 cents on the dollar," confirmed Judge Tuttle.

"Yes, that is correct."

"But it will take more than $8 million to do this."

"Yes, we understand that," replied Edsel.[12]

Although the Lelands were present through these negotiations and heard every word that was spoken, they later said that the Fords also agreed to pay back all of the Lincoln stockholders—an accusation Edsel vehemently denied.

That conflict would raise its ugly head later. For the moment, the Lelands and the Fords were the best of friends.

***

The auction—which was merely a formality—was conducted on the morning of February 4, 1922, on the front steps of the Lincoln Administration Building. A large throng had gathered, mostly Lincoln workers, to show their support of the men who would save the company and their jobs. Henry and Wilfred Leland arrived first, then Henry and Clara Ford, and finally Edsel and Eleanor, the younger Fords driving up in a custom-bodied Lincoln sedan. They all stood outside the Lincoln offices for photographs.

The proceedings started at precisely 10:00 o'clock. Harold Emmons, representing the Fords, immediately bid $8 million, the previously agreed to price. There were no other bids. After the obligatory last chance was offered, the gavel came down. "Sold to Mr. Emmons!" The crowd let out a loud cheer, and automobile horns began blowing...[13]

More photographs were taken, and then the Fords and the Lelands went inside to sign the contract. With Edsel and Wilfred sitting at a small table, and the two Henrys standing behind them, ownership of the Lincoln Motor Car Company passed from the Lelands to the Fords. Wilfred signed on behalf of the Lincoln Motor Car Company, and Edsel, as president of the Ford Motor Company, affixed his signature to the contract. The entire ceremony lasted only a few minutes. Then the two families stood up, walked out of the Lincoln offices, got into awaiting Lincoln motorcars, and drove away.[14]

***

It seemed as if it were a marriage made in heaven. Although the Fords now owned Lincoln Motor Car Company outright, they retained the original management structure as part of the agreement with the Lelands. Henry Leland remained president, and Wilfred Leland was vice president. To represent the Ford interests, Edsel was appointed second vice president and Bert Craig was appointed treasurer.

The industry was also relieved that Lincoln had been saved. "We are very glad that you have the Lincoln car, and that this fine car is still to be built and sold," wrote Hermann Brunn to Edsel. "They cannot fail to be a large and profitable business."[15]

Unfortunately, the honeymoon did not last long, and operations at the reorganized Lincoln Motor Car Company quickly deteriorated. The first disagreement occurred when the Fords tried implementing proven Ford manufacturing techniques to reduce costs and increase production, and to expand sales through established Ford dealers.[16] The Lelands thought that the quality of their cars was more than adequate and that selling Lincoln motorcars through Ford agencies was similar to selling expensive jewelry in a general store.

By the middle of May, it became evident to the Lelands that their sovereignty in running Lincoln was being restricted more and more each day. Sorensen and his manufacturing men invaded the company and began changing production methods and material specifications. The Lelands protested, but they were ignored.

Frustrated, and sure that things would only become worse, Wilfred Leland wrote a letter to Henry Ford at the end of May, saying that he and his father wished to purchase the company back from him. "While with all your vast interests the Lincoln may not be of great importance and may easily be swallowed up within the Ford organization," wrote Wilfred, "we assure you that with us it is the most important thing in the world."[17]

Henry never replied to Wilfred's letter, or to two others that Wilfred subsequently sent. Likewise, when Wilfred tried to contact Henry by telephone at Fair Lane or at the plant, Henry either refused to answer the phone or told his secretary to tell Wilfred that he was not in the office.[18] Realizing that he could not play opossum forever, Henry finally took one of Wilfred's calls and told him to meet him at Fair Lane to discuss the matter. "Drive through the rear gate," Henry said.

Wilfred did as he was instructed, but when he approached the mansion, he was confronted by one of the guards "who pointed a rifle at me and ordered me to stop," Wilfred recalled.[19]

Regardless of what Wilfred told the guard, the guard would not let him pass. As voices began to rise, Henry called the guard to let the young Leland pass.

Surprisingly, Henry listened to Wilfred for quite some time but eventually put an end to the discussion and to any hope the Lelands had in reacquiring the company. "Mr. Leland," bellowed Henry, "I wouldn't sell the Lincoln plant for $500 million! I had a purpose in acquiring that plant, and I wouldn't think of letting it go!"[20]

Shortly thereafter, both Henry and Wilfred Leland were fired.[21]

Henry Ford's purpose in obtaining the Lincoln Motor Car Company in the first place was not to help the Lelands, but to "give his boy something to play with—to keep him off his back," recalled a Ford executive.[22] However, Edsel wanted the company as much as Henry wanted him to have it. Although he was president of Ford Motor Company, Edsel was already feeling resistance from his father in allowing him to run the company as he saw fit. As Edsel saw it, Lincoln could become his own little fiefdom to run as he pleased.

***

Soon after the Fords had taken over Lincoln, both Henry and Edsel began receiving letters from the old Lincoln stockholders who were asking for the return of their money. So great was the influx of letters that Edsel had Lepine reply by form letters to the stockholders, saying that the company had not worked out any plan for taking care of the previous owners in the Lincoln Motor Company. Edsel hoped that the previous stockholders would eventually give up their quest for reimbursement. That may have happened, but then Henry Ford made a major blunder.

During a celebration at the Lincoln factory for Henry Leland's seventy-ninth birthday, Henry Ford presented the elder Leland, in front of hundreds of Lincoln employees and former stockholders, with a check for $363,000. "Show it to them, Mr. Leland, show it to them," exclaimed Henry Ford. "Tell them that they are all going to get them just the same." There is some question as to what the check was for. Leland thought it was for his Lincoln stock; Ford said it was a bonus for starting up Lincoln production under Ford ownership. Obviously, the others thought that they also would soon be paid for their old stock.[23]

Edsel was furious. His father had done his share of upsetting things over the years, but nothing as blatant at this. If Edsel had been present, he would have been quick to deny that the check was in any way related to Leland's old stock. Later, Edsel issued a statement, saying that the check was for services rendered by Leland during the transition period.[24]

Meanwhile, the Fords had made good on their promise of paying all of Leland's creditors. By March 1923, a little more than a year after purchasing the company, the Fords had paid out more than $4 million to creditors of the old Lincoln Motor Car Company. As far as Edsel was concerned, this fulfilled their legal obligation.

The now ousted Lelands felt differently. They believed that the Fords still owed payment to the former stockholders. Wilfred, in particular, took up the fight, and on March 19, 1924, he issued a public letter to Henry Ford in *Detroit Saturday Night*, hoping that public humiliation would force Ford to act. When that did not work, Wilfred and 1,800 stockholders filed suit. The suit asked for repayment in full to all stockholders of the old Lincoln Motor Car Company.

After more than three years of litigation, the Lelands lost their fight in getting the Fords to reimburse the stockholders in the original Lincoln Motor Car Company. In two trips to the Supreme Court of Michigan, the court repeatedly found in the Fords' favor. The Lelands failed to prove that the Fords had promised to pay the Lincoln stockholders as part of the purchase agreement.

Henry Leland died a few years later, in 1932, fully aware that many men had lost money in his venture.

***

Even at $12 million, everyone, including the Fords, thought that the Lincoln Motor Car Company purchase was made at a bargain price for such a high-quality plant and nameplate. Regardless, when the Fords took over the company, it was far from being a going concern and desperately needed resuscitation. "It was shut down tight and gave the appearance of a very stagnant organization," recalled Edsel. "At the time of the receiver's sale, there were 241 finished cars on hand and the price on most of these models had to be cut in order that they might be moved, the reason for this being that the body types were obsolete and old-fashioned. We found that the selling organization was stagnant, and that while it did have good machinery and a good nucleus of a manufacturing organization, it required a lot of bolstering up before it produced quality cars on any sort of economical basis."[25]

To revive the company, Edsel ordered an expansion of the facility and spent hundreds of thousands of dollars on new machinery. Soon, he had the plant producing 50 finely crafted Lincoln motorcars per day.[26] Even then, it was not enough to satisfy the demand for Edsel's stylish Lincolns. Gaston Plaintiff, the Ford branch manager in New York City and a close friend of the Ford family, wrote to Edsel to plead for more cars: "The Limousine is really our biggest seller, and we are not receiving anywhere near what we require from the factory; in fact, we are so short that we can't make deliveries of this model. Won't you see what you can do about helping us out?"[27] Plaintiff also cautioned that both Packard and Cadillac had built beautiful salons in the city and that every effort should be made to ensure that Lincoln quality is maintained.[28]

As Plaintiff would soon realize, Edsel would do more to boost the prestige and sales of Lincolns than he could ever have imagined. Edsel would do this by tapping into his creative side, adding stylish bodies to the well-built Lincoln motorcars.

Chapter 19

Carrossiers

"The motor body designer adopted streamlining not... as an aid to speed and economy, but as a beauty factor."
Edsel Ford[1]

When the Fords took over the Lincoln Motor Car Company from Henry and Wilfred Leland in 1922, no one could fault the mechanical integrity of the motorcars they built. The Lelands had always produced quality automobiles, and by the early 1920s, their company was producing almost perfect machines. Their penchant for workmanship was clearly visible in the finish of the engine, the part essential to the motorcar, but rarely subjected to aesthetic attention. First, they had the outside of the engine blocks and heads polished, then beautifully painted, which made some sense, because those parts would be visible occasionally. However, what made no sense was that they also insisted on making the *inside* of the engine as attractive as the outside. They actually ordered their journeymen to machine and polish the crankshaft and connecting rods all over, rather than leaving them partly rough, even though such extremes did nothing to the affect the performance or quality of the engine.[2]

Through such dogged insistence, the Lelands achieved their goal of mechanical perfection. Unfortunately, it came at the expense of aesthetics. According to many designers at the time—and apparently by many others, as evidenced by the dismal sales figures for Lincoln—the Lincoln motorcar was nothing more than a "box on wheels."[3] At the time, Lincolns had power (all had large V-8 engines), speed (most could easily cruise at 75 miles per hour all day long), and quality (each vehicle was exhaustively tested before it left the factory). However, the cars desperately lacked aesthetic appeal—"window dressing," as the designers liked to call it.[4] In their blind pursuit of quality, the Lelands simply over-emphasized the non-visible features of their fine automobiles and under-emphasized the visible. That is what eventually caused Lincoln customers to drift to other makes,[5] and that was the primary reason Edsel was able to acquire the company at below market value.

***

Although the Lelands were zealous in their pursuit of quality, Edsel understood that quality was the foundation of any fine motorcar, and he made sure that the Lincoln quality was not compromised when he took over the company. In fact, this was one area where Edsel actually took cues from his father and instructed Lincoln engineers to design in quality only where it made sense (and cents) and only where it counted in the eye of the consumer. "There was absolutely no question of cheapening the quality of the Lincoln car," recalled Harold Robinson, Lincoln plant manager. "In fact, I know that Edsel Ford insisted that we hold our limits closer so that the Lincoln car would be the finest precision-built car that there was. He ordered that the precision be maintained and even increased."[6] Edsel stopped the unnecessary machining of internal engine parts immediately, and he implemented other Ford-proven methods that reduced costs while maintaining quality.

What had worked so well for his father also worked for Edsel, and the perception of Lincoln quality actually improved under this strategy. "I wish to compliment you and your organization very highly on the splendid appearance of this Lincoln chassis you sent to us," wrote custom body designer Hermann Brunn to Edsel. "It is certainly a beauty and gives evidence of more careful attention to detail than any chassis we have ever received from the Lincoln Company."[7]

During this early stage of managing Lincoln, Edsel also felt that it was far more advantageous to insist on quality manufacturing rather than tout "Quality Lincolns" through advertising. Because Lincoln quality was already well known, all Edsel had to do was to maintain it, and that fact would continue to find its way to the buying public. Fully aware of Edsel's reticence to over-sell his products, Arthur Brisbane, editor of the *Detroit Free Press*, pleaded with Edsel to simply state the facts. "I dictate this letter to the phonograph in one of your Lincoln cars," Brisbane admonished, "and I wish the Lincoln car were more aggressively advertised. It seems to me that the best car made in America or anywhere else for that matter ought to be advertised as 'THE BEST CAR MADE.' "[8] Edsel appreciated Brisbane's support but remained reluctant. Besides, why spend a lot of money on advertising a product that was only partially complete? As far as Edsel was concerned, before he could embark on a glamorous advertising campaign, he needed to add some panache to his Lincolns.

Producing quality products, of course, was nothing new to the Lincoln Motor Car Company or to Ford. What was not second nature to either company was styling. It was the one element that needed to be developed and honed, both at Lincoln and at Ford. Edsel was determined to attain it at both. He told custom body designer Ray Dietrich that he was willing to invest time and money to develop the Lincoln so that it would take its rightful place in the prestige field.[9] If Edsel could accomplish that, he left little doubt that he would turn his efforts to his father's products. He intimated to Ford executives that he hoped a certain amount of Lincoln design would "dribble down" into the Ford line.[10] First, however, Edsel would concentrate on improving the styling of Lincolns. Then, when appropriate, he would transfer that knowledge to Fords. For the foreseeable future, Edsel would devote most of his time and energy to that end.

***

The Lelands should not be overly criticized for emphasizing quality and understating aesthetics. Similar to Henry Ford, Henry Leland became involved in the automobile industry in its infancy. At that time, the top priority was making automobiles that worked—design and aesthetics were barely a consideration. By 1922, as the automobile industry was about to enter what is now considered the Classic Era, the skills and technology to manufacture a quality motorcar were in abundance. The proof of that could be seen in the thousands of reliable, albeit homely, Model T's being manufactured every day at the Ford factories. After the physics of the machine were conquered, the public began forcing the rapid implementation of stylish design. No longer were customers satisfied with an automobile that simply worked. (That also would become evident in the decline of Model T sales.) They now wanted a machine that looked good, too. It would take some time before the mass producers such as Ford and General Motors would incorporate stylish designs into their production automobiles. However, the luxury motorcar builders such as Packard, Duesenberg, and Franklin understood the changing market and began placing emphasis on body design in addition to mechanical integrity. The Lelands' reluctance to understand or accept this new strategy assured their fate. When Edsel took over their company, he not only began emphasizing stylish designs as did the other prestigious automobile manufacturers, but he also quickly pushed his company into the forefront of the new trend.

***

With this new direction came custom body builders—coachbuilders, as they were commonly called, or "carrossiers" for the more sophisticated. These were companies that specialized in handcrafting luxurious, specially built automobile bodies to place on expensive, powerful chassis. They were also the first to institute automobile design and styling as a separate and distinct process in the manufacture of automobiles. Eventually, they would pique styling to a point where it would become the most significant factor in the purchase of an automobile, a factor that continues to this day.

Similar to many automobile companies, the custom body makers evolved from early-day carriage makers and simply transferred the skills they had used to build buggies, coaches, and broughams to constructing phaetons, roadsters, and sedans. Each motorcar body they built was painstakingly hand made by craftsmen who spent the good part of their lives making certain parts or performing certain functions in the drawn-out process of automobile body manufacturing. Some specialized in metal forming, others in woodworking, and others in upholstering. The metal craftsmen hammered, formed, and stretched thin sheets of aluminum into fenders, doors, and body panels. Woodworkers cut, glued, and steam-bent lengths of oak and ash into intricately shaped body framework. Top grain leathers and fine fabrics were cut, sewn, and stuffed into plush, smooth seats by upholsterers. Hardware items such as door handles, latches, and window stanchions were cast in solid brass and then highly polished—all by hand. The finishing work—polishing wood, painting the body, and fitting the top—was done by other specialists and could take weeks to complete. With the patience of Job, these craftsmen varnished and painted until every body part attained a lustrous finish. Finally, when all hardware was cast, when all sheet metal panels were smoothly formed, finished, and fitted together, and when all upholstery was plumply rolled and precisely tucked, the body was fitted onto the precision chassis and made into a whole lovely creation.[11]

Carrossiers also needed two other specialties to build automobile bodies: body draftsmen and body designers.

Instead of using computer terminals to develop a body layout, as is the practice today, the body draftsmen in the early years used ruling pens, India ink, and tracing cloth. Andrew Johnson trained many of them at New York City's School of Carriage and Automobile Design.

Johnson learned his craft of surface geometry—the process of "projecting" three-dimensional forms onto a two-dimensional piece of cloth, called working drawings—from shipbuilders in Bordeaux, France, in the early 1900s.[12]

The task of the body draftsmen was to develop and communicate the compound and convoluted shapes of the refined automobile body to craftsmen through these working drawings. Without these drawings, it would have been practically impossible for the craftsmen to construct the various parts of the automobile body from a preconceived design. These men who "laid the ink" never received the accolades and recognition afforded later to the automobile designer—even though their working drawings were meticulously detailed and exquisitely executed—simply because they failed to cross the Rubicon from refinement to aesthetics. They made great strides in transforming the already-existing body building process by evolving the customary rectangular and square carriage forms into more symmetrical surfaces and graceful curves. They were even daring enough to smooth the abrupt differences between the engine hood and cowl, domed fenders, and roof panels, and managed the proportions of windows and doors.[13] However, that was not enough. Their efforts only provided a more propitious layout, making the craftsmen's job easier. The creative eye of the designer would be needed to complete the journey.

If it was any solace to the body draftsmen, the custom body era could not have occurred without their skills. However, it also could not have occurred without the body designers, for only they had the inimitable capacity of blending, shaping, and transforming the more graceful lines and congruent surfaces started by the draftsmen into automobile designs that could be considered works of art. "The difference between a body draftsman and a body designer," explained an early designer, "is that the designer has the conception, and the draftsman interprets it into reality."[14] There was little correlation between the two skills, and thus few body draftsmen became body designers. In fact, some of the great automobile designers could not draw well.

***

Obtaining a custom made body from a coachbuilder was more involved than merely placing an order for one. When a person decided to buy a custom motorcar, he started a process that would take thousands of man-hours to complete, and practically drain his bank account by the time it was done. It was an arduous task for the generalist, but an adventure for the perfectionist, because the process of buying a custom-made automobile required countless detailed decisions, months of perseverance, and at least $10,000 in cash.

Choosing an appropriate chassis from the myriad of offerings was the first decision a customer had to make. He could choose from one of the many American-made chassis, such as Duesenberg, Packard, Marmon, or Lincoln—the chassis Edsel wanted everyone to order. Any of these chassis would be delivered to the carrossier complete with engine, frame, wheels, radiator shell, front fenders, and hood—everything, in fact, except places to sit and surrounding bodywork.[15] Or, the customer could order a foreign chassis, such as an Italian-made Isotta-Fraschini or Spanish Hispano-Suiza, which came with everything the domestic chassis did except front fenders and hood. On these imported chassis, the designer had to develop what he called the "running gear": front fenders, running boards, splash aprons, and hood.[16] Of all the chassis available, both domestic and foreign, the designer loved the more exotic foreign chassis the most because it gave him the freedom to design the entire outer surface of the automobile—from the hood to the spare tire mount, and from the top of the running boards to the long, flowing fender treatments.

If obtaining a chassis seemed unduly complicated, that was actually the easy part. Once that was done, the customer now had to select a carrossier to design and build a body to match his exquisite chassis. Again, he could choose from a number of coachbuilders, a task even more onerous than choosing a chassis. In terms of quality and performance, one chassis was as good as another, and one Lincoln or Isotta chassis was identical to every other Lincoln or Isotta chassis. The body, however, was different. Because it was individually designed and hand made, the body became a one-of-a-kind masterpiece, unique among all automobiles. The customer had to take great care and forethought in selecting a design from the carrossier's proposals, for he ultimately would be responsible for its outcome and would have to face the results every time his chauffeur pulled the car up to the front door.

After choosing one of the designs, the intrepid customer had to analyze and approve its every line, contour, and moulding; every knob, trim piece, and ornament; and the upholstery fabric and body color. The carrossier would not begin building it until the customer had approved each of these items in advance.

***

No one enjoyed this protracted process of custom body making more immensely than Edsel. He followed every custom-made Lincoln body from inception to completion as if each one were for his own personal use.[17] Edsel's eye for detail and exacting personality compelled him to analyze every new design down to the minutest detail. "We are wondering," wrote the president of Packard, "if there is any particular reason for your desire to have the extra tires for the chassis you ordered rib tread? The standard rear tires on our cars are Goodyear cord, All Weather tread, and all our customers so far have desired this tread for their extra tires."

"I would say," replied Edsel, "that the specifications as made out are all right; that is, rib tread is desired, and I would not care for the All Weather."[18]

Only after spending countless hours specifying and approving every detail such as this on each job would Edsel approve any of the carrossier's design proposals.[19]

Such in-depth analysis may seem like indecisiveness, but it was actually one of Edsel's greatest strengths as a design critic. Such painstaking examination—reviewing every line and every contour of every design—enabled Edsel to pick out the nuances of a proposal that could actually change a mediocre design into a classic one. However, once Edsel visualized the overall design or had a specific detail in mind, his internal deliberations would end, and he would pursue the design or detail doggedly to a conclusion.[20]

Automobile designer Raymond Dietrich soon learned of Edsel's resoluteness when he received a letter from Edsel about one of their projects. Clearly, Edsel was upset. "First of all," Edsel wrote, "I do not think that your production model is anything like the first sample that was built... The detail is quite different on the interior and to my mind the job is a very inferior model and does not in any way come up to the standard that we set in the Lincoln body types."[21]

When Edsel gave a body designer specific instructions, he fully expected them to be carried out, exactly as he specified. He ensured their compliance by demanding that he approve every step of the design process. "I think you are working entirely too fast on this job," Edsel wrote to Hermann Brunn about a new Lincoln design that Brunn was developing. "On the first sketch that was submitted, you had drawn up the working drawings before I had a chance to approve or disapprove it, and now you have the working drawings finished again...

"This all costs money and seems foolish until I am satisfied with the sketch. I have not given you the go-ahead, and please do no more until I do so."[22]

Edsel never backed down in demanding only the highest quality work from the custom body houses, but he also understood the coachbuilders' plight. Hermann Brunn was in Edsel's office one day, finishing the details on an order that Edsel was placing for 50 custom Lincoln bodies. Brunn was overjoyed with the order, but he was concerned about payment. "I almost lost my business once, by a similar production order, when it was cancelled about halfway through," related Brunn. "I hope you will not be offended by my making this statement, but I really almost rather have half the number and be sure they were going to be taken. Look, you might be over in Europe and the bottom fall out of the market, and we'll get a telegram to stop working, and there I'll be again."

"Well, Mr. Brunn," Edsel replied, "I understand, but I'll tell you what I'll do. I'll make the contract non-cancelable, no deferred shipments, and I'll sign it myself. Now, how do you feel about that?"

"Well, Mr. Ford, that's wonderful."[23]

Through such amenable business relationships, Edsel was able to secure the select services of Brunn & Company and other custom body makers.[24]

***

The demand for expensive and luxurious motorcars equipped with bespoke bodies exploded with the economy throughout the 1920s. Custom body shops sprang up in the wealthiest parts of the country: New York, Massachusetts, and California. Wherever there were rich or *nouveau riche*, there were custom body builders. They were in Hollywood to build cars for movie stars. They were in Boston to serve its aristocracy. They were in New York City, catering to paper-rich stockbrokers.

Le Baron Carrossiers was one such body builder. Started by a trio of energetic and enterprising men, it quickly became the premiere design studio in the country. Their story is significant not only for that, but because Le Baron designed automobiles for Edsel Ford.

The idea to go into business for themselves was the brain child of Thomas Hibbard and Raymond Dietrich, two highly competent body designers of the Brewster Company of Long Island City, New York, which was considered by many to be "the greatest coachbuilding firm in America."[25] Hibbard had once worked for the Rubay Company, a Cleveland, Ohio-based body builder, and then went to C.P. Kimball and Company in Chicago, where he was chief designer before joining Brewster in 1918.[26] Dietrich, on the other hand, arrived at Brewster for the first time five years earlier, left in 1917 to work for Chevrolet, and returned to Brewster late in 1918. Meanwhile, Ralph Roberts, a Dartmouth student who had been interested in automobile design since boyhood, decided to pursue that field instead of biology. Roberts loved to draw automobiles and had an exceptional sense of their design. While still at university, he sent a portfolio of sketches to Fiat of America, hoping it would recruit him. Fiat was not interested. Shortly thereafter, Roberts visited Rubay, not in search of employment but simply to obtain a firsthand account of how automobile bodies were built. Hibbard gave Roberts a plant tour.[27]

Roberts returned to Dartmouth and continued his studies, but began corresponding with Hibbard. Shortly after Roberts' visit to Rubay, Hibbard left the company, went on a stint to France for the Army, and returned about a year later, accepting a position at Brewster.[28]

During their years at Brewster, Dietrich and Hibbard witnessed the growth of the custom body business to the point where it was even greater than what Brewster could easily accommodate. They believed that much of that growth was due to their combined talents. Dietrich and Hibbard became weary of developing new designs for a small wage and decided that there was plenty of room for another design house. Thus, in 1921, they left Brewster and started their own design studio in New York. Dietrich and Hibbard asked Roberts, who had since graduated, if he wanted to join them in their new venture. In a decision that would dictate the next two decades of his life, Roberts agreed and met them in New York to work out the details.[29]

Their first task was to find a name for their new firm. "It has to be a French name," demanded Hibbard. "Everybody knows that anything French is stylish!" However, not knowing the French language, Hibbard picked up the New York telephone directory and began thumbing through it looking for French-sounding names, eventually coming across the name *Le Baron*. "Here's one," he said. "It sounds French enough! Let's call our company *Le Baron*. No," he hesitated. "Let's call it *Le Baron Carrossiers*. That really exudes style!"[30]

After settling on a name, they were fortunate to secure some office space on "automobile row" in New York at 2 Columbus Circle, south of Central Park near 58th Street and Broadway.[31] They rented some space on the fifth floor, tore out the partitions, and made one big reception office-drafting-drawing room. The layout of their new studio was functional, its furnishings austere. Upon entering, Roberts' office desk was immediately to the left and farthest from the windows, which lined the walls. Behind him, toward the center of the room and located in an area drenched by natural light, was Dietrich's drafting table, where he produced full-size working drawings. Next to him was Hibbard's office desk, where Tom did all of his design work—he never did any work on a drafting board.[32]

As business developed, Roberts recruited an office secretary, an artist, and an apprentice designer. The secretary sat next to Roberts. Roland L. Stickney, an artist Roberts recruited from the Locomobile Company because of his expertise in creating "picture perfect" watercolor renderings for promotional catalogs, sat near one of the windows overlooking 59th Street. Hugo Pfau, then a part-time office boy and budding apprentice, sat next to Stickney. Along a back wall stood a specially made filing cabinet with large, flat drawers to hold sketches, blueprints, photographs, and renderings.[33]

***

The three men divided ownership of the company into three equal shares, but on an unequal basis: both Hibbard and Dietrich received one-third of the stock for free because it was their expertise upon which the company was founded. Roberts, on the other hand, had to pay for his share, and borrowed $700 from his mother to buy his one-third ownership.[34]

***

Le Baron Carrossiers was unique in a unique industry—not because it was new, but because of the way it was organized and run. Unlike all other coachbuilders, the trio did not manufacture custom bodies—they did only the design work. They consulted the very wealthy who were interested in luxurious and expensive one-of-a-kind automobiles, and then they developed motorcar body designs that suited their elaborate tastes. Working in the same way that an architectural firm operates, they would develop a design on paper, have the client approve it, and then contract with an outside company to construct the coachwork. Upon supplying a complete set of blueprints and specifications to a bodybuilder, they would oversee the entire project from drawing board to delivery.

Dietrich and Hibbard did all of the design work. Roberts, while having an excellent sense of automobile styling and design—a "car feel"—that was similar to Edsel's, was not as deft on the drawing board as his two partners and thus assumed the responsibility of running the office.[35]

On their own, the trio submitted designs to anyone who would look at them, and for the first year or so of the studio's life, they barely generated enough income to feed themselves.[36] Money was so tight during these formative years that Dietrich's wife had to make pants for their little boy out of Ray's worn-out ones.[37] However, as one design contract begot another, and as each boastful owner showed off his graceful Le Baron design to his affluent friends, the firm soon was offering its expertise to the likes of Mrs. Flo Zigfeld, actress Marilyn Miller, philanthropist Maurice Heckscher (in whose building the Le Baron studio was located), and singer Al Jolson.[38]

One reason the Le Baron team succeeded was that they never turned down a request, no matter how outrageous, and they had the uncanny ability of incorporating any idiomatic feature a customer requested into a beautiful bespoke body. Ralph Roberts tells the story of Al Jolson, for whom Le Baron designed a car to fit his personality. Jolson had always owned Lincolns, even when they were not as refined as other makes; however, with his next one, he wanted a Lincoln design that was conspicuously stylish. "When I drive down Fifth Avenue," instructed Jolson in his jocular manner, "I want people to say, 'There goes Jolson!' "[39]

Not to be thwarted, Roberts explained Jolson's request to Dietrich and Hibbard. The design they developed had the lowest silhouette they had ever conceived, thinking that that would surely please the entertainer. Unfortunately, when Roberts showed Jolson their idea, the singer was not satisfied. "I want it lower!" remonstrated the singer.

"But you wouldn't be able to wear a hat in it now," retorted Roberts.

"I don't care if I have to crawl in it on my hands and knees!" replied Jolson. "Make it lower!"

Roberts placated Jolson by saying that they would take care of it. The next time Roberts saw Jolson was when he came to pick up his new Le Baron-designed Lincoln. Jolson was very pleased with it, although it was not an inch lower than the original proposal.[40]

In this car, Jolson's chauffeur would drop him off at Boston's Back Bay railroad station, speed over the narrow and winding roads between Boston and New York, and be waiting for the singer when he arrived at Grand Central Station. That was at a time when trains were fast and cars were slow.[41]

For composer Irving Berlin, the Le Baron team built a four-octave, imitation keyboard into the back of the front seat of his bespoke body. "It doesn't have to be functional," Berlin told Roberts. "I just want to be able to pantomime the tune as it comes to my head. That way, I will be able to remember it easier."[42]

These are only a couple examples of the extremes to which Le Baron went to please its clientele. Even the "standard" designs from Le Baron were customarily equipped with luxurious amenities, such as built-in bars made of silver maple cabinetry and a zinc-lined icebox. They came replete with a set of silver tumblers and cut-glass decanters. If not top-grain leather, the interiors of Le Baron coachwork were often upholstered in needlepoint embroidered broadcloth.[43]

Business for Le Baron was developing quite nicely (at least its founders were eating regularly now) when Edsel Ford heard of their expertise. What began as another of Edsel's routine carrossier evaluations developed into a unique relationship between himself and Le Baron that ultimately changed the nature of the automobile industry.

***

Edsel was well acquainted with all of the prestigious custom body houses: Brewster, Brunn, Judkins, Murray, Locke, Willoughby, and Kimball of piano fame.[44] Over the years, he had either contacted each of these firms or had purchased bodies from them. In the early years, Edsel purchased most of these bodies for his own personal use. However, now that he owned the Lincoln Motor Car Company, most of his orders were for Lincoln customers.

An inventory of Edsel's motorcar stable in 1923 reveals not only his eclectic taste, but also a collection worth more than $100,000. He owned everything from a $900 Ford delivery truck to an exotic Hispano-Suiza costing nearly $14,000. No less than seven Lincolns—sedans, coupes, phaetons, and limousines—all replete with custom-made coachwork, were available at his beck and call.[45] One of Edsel's Lincoln coupes had a graceful Brunn body for which he paid nearly $5,000. Another—which was for Eleanor's use—had a Brewster body that cost more than $3,000.[46] Four years earlier, Edsel purchased a $5,100 Packard chassis and had it shipped to New York to be outfitted with Holbrook coachwork.[47]

Edsel owned more cars at one time than a group of men would own in their lifetime, but he was not frivolous with money, and he scrutinized each purchase. Enamored over the quality and workmanship of the Hispano-Suiza, he ordered one but cancelled it a short time later after the company tried to double the price. "The only thing wrong with the whole thing is the matter of price," he wrote to Brewster. "I do not feel like putting such a large amount of money into an open type of body..."[48] Edsel also did not like to be taken advantage of simply because he enjoyed luxurious cars and had the wealth to buy them, and he made efforts to minimize such possibilities. Instead of shopping for cars himself, for example, he often would ask his business associates to make inquiries on his behalf. "Please look up German Mercedes agency," he telegraphed Gaston Plaintiff, the Lincoln sales manager in New York. "Send me catalog latest 6-cylinder car with supercharger. Get price, but do not reveal for whom."[49]

***

Unbridled by cost constraints and concentrating purely on aesthetics, Edsel realized that the custom body designers were setting the styling trends for the future.[50] However, he was still a novice in the design of automobiles, and, not having any expertise at Ford or Lincoln, he learned to leverage the strengths of these carrossiers to full advantage. Through them, Edsel would hone his already ingrained styling sense to a sharp edge and pull his stodgy Lincoln into an enviable position, to the surprise and expense of his rivals.

Edsel's first step in raising the prestige of the Lincoln was to contract with many of these custom body houses for one-of-a-kind bodies. Sometimes, he would select a design from the carrossiers' portfolio; other times, he would offer his own design ideas for them to develop. Wrote custom body designer Hermann Brunn, "It is also understood that the writer is to design a speedster body for your personal use to be mounted upon a Lincoln chassis carrying out your ideas of design as expressed...while in your office."[51]

Regardless of whose idea was used, Edsel worked with all custom body builders on the same, equitable basis. For a predetermined flat fee for each new design, the carrossier would submit design sketches for Edsel to evaluate, change, and approve. After he had made all of the changes to his liking, he would return the design to the coachbuilder, from which a complete set of working drawings and specifications would be made. Before releasing these drawings and specifications to the craftsmen to begin construction, the coachbuilder would have Edsel approve the revised design one last time.[52]

The Lincoln Motor Car Company was making the most of its own bodies when Edsel took over the company. However, as Edsel slowly began implementing "special cuttings" (a design that could be used in a semi-production run), the number of Lincoln-made bodies decreased, and the number of supplier-made bodies increased.[53] When Edsel approved a new design and it was transferred into working drawings, he was free to have the bodies built either in his own plant or by an outside supplier. If one of the body builders supplied the bodies—which was typical, particularly between 1925 and 1938—they were usually finished "in the white," meaning that they were supplied as raw aluminum shells, or with only a light-colored prime-coat of paint that made the bodies appear white. The bodies had no color-coat or trim pieces.[54]

Not all of the custom designs were furnished without paint, however. Many times Edsel would seek advice on color schemes for all of his vehicles—customs, production Lincolns, and even his mass-produced Fords. He knew that if he relied on his own severe taste, all of these vehicles made under his direction would have the same satin finish—not the typical shiny finish—and be painted his favorite gun-metal gray.[55] Even when Edsel was forced into the polychromatic realm, the color combinations he developed were quite austere. "With regard to the color of the new job," Edsel explained to Brewster, "I have been unable to make up my mind, but have an idea this time of two schemes: one is a combination of two tones of tan or gray-brown; the other, two tones of gray-green, more commonly known as sage brush green."[56]

To overcome this conservative bent, Edsel would seek color ideas from his custom builders, production suppliers, and even color consultants.[57] The custom builders were particularly eager to provide to Edsel their ideas on color schemes. Edsel had Dietrich develop some color proposals on several Lincoln models. The results—canary yellow on black and teal green on red—were probably too flamboyant for Edsel, but they hit the mark with the Lincoln sales group.[58] "Mr. Dietrich worked up some wonderful color work on Lincolns, especially the cabriolets and sedan models," wrote Edsel's New York sales manager. "I saw some the other day and believe we could sell quite a few of them."[59]

When Edsel became more involved in the design of Ford automobiles, he also would solicit color ideas from Ford's large paint suppliers. The initial approach for these companies was to obtain a number of production Fords and paint them in their proposed color themes. When the painting was completed, one of Edsel's engineers would bring the freshly painted Fords to

Dearborn for Edsel to evaluate. Being able to see the actual product in its new coat, Edsel would make his choices.[60] However, that system of "proofing" became so expensive and time consuming that the paint suppliers devised a more expedient method. First, they would develop a variety of colors and place them on metal panels. Then Edsel would choose the colors he wanted to evaluate more closely. Next, the paint manufacturers would paint miniature Ford models made from metal stampings in the colors that Edsel wanted to see. "Mr. Edsel Ford would then look the jobs over" and decide which colors to use, if any, in Ford production.[61]

Even with these procedures and Edsel's keen awareness of the importance of color in the design of automobiles, it was not until 1933 that Ford automobiles had the same color on the fenders as on the body. "Prior to that," recalled a Ford engineer, "all fenders had been painted black. Period. The change was brought about by sales demand and Edsel's influence."[62]

"But today, form and color are among the prime considerations of the builder," wrote Edsel. "Along with streamlining had come a burst of new automobile colors—yellows, tans, and grays that would have seemed rather garish but a decade ago, but which today are accepted as distinctly fitting."[63]

Although Edsel's penchant for design tended toward the conservative side, his design philosophy was, nevertheless, widely accepted, for there were abundant design ideas and perspectives that were quite extreme. One tire manufacturer, thinking it could capitalize on the increasing use of color by car designers, hoped to convince Edsel to use its new "designer tires" on his Lincolns. "The Armstrong Rubber Company has taken a forward step in giving you tires that harmonize in looks with your new Lincolns," wrote the company president. "Will you please send us literature showing the different new models in their correct and true colors?"

"We will match these colors either in the same shade or in beautiful contrasting colors, and these harmonizing tires will not be painted or gilded on the surface but will be colored right through and through, including the tread portion as well as the sidewalls."[64]

Edsel's artistic sense never allowed him to shut out new ideas, but he replied to this radical idea in the same diplomatic manner as he would to any similar suggestion: "I don't want to move that fast."[65]

***

Edsel's first visit to Le Baron Carrossiers was in the fall of 1922 after attending the second annual New York Automobile Salon.[66] From the numerous offerings displayed at the salon, Edsel was most impressed with the Lincoln sports phaeton entered by Le Baron, which had been designed for Milton Budlong, manager of the Lincoln agency in Manhattan. Never before had Edsel seen such graceful and congruent lines on one of his precision Lincolns. He could hardly wait to visit the Le Baron studio and meet the men who designed it. The Le Baron members were not completely obscure to Edsel—he had met both Hibbard and Dietrich while they worked at Brewster.[67] However, the designs Edsel saw then were graceless compared to what the two had produced for Budlong.[68]

Budlong made arrangements for Edsel to visit the Le Baron studio as soon as the salon ended. Arriving early one morning, Budlong introduced Edsel to Dietrich, Hibbard, and Roberts, and Dietrich and Hibbard told stories about how they had met when Edsel had visited Brewster. After these formalities, the trio showed Edsel some of their design work.[69]

To say the least, Edsel was ecstatic with their offerings and talent, and he immediately began ordering "custom jobs" in earnest from them. Starting with a couple of Le Baron designs for his own stable, Edsel quickly began ordering more for his various East Coast Lincoln agencies.

As soon as Le Baron had completed a car to one of its new designs, Edsel would advertise the car in various *bon ton* magazines such as *Spur* and *Vanity Fair* and in his own in-house product catalog called *The Lincoln*. The combination of Edsel's quality Lincoln chassis and Le Baron's exhilarating designs was an instant success. Wealthy customers in the market for a new motorcar were enticed into Lincoln showrooms with the new offerings. At one point, Edsel became so enthralled with the new activity that he pressed Dietrich for another new design so that it could be offered in upcoming monthlies. "I'm working on one now, a convertible sports sedan," Dietrich informed Edsel, "but it's just a rough sketch. I don't think I can get it done in time."[70]

"Don't worry if you don't have all the details worked out," Edsel hastily replied. "Just get the design ready for an ad. We'll worry about the details later."[71]

Under Dietrich's direction, Stickney took the design and brushed one of his impeccable renderings, just in time for publication. The model was an instant success. Only days after the ads hit the newsstands, customers and dealers began calling the Lincoln sales office, bombarding it with requests for information about the new offering and asking when they could purchase one.[72]

***

Although this direction produced very expensive Lincolns, it was, nevertheless, the correct approach. People who could afford custom-made motorcars cared little about cost, as long as the prestige and styling were there. Edsel's Lincolns already had an irrefutable reputation for quality, but when he combined that attribute with the fashionable designs of Le Baron, "Ford-Lincolns" soon became all the rage. That, combined with Lincoln affordability (a custom-designed Lincoln usually cost less than half as much as a Rolls-Royce or a Hispano-Suiza), they were veritable bargains, and Lincoln sales increased significantly.

To Edsel's delight, the demand for Le Baron-bodied Lincolns grew to a point where he had to either change the way he was building Lincolns or limit the number of Lincolns he could offer. Being a Ford man at heart, the last thing he wanted to do was to limit production. Instead of ordering one bespoke body at a time, Edsel shortened the lead-time of building custom-made bodies by ordering a series of five or six bodies of Le Baron design at one time. This was a boon for Le Baron because now they were both designing and building bodies. In 1924, they purchased the Bridgeport Body Company in Bridgeport, Connecticut, a firm with which they had worked since starting their studio.[73]

Le Baron would develop a series of designs for Edsel, have him approve the designs, and then send the designs to Bridgeport for construction. As soon as construction was completed, Edsel would have the bodies delivered to his strategically located agencies, with Budlong's Lincoln salon at the top of the list.[74]

This process turned out to be both productive and innovative. Through it, Edsel was able to provide more custom Lincolns to meet the growing demand. Likewise, because each car could be painted and upholstered to the exact specifications of the customer, the customers felt they were buying a one-of-a-kind vehicle. This idea was expanded until Edsel was making 50 to 200 special cuttings at a time.[75] These short production runs were miniscule, compared to what was being turned out at the Ford plant. However, its significance was no less remarkable. For the first time, Edsel began the process of designing cars initially for aesthetic appeal and then for production constraints. He soon proved that a stylish automobile could be produced as efficiently and voluminously as a homely one. In doing so, Edsel changed forever the process of building automobiles.

* * *

One time, Edsel asked Ralph Roberts and his team at Le Baron to design a coupe that could be used as a special cutting. Dietrich set to work and developed a design that he believed was as stylish as any of his previous designs. The only difference was that it was suited for volume production. When the design was completed, Roberts sent a rendering of it to Edsel, with their proposal for building one sample body that Edsel could use for evaluation, at a cost of $3,000. If Edsel liked it, they would sell him the design for $10,000. With that, Edsel could manufacture as many bodies to that design as he desired.

Edsel reviewed the design and proposal, and he wrote to Roberts that the design was intriguing but "the price is kind of high."

Roberts was somewhat dismayed at Edsel's reply because the prices he quoted were well within the range that he had charged Edsel on previous projects and were, in fact, well below their typical fee of $5,000. He was ready to forget the whole matter when, after talking it over with his partners, Dietrich suggested that instead of losing the order altogether, Roberts should write to Edsel one more time and ask him openly what he thought a fair price would be for the project.

Roberts hesitated momentarily, arguing that the project was well worth the asking price, especially because Edsel could spread the cost of the design over as many Lincolns as he wanted. Roberts certainly wanted the order, but he knew from previous dealings with Edsel that Edsel was not much akin to price negotiation. Edsel either thought the price was fair and gave you the order, or he thought the price was inappropriate and told you so. Nevertheless, Roberts followed Dietrich's suggestion and sent a letter to Edsel.

A few weeks later, Edsel's reply arrived: "Thank you for your letter concerning the coupe project," he wrote. "I very much like your design, but would consider a fair price to be $9,000."[76]

* * *

Whenever Edsel visited the Le Baron studio, he first telephoned Roberts to let him know that he was coming. Edsel would spend much of the first part of the day with his New York sales staff, and then he would drive to the Le Baron studio after lunch. Greeting Roberts as he walked into the studio, the two would exchange pleasantries, and then Edsel would proceed directly to Dietrich.

Dietrich would start the discussions by reviewing all of the designs on which he and Hibbard were working for Edsel. Taking sketches and drawings one by one out of the "Lincoln drawer," the two would analyze each design, first on its own and then collectively. Hibbard occasionally would interject an opinion or judgment, but normally these dialogs occurred strictly between Edsel and Dietrich. Some have said that the reason Edsel worked more closely with Dietrich than with Hibbard was that he was more impressed with the line, beauty, and luxury of Dietrich's work. To Edsel, Hibbard's styling seemed a little too inelegant. However, the reasons were more innocuous than that. First, Hibbard was with the firm only a short while—he headed to France to start his own studio in 1923—and thus never had a chance to build a rapport with Edsel. Second, Dietrich was more outgoing and "easier to talk to" than Hibbard. Naturally, Dietrich would take control of the conversation whenever Edsel visited. Regardless, both Dietrich and Hibbard "were excellent designers," and Edsel's appreciative eye and the way he became thrilled by a beautifully executed design always inspired them.[77]

Dietrich would then move briefly to designs on which Le Baron was working for other clients. By this time, Edsel had propped himself up on the corner of Dietrich's drafting table, and, for the next hour, their conversation centered on the designs coming from Europe, especially those from England, France, and Italy.[78] An abundance of ideas and inspiration evolved from discussing these "continental cars," as Edsel liked to call them.

***

These visits and discussions would occur throughout the year and anytime that Edsel visited New York, which was often. However, their elaborations never became more intense or more impassioned than during the annual Importer's Salon. Begun in 1921, the salon was held every year for the next decade. Limited initially to only European manufacturers, it was opened soon to American custom body builders as well. These salons were distinct from the annual automobile shows that were held by the numerous automobile manufacturers. The salon had restricted and limited attendance, whereas the general shows were open to the public. The two forums did have one thing in common, however, and that was to sell cars—the former to the elite of society, the latter to society.

The salon was held every December in the luxurious ballroom of the Commodore Hotel in downtown Manhattan, then at The Drake in Chicago in January, and finally, in February, at the Biltmore in Los Angeles.[79] The New York Salon, however, was *the* show to attend, simply because it was the first of the tour and because the "Who's Who" of automobilists and society attended this first leg. Admission to the gala, black-tie affair was by invitation only and was strictly administered by the czar-like director of the salon, Pop Lazcarus. He ensured compliance by placing Pinkerton guards at all ballroom entrances. Lazcarus himself issued a select number of engraved invitations to all of the exhibitors and to the automobile executives of Detroit, such as Henry and Edsel Ford.

On the times that Henry Ford did attend the salon, he attended only to be seen and did not spend much time in reviewing the motorcars on display. "He really didn't seem interested," said Roberts, "except in new engineering disclosures."[80] Edsel would tag along courteously, preferring instead to avoid the limelight and attend the spectacle before hours so that he could analyze, critique, and evaluate every offering in his typical fashion—in detail.

Accompanied by Gaston Plaintiff and Milton Budlong, Edsel would visit the spectacle the day before the salon officially opened. Attired in business suits, Roberts would meet Edsel and his associates outside the hotel and escort them into the display area, where they would meet with Dietrich and Hibbard. Free from distractions, the troupe walked unimpeded up and down the aisles as Edsel pointed out what he liked or did not like in each showpiece.[81] Together, they would express awe at the magnificent workmanship of Rolls-Royces, Hispano-Suizas, and

Isottas, and the graceful lines of the carrossiers' offerings. They would discuss the attributes and frailties of each motorcar they saw, and how they might incorporate some of the best qualities into Edsel's Lincolns.

According to Dietrich, Edsel looked at every motorcar on exhibit, and he did not miss a single detail, no matter how trivial, that might distinguish one elegant design from another. In addition to the details, Edsel would also view each motorcar from a distance, similar to an art critic examining a fine work of art. Then he would walk up to it and touch its surfaces, feel its quality, and study its workmanship. Frequently, Edsel would ask Dietrich or Roberts or Hibbard to climb into one of the vehicles with him to feel the depth and comfort of the cushions, the location of the controls in the driver's seat, or amenities available in the passenger area, never hesitating to compliment or offer constructive criticism on his findings.[82]

Edsel worked with many carrossiers, but his interaction and ultimate friendship with the Le Baron designers honed and polished his sense of design and feeling of form. Over the years, they collaborated on the design of dozens of bespoke motorcars, each one more refined and dignified than the previous one. With each new project, Edsel increased his understanding of the design process and refined his sense of styling.

Not only were the Le Baron team and many Lincoln customers aware of this, the international community—the sphere Edsel so desperately wanted to emulate—also realized Edsel's new-found talents. In 1926, at the first International Salon of Beauty in Automobiles, a design competition held in Milan, Italy, the design talents of Edsel were first recognized. His Lincolns won first place in the Cabriolets and Convertibles section out of 15 competitors and took second place in Open Torpedo and Sport Types out of 13 entries.[83] Edsel was ecstatic. All of his hard work and determination had paid off. Now, the competition had to worry about Edsel and his Lincoln.

* * *

Two coincident events that occurred at the close of the 1920s triggered the demise of the custom body era and the development of design studios at the mass producers in Detroit. First and foremost was the economic collapse in October 1929. Expensive as they were at the outset, not even the *haught mond* were too excited about spending $10,000 or more for a new automobile, no matter how elegant. After the onslaught of the Depression, the demand for custom-made motorcars dropped precipitously. The industry had all but disappeared by the early 1940s, although a limited number of custom-made automobiles could be obtained through World War II. The other incident was the incorporation of stylish designs into production vehicles. The process, initiated by Edsel Ford and Le Baron, was skillfully and rapidly being implemented at General Motors by Alfred Sloan and Harley Earl.

Many of the custom body companies were located on the East Coast, far from the monotonous production lines of Detroit. As the demand for custom designs waned, the custom builders had to find ways to maintain their businesses and thus came courting the mass-producers, especially Edsel. They began bringing design sketches of various ideas and even complete bodies to Dearborn for Edsel to evaluate, criticize, and hopefully buy.[84] "Whenever you are ready to take up the matter of Brewster & Company bodies for the Lincoln car," wrote the company president, "we shall be glad to consult with you in Detroit...and show you bodies finished and in construction that are splendidly adapted to the Lincoln chassis."[85]

The difficulty for Edsel was that every time he wanted a special cutting, he had to manage the unwieldy process of getting the run produced. First came the correspondence with and travel to a distant design firm—usually Le Baron in New York. Then Edsel had to bring the design to Detroit so that either Lincoln or Murray Corporation could produce it in quantity. If the manufacturer had questions about a certain aspect of the design, or if problems arose during its production, Edsel had to act as middleman, again becoming bogged down in more correspondence and travel. At the rate Edsel was now introducing special cuttings, it did not take long for this arrangement to become a logistical nightmare.

To make the entire process more efficient and timely, between 1924 and 1925 Edsel began pressing the Le Baron team to move their studio to Detroit. It seemed nonsensical for them to remain on the East Coast, when nearly 50 percent of their workload was now coming entirely from Edsel.[86] Additionally, the arrangement Edsel offered Le Baron was a liberal one. The cash amount was never disclosed, but Edsel would allow Le Baron to solicit other business as well. His only desire was to have them located close by him so that they could work together more effectively on his growing requirements. However, Dietrich and Roberts were not interested in relocating.

Edsel did not give up. In 1925, he decided to take a different approach. Instead of soliciting them directly, Edsel began putting pressure on Allan Sheldon, president of Murray Corporation, to retain some professional body designers of his own. "You do a substantial part of our body business—and that's about one third of the entire car—and yet I pay for all of the design work and advertising," Edsel explained to Sheldon. "In order for us to remain competitive, we must continue to advertise and develop new designs, but we cannot do both ourselves. You need to provide design expertise, and I know where you can get it."[87] Edsel then proceeded to educate Sheldon on the skills and expertise that the Le Baron team offered.

Sheldon understood, and he contacted Dietrich immediately. However, he had no more success than Edsel had. Regardless, Sheldon persisted and, after months of effort, made a final appeal. He called Dietrich and pleaded with him to simply come to Detroit "just to see what we would do for you and Le Baron."

Dietrich agreed and went to Roberts' home, where he was recuperating from an illness, to let him know what he was going to do. "That's fine," Roberts replied supportively. "But let's confirm what we have always said to each other: 'We are a team, and neither one of us will make a decision without the other's agreement.' "[88]

"Don't worry," Dietrich said consolingly. "I just want to see what he has to offer. I won't do anything foolish." But he did. Dietrich telephoned Roberts immediately after meeting with Sheldon and told him that he was moving to Detroit—alone.

That was fine with Roberts. He preferred living in New York anyway and maintaining his and Le Baron's independence.[89]

With financial backing from Murray Corporation and Edsel's assurance of continued orders, Dietrich came to Detroit and set up his own design studio and body manufacturing business, which he called Dietrich, Inc. He even carried his brashness one step further and took up residence on Jefferson Avenue, next door to Edsel.[90]

Sheldon's offer was not the only one Dietrich received during this period. He was also being wooed by Larry Fisher of General Motors, and he visited Fisher while in Detroit.[91] Dietrich, in effect, could have been General Motors' first design chief had he accepted Fisher's offer at the time. However, he turned it down because he wanted the freedom that the Murray/ Ford deal afforded him.[92]

***

Dietrich's departure left Roberts in the lurch, at least for the immediate time. What eventually transpired, however, turned out to be both ironic and poetic justice for Roberts. As for Edsel, he got what he was seeking also.

Unable to make adequate design sketches himself, Roberts leaned heavily on Roland Stickney to keep the design firm in business. Not unlike the relationship that developed between Edsel and his design chief in later years, Roberts and Stickney became inseparable. Roberts, too, had developed his innate design talent over the years by working shoulder to shoulder with Dietrich and Hibbard, and he had always been skillful at differentiating a good design from a poor one. Stickney, on the other hand, was not an astute designer, but he could take someone else's idea and develop it into a beautiful and saleable rendering, something that Roberts could provide to any custom body shop to have constructed. Together, Roberts and Stickney became a formidable design duo—each needing the support of the other, and each contributing to the success of the other. The rapport between the two became so piqued that, after meeting with a client, Roberts could call Stickney and describe to him over the telephone what he envisioned for a body design. By the time Roberts returned to the studio, Stickney would have some initial sketches ready for Roberts to review. "And they wouldn't need much changing either," exclaimed Roberts.[93] Before passing their design to the customer for approval, Roberts would sometimes have his wife, Dorothy, critique the design one last time to catch any subtleties he or Stickney might have missed. Everybody at Le Baron needed the assistance of each other, explained Roberts. "As a group, we designed."[94]

Roberts and Stickney survived, as did Le Baron. During the intervening months, Grover Parvis, the manager of the local Packard agency, and Hugo D'Annunzio, manager of the city's Isotta-Fraschini dealership, ordered bodies from Roberts until the new Le Baron team built up its own clientele, and until former Dietrich and Hibbard customers reordered new designs.

Business became so brisk that in 1927, Roberts called his absentee partner and asked if he would be interested in selling Roberts his interest in the firm. Dietrich was surprised at the offer. He knew that Le Baron was obtaining orders, but he could hardly imagine that they were receiving enough to precipitate a buyout. In his mind, the stock had to be worth next to nothing; therefore, Dietrich gladly sold Roberts his shares in Le Baron for book value.

What Dietrich did not know was that Roberts was now being wooed by a large Detroit manufacturer himself, again at the request of Edsel Ford. This time it was Briggs Manufacturing.

Essentially, Edsel gave Walter Briggs the same ultimatum that he had given to Allan Sheldon a few years earlier. With Briggs building the preponderance of bodies for Ford Motor Company, Edsel wanted the company to start contributing to the design of the bodies as well. Thus, Edsel suggested to Briggs that he try to bring Roberts to Detroit to not only head the Le Baron studio but to design production Ford bodies as well.

This time, Roberts was more than receptive to joining with a Detroit manufacturer. He was already beginning to see the decline in custom body orders and the rise in the so-called "semi-customs," initiated by Edsel. If he were going to sell out, now was the time to do it.

Unlike Sheldon, who solicited the Le Baron team directly, Walter Briggs sent his attorneys to work out a deal with Roberts. They met regularly, but it took longer than Roberts had anticipated to finalize the details to bring him to the Detroit manufacturer.

Shortly after the first of the year, during a meeting with Dietrich to discuss future body requirements, George Walker, the Lincoln purchasing manager, abruptly told Dietrich that Walter Briggs had just bought Le Baron "for a considerable amount of money."[95] Dietrich was dumbfounded, and Roberts now found himself indirectly in charge of all Ford design.

Roberts' relationship to Briggs was more confining than Dietrich's was to Murray (that is, Roberts worked directly for Walter Briggs), but Roberts ended up with an unexpected arrangement. Through some verbal gymnastics, Roberts was able to convince Briggs not only to open a Le Baron studio in Detroit, but also to retain the New York office as well. His argument was sound, for it was important to the prestige of the studio to be able to say that it had offices in Detroit and New York.[96]

Briggs turned over an entire plant to Roberts, and for nearly two years, Roberts managed both design studios, commuting regularly from New York City to Detroit. In winter, he took the New York Central Detroiter from city to city. When weather permitted, he drove the 700 miles from office to office in his own custom-made Chrysler Imperial convertible.[97] Taking the scenic route through upstate New York, he would catch the Buffalo & Cleveland sidewheeler *The Greater Detroit* at Buffalo and sail westward to Detroit, passing the mammoth Ford Rouge manufacturing complex as the steamer chugged up the Detroit River.[98]

Roberts lived to be 100, and he attributed his longevity to leaving Detroit and getting away from the bustling automobile industry. Every time he approached the Ford factory during these trips, he said that his "stomach got filled with butterflies and anxiety," just thinking about the job he had to do for Briggs and for Ford. However, tears came to his eyes when he began reminiscing about automotive giants such as Henry and Edsel Ford, Walter Briggs, and Walter Chrysler—all of whom he knew personally—because "these were great men, and they could do anything they set their minds to."[99]

***

The success of the Lincoln, both mechanically and aesthetically, gave Edsel confidence in his managing and design skills. Where he once constantly talked about the style, grace, and mechanical superiority of European cars, by the mid-1920s he could honestly say that his Lincoln had become their match, if not their master.[100] In a rare article he wrote for a Detroit newspaper, Edsel boasted of this new reality[101]:

> *While for many years the average American car owner looked up to the foreign built automobile as being the finest example of coach work and automobile production in the world, the United States is now recognized as being the home of finer automobile creations than any other country may point to.*
>
> *The Lincoln car is a typical example of this fact. It is the policy of this company to build the finest car that the tremendous resources of the Ford Motor Company will allow. This means that the finest of materials and engineering talent will go into the making of the Lincoln, and this fact is being recognized abroad, as well as in this country.*

European manufacturers and engineers are beginning to come to America to get patent rights and body designs for their cars. Hardly a week passes but what the Lincoln factories in Detroit are visited by some of the leading foreign engineers. Practically every one of these visitors marvels at the manner in which the Lincoln is built, and many express amazement at the fine manufacturing limits which it is possible to obtain in machine work.

The market in the United States for foreign-built cars is slowly passing. Today, the vast majority of Americans are forsaking the European cars and turning to homemade products.

In the same way, styles in American coachwork are far superior to those of foreign-made cars. It is possible to compare foreign built and American products in motor car values, and it can easily be seen that they are behind this country in styles.

All these factors combine to give value in the car made on this side of the Atlantic, at a lower price. It is true that there are many excellent foreign-built automobiles, but comparing them for appearance, value, and cost with such a product as the Lincoln, will show anyone how far superior American automobile manufacturers have come.

From the time that Edsel acquired the Lincoln Motor Car Company to the early 1930s, he continued to hone his design skills and developed his own unique perspective as to how an automobile should appear. "In their coherence and satisfying blend of solidarity, dignity, and grace," recalled an automotive writer, "many of the body designs for the 1932 through 1937 Lincolns may well be said to have been perfect examples of Edsel's maturing taste."[102]

Chapter 20

Felicitous Eye

"There is a tendency to keep monkeying with styles and to spoil a good thing by changing it."
Henry Ford[1]

"It is the well-designed object which...wins the race for public favor. It is their appreciation of—and consequently the selling power of—stylish designs."
Edsel Ford[2]

When Henry Ford and his engineers developed a new automobile, they would literally hammer together a prototype to evaluate. They looked at a car as a conglomeration of individual pieces and designed each of those pieces separately from all others. It was common, for example, for Henry to have Charles Sorensen make wooden models of various parts so that he could tangibly evaluate them in three-dimensional form. If the part was "goddamn good," they made it. If the part was "no damn good," they would start over.[3] When each part was "damn good," Henry and his men would assemble the parts as best they could, using brute force if necessary, into what they considered a complete automobile. It was then produced without further experimentation or regard to its aesthetic appeal.

"We couldn't style the car," recalled Joe Galamb. "Henry insisted that everything be square so that it could be made easier—interchangeable doors from left to right, for example. When everything is square, you can't get any slant to anything. Henry Ford was insisting on that fully."[4]

This approach to automobile design, as rudimentary as it seems, was not unique to Henry Ford or the Ford Motor Company. The early automobile pioneers were having a difficult time making machines that worked, let alone ones that were appealing to the eye. Therefore, aesthetic appeal was the last thing on their minds. As Edsel once explained it, "the problems that gripped the industry as a whole and forced body design into a place of secondary importance were those of speed, reliability, and economy, and development of the mechanical requirements to attain these ends."[5]

The little horseless carriage that Henry built in his coal shed so many years earlier exemplified this methodology. Understandably, it was austere and simple in design—after all, it was nothing more than a buckboard powered by the newfangled internal combustion engine. It would be difficult to imagine Henry coming up with anything more exotic. Even if his horseless carriage did not have the sleekness of a sleigh or the opulence of a brougham, it had one redeeming attribute: it ran, and it ran well. That is all Henry intended it to do—nothing more, and nothing less.

Functionality—that was what Henry believed in, that was his goal, and that was what he ultimately achieved. Unfortunately, he would carry this objective to the point of absurdity.

***

When Henry founded the Ford Motor Company in 1903, his primary objective was to build durable automobiles that worked. The first car he designed for the Ford Motor Company followed his usual tack: it was simple and spartan, made of the finest materials available, and it worked. Even the name of the car—Model A—was simple and illustrated Henry's bent for simplicity. Although it worked well and sold well, the Model A was not simple enough nor durable enough for Henry. He continued to modify it, redesign it, and simplify it over a five-year period, through models B, C, F, K, N, R, and S. From these precursory models, Henry culminated his years of experience and solidified his design philosophy in a car he named after the next available letter—T.

Similar to Henry's previous models, the Model T was a homely contraption. Its four large, spindly wheels appeared barely strong enough to support their own weight, making the car seem rickety and unsteady. From the front, its fenders were flat and high above the tires, resembling the wings of an airplane rather than the parts of an automobile. Also, the headlights were out of proportion to the remainder of the car, giving it a comical, frog-like appearance.

All of Henry's earlier cars were functional, but not in the same way as the Model T. It was light and powerful, rugged and durable. It was, in fact, the perfect car for the times. The public needed the durability and utility of the Model T. At the rock-bottom prices at which the Model T was offered, only Henry's contrivance fulfilled that need. At that time, few paved roads existed, except those in the city. Likewise, the dirt roads that did traverse one town to another became rut-infested quagmires with the slightest rain. Because of its simplicity and durability, only Henry's Model T could negotiate the muddiest roads and the deepest ruts without hesitation. The utility of the Model T—not its beauty—made it popular with the public and made it sell. That would remain steadfast in the mind of Henry Ford.

***

During a 20-year period, from 1908 to 1927, Ford Motor Company made only the Model T. The vehicle came in a variety of body styles—two-door, four-door, roadster, touring, and, for a short time, a town car. However, all were still Model T's.

What is more astonishing than the longevity of the Model T was it dominance in the market. Between 1921 and 1923, Ford controlled 60 percent of the entire American automobile market with only this one vehicle.

Unfortunately, and probably inevitably, Henry became blinded by the phenomenal success of the Model T. He fervently believed that if changes were made to his beloved creation, his formula for success would vanish. He insisted that the only way to entice the public to buy thousands of his motorcars was to make them functional, make them cheap, and make them the same. As far as Henry was concerned, aesthetics had nothing to do with building automobiles. He was right, of course, in a perverse kind of way. Aesthetics had nothing to do with the success of the Model T because the Model T had no aesthetics. The public bought Henry's car for its functionality, not for its appearance. However, the Model T probably would have been equally successful had it been *designed* rather than seemingly thrown together.

Henry's philosophy was effective in the early years of the industry, but progressively it became more of an aberration than a truism. When it came to meeting public demand, Edsel warned his father many times that what was working today may not work tomorrow, and certainly what was stylish today would not be stylish tomorrow. For Ford to remain a viable business, cautioned Edsel, the company would have to take steps to meet the changing market. Henry would not listen. His stubborn resistance to change his Model T to meet changing customer needs would eventually cause him to lose his commanding share of the automobile market.

***

A dichotomy existed between Henry and Edsel from the beginning. Henry, of course, was one of the earliest pioneers in the automobile business, and he did what he had to do to survive and succeed. He did not simply make the quadricycle one year and then form the Ford Motor Company the next. There were years of trials, tribulations, and adversity before Henry developed the Model T and reaped the riches of its success.

These various attempts and failures at trying to build a utilitarian automobile hardened Henry's resolve, and he became even more unyielding when he eventually created the Model T. When critics scrutinized the Model T, they were, in effect, criticizing Henry, and any affront to his car was an affront to him. Henry's sheer determination had taken him through years of anguish to develop the Model T, and he would listen to no criticism of it, even from his only son.

Edsel, on the other hand, grew up with the automobile and gained an entirely different perspective of it. As the automobile developed, he developed. As it matured, he matured. Eventually, Edsel and the automobile became inextricably intertwined. Because Edsel did not have to fight for his existence within the industry as his father had, he was more receptive to the demands of the market.

Edsel's immersion in the automobile business literally from birth developed within him an uncanny "auto sense" and a disregard for the status quo. This gave Edsel the ability to critique an automobile quickly and objectively from its outer skin down to the nuts and bolts that held it together. From his range of experiences, Edsel understood the utilitarian features of his father's Model T but appreciated the refined qualities of more expensive cars such as a Stutz or Duesenberg. He wondered why the two approaches could not be melded together, and he relentlessly worked toward that end.

Because of his receptivity and exposure to all makes and models of automobiles and his growing understanding of the automobile industry as a whole, Edsel soon found the Model T boring and lackluster. He was fully aware of what had made his father's contraption so successful, but he also was convinced that a purely functional motorcar could have lines pleasing to the eye. There was no reason that functionality and aesthetics had to be mutually exclusive. Edsel knew that automobiles must be dependable and reliable, but he also believed that styling would become a significant variable in the formula for success in the future.

***

Edsel's involvement in styling occurred for a few interweaving reasons. Being forced to analyze the various aspects of the company due to Henry's constant interference, Edsel realized that in the wide scheme of automobile manufacturing, his father "didn't care what a car looked like as long as it did what he wanted it to do."[6] To Henry, form followed function, and the function of the early automobiles, exemplified in his own beloved Model T, was ruggedness, durability, and reliability. The only worthwhile aspect of his company was manufacturing—building a quality automobile, in high volumes, at low cost—not whether his automobiles were appealing to the eye. As far as Henry was concerned, the voluminous sales of the Model T were enough proof that an unglamorous but functional automobile was what the public wanted, and that was what he was determined to produce.

So adamant was Henry against styling that he barely changed the appearance of the Model T over its 20-year production span. Its monotonous and humdrum design became the brunt of many jokes. During one presidential campaign, Will Rogers wittingly said, "Ford could be elected president all right. He'd only have to make one speech, 'Voters, if I'm elected, I'll change the front end.' "[7]

Edsel had loftier aspirations. He wanted to utilize his knowledge and understanding of automobiles and the industry to move the company into a more progressive stance by combining the artistry of custom-made motorcars with the severely functional requirements of mass production. In other words, Edsel "believed that mass production could turn out beauty at the same time it turned out volume."[8] If he could capitalize on his father's apparent indifference to styling, he might be able to develop this aspect of auto making free from his father's interference. By proceeding one step at a time, and by gently nudging his father toward more stylish cars, Edsel hoped to obtain control over at least this one aspect of his father's empire, perhaps contributing to its development and success as well.

However, the most significant reason Edsel pursued styling was for its intrinsic value. It provided him with an "indispensable outlet" for expressing himself creatively and artistically.[9] Eventually, Henry turned over complete responsibility of styling to Edsel (one of Henry's few benevolent acts toward his son) and, in so doing, nourished Edsel's self-confidence by enabling him to combine his artistic ability with his business responsibilities.

***

Edsel's felicitous eye and ability to conceptualize a form was unique within Ford Motor Company (at least until he recruited other artistic fellows similar to himself). His approach to automobile designing was more cultured than that of his father, but no less successful. Unlike Henry, who started each part and worked toward an undetermined final form, Edsel started with a conception of the final form and worked "backwards," developing the individual parts that would make up the predetermined whole. With this approach, Edsel placed form over function, leading to the creation of motorcars unparalleled in elegant composition.

***

Edsel was a born with an acute sense of style. As the automobile became more refined mechanically and aesthetically, so did Edsel's design acumen. He had a natural ability, but he developed his talent by poring over automobile catalogs and magazine advertisements that expounded the virtues of all types of motorcars. From these ads, Edsel studied every design in detail, assimilating the best attributes of each. Early in his teens, he began a collection of scrapbooks, saving within them pictures and clippings of automobile designs he thought the most beautiful and dramatic. (His initial entries were of Model T's—equipped with special coachwork.) His taste became more refined as he grew older, and instead of saving pictures of fancy Model T's, he began collecting images of exotic American-designed cars such as George T. Brokaw's special Marmon *Canoe Roadster* and William Wrigley's *Semi-Touring* Auburn. William Vanderbilt's custom Locomobile and his Hispano-Suiza special phaeton also captivated Edsel.[10] Edsel exhaustively examined pictures of these outstanding motorcars after pasting them in his scrapbook.

Edsel evaluated each of these motorcars—as well as hundreds of other designs—for special nuances. For instance, he would take a picture of a Vauxhall and draw a circle around its graceful fenders, or he'd highlight the split windshield of a Grand Prix race car. Once, when he was in an especially insolent mood, Edsel inserted a picture of a Model T with an English body alongside a Rolls-Royce. From his perception, "except for size, they looked very much alike."[11] He'd spend hours analyzing each pitch of the fender, each slope of the rear bodice, and each angle of the windshield for each of the automobiles in his scrapbooks. In this way, Edsel not only absorbed the appearance and construction of each vehicle, but he also assessed their characteristics and personalities. This in-depth analysis of various sizes, configurations, and classes of motorcars cultivated within Edsel an uncanny sense of design and styling, and augmented his intuitive talent for design. Over time, Edsel formulated his own personal theory and approach to automobile styling, eventually developing an "unusual sensitivity and a rare feeling of form, composition, and color."[12]

By the time Edsel was in his late teens, his father had amassed enough wealth to enable Edsel to build special vehicles according to his own specifications and designs—a privilege accorded to only the wealthiest in Detroit. At a time when the automobile remained a luxury, being able to design and build automobiles, not only owning them, exemplified the status his father had already attained.

Most of these early experimental designs were merely stripped-down Model T's that Edsel called speedsters. Resembling production Fords before their bodies were attached to the chassis, these speedsters had a monocle windshield, dual bucket-type seats, an exposed gas tank, sporty wire wheels, and swooping fenders. One of Henry's employees remembered Edsel driving to his father's mansion in Dearborn: "I know he had these sporty automobiles, and that's what I liked about Mr. Edsel coming out—seeing these nice new cars. They were Fords, but they were fixed up just a little bit different. I remember that little six-cylinder Ford he had with a yellow body."[13]

Indeed, Edsel's speedsters were much more than souped-up Fords. Compared to regular Model T's, which had square, top-heavy bodies mounted on top of black-painted wooden spoke wheels, Edsel's personally designed speedsters were exotic and racy in appearance.

We get a rare glimpse into the intensity Edsel placed in these projects, as he did in most of his endeavors, from a notebook he kept. In it, he listed the things he had to do day by day to complete one of his speedsters: "separate engine from muffler..., lengthen steering post..., turn over steering wheel..., draw hub caps..., make bearing retainer..., shorten running boards..., install rubber fenders..., remove 2 leaves out of rear spring..., color: gray, white pin stripe."[14]

These modifications were relatively easy to incorporate into a production car, but Edsel went further later. In one car, he had Ford engineers design and build for him a pair of bucket seats, a set of wire wheels, a V-shaped radiator, and a complete six-cylinder engine to place into one of his speedsters. "Mr. Edsel Ford often came into the shop and watched us work on it," recalled one of the engineers assigned to the project. "He was a quiet man, very retiring, but he had a real smile on his face as he watched us work on his car."[15]

According to Ernest Kanzler, Edsel owned 25 to 30 beautiful and exotic hand-made vehicles over the years.[16] However, those were not the only unique motorcars Edsel owned. Although Edsel still lived at home, Henry indulged his son's penchant for stylish cars by buying him sporty Mercers, elegant Locomobiles, and dashing Delages. After getting married, Edsel continued to buy other expensive, exotic motorcars to fill his stable. It actually became more common for him to be seen driving one of his original and distinctive automobiles than one of the manufactured contraptions churned out by his own company. "Every once in a while," recalled Edsel's son, William, in an understated account, "a new car would appear in the driveway which was clearly not a production car."[17] Ordered and built to his exacting requirements, Edsel's variety of unique motorcars represented the evolution of his design philosophy and provided a means for him to season his "natural talent for styling."[18]

An example of Edsel's eclectic taste was the eight-cylinder Grand Prix model Bugatti he bought in 1926. "I have been fussing with the Bugatti quite a bit since it arrived here," he wrote to a friend, "and have been able to make it run very well, due to checking over the distributor coils and replacing the spark plugs with some which will not corrode from excessive oil. I have also put some baffle plates in the muffler, which has deadened the exhaust, and am just about to put the car in the paint shop and try and make it look like something."[19]

***

Of all the personal cars Edsel owned over the years, two of them that were built in the early 1930s exemplified many of the design ideas he had been cultivating with his earlier models. The first one was a boat-tail speedster built from a production 1932 Ford. The little car had many radical features for its time: all-aluminum body, no running boards, a raked split windshield, smooth exterior, and a V-shaped grille shell. Taking advantage of the vast manufacturing capabilities of the company, Edsel had the body and fenders of the car built at the Ford Dearborn Engineering Laboratory—the trim, paint, and final finish were done at the Lincoln plant.[20] The second prototype was also a boat-tail speedster, but it was built two years later from a production 1934 Ford. Similar to the 1932 prototype, this special had a body made of aluminum panels that were shaped over a frame made of airplane-type tubing. Both the radiator ornament and filler cap, so prevalent on automobiles at the time, were "designed out" of this car. Running boards were not included on this model either.

Both of these specially designed roadsters were expensive to build—each costing more than $100,000. However, they helped solidify some of the most innovative concepts that Edsel had been formulating in his mind—design ideas that eventually would be incorporated into production Fords. The streamlined surfaces that the tubular framing construction afforded and the elimination of superfluous features such as exposed running boards, radiator caps, and other ornamentation would eventually be incorporated into all of Edsel's designs and were to become the hallmarks of his styling approach.

***

After all the years of building and studying exotic-looking motorcars, and with his penchant for styling, "Edsel was not a designer in the strict sense of the word," explained Bob Gregorie, Edsel's chief designer. "But he had a sense of what looked right, and that made him an excellent critic. He could control the design by being a critic. He understood what I was talking about, and I understood what he was talking about."[21] It also enabled Edsel to direct and guide the styling team, which he would eventually develop, toward handsome, timeless automobile designs. Edsel never professed to be a designer, but he admitted that he was able to discern what did or did not look right—what was or was not a good design. "I do not know the manufacturing end so well as I know the commercial end," Edsel said in a rare interview during the development of the Model A, "but I think I can judge design, for on that I have had long experience. But I can only judge."[22]

Raymond Dietrich attributed Edsel's design talent or affinity for style to Edsel's "discerning eye." Edsel understood the difference between the "ordinary line and the lines of the classics," Dietrich explained.[23] "He had an instinctive liking for dignity in an automobile...dignity that reflected a purpose."[24] To Edsel, every line, every contour, every angle in an automobile design must have a purpose and must be simple. "He had a very keen feeling of the pure sensitivity of a line and a feeling of lightness," recalled his chief designer. "But above all, Edsel was a stickler for fineness. He liked a trim, delicate effect of a car. He didn't care for a big, bulbous sort of design. He wanted delicate bumpers and delicate doors handles and things like that."[25]

When Edsel said that he wanted to make the best car in the world, he meant not only tested mechanical qualities but also superior aesthetic appeal. He wanted the best styling in the world, and he knew that the only way he would get it would be for him to set up his own design department within Ford Motor Company.

***

In January 1931, Edsel took the first major step in aligning Ford with the design initiatives of General Motors and Briggs when he hired Eugene T. "Bob" Gregorie, the first true automobile designer for the company. Ford did not have a design department before then, and it would be another four years before Edsel had a complete design team working for him. For now, Gregorie would assist Edsel in transforming Henry's "standard practice of engineering a car and then covering it with sheet metal as the occasion demanded" to "get some style into the cars, and to try and sell more by styles" instead of pure utility.[26]

Gregorie came to Ford Motor Company and gained Edsel's attention through an uncanny and prophetic chain of events. Born on Long Island, New York, in 1908, and raised there, Gregorie had an instant affinity for water and spent most of his spare time as a boy "fooling around with boats, especially yachts." He gained no formal education except what he learned around the docks, but that was sufficient for him to land his first job as a marine draftsman with the Elco Works, a boat building firm in New Jersey.

Gregorie's true love would always be the elegant lines of a graceful yacht, but he developed an appreciation for automobile design through the number of European motorcars that his father owned over the years. This appreciation soon led Gregorie to the Long Island design firm of Brewster & Company in 1929. Brewster had been designing and building custom automobile bodies since the beginning of the twentieth century and had been a custom carriage maker since 1810. Edsel Ford had purchased several Brewster bodies for his personal Lincolns. There, Gregorie learned the basic skills of automobile body design.

Young and restless for adventure, Gregorie left Brewster after less than a year and headed for the large production companies of Detroit in the fall of 1929. Upon arriving at the motor capital, he proceeded directly to the General Motors headquarters on West Grand Boulevard, showed them his credentials, and immediately landed a designing job in the newly established Art & Colour Section of the company. Gregorie's skills won him the job, but the economy quickly stifled his career at the large automaker. Three weeks after he had started work at General Motors, the stock market crash hit Wall Street, and Gregorie was laid off.

Undaunted, Gregorie went looking for another design job with one of the other automobile companies in Detroit, including the Lincoln Motor Car Company. Unfortunately, all of the other companies also were reeling from the crash and were not able to hire anyone. Relying on his naval architectural experience, Gregorie instead landed a job helping to finish the design of a yacht for a local entrepreneur. That job lasted until midsummer, and when it ended, Gregorie returned to Long Island.

In the fall of 1930, Gregorie went with this family on holiday to South Carolina. Shortly after New Year's Day in 1931, he received a telegram from the chief body engineer at Lincoln, Henry Crecelius, whom he had met after he was laid off from General Motors. The telegram said that Lincoln was in need of a body designer and that the position could be Gregorie's if he wanted it. He could start immediately.

Gregorie took the first available boat out of Charleston and headed to New York, where he then took the train to Michigan. Tired of pacing up and down the decks of the ship, Gregorie decided to pass some time by allowing a gypsy to tell his future. "A gal on the boat was reading tea leaves," recalled Gregorie, "and she said to me, 'I can see a small, dark-haired man, and I can see a long, low, white building. He's going to be an important part of your future.' "[27]

"Well, the man was, of course, Edsel Ford, and the building she described was Ford's Engineering Laboratory."[28] Gregorie would spend the next one and one-half decades at the Engineering Laboratory, designing Ford automobiles under the auspices of Edsel Ford.

When Gregorie arrived in Detroit for the second and last time, Crecelius put him to work developing body designs for Lincolns. Soon, Edsel became acquainted with Gregorie's work and realized that Gregorie possessed the talent and personality he was seeking to establish his own in-house design department. Within a few months, Gregorie was reporting directly to Edsel.

"Gregorie was a smart, cocky young fellow who knew it all," recalled Laurence Sheldrick, "and he wouldn't take anything from anybody."[29] Normally, Edsel shied away from overbearing people, but he and Gregorie seemed to hit it off from the start. Besides, Edsel knew that if he were going to set up a design department within the company, he would need somebody like Gregorie, whose talent and temerity could stave off the interference he doubtlessly would receive from Henry's men. Regardless of Edsel's innocuous intentions, he knew that the mere existence of a design department within the Ford Motor Company would threaten the production men of the company, such as Martin and Sorensen, and even Henry himself. They would consider the establishment of a design department as an attempt by Edsel to usurp power from them, not as a means of making the company more competitive. Although Edsel would have preferred a more reserved person with whom to work, Gregorie's self-assertiveness, confidence, and talent fit Edsel's needs perfectly. "Edsel thought an awful lot of Gregorie," recalled a Ford employee.[30]

After Edsel had brought Gregorie under his tutelage, he had the makings of a full-fledged design department—of only three men. Edsel, of course, was the overall director of the department, providing them with inspiration. John Crawford, Edsel's longtime executive assistant, was his production expert, and Gregorie, only 22 years old, was the designer.

Similar to Gregorie, Crawford was a crusty, forthright type who had worked for Ford since Edsel had become president. Because he was both an attorney and an engineer, he was especially helpful to Edsel and would advise him on both technical and business matters. In Edsel's embryonic design troika, Crawford was the "modifier." "I had to figure out how or if a car could be built," Crawford recalled. "Could we form a sheet of steel into the desired shape? Was chrome trim practical?[31]

Gregorie's role was to "put on paper Mr. Edsel Ford's thoughts on design."[32] However, his cocksure attitude "kept him in continual hot water," especially with Henry Ford.[33] However, Gregorie reported directly to Edsel, and thus he never suffered any retribution from Henry as so many others had endured. Occasionally, Henry Ford would come into Gregorie's office to see what Gregorie and Edsel were working on. Trying to be polite, Gregorie would begin to explain a particular design or project, but Henry would stop him short. "That's between you and Edsel!" Henry would blare. "That's between you and Edsel!"[34] Then he'd walk out of the room.

***

During the 12 years that Edsel formally directed styling at Ford Motor Company, he kept the size and atmosphere of the design department very simple. It was always a small staff—never exceeding more than 100 people while Edsel was there. "He [Edsel] knew most of the boys in there by their first name," explained Gregorie. "It was just a friendly, folksy arrangement."[35] The men called Gregorie by his boyhood nickname, Bob (some called him E.T.), and he enjoyed giving them nicknames of their own.[36]

Knowing the demeanor of artists and other creative people, Edsel managed his department in a similar free-flowing arrangement. "He set up no rules. I didn't even keep regular hours. Sometimes I'd just take off for a long trip in my car to clear my brain," recalled Gregorie.[37]

Although Edsel's design department was simple and probably understaffed, it had the design talent and capabilities equal to any of its competitors, especially General Motors. The only similarity between the General Motors Art & Colour Section and the Ford design department was that each was created specifically to design automobiles. The General Motors department, supported munificently by Alfred Sloan, began small (it had 50 people shortly after it was established) but quickly grew. The Art & Colour Section was separated into different, smaller departments, each specializing in one of the five car divisions of General Motors: Chevrolet, Pontiac, Oldsmobile, Buick, or Cadillac. At Ford, all of its designers designed all of its products—Fords, Lincolns, Zephyrs, and Mercurys. There was no differentiation regarding who worked on what. Harley Earl, an imposing, boisterous Californian, ran the General Motors design department, whereas Edsel, unpretentious and gracious, ran the Ford design department.

Both Edsel and Earl were extremely talented men when it came to understanding automobile design, but each man ran his design department with diametrically opposed methods. Edsel was democratic and paternalistic, whereas Earl was dictatorial and authoritative. This disparity in personalities and management techniques made the design studios of General Motors world famous and the one at Ford essentially unknown. Earl, with his flair for notoriety and need for recognition, made sure everybody was aware of him and his design department. Edsel, on the other hand, was content in developing his designs out of the limelight and in the seclusion of his Dearborn studio. Ironically, Earl's Art & Colour Section would become the epitome for automobile design studios, but the designs created by Edsel's diminutive staff were to become the envy of the industry.

Although Edsel held the title of president and Earl the title of director, Earl was probably Edsel's closest counterpart in the area of automobile design. Only 16 days apart in age, the two men would direct their respective design studios as though they were generations apart. Both used energy and talent, but Edsel used progressive techniques in managing his studio, whereas Earl used dictatorial methods so typical of that time. Edsel and his team of designers developed their models within a group atmosphere, but Earl used fear and intimidation to obtain designs from his men. This dichotomy was due to the different personalities of each man, but the organization in which each worked also played a part. Edsel himself had the final say in all Ford designs, but Earl had to sell his ideas to divisional heads and to Sloan himself. "It was not a question of selling or intimidating a policy committee," confirmed Gregorie, "but of working together with Edsel toward a common goal."[38] Even Gregorie, who tended toward independence, supported Edsel in his team approach. "There are not going to be any prima donnas here," Gregorie told his staff. "We are all working boys. Let's do a good job. Let's please Mr. Edsel, and if there are any problems, let's sit down and talk about them."[39]

That was not the situation at General Motors. Because of his domineering personality, Earl set up an elaborate hierarchy within his studio, which deliberately put distance between himself and his designers. At Ford, however, there was but one layer of red tape—Gregorie merely had to convince Edsel. Indeed, the bureaucracy at Ford was so slimmed down that creating and selling the design of the product became indistinguishable.[40] For Earl, it was a continual selling process. When he had design ideas to propose, he would have to present them at an executive showing, where all of the General Motors divisional general managers and other executives could either approve or veto the design. "Earl was always kicking against the pricks to get his designs approved," recalled Frank Hershey.[41]

***

Day after day, Edsel would devote an hour or two after lunch to the work of the Design Department. He would leave the Round Table luncheon in the northeast corner of the Engineering Laboratory and walk, usually alone, down the length of its immense corridor, and then enter the design studio. Edsel would meet with Gregorie, and the two men would look over the clay models, make changes, and make decisions.

Bob Thomas, a young design apprentice at the time, explained how easily decisions were made, and how unbureaucratic Ford's design studio was[42]:

> *They needed some ideas for a rear deck handle, so I made some sketches. Gregorie liked one of my designs, so he told me to draw it up and have it made at the Rouge plant.*
>
> *About two weeks later, this beautiful chrome-plated model appeared in the studio. The thing was cast in solid brass and weighed a ton. Anyway, when the time came to choose a rear deck handle, we all had the proposals laid out on a big table covered with black velvet. As was general practice, Edsel Ford came into the studio after lunch in the lab. I remember waiting breathlessly as he picked up my design and said, "This is it."*

Despite their disparity, both Edsel and Earl shared one thing—they both knew design. When the prolific custom body designer Ray Dietrich first met Edsel, he was "thrown completely off guard by his extensive knowledge of the entire field of prestige cars, both European and American." When Dietrich asked Edsel questions about various aspects and qualities of different automobiles, "Edsel referred to the different body styles by correct type of name according to the Society of Automotive Engineers' standard body nomenclature."[43] Earl likewise understood design, but he was not as articulate as Edsel. In fact, Earl literally invented words to represent the visual forms he had in his head, and he would force his designers to understand and use his fabricated vocabulary. "I want that line to have a *defluncky*, to come across, have a little hook in it, and then do a little *rashoom* or a *zong*."[44]

"When Earl called something a *blitz line*, that became a blitz line," recalled Frank Hershey. "Then anytime he said blitz line, you knew what he was talking about. Out of abject terror, Earl's words became a permanent part of everybody's vocabulary."[45]

Although the managing styles of Edsel and Earl definitely differed, the most obvious contrast was in the way they dressed. Edsel dressed neatly, but he was conservative. "He didn't care for flamboyance in his clothes, or lifestyle, or in his automobiles, or in anything else," recalled Gregorie. "Whatever Edsel wore, drove, or owned had to be simple, plain, and in good taste—immaculately good taste."[46] Earl, on the other hand, thrived on extravagance. He too was a fastidious dresser, but his taste lay on the other end of the fashion spectrum from Edsel's. According to those who knew him well, it was common for Earl to wear purple shirts, bronze-colored suits, and suede shoes.[47] The macho-men of Detroit, such as the Fisher brothers, for example, used to make fun of Earl's flamboyant suits, shirts, and neckties.[48] "He was showing them that he was a designer by wearing all this stuff," explained one of Earl's designers. "He was a big man (he stood 6-4), he should have dressed very conservatively. But he didn't. He had an image."[49]

Edsel and Earl had one thing in common, however—they both used a hands-off approach in managing their studios. In other words, each man directed his designers toward designs that they—Edsel and Earl—envisioned, through verbal cues, as opposed to literally showing them what they wanted by making a sketch or drawing. Earl's reason for using this approach was quite different from Edsel's, however, and very enlightening. Edsel could have made a sketch or a drawing—even a complete rendering—to describe a particular shape or form that he had in mind to Gregorie, if he wanted to do so. However, he never used his artistic skills in that manner. Instead, he kept his illustrative skills to himself. On the other hand, Earl would not sketch a form or design idea simply because he could not draw.[50] This lack of artistic skill, combined with Earl's difficulty in articulating his desires to his staff, made him practically inexpressive and made working for him frustrating. "He knew what he wanted," recalled one of his designers, "but he couldn't draw it for you. And he'd get so impatient with you if you didn't get what he was trying to describe, that he'd start yelling. 'Now, goddamn it! I want it like this!' "[51] Typically, Earl would sit in his Hollywood director's chair and instruct all of his designers to sit around him. "Now, move this back. Now, do this," he would command. "He'd sit there and everyone would run around him like a bunch of monkeys," recalled Bill Mitchell. "That was the way cars were designed at General Motors when he was there."[52]

Edsel would never become upset if a design were not to his liking. He would work with Gregorie and his designers and elicit their ideas using a civil, intelligent method. He would go to the designer's drawing board and discuss the design with him. "I like that," he would say in encouragement, or "I don't like that," in disappointment. If it were a particularly poor design that he was evaluating, or if changes were made at his direction that did not turn out quite as he had envisioned, Edsel would not chastise the designer. Instead, he would say simply, "Oh, gee, that looks pukey, doesn't it?"[53]

As an overall design came closer to what he envisioned, Edsel would suggest minor changes until he was completely satisfied with it. "Oh, how about making this a little longer," he would say. "Make this a little thicker here. Would it do any harm," he would question rhetorically, "to round that a little more?" When he was satisfied with the design, he would confirm the idea with Gregorie. "Does it look good to you? Okay? All right, let's do it that way, huh?"[54] During these discussions, Edsel "had a quick way of moving his hands when explaining something—moving them downward or upward or in a swooshing motion to indicate height and width and shape."[55]

Through this artistic exchange—offering criticism and taking suggestions, offering direction and receiving rebuttals—Ford cars were designed when Edsel headed the Design Department. He would set the direction for the team, and they all pursued it until an acceptable design came to fruition.

***

Throughout this interactive process, Edsel was trying to achieve some specific designs that he had in mind. He was never arbitrary or discretionary—he always had a reason for his suggestions, a purpose behind his intent. However, "when Edsel made up his mind, he wanted something, he wanted it to look like he wanted it to look. He was obstinate as hell," recalled Gregorie.[56] That, of course, was Edsel's prerogative and his responsibility—to get the lines, the flow, the styling that he envisioned from his designers.

The most explicit description of how Edsel operated, how he got the best from his designers in an unobtrusive and intelligent manner, was eloquently described by industrial designer Walter Dorwin Teague. Throughout the 1930s, Teague designed several exhibition buildings for Edsel. While his direction explains the process he and Edsel used to develop architectural schemes, it is not difficult to envision that Edsel used a similar process in the design of automobiles[57]:

> *It was a thrilling experience to submit a complicated project to him—for a major exposition building, say, where imagination and invention had scope. With keen perception he would take in the dominant motif of the scheme, and then follow*

it tirelessly down through the entire composition. No detail escaped him: every aspect, every feature was fitted into the complete visualization he was building up in his mind. He needed to know the proportions of every opening, the color of every wall, the tone and texture of every floor. Throughout, he never disturbed the plan by injecting an alien preference of his own. But from time to time, he would put his finger on an element that was out of key, in a big or little way, not fitted perfectly into the designer's major intention. So the design would grow, become more refined and clarified, more perfectly itself. And when the session or long series of sessions had ended, the designer would feel that he had been assisted by a master.

***

In the process of designing a new Ford, Edsel would first ask for design ideas in two-dimensional form. "He loved to come in and sit on one end of a drawing board," explained Gregorie, "and have me pick up an old blueprint and sketch something on the back of it—a hood ornament or some damned thing."[58] They would chat about automobile designs, and Edsel especially liked to tell Gregorie of the graceful cars he had seen in Europe while on holiday there. Gregorie and Edsel thought similarly, and during this stage of design, they worked best together. Both men liked plain and simple designs, "so that is one reason Edsel and I hit it off right away," says Gregorie. "He also liked the idea that I designed yachts. He loved yachts." Occasionally, Edsel would bring in a clipping of an especially pleasing boat design, and "we'd talk back and forth" about its salient features.[59] Usually, Edsel "had no time for small talk—absolutely none," continues Gregorie.[60] Therefore, these extraneous discussions happened only on rare occasions.

Gregorie's experience in yacht design gave him an uncanny sense of proportion and feeling for a beautiful line.[61] "I can look at a line and tell whether it is fair or should be tucked in here or tucked in there," explained the designer.[62] Gregorie would dissect conversationally the design of a car while Edsel stood by and pondered it. Edsel would seek ideas in similar fashion from all his men—from his engineers, body engineers, and, above all, his stylists. Each part of the car, and eventually the whole car, was designed in this manner.

After the overall design was established, the concept was transformed into a three-dimensional clay model. In the early years of Edsel's design team, their studio was not equipped with the proper tools and equipment to efficiently build these models. As they became more involved in projects, and as the need for such equipment increased, Edsel installed a metal shop and a wood shop. People from experimental engineering, pattern makers, and plaster sculptors were brought in, so that the entire design process could be implemented within Edsel's purview. Before the styling group was formally established, Gregorie would create various styles and have them developed into three-dimensional clay models. Joe Galamb, head of the Ford body engineering group, would then analyze each design for manufacturability. Gregorie merely made the presentations of his designs. "He was the artist," recalled Laurence Sheldrick, "and the other group determined the body styles that they would use."[63]

When a model was complete, "Edsel liked to criticize the first model in clay. If he didn't like the shape of the grille, or the back end, or the fender," he would ask his designers to change them. When he "made suggestions, we didn't have to make an entirely new clay model, we just made the changes on the ones we had," recalled Joe Galamb. "He would come back two or three days later to inspect it again. Everything was ready after he okayed the clay model to make up the final drawing of a sample. He never made any changes after we had our sample made up."[64]

During these critique sessions, Edsel sometimes could not express why a particular aspect of the design appeared wrong. He was usually quite articulate about explaining a design, but he had an "active mind" that could synthesize a design faster than he could describe it.[65] Although he had pursued both painting and sculpture as a hobby for years, Edsel never did sketch on paper an idea that he had and give it to his designers.[66] Instead, he would describe what he wanted as best he could by moving his hands in quick motions to "explain" something. "But even if he couldn't explain exactly why he didn't like a particular design," explained Gregorie, "I had enough appreciation of his judgment that I would modify it to a point so that it pleased him."[67]

***

Such was the rapport Edsel had with his men, and the respect that the men had for Edsel. He did not have to resort to scare tactics or intimidation to effect requisite designs, as Harley Earl did. Edsel simply managed and directed more effectively. "I've gone into the design department with Edsel many times," recalled Henry Doss, a Ford executive, "and he was definitely in charge of that."[68]

Chapter 21

Edsel's Sanctuary

"We didn't call it 'Styling Department,'
because 'styling' smacked of women's wear."
E.T. Gregorie[1]

"Edsel always liked swanky automobiles,
good-looking cars."[2]
Harold Hicks

After the Model A was well into its production cycle, Edsel began thinking about the next model change. Henry was probably thinking that he would produce the Model A for years and "jam it down the public's throat" for at least as long as he did with the Model T.[3] However, Edsel realized that styling changes would have to be done more often, and he knew that the Model A was only an interim step in moving Ford toward the realities of automobile manufacturing being championed by Alfred Sloan at General Motors.

Three years after the Model A changeover, Edsel began development of the Model B.[4] With this car, Edsel wanted to add more of the refinements that he did not have time to incorporate into the Model A. "The efforts on the Model B were to make a car that was longer looking and decidedly lower looking," recalled the chief engineer at Ford. "He wanted the Model B to have more of a big appearance."[5]

To help on the project, Edsel called upon the services of Briggs Manufacturing and its newly recruited design chief, Ralph Roberts. Edsel would eventually bring E.T. Gregorie into the process, but that would not occur until a year or two later. For the initial years of the 1930s, Roberts and his staff did most of the design work for Ford, through Edsel's direction.

***

Roberts had a number of talented designers working for him, John Tjaarda and Al Prance among them. Tjaarda was a Dutch engineer who had come to America in 1923 and was involved in several design stints before coming to Briggs. He initially went to California and designed special-bodied cars for movie stars. From there, he went to work for Locke & Company, coach builders in Rochester, New York, where he designed special bodies for Pierce-Arrow, Hispano-Suiza, and other luxury car manufacturers. He spent another two years with Duesenberg and then went to the General Motors Art & Colour Section in 1930. Within two years, Tjaarda was working at Briggs.[6]

Al Prance had worked with Roberts while he was at Le Baron, and he designed most of the interiors for their custom body designs. Edsel knew of Prance's work with Le Baron and thought he was a "tremendous designer."[7]

Roberts' job was to have his men make sketches and clay models and present them for consideration to Ford, Hudson, Studebaker, and other manufacturers for which Briggs made bodies. "The idea was to sell these companies on specific new designs and body programs," recalled Roberts.[8]

The design process used by Briggs was similar to that used at General Motors, but instead of building full-size clay models from design sketches, Briggs built its full-size mock-ups from wood. The car manufacturers would then approve or reject these design concepts, based on their evaluation of the mock-ups.

* * *

When Roberts was working on new design ideas for Ford, Edsel would usually drop in at the Briggs plant on his way to his office at the Rouge Administration Building. (The Briggs plant was located approximately halfway between Edsel's Gaukler Pointe mansion and the Ford factories in Dearborn.) Edsel would visit two or three times a week, around 9:00 in the morning. Most of the time, he would visit Roberts alone; however, when the need arose, he would ask Joe Galamb to meet him in the lobby, and they would both visit Roberts.[9]

Upon his arrival, Edsel would walk through Briggs' elegant reception area, replete with marble fireplace and stately curtains, and proceed to Roberts' studios located on the fourth and fifth floors of the company's Mack Avenue plant. "There wasn't a lot of preparation whenever Edsel would come over," explained Roberts. "He just came up to our studio and reviewed what we had. Our different designers would submit sketches of suggestions, which I would go over with Edsel. Eventually, we would come up with one he thought we could develop."[10]

"Once we settled upon an idea," continued Roberts, "we would make a scale model, a full-size blackboard drawing, and then change these to the satisfaction of Edsel. Then we would get Joe Galamb to approve them, more from a technical side than from the design side. The design was then forwarded over to our body engineers at Briggs for their evaluation. After the production engineers approved the design, a wooden, full-size mock-up was made, entirely complete: exterior, interior, doors that opened... Slight changes could still be made at this point..."[11]

These wooden models were extremely realistic, even more so than the clay models that superseded them, primarily because the wooden models had functional interiors, whereas the clay models represented only the exterior form of a car. The wooden models also proved more useful to the manufacturers because those models could be exhibited at automobile shows. "One time," recalled Roberts, "while a mock-up was in our executive garage ready to be sent to Edsel in Dearborn, our CEO, Walter Briggs, in a hurry and filled with anticipation of another new Ford order, rushed down to thc garage, jumped into the mock-up, pumped the accelerator and turned the key switch... No go!"[12]

"Once a mock-up was ready, we'd send it to Dearborn—nobody was going to get Henry Ford to leave Dearborn!—where Edsel would evaluate it and show it to his father and the other executives."[13] Because Edsel had been closely involved with the design from its conception, this stage was essentially a formality.

Through this procedure, this process of give and take, and of offering suggestions and constructive criticism, the 1932 Model B was created. Although the Model B was initially produced with a power train almost identical to that of its predecessor, its outward appearance was a complete makeover. With its subtle curves and graceful lines, the new Model B was longer, heavier, and roomier than the Model A.

Slowly, Edsel was edging Ford Motor Company toward Sloan's idea of yearly model changes.[14] It took Edsel 20 years to implement a change from the Model T to the Model A, but only 4 years to change from the Model A to the Model B. Edsel's next change would take him only 12 months.

***

While Edsel was working with Briggs on the replacement for the Model A, he was also working with Gregorie on a vehicle design for the European market. This car, known as the Model Y, would be the first Ford vehicle designed by Gregorie for Edsel.

The Model A was a resounding success in the United States but a dismal failure in England. (This was the car that Kanzler was criticizing in his memo to Henry.) According to Ford's British managers, the Model A was "too expensive, too costly in taxes and gasoline, and too large for narrow streets." Additionally, two of Ford's rivals in Europe—Morris Motors and Austin—were threatening to run away with this market if Ford did not act quickly.[15]

In transatlantic telephone calls, Percival Perry, the British manager for Ford, explained the predicament to Edsel and then sent him a list of what he thought should be the requirements for a competitive car.[16] The list exemplified the great disparity between what the European market demanded and what the Model A offered. "The car must have a small-bore engine of about 8 English horsepower, a 90-inch wheelbase, a narrow tread and limited weight."[17] The Model A, in comparison, had a large bore, 40 horsepower engine, 103.5-inch wheelbase, and wide tread.

The mechanical aspects for the new "Y" were assigned to Laurence Sheldrick, a graduate engineer from the University of Michigan, and Ford's power train expert. The responsibility for engineering the body was given to Eugene Farkas. "I got acquainted with a young fellow who got started with Ford maybe a month or so before we started this project," explained Farkas. "His name was Gregorie. I got acquainted with him and he showed me some advanced sketches of different body designs that he had developed."

"I said, 'You're just the man we want. We want you to help us out on this little English job. Design us a nice, up-to-date body for it.' "

"That's what Mr. Ford has asked me to do," replied Gregorie.

"He was delighted," recalled Farkas. "Gregorie was just a young fellow with not too much experience, but he had talent. So he made some sketches of the little car."[18]

Coincidentally with Gregorie's work, Ralph Roberts heard about the project and decided to pursue some design ideas for the new car on his own, hoping he could sell one of them to Edsel. As soon as he had some ideas ready, Roberts took them to Dearborn to show Edsel.

After reviewing his designs, Edsel told Roberts, "I wouldn't care for it, but if that's what Perry wants, okay."

Roberts took that as a cue from Edsel, thinking that if he could sell Perry on his design, he could capture this new Ford business. So he sailed to England and showed his "pre-authorized" design to Perry and his staff. "That's not what we want," Perry said, after seeing Roberts' proposal. "But if that's what Edsel wants, I guess it's okay." Roberts knew at that point that he had lost the contest.[19]

Edsel chose Gregorie's design over Roberts' design solely on its merits. Gregorie's design was well proportioned for such a small vehicle. It had distinctive but simple lines—attributes that appealed to Edsel's desire for having a "trim, delicate effect on a car," as Gregorie explained it.[20]

* * *

Actually, Edsel thought so much of Gregorie's design that he ordered a complete redesign of the recently completed Model B. He took the body drafts of the Model Y to Clare Kramer, a body draftsman who worked for Joe Galamb, and told him to "step up" the proportions of this car to an American-sized car.[21]

A week or so later, Kramer cornered Edsel and showed him the profile that he had developed. It was a stunning transformation. The larger proportions of the 1933 American Ford transformed the proportions of Gregorie's handsome Model Y into a strikingly beautiful automobile.

"Roll it up for me," instructed Edsel, pointing to the drawing. "I want to take it with me."

With drawing in hand, Edsel went immediately to Gregorie's drafting board and showed him the design. "How do you like it?" Edsel asked.

"Oh, fine," replied Gregorie.

"Do you recognize it?"

"Yes, of course, Mr. Ford."

The entire thing completely surprised Gregorie, but he was proud of the outcome. "Body Engineering handled the whole process at the order of Edsel Ford," explained Gregorie. "They stepped up my Model Y design a certain percentage to produce the smart 1933 Ford line."[22]

The 1933 Ford was the second major redesign of the Ford line in as many years. The 1932 Ford had been a crash program, from the chassis and new V8 engine developed by Henry, to the body design developed by Edsel and Roberts. However, as far as Edsel was concerned, it was only an interim step toward building more graceful motorcars. The 1932 Ford was still too stubby for Edsel's liking. For the next new Ford design, Edsel wanted a more graceful design on a longer chassis.

Of course, Henry was reluctant. He was not used to making major changes so often. He changed from the Model A to the Model B only because of pressure from the market and his desire "to put the competition in their place" by developing the first low-cost V8 engine in the industry. "Edsel liked the styling of the English Ford so well," recalled one of the engineers, "that he sold it to Mr. Ford to redesign the '32 model for 1933. In such a short time, the body change was very costly, but Gregorie's design appealed to him that much."[23]

Edsel won this battle because it would not cause any grief on Henry's part. Edsel already had the body design done, and all he needed from his father was a chassis with a longer wheel-base. After considerable discussions, Henry agreed to add six more inches to the wheelbase of the 1932 Ford. Although not a considerable amount for the engineers to change, the added length helped Edsel tremendously in developing automobiles with a more graceful appearance. That is why transforming Gregorie's Model Y onto the larger wheelbase turned out so well.

Now that Henry had allowed two major changes in four years, Edsel was confident he could convince his father to make changes more often in the future.

***

While Roberts and his designers were hard at work developing the next generation of Fords, the only thing that Gregorie could do was work on new ideas for Crecelius and Edsel, most of which never saw the production floor. However, that was about to change. In February 1935, Edsel took his annual winter vacation to his Florida home in Hobe Sound, and there he finally decided to establish a formal design department within Ford Motor Company.

Edsel had wanted his own design department for years, but his father's reluctance toward styling and the lack of an adequate design staff prevented that. "Edsel recognized that the establishment of the Design Department was a likely candidate for becoming just another episode in this on-going battle with his father," explained Emmett O'Rear.[24] With the success of the Model A, the 1932 Ford, and more recently Gregorie's designs of the Model Y and the 1933 Ford, however, Edsel sensed the timing was right to make the move. Gregorie's skills played no small part in Edsel's decision. He had seen how well Gregorie could design automobiles to his liking and how forcefully Gregorie stood up to the sniping and insults from the manufacturing men. Consequently, Edsel felt that Gregorie had the wherewithal in all areas to direct a design department.

***

Nobody knows if Edsel discussed the decision he was about to make with his father. If he did, he probably did it over the telephone while in Florida, relaxed, with his argument prepared in detail beforehand, and while his father was at home relaxing next to the fireplace. It's not that Edsel was afraid to make a decision. There were many incidences over the years where he made quick, decisive, and proper decisions. Similar to managers of today, however, Edsel preferred making decisions only after gathering sufficient information, rather than in the heat of the moment. He detested confrontation and "never cared to get involved in the heat and squabble of any kind," explained Gregorie. If Edsel were forced into a position to fight for his point of view, he would simply walk away. That is why Edsel made the decision to establish the Ford Design Department while in Florida, rather than face to face with this father or in the company with Sorensen and Martin.[25]

It happened in the afternoon, shortly after Gregorie had returned from lunch. He was working on some design ideas at his desk when the telephone rang. "Mr. Gregorie," he answered.

"Hello, Mr. Gregorie. This is Edsel Ford."

"Hello, Mr. Ford. What may I do for you?"

"Well, I've been thinking about our design activities, and would like to ask you if you would like to head up your own Design Department?" explained Edsel. "I've already talked it over with Charlie Sorensen, and he is prepared to revamp the Engineering Department however you would like it. How does that sound?"

Gregorie, taken aback by the offer, hesitated a moment, but finally said, "I'd be overjoyed, Mr. Ford."

"Fine. Fine." replied Edsel. "Why don't you get with Charlie as soon as you can and get things going."[26]

With that one phone call, Gregorie went from being a lowly "draftsman" to a high-ranking Ford executive on the level of Sorensen, Martin, or Galamb. Gregorie had been reporting directly to Edsel for about a year, but the others considered him more as Edsel's consultant than as an executive. Now, Gregorie was assigned the responsibilities of managing a full department, one that would eventually take on the importance of purchasing and engineering. Gregorie was only 27 years old.

"It's a strange thing how it happened that way," explained Gregorie. "Edsel gave the order, and things had to happen. He was, after all, president of the company. He wanted things moved in that direction, and he didn't want anymore dealings with Briggs as far as design was concerned. Everything was closed off at that point."[27]

Interestingly, Gregorie did not have to search for Sorensen that day. "Charlie Sorensen got his orders from the bossman, and he came to me and told me that he was at my beck and call to establish the Design Department with full steam ahead, no strings attached," explained Gregorie. "And, oh, he was sweet as honey! He was usually hard-boiled and then some, and boy, syrup couldn't have poured out of his mouth any sweeter than he was! 'Anything the boy wants,' he told me. 'Anything the boy wants. Why, you just tell me, and I'll fix it up for you.' By the time Edsel came back a month later, why, we were going full bore."[28]

***

Edsel appointed Gregorie as the first chief designer at Ford because of Gregorie's extraordinary design talent. However, Edsel also made him chief designer because of his integrity and fortitude. "He knew I could handle my end if he backed me up," explained Gregorie. "I couldn't have managed things without his backing. But he needed me, too. We had an unspoken working arrangement between us: I expected his backing where it was needed, and if he needed my backing, why, I'd try to go to bat for him. Many times, I would get into a heated discussion about this issue or that issue, and Edsel would stand back and nod his head, and things happened. He apparently went to his father and just told him that's the way it had to run—that was the new policy. The old man could see what was taking place—that styling was a big issue. So that left the chassis people to their own P's and Q's. Mr. Edsel Ford very subtlely let these people know that I wasn't to be messed with. They were to work with me or else. It was an unwritten piece of judgment. That issue was all it took to get the respect that I needed to establish the department and get respect for the way it was operated. So that was the break-off point. Oh, I had to stand up for the department. I took a chance, and it worked out."[29]

Whether intentional or coincidental, Sorensen told Walter Briggs about the formation of a Design Department within Ford Motor Company. Sorensen ran into Briggs in Palm Beach and told him that Edsel had decided to establish his own in-house Design Department. The company would continue to buy bodies from Briggs, but as far as body design was concerned, Briggs' services would no longer be needed.

The next day, Briggs called Ralph Roberts and asked him how much work he had done on the Ford account. "Well, we have quite a few design concepts and several wood models completed and several more well under way," explained Roberts.

"Gather up whatever you have and take it all over to Dearborn," ordered Briggs. "We have just lost the Ford account."

"It was just that quick," explained Roberts. "By the time we had collected all of the materials and models, as W.O. had directed, we had about six truckloads of material to drop off at Ford."[30]

Edsel did not have any animus toward Briggs or its designers. He pulled out of Briggs "because he wanted to make his own design decisions in the privacy of his own quarters," explained Gregorie. "Over at Briggs, his decisions were exposed to a lot of other people who were heavily involved with design work for other companies, not just Ford, and he didn't like that. (He had been cautious of this for years. Back in 1925, he had learned that Ray Dietrich had been showing designs that he had worked up for Edsel to the chief body engineer of the Marmon Motor Car Company.[31]) Edsel was glad to pull away from Briggs. That is the way he wanted it. He was a lot more comfortable that way."[32]

***

Sorensen and Gregorie decided to set up the new Design Department in the southwest corner of the Engineering Laboratory, which means that Edsel must have received permission from his father before making the momentous decision. The southwest corner was where Henry used to conduct his old-fashioned dances, forcing his executives to learn polkas, cotillions, and quadrilles. "To some of us," said one of the old-timers, "that was sort of sacrilege and a sign of changing times, when styling could take over Mr. Ford's dance floor."[33]

Years before this occurred, Henry had a dance hall constructed on the grounds of Greenfield Village and had his old-fashioned dances conducted there. (Ford's executives did not enjoy going to the dances at Lovett Hall any more than they did when the dances were held in the Engineering Laboratory.) When winter arrived, the executives could count on receiving an invitation every weekend. Everyone complied, of course, but when Clara told Henry that she was bored, Henry tried livening up these parties by inviting Edsel's Grosse Pointe friends.[34] That helped a little, but because smoking and drinking were forbidden at these dance parties, dancing old-fashioned dances to liven the spirit left much to be desired. After a while, Edsel and his friends would have a cocktail party first, then come to the dances.[35] During the dances, Edsel and his friends would sneak down to the downstairs lavatory and have a smoke, washing their mouths out with mouthwash before returning to the dance floor.[36]

When the Grosse Pointe crowd first started attending the dances, they felt lost. None of them knew any of these old-fashioned dances; rather, they were used to the two-step or the fox trot. To make them feel comfortable, Edsel would step in and ask the instructor to play a two-step. When the dance ended, he would say, "That was a beautiful waltz, wasn't it?"[37]

"I went to parties at Lovett Hall when the Dearborn crowd was there and when the Grosse Pointe crowd was there," recalled Catherine Ruddiman, "and they were definitely much more fun with the Grosse Pointe crowd."[38]

Although Henry Ford had discontinued his folk dances, he still had a gypsy orchestra of six musicians play about three times a week. They would play for an hour, starting at 3:00 in the afternoon, right over the wall from the Design Department. "Every now and then, one of those musicians would hit the cymbals or the drum just right and...BONG!...nearly startling me and my boys off our chairs!" recalled Gregorie.[39]

***

Nevertheless, the location of the Design Department was ideal. With skylights for natural light, a teak floor, and floor-to-ceiling windows on the west side overlooking a man-made pond, the environment was perfectly suited for creating. The entire Design Department was always kept immaculately clean. No matter how much clay was thrown during the day, a cleaning crew would come at night to sweep and polish the floors, returning the department to it pristine condition before the next business day.

No permanent alterations were made to the holy area, however. The new Design Department was merely cordoned off from the remainder of the Engineering Department with partitions made of yellow oak and frosted glass, with double doors at regular intervals so that large tables and models could be moved in or out as needed. The model shop was also cordoned off from the Design Department with these partitioned walls. Directly above the double-door entrance to the shop was a large stuffed black crow hanging from a wire from the ceiling. Nobody seems to remember why it was there, but it was probably hung there at the order of Henry Ford during the dancing days. Nobody had the nerve to take it down, for fear of reprisal from the old man. "It had a wing span of about three feet," recalled John Hay, a model builder in the shop. "They were strict about keeping the entire area clean, and every once and a while, my boss had me climb a ladder and dust that doggone bird off!"[40]

Gregorie's office was centrally located along the south side of the Engineering Laboratory. That wall formed one side of Gregorie's office. The other three walls were made of yellow oak, similar to the other partitions that made up the Design Department. However, instead of having frosted glass panels, Gregorie's office had clear glass panels, allowing him to "see the whole dang place sitting from his desk," explained John Hay.[41] Within his office, Gregorie had a desk, a drafting board where he would do his design work, and a small table on which he would make small clay models. His office had two doors: one to the Design Department through the east partition, and one that opened directly to the parking lot through the south wall. Gregorie would park his car next to this south-facing door. When Edsel visited Gregorie after lunch, he would walk down the long corridor from the executive dinning room, walk across the Design Department, and enter through the east door. However, if Edsel wanted to visit his chief designer in private, he would park his car next to Gregorie's and enter through the outside entrance.[42] Eventually, Gregorie had the periphery of his office surrounded by large, moveable blackboards, the kind on which the designers made full-scale drawings, so that he and Edsel could discuss designs without the whole department gawking at them.

***

While Sorensen was managing the construction of the Design Department, Gregorie went to work recruiting men to staff it. The first man he hired was Dick Beneicke, a master clay modeler who was working at Briggs Manufacturing. When Beneicke came to Ford, so did his close friend William M. Leverenz. Leverenz was also an architectural plasterer and had worked with Beneicke, decorating movie palaces.

Shortly after Gregorie had brought Beneicke and Leverenz on board, a number of other men joined his fledgling Design Department, and he placed them in charge of four separate sections: design, body layout/engineering, clay modeling, and the fabrication shop. Gregorie placed Beneicke in charge of clay modeling, with Leverenz as his assistant; Jimmy Lynch was in charge of the fabrication shop, with Bert Pugh as his assistant; and Martin Regitko was in charge of body layout and engineering. Gregorie separated the design activities into four separate activities and placed Bruno Kolt in charge of grilles/exterior trim, Walter Kruke in charge of interiors, John Walter in charge of instrument panels/steering wheels, and Willys Wagner in charge of lighting, bumpers, and wheels. Edward Martin assisted Gregorie in managing the design activities and was the department coordinator with engineering. All of these men came from diverse backgrounds, but under Gregorie's direction, they melded into a formidable Design Department.

Sometime in early 1937, Edsel came up with a practical idea that would not only help Gregorie staff his new department, but would help prevent Henry's interference as well. As one of the early designers explained, "It was common knowledge that Henry Ford never took any interest in anything Edsel did. Even to the extent of literally destroying Edsel's projects. That was the reason that Edsel suggested bringing in Trade School boys to the Design Department."[43]

The Henry Ford Trade School was located on the second floor of the "B" Building at the Rouge plant. There the students were taught math, geometry, trigonometry, mechanical drawing, machine shop theory, and how to operate all types of metalworking machinery. "We had outstanding instructors," said one of the first trade school recruits. "I learned how to run a lathe, shaper, vertical and horizontal mill, O.D. grinder and surface grinder, and spline mill." Many graduates went on to the Ford Apprentice School, where they were taught "advanced drawing, advanced math, and body surface layout," skills, which proved invaluable for those lucky enough to be assigned to the Design Department.[44]

"The Trade School was Henry's pet and he supported anything where 'the boys' were involved," explained Emmett O'Rear. "E.T. Gregorie was not liked by the other executives. They felt that 'styling was a lot of nonsense.' Henry Ford tended to agree with that approach, but tolerated it because of the boys. Therefore, if the Design Department was presented as a marvelous opportunity for students as they graduated, it would have a chance of survival."[45] At Edsel's direction, Gregorie interviewed a number of Trade School graduates and recruited five of the most talented: Bud Adams, Placid Barbera, Frank Beyer, and Emmett O'Rear were recruited as apprentice designers, and Francis Zawacki as Gregorie's secretary.

John Najjar and Bob Thomas joined the Design Department at this time also, and as an indication that Edsel's scheme of hiring "Henry's boys" was working, Henry Ford himself sent them to see Gregorie. At the time, Najjar was a student at the Ford Apprentice School, and Thomas was an apprentice of the resident artist at Ford, Irving Bacon. The last design apprentices recruited during this formative period were Tucker Madawick, a recent graduate of the Pratt Institute, and Duncan McRae, a designer who simply "knocked on the door" and got a job.[46]

Ross Cousins was the last of this initial group, but he was an illustrator, not a designer. Following his father's footsteps as a commercial artist, Cousins learned his trade from his father and by taking art classes at Cass Technical Institute in Detroit. Upon graduating in 1935, Cousins got his first job at a print shop that did a lot of automotive work. From there, he worked for another printing company and eventually ended up at Chrysler. While at Chrysler, he learned of a job at Ford and applied. Gregorie hired Cousins so that he could assist the advertising agencies in developing ads. "Ross could make nice drawings and show the car in front of a butcher shop, or a gal stepping out there and the guy putting groceries in the back and all that kind of stuff," explained Gregorie.[47] Many of the Ford advertisements showing illustrated scenes during the late 1930s and early 1940s were done by Ross Cousins.

"Those boys were 18, 19 years old when they came to work for me," recalled Gregorie. "I was only in my 20s."[48]

In addition to this diverse group of experienced draftsmen and neophytes, Gregorie also had several seasoned automobile designers on his staff. He is quick to point out that he never went out and solicited them. (He did not want to give Henry the impression that he was setting up a professional organization.) Rather, they all came to Gregorie, looking for a job. Martin Regitko was the first of these men. He came from Willoughby & Company, custom body builders in Utica, New York. Regitko's forte was laying out the body structure, from internal door pillars and structural members around window openings to exterior body surfaces. In the late 1930s, Hermann C. Brunn, who had learned the art of custom body design from his father who founded Brunn & Company, a notable custom body builder in Buffalo, New York, joined the Design Department. From the custom body house of Waterhouse & Company of Webster, Massachusetts, came Charlie Waterhouse, son of the firm's founder. John Dobben, formerly a designer and later chief engineer for Judkins Company, and George Tasman, body designer for Locke and later Le Baron-Detroit, were also members of Gregorie's staff. Edsel Ford had purchased thousands of bodies from these firms over the years and no doubt had known all of these men well before they joined Ford. However, it is unlikely that he encouraged any of them to move to Dearborn. After the war, Thomas Hibbard, who had once been a partner in the famous Le Baron body company in the early 1920s, joined Gregorie's staff from General Motors. "I'm not quite sure how Tom came to Ford," says Gregorie, "but he was working at the Rouge helping develop the Jeep. He came over to my department shortly after that."

"These men were body draftsmen for me," explained Gregorie. "They had no great knowledge of production bodies, but they were very knowledgeable people when it came to drafting standards and body construction. They had a good foundation in the business, and they didn't need re-education. We were fortunate to get them."[49]

***

By 1938, Gregorie had a staff of about 25 men, all making 65 to 75 cents per hour.[50] His staff eventually grew to about 100, which was similar to the size of one of Harley Earl's divisional design staffs at General Motors. One of the reasons that Gregorie could operate with such a small staff is that everybody was allowed to do almost anything they wanted. He did not delineate responsibilities to the point where only clay modelers could do clay modeling, nor were designers restricted to only designing. If a designer wanted to help on a clay model or a modeler had an idea for a hubcap or a grille, they were welcome to chip in. "The innovation could come from anybody and not necessarily all from Gregorie," confirms Adams. "We could make drawings, templates, cut up letters for instrument panels, even paint the clay models. We could bring in things that we had done at home and throw them into the pot for consideration."[51]

With input from all, Gregorie's men came up with ideas that transformed the industry. They were first to integrate the radio speaker with the radio and put it on top of the of the instrument panel. They originated large plastic panels on the dash, integrated the instrument cluster into the steering column, and were first to conceal the running boards. As Gregorie says, "There were always fresh ideas from our design group."[52]

***

In an era when dictatorial management was the rule, especially in the automotive industry and particularly at Ford Motor Company, Gregorie ran his department similar to a country store. "We were all a team," he proudly explained, and he treated his men the way he wanted to be treated. Gregorie related to his men in the same way that a teacher treats his students: directing, encouraging, and critiquing. Within the Design Department, he kept the atmosphere light.

Edsel supported Gregorie's team approach. He was never one for formality and did not care to discuss anything with a group standing around him.[53] He preferred talking with Gregorie one on one and deciding which way to go on a particular design in the privacy of Gregorie's sequestered office. Even with such talented artists as Cousins, Edsel preferred seeing one of Gregorie's quick sketches rather than fancy airbrushed drawings. "Edsel didn't care for flossy presentations," explained Gregorie. "He loved to be right there when I made a sketch—he liked to see it in the raw, so to speak. He'd always accept something that was done quick like that."[54] Because of this, there were never any formal design presentations given to Edsel by Gregorie and his staff, as there were at General Motors. All design decisions were made between Gregorie and Edsel, either in Gregorie's office or at the end of a drafting board in the Design Department.

Charles Sorensen could not get used to the democratic approach Gregorie was using in the Design Department. Although Gregorie was now part of the executive team at Ford, Sorensen never hesitated to criticize Gregorie's designs or throw him a barb every now and then to test his mettle. "I would stand back and watch Sorensen say something about raising a moulding one-sixteenth of an inch—a sixteenth!" recalled John Najjar.[55] Even the suggestion of moving the moulding such a minute amount was ridiculous—the change would not be discernable. Not all of Sorensen's criticisms were directed toward the design of the vehicle. He was mostly concerned with whether or not the new designs could be manufactured efficiently. If he thought the stampings were too deep or something else might cause a production problem, he would bring that to Gregorie's attention. "In other words, he was hedging against production problems," explained Gregorie. "He hadn't gotten used to changes every year."[56]

However, Emmett O'Rear vividly recalled a time in 1939 when Sorensen threw one of his famous temper tantrums against Gregorie. "Sorensen walked into the Design Department about one o'clock," explained O'Rear, "and since Gregorie had not yet returned from lunch, Sorensen walked up to my drafting board, pulled up a stool, and watched me as I worked on a new grille design." About 15 minutes later, Gregorie walked in, saw Sorensen, and came over. Gregorie said hello politely, but Sorensen ignored him for what seemed like hours, then finally asked, "Is this going to be a stamped grille or a die casting?" Gregorie began to explain that by designing it one way, it could be a stamping; by designing it another way, it could be a die-casting. Sorensen, not used to being lectured by anybody other than Henry Ford himself, became incensed. His face turned the color of liver, and he jumped up in disgust, knocking over the stool in the process. "When are you going to get it through your goddamned head," he yelled at Gregorie, "that I don't give a shit about your ideas! All I want to know is what does Edsel want?" Gregorie, not intimidated in the least, politely retorted, "I don't know what he wants. I haven't asked him yet."

"Then you do it, and be quick about it!" demanded Sorensen as he spun around and walked out the door, slamming it so hard that O'Rear was surprised the glass did not break.[57]

Gregorie simply grinned, shrugged his shoulders, and walked back to his office. He was not shaken in the least.

Gregorie's personality would have forced him to stand his ground regardless of the environment, but it was easier for him to do so because he had Edsel's complete confidence and backing. Sorensen and Henry's other men never accepted Gregorie as an equal; however, with Edsel's full support, they had no choice but to work with Gregorie.

***

Although it took Gregorie a couple of years to set up the Design Department, by mid-1937, he and his small staff had taken over all design activities for the entire Ford Motor Company. During the next 24 months, Gregorie and his men redesigned the entire Ford line—including commercial vehicles. They also developed the Mercury, created the Continental, and even designed a tractor at the bidding of old Henry. At the center of this flurry of creative activity was E.T. Gregorie. "He was *the* designer at Ford," explained John Najjar. "The rest of us did details."[58]

"Gregorie spent 50 percent or more of his time at his desk, cranking out ideas for every aspect of the car from a styling point of view," explained Emmett O'Rear. "He kept a ream (literally 500 sheets) of 8.5 × 11 white typing paper on the corner of his desk and grabbed off sheet after sheet, making these sketches. Shortly before lunch, he would come flying out with a handful of sketches and pass them out to Walt Kruke, John Walters, Bill Wagner, and Dick Beneicke. In my judgment, his ideas ran the gamut from bad to great, but the sheer volume was an unbelievable sight to behold. I've never in my lifetime seen a more fertile mind than Bob Gregorie's."[59]

When he was not creating a design on paper, Gregorie would be poring over an engineering drawing, arguing details with Sheldrick or Galamb, sculpting a scale model, or helping Beneicke work on a full-size model. "The clay modeling knife in Bob Gregorie's hand was a very artistic instrument," said O'Rear.[60] Gregorie would also do design work at home, especially if a project were falling behind schedule. "I kept a pad at home, and I'd occasionally spend an evening sketching certain features that I had in my mind from a day at the office. I'd take a sketch pad and draw four or five ideas of something that was going on during the day, and take them in the next morning, and give them to one or two of the boys, and say, 'Pick up from here. See what you can do with it.' "[61]

The one aspect of design that probably interested Henry more than Edsel was the commercial line of Ford. Henry's pickups, trucks, and delivery vans were all highly functional; however, after Gregorie got through with them, they were aesthetically pleasing as well. "We were always working on commercial stuff," recalled Gregorie, "but there was very little artistic effort put into it, if I can put it that way. It was usually pretty well controlled by a basic set of dimensions: the width of the body, wheel clearance, accommodating various size tires, and overhang, so we didn't have much leeway. But they always turned out passably decent looking. Our efforts were primarily on surface detail. We usually made quarter-size models of the pickups, where we had the most amount of leeway and control."[62]

***

The design process implemented at Ford by Edsel and Gregorie was unique in the industry, and it enabled them to produce their landmark designs. The first step in this process was to develop an initial design concept, which Gregorie always did himself. Then a one-tenth or quarter-size clay model would be made. In the early years, Gregorie made these himself. However, as his apprentices became more experienced, he delegated much of this work to them. If the small-size clay model appeared promising, then a full-size clay model would be constructed. Edsel was involved in all of these steps, but he usually waited until the full-size clay model began to take shape before providing his input. "In many instances, Edsel would take his hand and move it over the body design as though he was smearing the clay on to make his point," explained Gregorie.[63] As soon as the full-size clay model was to his liking, Edsel would officially approve it, and the body engineers would take over from there.

"All my marine architecture training was wonderful for the development of automobile shapes," explained Gregorie. He used modeling bridges and Table of Offsets when he had designed boat hulls. "We had water lines, deck lines, and sheer lines, and sections through the hull at various places. I just applied this process to the automobile business. I mean, designing automobile bodies was duck soup compared to the outer body of an 80-foot yacht!"[64]

At this point, Clare Kramer and his group of production draftsmen became involved. They were located in one of the old Ford Tri-motor division hangars, and their responsibility was "to put these sections on a full-size engineering draft, clean up the lines, and get the surfaces defined and trued up, so that they could take the draft back into the body engineering activity and do all the structure and the hinges and everything else that goes into a body," explained Adams.[65] When the engineering drawing was completed, it was turned over to the pattern makers and then to the die makers.

This fundamental design process is what differentiated Gregorie's designs from those of Harley Earl. Instead of following Gregorie's design process of developing a design concept, starting a full-size clay model at the earliest convenience and then working out the final shapes in clay, Harley Earl and his men took the opposite approach. They started with a full-size blackboard side view and, perhaps, end views or some cross sections. They would then complete the shape on the drafting board using various trade secret 'keys' to handle transitions in shape—to define the shapes between what they had established as basic sections. After completing these two steps, the design would be transferred from the two-dimensional drawing into a three-dimensional clay model. "[General Motors] usually had a good line flowing from their rear fender to the rear door, to the front door, but then it broke in plan view, the cowl and hood line—a sharp break inboard," explained John Najjar.[66] "This type of approach," concurred Adams, "is what gave General Motors' products their architectural, hard feel, compared to Ford products. Our cars were generally [more] rounded and flowing and softer; not the stiff General Motors' type of appearance."[67]

***

A lot of time had passed since Edsel had first toyed with automotive design with his father's Model T runabout in 1911. Edsel would have liked to have started a Design Department within the company much sooner than 1935. However, there is little doubt that he had found in E.T. Gregorie the right man for the job. Edsel had considered Ray Dietrich years earlier but found him to be too arrogant and "too much of a playboy" for his taste.[68] Looking back, it is probably better that Edsel had not hired Dietrich, for after the custom body business went by the wayside, Dietrich never developed a suitable car for mass production—his forte seemed to have been strictly custom bodies. Gregorie, however, with his mechanical horse sense and his sense of style, had a knack for the production car.

Fortunately, Edsel and Gregorie quickly developed a mutual admiration for each other's personality and abilities. In the years the two worked together, for example, neither ever called the other by his first name. Gregorie always referred to Edsel as "Mr. Ford," and Edsel always called Gregorie, "Mr. Gregorie."[69] When it came to automobile design, the two were on equal terms as well. "Damn few people know how to design a car," exclaimed Gregorie. "You have to get the heritage of an old Duesenberg, an old Mercedes" to develop a good design.[70] With such requirements, Gregorie could not have described his patron more accurately. Edsel knew automobiles and he knew design, and his new chief of design would not have to explain to him the difference between a Duesenberg and a Mercedes, or between a cabriolet and a roadster, for that matter. In fact, Gregorie would learn plenty about styling and design from Edsel.

Edsel respected Gregorie's abilities, and he did not hesitate to show his admiration for him. Receiving such respect, although it was well founded, "was rare in the old automobile companies," confessed Gregorie, and he appreciated Edsel's support.[71] As for Gregorie, he had found in Edsel an interpreter for his design ideas. Together, Edsel and Gregorie were to become a formidable design team.

***

Edsel and Gregorie had many similar tastes when it came to automobile design. They both had a conservative bent—although Gregorie's was not as severe as Edsel's. They both appreciated the simple, straightforward line. Likewise, neither cared for a lot of ornamentation on a car. They both liked the design of a car to be "nice and trim and clean-looking."[72] Above all, they both had a deep-rooted appreciation for the basic form. For them, the form of an automobile was paramount in design.

However, this is where their tastes departed slightly. Gregorie preferred an "expressive, defined, and articulated" form. He liked the individual parts of an automobile body to flow together, yet retain their identity. Edsel liked forms with sharp contrasts and clear definition. Gregorie explained the difference this way: "It's similar to the difference between a sleek, delicate racing boat and a husky seagoing boat," he said. "Both have the ability to cut through the water...only one has a different purpose than the other."[73]

Their differences in design inclinations were slight, and any disagreements over which way a certain design should be developed were always worked out amicably between the two men. Edsel rarely gave a directive, preferring instead to solicit ideas and draw out new concepts through discussion. That is the way Edsel liked to work. He preferred to work in a casual atmosphere and without pretense. He came to the Design Department almost every afternoon, so he was thoroughly aware of the designs on which all of the men were working. Thus, if a decision had to be made, Edsel would make it right there, "just sitting at the end of a drafting board," recalled Gregorie.[74]

A project would usually be started by Edsel, but all of the design proposals and ideas would comc from Gregorie and his staff. Edsel, on his regular visits to the Design Department, would review their ideas, make suggestions, and guide the designers toward an overall design that was acceptable to him. That is probably why Edsel never offered a sketch of his own. If he did, it may have limited the designer's own talent, thus preventing the best possible design to unfold.

This process gave Ford designers much latitude. Edsel believed that automobile design was an artistic endeavor, and he treated all of his designers—especially Gregorie—as any beneficent patron would treat an artist. To Edsel, Gregorie was the artist, and Edsel merely gave him the means of creating. Gregorie appreciated Edsel's understanding of artists and the design process, as well as the support that Gregorie and his designers received from Edsel.

If Gregorie would have liked to change one thing in his relationship with Edsel, it was Edsel's introspection. "Edsel didn't have an expressive personality," explained Gregorie. "He was a very quiet, very reserved, unassuming man; always very solemn, very serious. He rarely exhibited enthusiasm, and he rarely exhibited distaste. It was very difficult to know, without very constant association with him, how his feelings ran. Maybe that was his privacy shield. I think he kept away from the masses that way. He didn't want to be influenced by people. He had his own way of determining whether a design or some aspect of a design met with his taste."[75]

Gregorie had his own ideas about how Ford design should evolve, but he was always reticent about proposing them to Edsel. "I had a feeling that he'd be offended if I suggested that sort of thing to him." Gregorie never hesitated in voicing his opinion about a particular design on which he and Edsel were working, but he was respectful and cautious as to how he presented his ideas. To help determine which design was better, for example, "I'd make a little sketch of what I had in mind on a piece of vellum and lay it over the design we were discussing so that Edsel could see what I was thinking." That way, both of their ideas could be considered simultaneously. On many instances, Gregorie had to concede that Edsel's idea was superior to his. "But Edsel was an excellent critic," explained Gregorie, "and he knew when something looked right. After all, a man doesn't have to be a designer to point out something and say, 'Look, the hood line doesn't look right,' or 'I think the window is too wide.' He doesn't have to have the ability to take a pencil like I have."[76]

This kind of interaction developed a mutual admiration between the two men. "If a design looked right to him [Edsel], I had a feeling it looked right to me," explained Gregorie, "and if he thought it looked right to me, unless there was something conspicuously wrong, why, it was all right with him." Edsel always had the final say, however, and Gregorie never tried pushing through an idea that Edsel did not like. That never would have happened anyway; the two men had "developed a good, balanced way of working things out" between them. "I got to know certain things that were acceptable to him," explained Gregorie, "and I could tell very quickly whether I could sell him on something or not."[77]

***

Edsel appreciated Gregorie's willingness to discuss design ideas with him and for not presenting arguments when none were needed. Within a short time, Gregorie had become one of the few people within the Ford organization in whom Edsel would confide without fear of being betrayed.[78] The only other company employee with whom Edsel had such a close relationship was his brother-in-law, Ernest Kanzler.

The combination of Gregorie's support and artistic abilities created an uncanny bond and relationship between the great designer and his patron—one teetering on the edge of the ethereal. "Edsel left the design intuition to me," explained Gregorie. "Even though there were two of us, we thought as one. My hands became his tools to develop designs. I'm sure that sounds odd, but the designs that we developed over the years were *his* designs, not mine. It's difficult to put into words our working relationship, but in essence, I was able to put on paper and into clay the designs *he* was visualizing in *his* head. And I could do it without a lot of discussion between the two of us. Edsel never took a pencil in hand and drew out an idea that he had—he depended on me to develop the design that he was visualizing. I sketched while he explained what he had in mind, and the design developed that way. Of course, we had similar tastes when it came to design, so what he liked, I liked, and what I liked, he liked. We never had any arguments over which way a design was to evolve. He'd make suggestions, and I'd make suggestions, and we'd alter the design until we were both satisfied with it. It was as if he were directing my hands towards the designs he wanted. I know that sounds bizarre, but that's the best way I can explain the way Edsel and I worked together."[79]

***

Henry Ford's office was in the Engineering Laboratory on the other side of the building from Gregorie's Design Department, whereas Edsel's office was four miles away in the Ford Administration Building on Schaefer Road. Edsel would drive to have lunch with his father and the other executives, and then go to see Gregorie. Occasionally, Henry would walk with Edsel to the Design Department, where the two men would go into one corner and talk for hours on end, Edsel sitting on a drafting board, his hands under his legs, feet dangling, and Henry sitting opposite him.[80] When that happened, Gregorie would instruct his "boys to stay clear and give them all the privacy they wanted, and observe the pantomime between the two of them."[81]

Most of the time, however, Edsel "came into the Design Department to deliberately avoid the old man," recalled Gregorie.[82] Because of Edsel's deep-rooted interest in design and his close relationship with Gregorie, he cherished the time he spent in the Design Department, and, to a certain degree, it became his sanctuary within the company. Whenever the designers were finished working on one of the design mock-ups, they would cover it with linen to keep off the dust and to keep the design out of view of prying eyes. On particularly wearisome days, Edsel and Gregorie would pull back the cover on one of these prototype cars, climb in, and talk. "It was cool in there," said Gregorie, "and when Edsel didn't want to be disturbed, we'd sit in there on those nice, big sofa seats and talk about various things. Their discussions usually centered around design, but Edsel occasionally would bring in some magazine clippings of boats—in the same way as Edsel used to clip automobile pictures from magazines when he was a teenager—and he and Gregorie would discuss their salient points.

At other times, Edsel would confide in Gregorie about the difficulties he was having in some other area of the company. Edsel never talked to Gregorie specifically about the problems he was having with his father concerning design, but Gregorie was able to read between the lines.[83]

"Old Henry Ford had absolutely no interest in the design and appearance of the car," explained Gregorie. "A lot of niceties that Edsel wanted built into the product, Henry felt were superfluous, and that the Design Department was nothing more than a hotbed for production problems. He never could understand why anybody would want to dress up in white smocks and build cars out of mud. And he tried awfully hard to stifle Edsel's involvement in design. For every new design that Edsel wanted to implement, the old man made him fight for it. That's what Edsel suffered from ever since he pushed through the Model A. But to Edsel's credit, he stuck to his guns and pushed through all the designs that he and I developed over the years. But Edsel paid dearly for it. After the meetings Edsel had with his father, he'd come over and talk to me, and he was all shook up. It would take him a half-hour or more to get his reasoning back. It was this constant wrangling with the old man that eventually wore Edsel down. It eventually killed him."[84]

That is why Edsel cherished his time in the Design Department. "He'd just like to come into the department and sit in one of those cars and rest himself—get away from the old man," recalled Gregorie. "He'd say, 'Gees, I wish I could sit here all afternoon.' Poor man, my God, I'm telling you..."[85]

After these respites, Edsel would discuss design ideas with Gregorie. "You'd be surprised at the decisions that were made just sitting at the end of a drafting board. Edsel and I would decide on this hood ornament or that hood ornament, this direction or that direction. There were no committees—nothing of that nature at all."[86] When it came to styling and design, recalled Charles Sorensen, "Edsel lived with that."[87]

***

Edsel took great pride in his design team. He directed them professionally and with great insight. They responded by putting his vision into sheet metal. Unfortunately, neither Edsel nor his team received much respect from Henry Ford. In fact, Henry would instruct his cronies—Harry Bennett and Laurence Sheldrick, in particular—to keep an eye on Gregorie and his men. "They used to drill holes in the partitions that separated styling from engineering and spy on Gregorie," recalled Cousins. Henry Ford's men would even go into the bathroom of the Design Department and "look under the doors and see if we were in there too long." They would also place watchmen at the gate and checked Gregorie's comings and goings, "just so they might find a reason to fire him. We had to almost fight for our existence."[88]

Henry and his cronies never got rid of Gregorie or any of his designers as long as Edsel was alive. The men in the Design Department all worked directly for Edsel—not for Henry—and Edsel protected them. They considered themselves artists and professionals. "The old man thought they were queer."[89]

Chapter 22

Design Synergy

"Edsel was a genius as far as design was concerned. And Mr. Gregorie was a master designer."[1]
Placid Barbera

By the 1930s, Edsel's design acumen had matured. During the next decade, it would burst into a period of immense creativity, manifesting itself into some of the most provocative automobile designs ever produced. In an article he wrote in late 1935, Edsel hints at what the designs of the future would be. "There can be little doubt," he wrote, "that, in the future, some form of streamlining distinctly suitable to the automobile will be developed."[2] No one could have imagined at the time that Edsel alone would soon successfully combine these ideas of the future with the pragmatic needs of the production line.

***

The idea of streamlining was nothing new to Edsel. Because of his involvement in the manufacture of airplanes, he was well aware of the effect the principles of streamlining had on the movement of machines. However, he had also experimented with the idea of streamlining automobiles during the days of the Model A. Knowing that the most streamlined shape is a teardrop design, he had his engineers set up a rudimentary experiment to validate this theory.

When viewed from the side—at least from the rear forward—the silhouette of a standard 1928 Model A two-door sedan approached a teardrop shape. Therefore, according to theory, it seemed logical that the basic Ford sedan would run faster (i.e., be more aerodynamic) if it were driven backwards. Edsel had his engineers rig up one to try the concept. To achieve the full effect of the experiment, the engineers reversed the steering on the car and tried to drive out the back window. In other words, they turned the car's seats around, moved the steering column from the front of the car to the back, and revamped the steering mechanism so that the steering wheel would turn the "back" wheels (which were previously the front wheels). Edsel drove the car first, "but that was kind of a bad deal," confessed one of the engineers who watched the ordeal. Edsel "got going with a pretty good speed...lost control and landed in a ditch. That stopped that development right there!"[3]

This unsuccessful and potentially tragic experiment proved one thing to Edsel, however. It demonstrated that the most effective way of testing the aerodynamics of an automobile design was exactly the same as that for testing an airplane design: running air flow patterns over miniature-scale models and then incorporating the data gathered in those experiments into new designs.[4]

Before he went headlong into experimentation, Edsel needed a base line from which to measure the aerodynamic effectiveness of new designs. He instructed his engineers to determine the air resistance on all current-production Ford vehicles. They did this in a most unusual manner. "We devised a way, whereby, we could hang a complete car up in a wind tunnel," describes Emil Zoerlein, the engineer in charge of the test program. "We made a frame which was streamlined...set the car upon this frame and then suspended the frame from the ceiling so it could swing." Attached to the back of the test car was a "take-off," which was connected to a spring scale. As air was blown toward the front of the car, the whole contraption shifted, moving the spring scale in proportion to the aerodynamic resistance of the design. This crude but effective method enabled Zoerlein to objectively measure and compare one Ford body style to another.[5]

After Zoerlein had compiled data on all production Ford models, Edsel had him perform airflow patterns on various experimental clay model designs—comparing those results with the previous production design data and determining if the new design had better or worse aerodynamic qualities. Initially, clay models of complete car designs were tested, but then the engineers determined that the design of individual body parts—a clay fender or a clay grille, for example—could successfully be evaluated independently. They also discovered that by covering the clay forms with metallic paper or tin foil, the "skin friction" of the clay surface could readily simulate that of a metal surface, thereby giving them more accurate data.[6] These innovative methods of obtaining data all contributed to Ford's wealth of information about aerodynamics and significantly reduced the amount of time and expense required to analyze various new designs.

Edsel never liked anything "heavy and bulldozish-looking," and these wind tunnel tests simply reinforced his intrinsic sense of the refined form.[7] From this point, Edsel would concentrate on the streamlined, aerodynamic form. When he began assembling his styling team, he directed them toward the coupling of these design concepts into stylish, congruent forms as well.

* * *

Although this was a noble styling objective, it was not necessarily easy for Edsel to implement—the production men had other thoughts. Unlike the stylists, the production men felt that more curved and pointed shapes of streamlining caused more manufacturing problems than styling benefit. "Just like the stylists of today," explained Laurence Sheldrick, chief engineer at Ford, in a somewhat insolent manner, "they forget that there had to be machinery put inside the car somewhere. They just draw these beautiful curved lines wherever they damn please and don't leave you any room to put the axles, steering gear, and other parts. That always results in a lot of pushing and hauling back and forth between the mechanical group and stylists."[8] Thus, every time a stylist became too extreme with a design, the production men would inflict some pragmatism into it, which immediately caused conflict between the stylist and the engineers. Whenever thcsc contentions arose, which was often, Edsel had to play the referee between the designers and the production men to get the design completed.[9]

This was a debate well understood by Edsel, for no one in the company had his grasp of both production and design.[10] He fully understood the manufacturing realities of the production floor and the aesthetic requirements of the showroom floor. Because of his in-depth knowledge of these two critical aspects of automobile manufacturing, not merely his ability to mediate their disparity, Edsel was able to direct his engineers and stylists into developing what the New York City Museum of Modern Art later called the "the first successfully designed streamlined car in America."[11] Known as the Zephyr, the car became one of Edsel's greatest design triumphs.

***

In 1934, sales of Lincoln automobiles plummeted to slightly more than 2,000 units from nearly three times that amount only a few years earlier.[12] This progressive decline was not caused by any misguidance by Edsel. Indeed, from the time Edsel took over Lincoln, he dramatically increased its sales, nearly doubling them in the first three years. Unfortunately, the onslaught of the Depression took its toll, even on affluent customers, and caused Lincoln sales to decline with those of all other car lines.

This was not necessarily detrimental to the entire Ford Motor Company because sales of expensive Lincolns never approached the voluminous stream of Ford sales. However, what was disconcerting to Edsel was the lack of anything else to offer from the company. A customer who would have purchased a Lincoln in 1929 may not be able to afford one now, but he may not want to buy a Ford either.

From Edsel's perspective, what Ford needed to shore up sales was not more Lincoln sales, but an automobile that filled the market niche between the low-priced Ford and the expensive Lincoln, a car that was more refined than the Ford but not as elegant as the Lincoln.

Both of Ford's major competitors had similar entries. General Motors had Pontiac, the next stepping stone for the Chevrolet owner. Chrysler had Dodge, a little more expensive than the Plymouth, but significantly cheaper than the top-of-the-line Chrysler. For Ford to remain competitive, Edsel wanted and needed a similar offering.

***

As the Depression took hold and automobile sales plummeted, Walter O. Briggs was lamenting the possibility of Ford reducing or even eliminating its body orders to his company—and justifiably so. Even during times of booming production, both Henry Ford and his production chief, Charles Sorensen, threatened more than once to drop Briggs altogether and bring the body business totally into their throes. They felt that Briggs was paying too much attention to its Chrysler business and not enough attention to Ford's, and they therefore would pay them back in kind by reducing body orders.[13] Ironically, Edsel forced this issue by contracting with his former high-school teacher, Clarence Avery, now president of the Murray Corporation, to build a portion of the vast body requirements for Ford.[14]

With production waning and the new threat from Murray, Briggs implemented a number of steps to reduce his vulnerability. First, he recruited Howard Bonbright, Edsel's close friend who accompanied Edsel on his cross-country trip more than a decade earlier. Bonbright was ostensibly hired to shore up Briggs' financial policies, but the real reason was to bring Edsel closer to the body manufacturer.[15] "If Bonbright had a business career," confirms Roberts, "it was not known to me. He was a prominent and wealthy Grosse Pointe socialite; probably Edsel's closest friend."[16] Second, Briggs began putting pressure on Ralph Roberts to come up with more new Ford designs to show Henry and Sorensen that Briggs was indeed dedicated to Ford Motor Company. Murray was not only causing competition in terms of manufacturing capability, but Sheldon hired Amos Northup to help in body design as well, which exacerbated Roberts' vulnerable position.[17] Roberts was already doing a respectable job for the body manufacturer, but when Briggs suggested to Roberts that the company needed new talent to stave off the Ford threat, Roberts immediately took steps to both satisfy Briggs' directive and safeguard his own position.[18]

Roberts, of course, was well acquainted with all of the custom body shops, and he knew of one exceptionally talented but frustrated designer at the New York firm of Locke and Company. When Roberts went courting John Tjaarda, neither man was fully aware of the other's motives. Tjaarda understood that Roberts was looking for additional help, but he was not aware of Roberts' predicament. Roberts, on the other hand, knew that Tjaarda wanted a more responsible and creative position, but he was unaware that Tjaarda's primary reason for being receptive to a move was not to gain a new position but to get into production an automobile design on which he had been working for the past six years. All Roberts knew was that Tjaarda was a designer of significant talent and reputation, and he felt fortunate and relieved to retain him.

When Tjaarda accepted Roberts' offer, Roberts immediately informed Briggs and at the same time convinced him that Briggs Manufacturing needed two separate design studios. In outlining his plan, Roberts said that he would retain his current position, staff, and location to work on production designs for various automobile manufacturers, but primarily for Chrysler and Ford. A second studio, headed by Tjaarda, should be established in the vacant design studio space initially walled off by Roberts when he first came to Briggs and would be used to work on more innovative and futuristic designs. Through this arrangement, argued Roberts, Briggs could offer the production expertise that Sorensen was seeking and the innovative design expertise Edsel was demanding. Briggs accepted the entire proposal, and, in the process, Roberts again saved his own hide.[19]

Roberts returned to his design studio and began working earnestly with Edsel on Ford designs, a responsibility he continued for only a short while. Starting in the late winter of 1933, Roberts began spending a lot of time in Europe, acting as liaison between Briggs Manufacturing in Detroit and its new manufacturing plant next to the mammoth Ford Dagenham, England facility. Roberts kept in touch with domestic design activities of Ford, but more and more, Edsel began absorbing that responsibility and handing it to Gregorie.

As for Tjaarda, he quickly took up residence in his new Briggs studio located at the opposite end of its Mack Avenue plant from Roberts. Shortly after settling in, he arranged a meeting between Bonbright, Briggs, and himself. Undoubtedly, Tjaarda was anxious to divulge his secret design to them.

Tjaarda started the design he was so anxious to present to the Briggs staff early in 1926. Since that time, he had been consumed with the development of a radical concept in automobile design and construction—so much, in fact, that he dubbed his project *Sterkenburg*, after his original Dutch family name of Tjaarda van Sterkenburg.

His initial Sterkenburg models incorporated many innovative concepts for the time, such as steel unit body and frame construction, exteriors made of either lightweight aluminum or fabric, independent four-wheel suspension, and rear-mounted engine. In later Sterkenburgs, Tjaarda retained all of these early concepts but added aerodynamic and streamlined features. By the time he came to Briggs in 1932, he had consolidated all of the independent aspects of the Sterkenburg models he had developed over the years and made them into a refined automobile.[20]

Tjaarda's dream now was to get his refined Sterkenburg produced. This is why he so readily accepted Roberts' offer.

After making the arrangements, Tjaarda prepared himself for the meeting. After years of development and futile attempts at selling his Sterkenburg design to other concerns, he knew that this would be his last opportunity to find somebody to back it. In preparation for the meeting, Tjaarda created an elaborate display, exhibiting a chronology of all his design sketches and scale models. With these props, Tjaarda gave the presentation of his life to Briggs and his staff. With as much gusto and enthusiasm as he could muster, and being careful to control his Dutch accent as much as possible, he described to them the advantages of his futuristic Sterkenburg design.[21]

Although many such projects ended in failure or were never completed, they never discouraged Edsel. He learned from all of them and always tried to retrieve their most embryonic ideas. If one of these ideas could not be developed or incorporated into a Ford-produced vehicle today, he thought, perhaps it could be at a later date. Tjaarda's rear-engine drive concept is a good example of this philosophy. Regardless of the public's negative reaction to it, Edsel thought Tjaarda's idea had merit and would someday be a commonplace design. "The next big development," wrote Edsel in 1935, "will be in rear-engine designs with all driving mechanism behind. That will prevent the drive-shaft from cluttering a low-frame car, and keep the noise and smell out of the front seat."[22]

Tjaarda probably could have sold his concept to Briggs with half the effort because Briggs thought Tjaarda's design had merit—but not the way it was. Instead, Briggs ordered Tjaarda to utilize some of his Sterkenburg concepts and develop an entirely new automobile, hoping that such an innovative design could replace the aging Lincoln body they had been producing for Edsel.

However, that was not the only reason. Late in 1934, Chrysler introduced a body design conceived in its own studio and built by the Budd Manufacturing Company of Detroit. Similar to Tjaarda's Sterkenburg model, the new Chrysler Airflow also had unit body construction and aerodynamic features such as fastback design, fender skirts, and a revolutionary down-curving front-end design, replete with "Art Deco grille."[23]

Sales of the ultramodern Chrysler Airflow began slowly and never gained momentum. During its first year, only 11,000 were sold. Critics began calling it the "Airflops."[24] After significant expense to save it—including a facelift by Ray Dietrich—the Airflow was discontinued three years later.[25]

Nevertheless, Briggs and Tjaarda pressed onward. Confined to his studio, Tjaarda worked in secret and was not allowed to discuss his new project with anybody except Bonbright, who became Briggs' liaison between Tjaarda and Edsel; Newton Manning, Briggs' production manager; and W.O. Briggs.[26] Not even Roberts was informed. "I didn't know a thing about it," recalled Roberts.[27] By 1934, the design was ready for Edsel.

Edsel was most intrigued when he saw Tjaarda's Sterkenburg-based design. "Here's something almost ready made," Edsel said to one of his engineers. "Let's take it and build an automobile under it."[28]

Although well aware of the anemic sales of the Airflow, Edsel was a firm believer that unit body construction was the only way of achieving aerodynamic shapes in an automobile. He even bought an Airflow and "went over it from one end to the other" to determine how it was constructed, and he had Zoerlein test its aerodynamic efficiency.[29] Besides, as far as Edsel was concerned, the extreme design of the Airflow was causing sales reluctance, not its method of construction[30]:

> *One daring designer, attempting to incorporate all these possible innovations into one car, has built a body bulbous in front, tapering toward the rear, with the sides shaped at the bottom into a distinct flare, the wheels sheathed, the lamps built into the fenders, the radiator and windshield having a radical slant, and details such as the door handles and other fixtures, which might offer wind resistance, concealed. In this car, the engine is in the rear. Yet, while this model is extremely interesting from the viewpoint of engineering and design, it is so radically different from our present accepted ideas of beauty and graceful, flowing mobility that it is difficult to believe it would receive general acceptance with the motoring public today.*

The design that Tjaarda had developed for Edsel was a masterpiece of aerodynamic form. By placing the engine in the rear of the car, Tjaarda eliminated the ubiquitous front radiator and grille and replaced them with an inverted boat stern shape, which ended at a one-piece windshield. From that point, the roofline of the car sloped downward in a dramatic but congruent fastback line. The fenders melded smoothly into the body, and the rear ones had flush-mounted fender skirts, which gave the car an even more cohesive appearance.

To test the public's reaction to the newfangled vehicle, a wooden mock-up was built and shepherded around the country with several other glamorous cars to see how the public would react. Edsel did not want a debacle with his design, similar to what was happening with the Airflow. During the shows, Briggs personnel mingled with the crowd, picked up any remarks and expressions being made about Tjaarda's vehicle, and collected them for analysis. The results were conclusive: "From a styling angle... 80 percent of the viewers liked the streamlining...but only 50 percent favored the rear-engine."[31]

Obviously, the public favored a front-engine design over Tjaarda's rear-engine one, so the decision was made to revamp his design, if possible, to accept a front-mounted engine. This would not be an easy task, for the major obstacle in developing a more aerodynamic design was always with those two parts. Incorporating a conventional front-end design, complete with radiator and grille, could all but destroy the streamlined design that Tjaarda had developed. However, if the car were to be a success, the challenge had to be met.

As had been the case with the Model A, Edsel realized that the front-end treatment of his new car would produce an acceptable design or a flop. After all, wasn't it the oblique and plunging front-end design of the Airflow that contributed to its demise? If an appropriate front-end treatment could not be created for the Zephyr, Edsel would have to scrap the entire project. It would be too risky otherwise.

It was, of course, Briggs' and Tjaarda's responsibility to salvage the design for Edsel. However, to shore up their efforts, Edsel reassigned Gregorie from working on Lincoln designs and asked him if he could provide some fresh ideas and design suggestions on Tjaarda's prototype. Edsel had confidence in Gregorie because Gregorie had done a marvelous job on the Model Y. Now, Edsel would challenge Gregorie's imaginative abilities again by coaxing him into designing an entire front-end treatment that would enable Tjaarda's design to accept a traditional front-mounted engine. According to Laurence Sheldrick, a Ford engineer who witnessed the entire Zephyr project, Gregorie "made his reputation with Edsel by refining and completing the styling begun by Tjaarda," which paved the way for Gregorie to become the first chief of design at Ford Motor Company.[32] Edsel had been considering both Tjaarda and Gregorie as his new design chief, but he settled on Gregorie.[33]

Tjaarda and Gregorie worked concurrently on this design problem, but neither was making much progress in its solution. As time passed and ideas were developed and discarded, Edsel became discouraged and doubted whether the design could be saved. In his usual, methodical manner, he ruminated the problem in his mind's eye. What could be done to Tjaarda's design, he thought, that would enable it to retain its aerodynamic appearance but accept a front-mounted engine? The Airflow had a conventionally mounted engine, but it was a styling disaster. However, the Airflow and Tjaarda's design had many features that Edsel wanted to retain. How could the two seemingly contradictory approaches be integrated into one exquisite form?

With the two designers at an impasse, and convinced that the entire project would be abandoned unless a solution was reached soon, Edsel decided to take a diametric approach to Tjaarda's bulbous design and instructed Gregorie to consider a front-end design that had sharp, vertical lines.

Gregorie was flabbergasted. Had Edsel suddenly lost all sense of styling and form? Edsel's directive seemed contrary to the purpose of aerodynamics, and Gregorie tried to persuade Edsel against pursuing such a nonsensical approach. However, Edsel "was very obstinate" about it, explained Gregorie. Edsel "loved pointy shapes," and he felt that if done properly, a pointy shape could be married quite handsomely to Tjaarda's otherwise bulbous shape.[34]

Uncertain of Edsel's genuineness, Gregorie began roughing out some ideas anyway. He made one sketch, then another, and then another. The more Gregorie sketched, the more ideas came to mind. He began to realize that Edsel's outlandish idea had more design sense than he had first realized.

After a few more iterations, Gregorie created a configuration that he thought would suffice. Working with Edsel's objective in mind, Gregorie discarded Tjaarda's front-end treatment entirely and sketched the prototype's front cowl, windshield, and front bumper on a piece of vellum as reference points. Then, starting at the top of the cowl, he drew an almost perfectly horizontal line forward, stopping it short of intersecting the vertical plane of the bumper. From that point, he drew another line downward in a severe, almost vertical angle, until it intersected the bumper. With these two lines, Gregorie defined the side elevation or profile of his design, which provided the area necessary for an engine. Switching to the front view, Gregorie shaped the grille. After several attempts, he settled on a sharp, "V" form that resembled an inverted prow of a ship. After combining all three features—the horizontal hood line, almost vertical front profile, and V-shaped grille—Gregorie's design provided the "pointy" shape Edsel was seeking.

Gregorie liked the design, but he felt it was too angular for Tjaarda's bulbous design. However, when the two seemingly discordant forms were mated together, they "clicked... They clicked, clicked, clicked."[35]

* * *

The Zephyr was not only a styling triumph, it was a production success as well, which, from a manufacturer's perspective, is the true worth of a design. During the year the Zephyr was introduced, more than 17,000 were sold. Selling at $1,275, it invaded the upper-medium-priced field—precisely what Edsel wanted—and essentially gave Ford dealers a third car to sell.[36]

Edsel received many accolades on his new Zephyr. Again, the Cartier brothers—overwhelmed by Edsel's earlier styling hit—each bought a new Zephyr.[37] "I bought myself a Zephyr several weeks ago," wrote a Zephyr convert, "and tried from the start to sell my wife on the car. She didn't tumble until yesterday when she whispered to me most confidently that she finally decided that after all, she did like the Zephyr better than her Cadillac!"[38] One of the first Zephyr owners wrote to Edsel, complaining moderately about some minor defects in his new car but praising its overall design. "I am glad that you liked the Zephyr," replied Edsel, "and hope that it is performing properly for you. Most of them are coming through in good shape now, but you know things sometimes happen."[39]

The Zephyr also achieved Briggs' objective. In fact, the car became Briggs' biggest moneymaker. "The reason," explained Tjaarda, "was that Briggs painted and trimmed the cars in their shops; they also made the hood and fenders and completed the assembly ready for installation of the mechanical units. Edsel Ford once laughingly told me that the Zephyr assembly line at the Lincoln plant was only 40 feet long because there really wasn't much left for Lincoln to do when Briggs was through!"[40]

The success of the Zephyr gave both Edsel and Gregorie a great boost of confidence. They realized they could produce a car that could simultaneously open a new market for the company and enable it to compete equally with its rivals in a new market segment. Additionally, the way Edsel directed the entire Zephyr project demonstrated that he had become "not a stylist, but a styling executive," capable of guiding even the most astute stylists into directions not even they thought possible. It also "revealed not only his [Edsel's] exquisite sense of line and form, but his innate directness and simplicity."[41]

More importantly, however, the Zephyr achievement confirmed what Edsel had hoped all along—that Gregorie was a designer among designers, capable of running the Ford design studio.

* * *

One minor flaw with the design of the Zephyr stemmed from the engineering of the vehicle, not from the body design. About a year after the Zephyr was introduced, the company started receiving complaints from dealers and customers in the Southwest, claiming that the new Zephyr overheated too easily. The condition was caused by a combination of mechanical design and styling. Because of Gregorie's front-end design, the engineers were forced to install a tall, narrow radiator behind the vertical V-shaped and pointy grille. This arrangement did not allow the engine fan, which was mounted low, right off the engine crankshaft, to pull in enough air across the entire surface area of the radiator to cool the engine. Frank Johnson, chief engineer for Lincoln, began going to see Gregorie about twice a week, asking him if he could do something to the front end to allow more air to pass through the radiator. (It probably would have been just as easy to relocate the engine fan, but Henry liked its simplicity—no belts, no pulleys—and thus that was out of the question.) Gregorie had never become used to the vertical front end he had designed for the Zephyr anyway, so he was more than happy to see what he could do to help.

After lunch one afternoon, Gregorie walked up to a Zephyr chassis that was sitting in the Engineering Laboratory and started taking measurements. He determined that the radiator would fit horizontally between the frame rails. Mounting the radiator on its side would allow the fan to pull more air across more area of the radiator, thus providing more cooling. "So, I had the boys change the radiator around and whip together some rough sheet metal work around the radiator and just piece it together with little pop rivets," explains Gregorie. "It was just a rough bunch of sheet tin. But the opening we cut in it to allow air to flow through was low and crosswise in front of the core. We put it in the wind tunnel, and we ran it, and gee, it cooled fine." When Gregorie showed Edsel his idea, Edsel said, "Well, let's go ahead and see what we can do with it." So, Gregorie had Dick Beneicke form a front end "with two grilles, quite low, and, gee, it turned out to be right nice-looking. Of course, Edsel okayed it right away for the '38 Zephyr."[42]

The original design of the Zephyr was beautiful in its own right, but Gregorie's new design was so stunning that it became an overnight styling sensation. Its two-piece grille had fine, horizontal bars, the first of its kind and the envy of the industry. Packard copied it immediately, and when Harley Earl, Gregorie's rival across town at General Motors, was shown a photograph of the new front end of the Zephyr, he became enraged. "My God," he remonstrated, "how'd we miss on that one? It's going to ruin us!"[43] "Yes," recalled Frank Hershey, "Harley exploded when I showed him a picture of the new Zephyr!"[44]

In response to Gregorie's styling coup, Earl ordered horizontal grille designs for all General Motors cars as quickly as possible. (His first design with a horizontal grille was on the 1939 Buick.) "All the other manufacturers thought it was a styling deal," explains Gregorie. "They didn't know I did it to correct this damn cooling problem!"[45] In fact, Gregorie reverted back toward a more vertical treatment on the 1939 Zephyr. He kept the basic outer shape of the two-piece grille of the 1938 Zephyr, but replaced its fine horizontal bars with fine vertical bars, giving the grille a more flowing effect, similar to a yacht cutting through water.

***

Throughout the whole period of Zephyr development, Gregorie was preoccupied with the Model K Lincoln. The "big 'K' Lincoln," as he referred to it, had been in production practically unchanged since early 1931 and was terribly outdated for being the most prestigious motorcar offered by Ford Motor Company. Add to that the fact that the Zephyr was not setting any sales records for Lincoln, and Gregorie was understandably concerned that Lincoln might be on its last throes. From his perspective—and it can be assumed from Edsel's as well—Lincoln needed a "spiritual boost" to lift it from its doldrums. What Lincoln needed, thought Gregorie, was a "classy" car, a car that surpassed even the streamlined Zephyr.[46]

As Gregorie was pondering what to do with the Model K, a concept popped into his head one evening in the fall of 1938. Because the new Zephyr had a unique chassis unto itself, with a shallow floor pan and low curb height, perhaps a new car could be designed around it that was sporty or elegant, or both, and could become the replacement for the Big K. If it could, then the suspension, engine, steering gear, and other mechanical parts of the Zephyr probably could be used as well. Gregorie thought about it most of the night, and when he came into his office the next morning, he sketched his idea.

Upon his arrival, Gregorie asked his right-hand man, Ed Martin, to pull out some one-tenth-scale drawings of a Lincoln-Zephyr four-door sedan. "As I sat beside him," recalled Martin, "he took out a piece of vellum, put it over the sketch of the Lincoln-Zephyr...and began drawing. After several minutes of silence, during which he kept drawing, he told me that since it was impossible to build a sports car on a Ford chassis, he would suggest to Mr. Edsel Ford that we build a special sports car on a Lincoln-Zephyr chassis..."[47]

Although the Zephyr was a comparatively long and low car, the design that Gregorie sketched appeared to be even longer and lower. It was not any longer or lower, of course, because the chassis of the Zephyr was being utilized. However, after Gregorie had moved the windshield backward, made the roof line lower, made the fenders longer and the hood longer and lower still, the car appeared significantly longer and sleeker than the streamlined Zephyr. In less than an hour, Gregorie had roughed out a design that eventually became the Continental. "When I tell people that I designed the Continental in 35 minutes or so, why, that's God's honest truth!" Gregorie proudly exclaimed.[48]

That afternoon, Edsel dropped by the Design Department as he normally did after lunch. As he approached his chief designer's office, Gregorie pulled the crude sketch off his drawing board and held it up for Edsel to see. "How do you like this?" he asked.

"Oh, boy!" exclaimed Edsel. "That looks great! That's it! Don't change a line on it! How fast can you have one built?"[49]

Edsel's Florida vacation was coming up in March, so that became the target date for completion.[50]

To ensure that none of the lines of the design were changed, Gregorie had an apprentice immediately take his sketch and glue it to a piece of masonite, cut it out on a coping saw, and set it up in a small modeling bridge. Then, in his office and with the help of Bud Adams, Gregorie modeled a one-tenth size, three-dimensional clay model. As work progressed on the model, "Edsel would periodically come in very quietly to review the model," recalled Adams. "If Mr. Gregorie wasn't there, I would scurry out and get him. I would then sidle up behind the two as close as possible so I could hear their comments firsthand."[51] The model was completed in about a week, and after having it painted Edsel's favorite gunmetal gray, Gregorie realized that he "had a real nice looking car."[52] So did Edsel, and the prototype was begun.

Before the craftsmen could start work, however, engineering drawings had to be made. Bud Adams transferred the profile of the clay model into a one-tenth scale drawing, and Martin Regitko used both the small clay model and Adams' scale drawing to develop a full-size engineering drawing that the model shop would use to construct hammer forms for the fenders and body panels. "Other details of construction were considered and worked out to meet any objections which might arise," recalled Ed Martin. "Men from the Lincoln plant, Harold Robinson, Walter Street, and Mr. Bartholomew were consulted."[53] At that point, the prototype car was built.

This was a highly unusual method of building a prototype. Normally, a watercolor rendering of an ideation was made to get a feel of the design proposal. If that appeared interesting, a number of additional sketches might be made to provide a greater perspective of the design. Then a full-size clay model would be built to see what the design looked like in full-size, three-dimensional form. All of that would have taken at least three months to complete, *if* everything fell into place. Even at this late stage in the process, the design could be significantly altered or even scrapped. The Continental bypassed all those steps. No watercolor renderings. No presentation sketches. No full-size clay model. Only the rudimentary profile sketch, the small clay model, and the abbreviated engineering drawing were made before the prototype was built. It wasn't until after the prototype had been built and Edsel's decision to put the car into production that a full-size clay model was made, with all necessary engineering drawings.

To meet Edsel's deadline, Gregorie took a production 1939 Zephyr sedan and had it revamped by hand into a new automobile based on his fresh design. The beautiful front end of the Zephyr that Gregorie had redesigned for the 1939 model year was left in tact, but many of the other body parts were either revamped or discarded. The cowl was retained, but four inches were sectioned from it and the doors to give the body a lower profile. The hood and all of the fenders were discarded and replaced with hand-fabricated ones, which were much longer than those of the standard Zephyr.

"There weren't months and months of drawing and building models," says Gregorie. "I put together so many cars like that, just out of odds and ends, so it became just a backyard project. The form was already precast: we happened to have a nice-looking front end, and the fenders were nice, and by the time we squished it down and raked the windshield a little more, why, you had your car! It just happened. It was a happy combination. When you stop to think of the agonizing that goes on when you attempt to do that from scratch with airbrush drawings and models and all that type of thing, I think we came out with something that was damn striking without all those shenanigans. It showed what could be done. Let's look at it this way: it was an unusual way to do things, and we at Ford at that time frequently did things in an unusual way and we frequently came out with some unusual cars. I don't think we would have gotten a car as attractive as the Continental turned out if we had hashed around with all sorts of meetings and conferences and all that other blah, blah. Edsel liked that. He had a sense of liking things done that way. In other words, the Continental could only have happened at Ford."[54]

Gregorie designed the Continental in the Design Department, but the actual car was built at the Lincoln assembly plant, 10 miles away from the Design Department. Edsel would digress from his normal routine and go to the plant several times a week to look at the progress of the Continental.[55] During these visits, two of the most striking and influential design features of the car were developed.

When he made the original sketch, Gregorie drew in an integral rear trunk, giving little thought to its functionality. By the time the prototype was practically finished, however, Gregorie realized that it was too small to hold a spare tire. His deadline was approaching rapidly, and Gregorie dreaded the possibility of having to revamp the trunk area at this late date.

Gregorie cornered Edsel as soon as he arrived and explained the dilemma. The two men talked for what seemed like hours but finally arrived at a solution. In what appeared to be a classic decision of placing function over form, it was actually a compromise to keep the project on schedule. Instead of redoing the trunk area so that it could accept a spare tire, Edsel and Gregorie decided to mount the spare tire on the outside of the trunk lid. They had no alternative;

the trunk was the only place available on the car on which to put the spare tire. "It surely wasn't a styling twist," explained Gregorie emphatically.[56]

The other design feature, which became a Continental trademark, also resulted from expediency rather than deliberate design intent. According to Gregorie, Edsel did not care for the use of a lot of chrome such as "big, grotesque hood ornaments or great big, thick bumpers" that were becoming prevalent during this time. In maintaining this philosophy, Edsel told Gregorie not to add any kind of trim to the body. He told Gregorie to leave off even the simplest chrome or decorative trim.[57]

Gregorie did not question Edsel's request because it made his task much easier. Thus, he had the car built completely bereft of any kind of body trim, keeping it "clean and neat—just fine surface lines."[58] This allowed the fine surface lines of the basic design to remain uncluttered by strips of chrome or stainless steel. The few chromed pieces on the car were the door handles, trunk latches, and hood ornament. Gregorie's illustrator, Ross Cousins, designed the beautiful hood ornament in an Art Deco motif. "1939 was the year of the World's Fair," explained Cousins, "and I got my inspiration for the sphere and lance design from brochures I had seen of the exhibition."[59]

After the Zephyr body was disassembled, cut, and sectioned, and the new body parts were added, all imperfections, rough surfaces, and transitional areas were filled in with lead and then smoothed. "I shudder to think how much the car weighed," explained Edsel's youngest son, Billy, who drove the prototype a few times. "I don't know how much lead was in the thing, but it was really a lead sled. It took about a mile to stop the car after you got it up to 65. The brakes were totally inadequate for its weight."[60]

Because the car was not built in the Design Department, the other Ford designers were completely oblivious to the progress of the project. The first time any of them saw the finished vehicle was when the prototype was finally completed and moved from the Lincoln plant to the Design Department. "We watched with amazement," recalls Bob Thomas, a junior designer at the time, "as the...car was moved into the studio. It was a puzzlement. What was the squared-off rear end with an add-on spare tire? That hood—Christ!—it was a foot longer than the Zephyr's! Has Gregorie lost his mind?" After overcoming his initial shock and analyzing the design more closely, Thomas and the other Ford designers realized that Gregorie had made "a beautifully proportioned car."[61]

"Edsel drove the prototype around for a few days and seemed pleased and proud of the result," recalled Harold Robinson, the plant superintendent. (His co-workers called him Robby.) "He came back to the plant one morning, however, to get the front back padding moved a little to better fit his back. He had a bad back."[62]

Edsel's chauffeur drove the car to Florida—it took three days and two nights, in those days—and met Edsel and Eleanor, who had taken the train, at the train station.[63] Edsel drove the stylish car around Palm Beach, much to the delight of onlookers, and it was the hit of the winter season among the Gold Coast *beau monde*. Edsel received such acclaim for the car that he telephoned Gregorie two weeks later, full of excitement. "Gosh," he exclaimed, "I've driven this car around Palm Beach, and I could sell a thousand of them down here right away. You'd better get over to the Lincoln plant and talk with Robby and see what you can do to set up an arrangement for limited production."[64]

Gregorie met with Robinson, and he immediately began work. "By the time Edsel returned from Florida," recalled Robinson, "we had bodies in paint, a good stock of parts made up, and several bodies on the metal finishing line. It was good that we proceeded with speed because upon his return, his first act of business was to come into the plant directly from the depot to see how we were doing."[65]

At that moment, Edsel was thinking of producing the new car in only a limited production run of 100 or 200 units, something he used to do with custom-bodied Lincolns more than a decade earlier. However, between the time of Edsel's consequential phone call and the time he returned to Detroit, he had decided to incorporate it as part of regular production at the Lincoln plant, commencing with the 1940 model year.

Initially, Gregorie referred to the new car simply as the "special convertible coupe," but before it went into production, Edsel decided to make it the top of the Lincoln-Zephyr line and christened it the Continental to depict its dignified but fast, sporty, and exhilarating design.[66] Formally speaking, the name of the new car was the Lincoln-Zephyr Continental. The hub caps on the prototype had Lincoln-Zephyr stamped into them. It did not lose the Lincoln-Zephyr moniker until 1941 when it became its own distinct model in the Lincoln line. With a price tag of $2,840, it was one of the highest-priced production cars available in the United States at the time. Only a scattering of Cadillacs and Packards were more expensive.[67]

Only 404 Continentals—350 cabriolets and 54 coupes—were built during the 1940 model year, and all of those were essentially built by hand. That volume was much too low to justify spending the hundreds of thousands of dollars it would have cost to tool up for the new model. Therefore, the Lincoln plant relied on its hundreds of craftsmen and artisans remaining from the custom-body days to build the limited number of Continentals. As in the old days of building custom bodies, they used wooden forms and Artz presses to form many parts. The roofs for the hardtop models were hand-made.[68] Production dies were made for the fenders, hood, cowl, and rear deck, and less expensive aluminum castings were used in place of typical steel stampings.[69] In areas that needed additional finessing or a smoother finish, lead was used to fill in any imperfections. It was not until the 1946 Continental went into production after the war that production tooling was used to make all parts of the luxurious model.

The production of the Continental was low, but it served an important purpose in the Ford lineup. From Gregorie's viewpoint, the Continental saved the Lincoln line. Sales of the Model K Lincoln had dropped dramatically after a high of about 3,500 units in 1931 to only slightly more than 200 units in 1939. "If it hadn't been for the Continental," said Gregorie, "the Lincoln name would have fallen into a precarious position after the exclusive K Model was phased out of the picture. With the K Model gone, there was nothing to carry the Lincoln name with any sort of prestige. The Lincoln-Zephyr never had prestige—it was a $1200 competition to Buick and Oldsmobile. So when the Continental came along, it gave Edsel Ford a good prestige boost with the type of car that he was primarily interested in, and it became an exclusive carry-over for the Lincoln name which carried it right up to the war."[70]

Gregorie had mixed emotions about the Continental. "Its profile, I feel, was just about perfect," he said. "It was just so simple, it couldn't help but look nice. We happened to have a nice-looking front end, and the fenders were nice, but I never prized it as a beautiful piece of work. I thought it was kind of weak in the rear end. I didn't care too much for the pinched-in rear end. It reminded me a little bit of a dog with his tail between his legs. I always thought it looked a little hungry. The bumpers were a little too delicate, too. I would have liked a little more beef in the bumpers. Of course, Edsel liked those delicate bumpers."[71]

Actually, Gregorie was surprised about all of the acclaim he received over this one car. "I mean, when you think of all of the other cars and trucks we designed, I don't understand why people single out the Continental. But it was spectacular, and it was more of a deviation from conventional design—let's put it that way—for anything that ever went into production. As it developed, it became very popular, and I guess it'll go down as one of my better-known activities."[72]

That is an understatement. Gregorie's Continental has been the only American luxury car honored for design excellence at the New York Museum of Modern Art. The brilliant architect, Frank Lloyd Wright, told a reporter in 1940 that he thought Gregorie's Continental was "the most beautiful car ever designed."[73] Few automobile designers can claim such distinction.

* * *

Thus, it is an automotive myth that Edsel Ford designed the Continental. Similar to all of the other motorcars created under Edsel's tutelage, his guidance and direction contributed greatly to its outcome. The Continental became his greatest styling triumph.

This was a vehicle Edsel wanted purely for himself, and he directed Gregorie to design it exactly as he wanted it, with no thought of the market. That is why Edsel did not hesitate to suggest that the spare tire be placed on the outside of the trunk or that the car should be finished without any kind of decorative trim. These were not bold steps perhaps, but they were fresh styling cues nonetheless. These features, and Edsel's penchant for simplicity and fine lines, enhanced Gregorie's design and transformed it into one of the most beautiful and admired cars ever built.[74]

Would Gregorie have come up with such a magnificent design on his own? Possibly. However, without Edsel's disregard for the marketplace, Gregorie's design probably would have been more in tune with what would have been acceptable to the public at the time. It would not have embodied Edsel's character and styling genius.

* * *

After the Continental was introduced, it was the only car that Edsel would drive. Occasionally, he would drive one of the new Ford V8's or one of the new Mercurys, but most of the time, he would be seen in a Continental. Now, whenever he wanted to impress somebody with a gift, as he liked to do occasionally, he offered the person a Continental.

Mickey Rooney, a young actor at the time the Continental was introduced, had just completed filming *Young Tom Edison*, a biographical film of Thomas A. Edison. Much of the filming was done on location in Michigan, and while there, Mickey visited Henry and Edsel Ford in Dearborn. They gave Mickey a personal tour of Greenfield Village, driving him around in a new Continental.

Realizing how impressed Mickey was with the new Lincoln offering, Edsel sent a memo to the Lincoln production manager, instructing him to build one "Lincoln Zephyr Continental for Mickey Rooney, specifications as follows: Lyon Blue paint, soft and very pliable leather interior, carpet dyed blue to match leather, tan top, radio and white sidewall tires."[75]

When Mickey received the car a few weeks later in California, he was elated. "I am enclosing some photographs taken several days ago at the MGM Studio when I delivered the Continental to Mickey Rooney," wrote Mickey's publicist to Edsel.[76]

> *I wish we had arranged for a movie camera at the time delivery of the car was made to Mickey. It would have been a masterpiece, even for him. He was like a wild Indian—jumping frantic with joy. I have never seen anyone in my entire experience so genuinely happy and enthusiastic about anything. He ran around the car two or three times, exclaiming "Zowie," "Hot-Diggety-Dog." "Can you mugs feature 'the Mick' behind the wheel of that baby?"*
>
> *He then jumped into the car and to the surprise of all of us, drove right out of the studio and around the block, and then back again. He had everybody in stitches. The people of the studio told me they hadn't been able to get Mickey to do much work for the past two or three weeks, as he talked about nothing but the new Continental Mr. Edsel Ford picked out for him.*

***

The Zephyr and the Continental were not only styling sensations, they were the manifestation of Edsel's dream of developing a line of Ford-produced cars that could compete with the entire array produced by General Motors. Ever vigilant and envious of Sloan's market coverage, when Edsel introduced these two cars, he filled two more gaps in the sparse Ford product lineup. For years, Ford had filled the low-end market with the Standard and DeLuxe Ford models, which competently competed against the General Motors Chevrolet and, to some degree, Pontiac. The Zephyr and the Continental covered the upper-medium-priced and luxury segments, offering Oldsmobile, Buick, and Cadillac appreciable competition.

Now, the major void in the Ford lineup lay in the increasingly popular and lucrative medium-priced market, where General Motors had significant strength. Pontiac and Oldsmobile were both formidable contenders. Gregorie drew up some charts comparing prices and weights of Ford products versus the competition, and the charts "plainly showed the hole in the structure that had to be filled," Gregorie said. "There was about a $500 spread—which was a lot in those days—between the Ford and the Zephyr, and there was nothing in-between."[77] Based on this information, Edsel and his sales manager, Jack Davis, decided that the new car should be larger and longer than the Ford and priced about $300 more.

Unlike the Continental, which was only an exercise in styling for styling's sake, Edsel approached this new project strictly from a business point of view.[78] Pontiac was beginning to take more Ford owners, and Edsel was determined to develop a Ford product to stop the defection. He instructed Gregorie to begin work on an offering that would compete head-on with the glorified Chevrolet of General Motors.

Meanwhile, Edsel began the arduous task of convincing his father of the need for another new car. That is when their on-again, off-again wars flared again. Even though Edsel had recently introduced two new models, he now wanted another one—a car that was slightly larger than the Ford, with a new name, to be built on the Ford production line. Henry did not put up much resistance with the Zephyr or the Continental, because both of these cars were being built at the Lincoln plant—Edsel's domain. But now that he wanted another new model built at the Rouge plant, Henry put his foot down. "The old man couldn't see why Edsel had to have two different cars with just a few inches in length and a different name," explained Gregorie. "Edsel never put it in as many words, but I could tell the resistance he was having with the old man. Edsel was trying to sell the old man on this car as a complete breakaway from the Ford, but the old man was only going to accept it if it smacked of Ford. He wanted to be able to point to the Ford insignia on the glass, and on the battery box, and all that kind of stuff."[79] Henry wanted the new car to have the same chassis, engine, and transmission as the Ford.

These requisites were all simply diversions. "Henry Ford was pretty cagey," explained Gregorie. The fact was that Henry did not want Edsel to add another new car to the Ford product line at all, and he made these demands in hopes of killing the project. What Edsel wanted was an entirely new car, but Gregorie could not understand why Edsel was leaning toward a Ford-looking car. Edsel wanted it larger, to be sure, and he gave Gregorie the added wheelbase that he had suggested. However, from Gregorie's perspective, "Edsel wasn't trying to step it up far enough from the basic Ford. He didn't want a new design. He wanted a Ford-looking design."[80]

On the other hand, Gregorie wanted to do something entirely different—something that would separate the new car from the basic Ford. He pressed Edsel to the point where it began to strain their relationship. "That was the thing that Edsel and I had a little funny feeling about," Gregorie tactfully explained. At the time, Gregorie was not aware of the difficulties Edsel was having in selling this new car to his father. Although Henry rarely came to the Design Department, he was smart enough to realize that Gregorie wanted an entirely new car and was attempting to influence Edsel in that direction. The irony was that Edsel also wanted an entirely new car but was reticent in explaining that to his father. "I don't think the old man would have stood for anything that was a definite breakaway in appearance from the Ford, because I sensed that Edsel had enough of a time putting the idea of the Mercury over with the old man." Edsel and Henry finally reached a compromise. Henry said that Edsel could have his new car as long as it was a continuation of the Ford appearance and it used his beloved buggy chassis.[81]

With Gregorie's desire to develop an entirely new look, he had to be careful in how he developed the new design. "We just weren't in a position to show Edsel anything that looked un-Ford-like," says Gregorie. "There was always that pressure, and I had to be awfully careful not to offend Edsel and his case and wants in dealing with the old man. I had to talk out of three sides of my mouth. You can't realize what I went through. I had to keep Edsel happy, and he had to keep his old man happy. I was beat going home some nights after spending hours with Edsel trying to work this thing out. The old man was smart enough to realize that I was pulling in one direction and I was trying to influence Edsel in a new direction. I could see that Edsel was caught in the middle. Edsel would come to me and in essence say, 'Dad can't see it this way, and I'm having a hell of a time satisfying him.' He couldn't put it in as many words, but he did express this fact to me very clearly."[82]

Realizing that he was putting Edsel in an uncomfortable position by developing unique designs for the new car, Gregorie finally acquiesced into developing a Ford-like design. "To have a complete breakaway would have been more than we could have asked for," says Gregorie. "It was that kind of feeling in all of my conversations with Edsel. I could see that he was up against a problem, and I had to go along with it. I was in no position to poke a hole in the idea and start a whole new complication." Edsel would have liked to have a car that was completely different from the Ford, but to satisfy his father, Edsel directed Gregorie to simply make a larger version of the Ford, which, for all intents and purposes, is the way it turned out. Gregorie says, however, that the design he developed "was enough of a change so that it didn't look exactly like the Ford, but for a new model, in my mind, I would have liked to have done something to give it an entirely different identity. But because of the pressures that were exercised in the development of the car, Edsel felt safer going along with his father."[83]

Based on Edsel's desires, the appearance of the 1939 Mercury followed the general theme and proportions of the 1939 Ford, but the body shape was a little wider and a little fatter. "In other words," describes Gregorie, "it was just a stylized Ford. We added a little more bulk to the car, and it showed up on the road as a little more important looking."[84] With its increased hood length and stubby-looking tail end, the new design was approaching Gregorie's idea of a well-proportioned automobile.

Tucker Madawick, an apprentice designer at the time, believes that the Mercury turned out quite well, given the constraints under which Gregorie and Edsel were working. "The Mercury was a real advanced car against the basic Ford," he says. "The car was *avant-garde*. And to get something like that into the Ford line was a real accomplishment. Remember, the old man still had a say in everything that Edsel did, and that was a big problem. We had more headroom in our cars than the other manufacturers, because the old man insisted that we sell these cars to farmers who went to the county fair wearing bib overalls—and they always wore their hats in the car. Harley Earl, with the full support of Larry Fisher and the top man, Alfred P. Sloan, never had this problem. He said, 'Shit, let them take their hats off!' The Italians said the same thing. You don't wear a hat in the car. Close the window! The contrast between Ford and General Motors was unbelievable!"[85]

When it came to naming the new car, Gregorie ran into trouble again. Edsel had chosen the name *Mercury* from a list of more than 100 others, which included everything from *Cyclops* to *Winged Victory*.[86] The name was not what caused the commotion, but Edsel's insistence in combining the name Mercury with the name Ford. Reminiscent of what he had done with the Lincoln-Zephyr, Edsel wanted to call the new car the Ford-Mercury. Gregorie was aghast. The design that was intended to be an entirely new car was already being diluted to resemble a Ford. By calling it the Ford-Mercury, there was almost no chance in separating its identity from the low-cost Ford.

Years earlier, Gregorie brought out a similar contradiction with the Lincoln. He mentioned to Edsel that it was incompatible to sell and service Lincolns at Ford dealerships and suggested that the two lines be separated. "Can you imagine a man paying $6,000 for a custom Lincoln and then having to take his luxurious motorcar into a Ford garage to get it serviced alongside an old dump truck or a tractor!" he explained. "Edsel got a big kick out of that." Joe Galamb and the other chassis people would never think of that, because they were not sales oriented. However, Gregorie was, and Edsel appreciated his sales perspective. When Gregorie mentioned to Edsel that the Ford name should be separated from the Mercury's name, however, Edsel "flared up." "What's wrong with the Ford name!" Edsel questioned heatedly. "We've spent 40 years building up the Ford name, and you're telling me it's not good enough to put on this automobile?"

"Boy, that was like a hot poker on Mr. Ford," recalls Gregorie. "He kind of had me on the hot seat, squeezing it for a minute!" However, Gregorie put forth his argument, nonetheless. "There's nothing wrong with the Ford name, Mr. Ford," Gregorie quickly replied, "but we need to get people to believe it's not a Ford so that they will move up to it, just like General Motors gets people to move up from a Chevrolet to a Pontiac to an Oldsmobile." Edsel left without saying anything further.

Meanwhile, Gregorie, who now thought his days at Ford were numbered, pursued his instincts nevertheless and had some of the sheet metal boys hammer out some hubcaps with only the Mercury name on them. At the same time, the Ford-Mercury was shown at the New York Automobile Show, and many of the Ford sales managers and dealers "climbed all over Edsel" when they saw the Ford name connected with the new Mercury. They explained to him the same thing that Gregorie tried pointing out a few weeks earlier. After the show, Edsel ran into Gregorie at the train station in Detroit and told Gregorie, "There's some things we have to go over. I'll be in right after lunch."[87]

During the train trip back to Detroit, Edsel thought about what the dealers had told him at the automobile show and changed his stance on naming the new car Ford-Mercury. Before the Mercury went into production, the Ford name was removed (but it was still on the glass and battery). Gregorie kept his job.

In the end, the Mercury "came out well enough" explained Gregorie. "In fact, we were tickled to death to get an okay on a car that was built as well as this one was."[88]

***

When Edsel was able to convince his father to switch from the Model T to the Model A, he opened an era unprecedented in Ford Motor Company until that time. From that point, Edsel was able to ramrod model change after model change within the company. It took Edsel 5 years to convince his father to change from Model T to Model A and 3 years to change from Model A to Model B. However, Edsel implemented model changes every year thereafter. Some of these changes were complete redesigns such as the 1933 Ford, and others were simply facelifts, such as the minor differences between the 1935 Ford and the 1936 Ford. Then Edsel also was able to introduce completely new model lines, such as the Lincoln-Zephyr, the Lincoln Continental, and the Mercury. These were major accomplishments for Edsel, particularly in light of his father's dictatorial control over the company and over him.

What Edsel wanted now was a line of entirely new automobiles, redesigned from the ground up, with a more modern power train, chassis, and body designs. After all, neither General Motors nor Chrysler was resting on its laurels. In addition to yearly model changes or facelifts in body design, both companies also were constantly updating the "hardware" of their vehicles. As far as Edsel was concerned, it would be only a matter of time before he would be able to do that as well. In fact, he and Gregorie began talking about a completely new line of Ford, Mercury, and Lincoln automobiles shortly after they had completed the new Mercury.

Looking back from a 60-year perspective on these product development ideas, one can see that Edsel and Gregorie would have taken the industry by surprise if the new cars that they had in mind had entered the marketplace in the early 1940s. However, the start of World War II in December 1941 all but put the brakes on their plans.

Chapter 23

The Ford Foundation

> *"Henry Ford was as selfish a man as God permitted to breathe."*[1]
> E.G. Pipp

> *"If the business should not require all the money it earns, then it would not be hard to find plenty of good uses for all that we have."*
> Edsel Ford[2]

With Edsel's growing responsibility with the design of Ford automobiles, and his burdening duties of running the mammoth Ford industries, he had little time to devote to philanthropic endeavors. Also, his experience with Admiral Byrd, combined with his intrinsic desire to be left alone, compelled Edsel to find a way to channel monies to charitable and worthwhile causes without having to be in the public eye and without taking up his valuable time. As one close associate explained it, "He was determined to handle the family philanthropies in a more business-like way and to be relieved from the day-to-day pressure and annoyance of many appeals."[3]

As president of Ford Motor Company and known as one of the wealthiest men in the world, Edsel received numerous letters every day, asking for donations to this cause or that cause, or simply contributions for personal hardships.[4] Most of these solicitations were unimportant or "bait," as Edsel liked to call them, and were from individuals trying to get a few bucks for free.[5] However, Edsel did not want to ignore the solicitations simply because of their volume, for there might be one in a thousand that had a legitimate cause. If there were, Edsel wanted to be notified of it. Thus, he had Lepine and his staff sort through all of the letters that came into his office, and they responded to most of them. At Edsel's suggestion, Lepine composed several types of form letters to send to solicitors so that "the writers would feel they had been well treated."[6] A typical letter would say: "I should be unable to offer aid from here, as Mr. Ford's existing commitments are such that it would not be practicable to attempt to aid all the individuals who apply here."[7]

***

Over the years, Edsel had been involved with so many philanthropic activities that he could have spent all of his time devoted to such endeavors. Starting with anonymous donations to the Detroit Institute of Arts, Edsel expanded his personal donations to organizations such as the Henry Ford Hospital, the Tau Beta Community House, The Edison Institute (popularly known today as the Henry Ford Museum & Greenfield Village), Children's House, and many others that will never be known.[8] "Ninety-nine percent of his [Edsel's] charitable and civic benefactions have never been heard of," wrote a reporter, "and yet one cannot move around this town without finding out where he has come in, usually to 'save the day,' but on the solemn promise of the sponsors that 'under no conditions' must it ever be known that he 'helped out!' "[9] But there were some that we do know of—some large, and some small.

For example, Edsel donated $750,000 to the YMCA, $150,000 to his alma mater, the Detroit University School, and $200,000 annually to the Henry Ford Hospital. On the smaller end of the spectrum, he made a one-time donation of $5,000 to the League to Enforce Peace (a League of Nations type of organization), and he paid for scholarships for two Mexican-American boys from Texas who had visited the Rouge plant.[10] Every year, Edsel donated approximately $75,000 to the Detroit Community Fund, a precursor to the United Way.[11] This latter charity became one of Edsel's favorites. During the Depression, the Detroit Community Fund was practically the only place where laid-off workers could obtain food, clothing, and shelter. Henry did not care much for community chests as a rule, but Edsel liked the way this group was organized and gave generously, providing $130,000 to the fund in 1932.[12]

In addition to monetary donations, Edsel also thought of ways to provide services to the needy. For example, he considered mass producing soybean milk for the poor.[13] One time, he had the idea to establish a restaurant where meals could be obtained for two cents, and he convinced the Fisher Brothers to donate one of their old factories for low-cost housing.[14]

William Cameron, Henry Ford's press agent, recalled a time when Edsel offered assistance to Detroit veterans returning after World War I. Hundreds of these men had ailments that had appeared during the war but could not be attributed to the war; therefore, they were denied government assistance. When Edsel was told about these men, he made arrangements to have every one of them receive medical care at the Henry Ford Hospital. "They were treated in all respects as private patients," recalled Cameron. "It was done without charge, and they were given private rooms. Some of them were in there for as long as two years."[15]

Because Edsel was president of the Henry Ford Hospital, he gave a lot of attention to its operation. Occasionally, he conferred with his father about hospital policy, but most of the time he bore the responsibility himself. To the management of the hospital, Edsel brought the same grasp of administration that characterized his work in the Ford Motor Company.[16] He and his mother redecorated the hospital, and he provided a smoking room for the nurses. Henry was not in favor of that, but Edsel made it clear that it should be done.[17]

Over the years, Edsel gave nearly $4 million to the hospital, and the money was used for everything from building improvements to medical research.[18] He outlined a broad research program, which included investigations in tuberculosis, artificial fever therapy, and the prevention of fluorine toxicosis.[19]

***

In the fall of 1924, Franklin Roosevelt, who had been disabled by polio in 1921, learned of a young man, also a polio victim, who had spent three successive summers bathing in the warm water pools of Warm Springs, Georgia. This young man, practically helpless, was helped each day into the pool. Over time, the man discovered that he could move his legs in the water; by persistent effort and exercise, he was able to strengthen them. By the end of the third summer, he was able to walk with the aid of a cane.

Roosevelt went to Warm Springs in the spring of 1925 and spent several weeks there, exercising and swimming in the pools. Although his condition did not improve, he noticed that other patients made remarkable progress. Roosevelt suggested that detailed reports be made on a group of these patients, and by the end of the summer, it was learned that many had made remarkable progress. At the close of the season, the report was submitted to three prominent orthopedic surgeons, and each concurred that a hydro-therapeutic center should be established at Warm Springs. To assist in the endeavor, Roosevelt established the Warm Springs Foundation in 1927 and immediately sought contributors.[20]

Edsel had known Roosevelt for many years, having first met him during World War I, and he was saddened to hear of Roosevelt's affliction. Thus, when Roosevelt asked for assistance, Edsel gladly obliged. He promptly sent a check for $25,000 to Roosevelt, which was used to build a glass-enclosed swimming pool at the resort. In subsequent years, Edsel contributed annually to the Warm Springs Foundation and participated in the President's Birthday Ball, which was established to raise money for polio research. When Roosevelt became President of the United States, he used his powerful position to solicit contributions from wealthy Americans for this worthy cause and in 1935 asked Edsel to serve as a member of this fund-raising organization. "I have followed with interest the splendid work which you have done toward the welfare of your fellow citizens," Roosevelt wrote, "and I can assure you that it will give me great personal pleasure to have you accept this appointment."[21]

Edsel could not refuse a direct invitation from the President of the United States, but he would have volunteered anyway, simply on the basis of their friendship. Edsel personally contributed $1,000 to the event and raised an additional $32,000 from his business associates. He continued to support Roosevelt's Ball in subsequent years but not in an official capacity.[22]

Along the lines of medical research, Edsel also developed an interest in cancer research. In 1927, he and two fellow Detroit businessmen, Richard Webber of the J.L. Hudson Company and Roscoe Jackson of the Hudson Motor Car Company, established a cancer research center in Bar Harbor, Maine. The three donors, who insisted on anonymity, pledged regular monthly contributions of $3,750 (of which Edsel's share was $1,875 a month) extending over a period from July 1, 1927 to June 30, 1932. These funds met the costs of the first laboratory and the other expenses of maintenance and salaries. In all, Edsel contributed approximately $200,000 to this fund.[23]

***

Sales of motorcars skyrocketed from 1910 through the 1920s, thanks in large part to the inexpensive Model T. However, the roads of the country were not keeping pace with the use of the automobile. Edsel well remembered his cross-country trip in 1915, when roads were nothing more than wagon trails and the most reliable routes were railroad tracks. Therefore, when the Lincoln Highway Association was established in 1913, backed by the leaders of the automobile industry, Edsel became one of its founding members and a director.[24] The objective of the association was impressive: to build a hard-surfaced two-lane road from New York City to San Francisco.[25]

Unlike the other automobile manufacturers who used their company's money to make donations—General Motors donated $100,000, and Willys-Overland donated $50,000, for example—Edsel donated his own personal funds.[26] Each year during construction of the highway, he donated $5,000.[27]

Although the entire project was interesting to him, Edsel became particularly involved in the so-called "Ideal Section" project. This section of the Lincoln Highway would be designed and constructed using the latest ideas in road construction technology and layout, and it would be used as a model for all future highways. Edsel appointed Jens Jensen, the landscape architect who had designed the grounds around Edsel's Grosse Pointe mansion, to design a 17-acre campground adjacent to the highway. "I feel that this contribution is one way of showing my interest in the Highway Association and the good work it is doing," said Edsel.[28]

In the end, the Ideal Section was indeed ideal. The road was several miles long, 40 feet wide, and made of concrete 10 inches thick. Sections were lighted, safety signs were installed, and a pedestrian footpath was constructed. A gas station was also incorporated into the design, and that structure was not approved until Edsel had approved it: "Mr. Ford's suggests that if the lavatory in the men's room was moved so as to be under the window, it would allow moving the partition so as to afford one additional toilet in the women's room." In total, the Ideal Section cost $74,000, of which Edsel contributed $33,000.[29]

***

Edsel's involvement in these philanthropic activities was pressing his time and imposing on his privacy. However, he knew these activities were important.[30] The first call on the money that a company earned properly belonged to the company itself, Edsel maintained, but humanitarian efforts had a claim on the funds that were not needed for operations, reinvestment, and reserves.[31] "If the business should not require all the money it earns, then it would not be hard to find plenty of good uses for all that we have. There is no end of opportunities for medical and scientific research, and then there is always education. It would be worth hundreds, perhaps thousands, of millions to get rid forever of any one of a dozen common diseases. And that is mentioning only one line of research."[32]

The money was not the issue. It was the time. Except on the occasion when Edsel was dumbfounded at the sight of $1 million of gold bullion that his father had given him on his twenty-first birthday, "wealth didn't seem to affect him [Edsel]," recalled a close associate.[33] "Money can hardly be a burden, unless one lives for money," explained Edsel. "I have never been able to see that the ability to earn money should be a curse or a burden, unless one chooses it to be so."[34] What Edsel sorely needed was an avenue through which he could channel his "broad humanitarian impulses."[35]

***

That way was through a formally organized foundation expressly established to distribute the vast amount of funds being generated by the Ford Motor Company and being solicited by individuals and organizations alike. In addition to this, Edsel was thinking toward the future and the continuation of the company, especially in light of the Income Tax Amendment passed by Congress in 1913 and the Estate Tax (commonly known as inheritance tax) of 1916. Although Edsel probably understood the effects these laws might have on him, his family, and the company, they were emphasized by his friends John D. Rockefeller, Jr. and Rockefeller's son Nelson. "Edsel had been distressed frequently by the plight of some of his friends or their heirs who were forced to liquidate at depressed values to raise tax funds and in the process divested their families of the control of businesses built up over many years."[36]

With the passage of the inheritance tax, it was mandated that all bequests of $50,000 or more were to be taxed up to a maximum of 10 percent. However, the super-rich realized that this was only a precursor of what was to come—10 percent now, and who knows how much later?

In 1924, Congress increased the tax to 24 percent on fortunes exceeding $10 million, and in 1932, under Franklin Roosevelt's New Deal, it went up again to 45 percent.[37] There was debate on the Senate floor before this latest tax increase was enacted, spearheaded by Michigan's Senator Vandenberg, who used Edsel's future inheritance as an example of the devastating effect the new tax bill would have on the rich constituents of his state. "The utmost to be anticipated would be that some part of the equity now represented by the common stock would be sold to other interests or to the public," argued counsel on behalf of the Internal Revenue Service. Vandenberg was quick with his rebuttal: "There need be no speculation as to what will happen to the great Ford industrial enterprise with the proposed tax confiscation," he said. "Bluntly, the Ford Motor Company will be driven into the hands of Wall Street, or its equivalent—and the money changers, who have striven vainly in years past to achieve this end and whom this administration says it proposes to drive from the temple, will have been handed the dominion which in no other manner would be obtained. If Mr. Ford left an estate of $300 million, the proposed inheritance tax on top of the existing tax would take $270 million of it."[38]

Senator Vandenberg's numbers were very close to being accurate. However, what he may not have realized, and certainly the IRS council did not, was that neither Henry Ford nor Edsel Ford nor the two men combined had that kind of money lying around.[39] To pay that exorbitant tax, the heirs would surely have to sell the company.

***

The Ford family was unique at this time. Unlike Standard Oil and United States Steel, which were owned by thousands of stockholders, the Ford Motor Company was owned by only three people: Henry, Clara, and Edsel Ford. The death of any of the Rockefellers or any of the Carnegies would affect the families only a small amount and would not affect their companies at all. With the Ford family, however, the sudden death of either Henry or Edsel would force the surviving members to sell off great blocks of stock to pay the federal government its due.

This became especially ominous in the summer of 1935. At that time, Edsel was listed as the richest man the world had ever known, and when he eventually inherited his father's portion of the company, he would become the richest man the world has ever known.[40] That August, the Wealth Tax was passed, implementing new and higher income and inheritance taxes. Estate taxes rose to 50 percent over $4 million and to 70 percent of all in excess of $50 million. These sums seemed staggering to the average Depression worker. However, to the Ford family, whose net worth exceeded $1 billion, they signaled the death knell to their great fortune and the existence of the Ford Motor Company. Without some kind of tax shelter, the death of either Henry or Edsel would force their heirs to pay at least $300 million dollars to the taxman, and that would have required them to sell Ford Motor Company.

Edsel knew he had to act, and act quickly. At the time of the passing of the wealth tax, Henry Ford was 72 years old. Although spry for his age, Henry could be taken suddenly. Plus, Edsel could not rely on his father to take the necessary steps to thwart the impending doom. Part of this was because Henry did not want to face his own mortality, but equally importantly, he loathed organized charity. From Henry's perspective, giving away a new Ford every now and then was sufficient.[41]

***

Through the advice of the Rockefellers and his own research, Edsel ordered Clifford Longley, the Ford family attorney, to establish a foundation.[42] This was done sometime between July 1935 and January 1936.[43] Running only three, double-spaced typed pages in length, the charter that created the Ford Foundation was "as simple and effective as the Model T."[44] Edsel was pleased and signed the charter on January 13, 1936.[45] Whatever Longley may have done to protect the Ford Motor Company and the Ford family during his long career, it would pale in comparison to what he had created in this document.

The purpose of the Ford Foundation differed from those of the Rockefeller and Carnegie foundations, for it was created to avoid income taxes. The other two foundations were not—neither of these taxes existed when they were founded.[46]

"Henry Ford did not see or anticipate the future of the Foundation," recalled Charles Sorensen. "Edsel Ford did realize the great possibility of the Foundation and it was anticipated he would carry out the future plan when the estates of Mr. and Mrs. Henry Ford were passed onto the Foundation. But the early death of Edsel delayed any positive plan."[47]

We do not know the specifics of all of the conversations that Edsel had with Longley, but we do know that it was Edsel's idea to incorporate the two most salient aspects of the charter, from a letter that Longley wrote to Henry Ford: "Sometime during the year 1935, we were asked to prepare wills for yourself, Mrs. Ford, and Mr. Edsel Ford. In my discussions with Mr. Edsel Ford, from whom we had our instructions at the time, two points of importance were emphasized... Mr. Edsel Ford desired...to perpetuate the ownership and control of the Ford Motor Company in the Ford family as far as it is legally possible to do so and to facilitate the raising of funds to raise estate and inheritance taxes."[48]

From this point, Longley took over, and he did his job well. Having the foundation pay for inheritance taxes could be accomplished quite easily. The legal term is *residuary legatee*, and this provision in law enabled the foundation to pay any taxes due by the heirs of the Ford fortune. In other words, the sum that the foundation received from the Ford estates was pared down by the claims of the government.[49] However, there was no legal provision that would allow a foundation to perpetuate *both* control and ownership of a company in the hands of the owners. Longley left the decision to Edsel as to which one he wanted.

In one of the more astute decisions in business history, Edsel chose control over ownership.[50]

***

Longley made the next shrewd decision. As a means of keeping family control of the Ford Motor Company, his recommendation was to reclassify the 172,545 shares of Ford Motor Company stock held by Henry, Clara, and Edsel into Class A non-voting stock and Class B voting stock. The two classifications of stock, Longley explained, would allow the Ford family to hold enough voting stock to ensure their control of the company, even though their ownership in the company would drop significantly upon the death of Henry or Edsel. (It was assumed, of course, that Henry would be the first to go.) The idea was to have the Fords bequeath their non-voting shares to the foundation and their voting shares to the surviving members of the family.

Edsel agreed to the proposal and discussed the idea with his father and mother. Clara quickly agreed, but Henry hesitated. First, he did not fully understand the necessity of a foundation nor why it was necessary to institutionalize his giving.[51] He also had his own ideas about how the foundation should be set up, but none of them made sense.[52] Most importantly, Henry did not like the idea of having to change his will, which would be necessary in order for him to bequeath his Ford stock to the foundation. "Mr. Ford became very upset over the whole thing," recalled Harry Bennett, "and, I believe, saw it as a plot to get everything away from him. At one point, Mr. Ford was so angry he told me he would be damned if he didn't spend every cent he had before he died."[53] When asked by a reporter what would happen to the vast Ford empire when he passed away, Henry replied, "There has been no plan for a Ford dynasty—rather to do the job as well as possible. I have just been telling Edsel that if we keep our noses pointed in the right direction, we will go along."[54] Charles Sorensen supports this statement. He knew that Henry Ford had made no provision as to the dispensation of the company after his death.[55]

Understandably, it took a lot of consoling from Edsel to convince his father that forming a foundation was in the best interest of the Ford family. Edsel explained to Henry that it was the only way the great industrial enterprise would remain intact, for the Ford family to remain part of it, and for the government not to obtain any piece of it. This last point, more than the others, was what finally persuaded Henry to accept the foundation idea.[56]

To Edsel's credit, he again faced down his father in order to do what was right for the family and for the company. "Edsel enabled the smooth transition of my father's plan," explained Longley's son. "Had my father had to work with Henry Ford to develop a scheme, something would have been done, but the likelihood of it coming out so smoothly is questionable."[57]

Before 1935 ended, Longley had reclassified the Ford family stock into 3,452,900 shares of the new issue, with 95% in Class A stock and 5% in Class B stock. These new classes did not affect the potential income from Ford dividends, because both classes would receive dividends on the same basis.[58]

The next step was to rewrite the wills of Henry, Clara, and Edsel. Longley did that in early 1936. Henry's will now stipulated that upon his death, all of his Class A stock would be left to the Ford Foundation, and his Class B stock would be divided equally among Edsel and Edsel's four children—none would be left to Clara or to Eleanor. Edsel's will was also revised so that his Class A stock (1,354,904 shares) would be left to the Ford Foundation, and his Class B stock would be divided equally among his wife and children.[59]

The genius of this scheme was that the Ford Foundation, not the United States Treasury, would be enriched with 95% equity in the Ford Motor Company, while the Ford family would remain in control of the Ford empire. At the time, it was called the "greatest tax dodge in history." However, although the Fords would be able to avoid paying huge sums in taxes, they would forever lose ownership in the mammoth Ford industries that Henry Ford had founded and Edsel Ford had developed.[60]

***

On January 13, 1936, Edsel, Longley, and Bert Craig, treasurer of Ford Motor Company, signed the simple, three-page document that established the Ford Foundation.[61] It started small—Edsel donated an initial sum of only $25,000, and its headquarters was set up in the Ford Engineering Laboratory. The purpose of the foundation was simple: "to receive and administer funds for scientific, educational, and charitable purposes, for the public welfare."[62] Said Edsel, "The foundation will take care of the various...activities that I don't care to do personally. It will be on a small scale, and I have no intention of making it larger."[63] For years, the foundation was a minor concern, even to the Ford family.[64]

The Ford Foundation was doing what Edsel had wanted—taking over the philanthropic activities that he supported but with only minimal involvement from him. All he had to do now was attend annual board meetings and make contributions if and when he saw fit, which must have been often; by the end of the year, Ford Foundation assets had already reached nearly $2 million.[65] The Ford Motor Company also began donating funds to the foundation, ostensibly to be used for any worthwhile endeavor. However, most of the funds were funneled to the Henry Ford Museum and Greenfield Village. Between 1938 and 1944, Ford Motor Company gave more than $10 million to the foundation.[66] Therefore, the assets of the fledgling foundation began to grow, allowing it to dole out donations to various philanthropic activities, with little or no input from Edsel.

***

Edsel could now put his mind at ease. No longer would he have to wrestle with countless pleas for contributions or worry about what would happen to the Ford Motor Company in the future. With Henry, it was another matter. Although he acquiesced to the idea of the foundation and changed his will to accommodate that, he was beginning to have second thoughts, the ramifications of which would not be known for a few more years. That is why Henry refused to be one of the incorporators of the new foundation.[67]

Chapter 24

Bennett Motor Company

"Edsel disliked Bennett very much."[1]
William Cameron

"The relationship between Edsel and Bennett was like a cat's claw covered by fur."[2]
Albert Wibel

Until this time, Edsel Ford had been president of Ford Motor Company for more than a decade and wielded power in all areas of the company except engineering and labor. His father had always dominated engineering, and Henry had entrusted labor to a Napoleonic character named Harry Herbert Bennett.

Born in 1892 and raised in Ann Arbor, a college town 30 miles west of Dearborn, Bennett was about two years older than Edsel. He joined the Navy in 1909 and had traveled the world during his three-term stint.

Bennett was a short, stocky man who developed a braggadocio and physical build as a means of protection. In the Navy, he was, among other things, a deep-sea diver and prize-fighter—his ring name was "Sailor Reese."[3] Bennett could talk fast and was quick-witted, and he could tell a joke with the best of them. If he could not talk his way out of trouble, he never hesitated fighting his way out.

Informally, Bennett was fun to be around. However, when it came to work, it was best to either be on Bennett's side or stay away from him. To one Ford executive, Bennett was "very unimpressive."[4]

***

While on leave in New York and contemplating whether he should re-enlist for a third tour of duty, Bennett and a friend got into a brawl with customs officials at Battery Park. The reason for the scuffle is lost to history, but *The New York Times* journalist, Arthur Brisbane, saved Bennett and his friend from spending some time in jail. By coincidence, Brisbane had watched the entire incident unfold—seeing how well the diminutive Bennett handled himself against much larger men. He intervened on Bennett's behalf and cleared up the matter with customs.

Bennett thanked Brisbane for his help, and before he could excuse himself, the journalist asked Bennett if he would like to meet Henry Ford, who was in town visiting his New York distributor. Bennett never knew why Brisbane wanted him to meet the automobile magnate—all he could think of was that Henry Ford was looking for a young man who could take care of himself.[5]

"Can you shoot a gun?" Ford asked Bennett.

"Sure can," Bennett replied.

"Good. I can use a man like you at the Rouge. The men out there are a pretty tough lot, and I haven't any policemen out there."

"Well, I won't work for the company, but I'll work for you," Bennett replied.

"Fine," said Henry. "Go home, and I'll send for you there."[6]

It was months before Henry got in touch with Bennett, and when he did, he sent Bennett to the Rouge plant as promised. "I'm sending you over there to be my eyes and ears," he told Bennett. Then he added, with a tone of warning, "There may be a lot of people over there who want to fire you, but don't pay any attention to them. I'm the only one who can fire you."[7] Bennett took that last statement to heart, and he would lean on it for the next 30 years.

***

For years, Ford Motor Company had a security force called Ford Service. The job of Ford Service was to guard the factory gates, prevent theft, and keep order over the 100,000-men workforce at the mightiest manufacturing plant in the world. A facility as large as the Rouge—which covered more than 1,000 acres and had more than 93 separate structures—needed a policing force.[8] However, what began as a business necessity turned into an arm of tyranny.

Bennett took over as head of Ford Service from Pete Kelly, who had fallen out of favor with Henry Ford. According to Bennett, Henry wanted Kelly to use some of his men to spy on Edsel, and when Kelly refused, Henry fired Kelly.[9] That is why Henry was looking for somebody when Bennett presented himself in New York.

When Bennett learned the reason for Kelly's dismissal, he quickly assessed the situation at Ford Motor Company: do what Henry Ford asks, or find yourself out on the street. To Bennett, it was really no different than military life. After six years in the Navy, he had learned discipline and was used to taking orders and carrying them out without questions. No matter what Ford asked of him, Bennett would do his best to accomplish it and usually succeeded. Henry's usual practice was to test someone's loyalty by giving him some outlandish assignment that was sure to fail, simply to see how he would handle it. In most cases, the stooge would argue why it could not be done or why it should not be done, only to receive Henry's wrath.

With Harry Bennett, it was different. "I was Mr. Ford's aide, his master of all work," recalled Bennett. "I learned to know Mr. Ford. I learned how he thought, how he felt, how he operated. I also learned what he expected of me."[10] Thus, when Henry asked Bennett to do something, Bennett would get started so quickly that a lot of damage would be done before Henry could stop him.[11] Henry loved that. He loved it when he did not have to explain his every move to someone, and he loved it when someone would carry out his directives without hesitation and without question. Bennett loved having the power to wield around the company to accomplish Henry's orders.

One time, a reporter asked Bennett in Henry's presence if there were anything Bennett would not do for his boss. "If Mr. Ford told me to blacken out the sun tomorrow," Bennett replied with all seriousness, "I might have trouble fixing it. But you'd see a hundred thousand sons-of-bitches coming through the Rouge gates in the morning, all wearing dark glasses."[12]

***

The obsequious relationship that developed between Henry Ford and Harry Bennett sneaked up on Edsel. When Bennett joined the company in 1915 and became manager of Ford Service, Edsel was still a young man learning all aspects of the company under the watchful eye of his father. As Henry's son and heir apparent to the Ford empire, Edsel need not fear any man, let alone an ill-educated sailor watching over plant security. However, as the years passed, Henry watched both men mature and perform their duties in completely opposite ways. Edsel, of course, carried out his work in a quiet, unassuming, and democratic way, whereas Bennett carried out his work in the manner of the roughneck that he was. Henry liked that.

This became particularly evident after the change from the Model T to the Model A in 1927. Henry had never forgiven Edsel for taking a stance against his beloved Model T. To Henry, it was insubordination of the highest degree. Harry, Henry thought, would never have questioned the viability of Model T as Edsel had done.

Eventually, Henry began treating Bennett as if he were a son, and treating Edsel as if he were a hired hand. Bennett would pick up Henry at Fair Lane in the morning, and the two men would drive to work together. If somebody wanted to find Henry during the day, he could usually find him in Bennett's office located in the basement of the Ford Administration Building. "Henry Ford would go over there quite often," recalled Bob Gregorie. "He'd go in and sit down and place his feet up on Bennett's desk, and the two would bullshit for hours on end."[13] As the years passed, Henry often met with visitors in Harry's office instead of his own. This allowed Henry to get up and leave without being beholden to the visitor.[14] After work, Henry would call Bennett at home every night at precisely 9:30. Each had radiotelephones in their automobiles with which they could call each other—Bennett's number was 9, and Henry's was 10.[15]

Edsel likewise had a direct telephone line from Gaukler Pointe to Fair Lane, but he and his father rarely used it.[16] Henry never went to Edsel's office at the Administration Building, although it was directly above Bennett's office, nor did he pick up Edsel at his home and drive to work together. The only time that Henry allocated to Edsel was their regular luncheons in the Engineering Laboratory or during impromptu meetings in the design studio or somewhere in the plant. These meetings were awkward for both Henry and Edsel—they were not the congenial, chummy talks that were typical of those between Henry and Harry. Bennett appeared to have become the son that Henry had always wanted.

Edsel and Bennett were never on completely friendly terms. During the early years, Bennett was a lowly supervisor, and although Edsel treated all Ford employees graciously, there was no reason for the two to fraternize with each other. Bennett simply carried out projects for Henry and never interfered with Edsel's responsibilities. Later, however, when Bennett began to benefit from the graces of Henry, tension developed between Bennett and Edsel. As one executive described it, "they got along together like a couple of strange bulldogs."[17] Periodically, Henry would make efforts to bring the two men together; however, when Edsel and Bennett appeared to be making progress, Henry would break it up. "Now, Harry, you think you're getting along all right with Edsel, but he's no friend of yours."[18] The same thing occurred between Bennett and Charles Sorensen. Initially, the two men got along as well as any other pair of Ford executives, but Henry "generated conflicts" between the two men.[19]

It took Bennett 25 years to become Henry's right-hand man, but when he arrived, Henry entrusted him with so much power that even Edsel was forced to yield to Bennett.[20] "Bennett's orders can be rescinded by no one save Henry Ford," wrote one reporter.[21] So great was Bennett's power in the Ford Motor Company that in the late 1930s and early 1940s, Ford workers began calling the company "The Bennett Motor Company."[22]

***

Bennett took full advantage of his position as head of Ford Service. Although Henry did not pay Bennett much in salary, he gave Bennett an open checkbook for his personal expenses. "He [Bennett] always carried a roll of cash in his pocket," recalled a Ford executive.[23] Bennett used Ford Motor Company men and materials to remodel his house. He bought appliances on Ford's account, and he even bought a 75-foot yacht on which to "entertain" Ford Motor Company guests.[24] (Bob Gregorie designed a yacht for Bennett, which he was going to have built in the Rouge locomotive repair shop but instead had it built at a Detroit boat works.[25]) Bennett kept the yacht docked at the Grosse Ile yacht club, the same place where Bob Gregorie moored his boat. Bennett also had a summer cottage built on Grosse Ile on land deeded to him by Henry and Clara.[26]

Many times, Henry told Fred Black "that the Ford Motor Company would be carried away if it wasn't for Harry Bennett."[27] Although there is some truth to that statement—after all, Bennett was in charge of plant security—Bennett himself partook of many Ford Motor Company assets, all with Henry's approval. One day, for example, the Accounting Department contacted Bennett regarding an order of building brick that he had requisitioned. The brick company had not been paid for the shipment and was asking the Ford Purchasing Department for payment. However, the Purchasing Department claimed that it did not have an order for bricks. The brick company replied that it had a receipt stating that the shipment was delivered to Ford Motor Company and was signed for by Mr. Bennett. When the Purchasing Department learned of this, one of its managers contacted Bennett and told him, "You ordered it, you pay for it!"

Bennett replied, "Who in the hell is worried about a shipment of brick? You go and pay the bill and forget about it!" The manager backed down, and nothing more was said.

The truth was that the shipment of brick did indeed come into the plant and was signed for by Bennett. However, Bennett then ordered the shipment to be taken to a friend's farm for use in building a dairy barn.[28]

***

Although Bennett's position was lowly when he started with the company, he successfully developed it into a formidable power base—all with Henry's approval. He staffed it with former football players, boxers, wrestlers, and ex-policemen. He also hired hardened criminals. On Bennett's payroll, for example, were noted gangland characters such as "Legs" Laman, Joe Adonis, Joe Tocco, and Chester LaMare, known as the "Al Capone of Detroit."[29] Both Tocco and LaMare were part of the Italian underworld and on Ford premises were seen carrying .45's in their belts. "You'd see these Italian guards at the Administration Building, milling around and keeping their eyes peeled all the time, many working on the lawns," recalled Irving Bacon.[30]

Henry Ford began the practice of hiring ex-criminals long before Bennett started working with the company. As Bennett relates, "Mr. Ford had a profound morbid interest in crime and criminals.[31]

Bennett may not have started the practice, but he certainly elevated it to new heights. By the end of 1940, it was estimated that 20 percent of the employees working in the Rouge plant were ex-criminals hired by Bennett to work as Ford Servicemen.[32] "They're a lot of tough bastards," explained Bennett, "but every goddamn one of them's a gentleman."[33]

* * *

The reason Henry Ford encouraged Bennett to fraternize with organized crime was that Henry was deathly afraid that his grandchildren would be kidnapped. By having Bennett keep in contact with these underworld elements, Henry hoped to prevent any such attempts.[34] According to Walter Doehler, Henry's close associate, Bennett did just that. "Edsel and his children would have been kidnapped if it had not been for Bennett's vigilance," Doehler said.[35]

Doehler's perception is somewhat exaggerated because Edsel had his own team of watchdogs. However, Bennett wanted everyone to believe that the only reason Edsel and his family did not succumb to evil doings was because of him. Early in their association, Edsel did not mind Bennett's exaggerations and, in fact, consented to allowing Bennett's men to tail him and his family wherever they went, be it shopping, to work, to school, or to a baseball game.[36]

Being head of the richest family in the country, Henry was justified in worrying about the security of his grandchildren, especially after the kidnapping and murder of the Lindbergh baby in 1932. However, Edsel had also taken steps to protect his family, for he was accustomed to threats of extortion.

Every day, "crackpot letters" came into Edsel's office, and Lepine read through all of them, passing along those that seemed most menacing. Edsel would read them, writing in the margin, "Forward to Harry Bennett."[37]

One time when Edsel's daughter, Josephine, was attending Chatham Hall in Georgia, a scary incident occurred. Two days before the end of the school year, a young man wearing what apparently was a uniform of the Civilian Conservation Corps asked one of the girls in the school if there was a girl named Ford who attended the school. He told her that he knew some of Josephine's friends up north.

After hearing about the incident, the headmaster of the school informed the sheriff and demanded that his deputies step up their surveillance of the school. The headmaster also hired an additional guard and took special precautions to ensure that no strangers were on the school property.

When Edsel was informed of this incident, he immediately informed Bennett, whose goon network stretched as far south as Miami.

Fortunately, nothing further happened. The authorities picked up a man who fit the description and found him to be harmless. He was part of a nearby C.C.C. camp and seemed to be of "little intelligence." Evidently, he heard in some way that Josephine was there and simply wanted to meet her.[38]

Edsel had 18 bodyguards to protect him and his family 24 hours a day. There were six armed bodyguards at his residence eight hours a day for three shifts—one in the guard house at the entrance of the estate, one in the kitchen, one in the boathouse, and three roaming the grounds—providing constant surveillance.[39]

This sense of security was one of the reasons Edsel kept himself from the limelight as much as possible. Because Edsel limited his public appearances primarily to company functions within Dearborn, many people in both the company and the dealer organization would not recognize him if he walked in the door. "Will you please send us a photograph of Mr. Edsel Ford," wrote the New Orleans branch manager. "So many of our dealers visiting the Branch inquire about Mr. Ford, and we have been embarrassed several times of late when we were unable to show them his picture."[40] However, the New Orleans branch manager did not receive a photograph, nor did anybody else.[41]

Edsel's chauffeurs were not only his drivers—they served as his bodyguards as well. They were armed and trained in evasive driving techniques. Whenever they drove Edsel or members of his family, they would drive through red lights after nightfall to avoid the risk of kidnappers who might be waiting for Edsel's car to make an expected stop.[42] These were simple tactics, to be sure, but effective nonetheless.

One evening in 1933, one of the Ford chauffeurs was put to the test. That evening, Eleanor left her Grosse Pointe estate to attend a party. When her Lincoln picked up speed along Lake Shore Drive, another dark-colored Lincoln sped up to hers and motioned her driver to pull over. Believing that it might be a holdup, Eleanor's chauffeur pushed on the gas and headed for the nearest police station. In the ensuing minutes, both cars sped down the winding road, each one crashing into the other, with the Ford car tearing off one of the fenders of the assailant's vehicle.

During the ruckus, Eleanor's driver realized that the pursuers were policemen, not hoodlums, and pulled over. The whole matter turned out to be a misunderstanding.

When the Ford limousine pulled out of the estate, the police thought that Edsel was in it and tried to chase him down to serve him a subpoena to appear at a business case. When that turned out not to be the case, the police apologized to Mrs. Ford and allowed her car to leave.[43]

These precautions were taken in addition to Edsel's already fortified estate. A high iron fence separated his property from Lake Shore Drive, and high stone walls topped with barbed wire ran down both sides of the property from Lake Shore Drive to the lake.[44]

A few years after the Edsel Fords moved to their Gaukler Pointe estate in 1929, it had become a veritable encampment. Anticipating the worst with labor riots that sprung up in the 1930s, tear gas reserves and machine-gun encampments were set up at various locations along the perimeter of the property to protect Edsel and his family from possible siege.[45] Edsel was not alone in this regard. Henry's chauffeur carried guns under each shoulder and had two revolvers built into his car.[46]

William Rinehart was a mechanic who worked on Edsel's private boats, which were kept in the boathouse in the back of Edsel's Jefferson Avenue residence. Rinehart would go there during the day and enter the boathouse by parking on Jefferson Avenue and walking through a vacant lot that separated Edsel's home from a luxury apartment building. Each day, he would take this same path, unaware that Edsel's bodyguards watched his every step. Although the guards knew why Rinehart was there, they approached him one morning and told him to stop taking the shortcut. "Say, after this, come in the residence drive past us," ordered the guards.[47]

At both the Jefferson Avenue and Gaukler Pointe residences, guards patrolled the grounds and shoreline constantly. If the family ventured out onto the grounds for even a brief walk, their sojourn was monitored through binoculars.[48]

Many times Edsel asked his assistant secretary, Arthur Backus, to escort Henry Ford II and Benson on their trip home from The Hotchkiss School in Connecticut. While Backus was watching the Ford children, Bennett had a couple of men on the train as well, watching all three of them. "I distinctly remember one occasion when we left New York's Grand Central Station by automobile," Backus recalled, "and Henry Ford II, looking out the back window, remarked 'Here come those rats now,' referring to Bennett's men who were tailing our car."[49]

On another occasion, Backus was even asked to stand in as Edsel's proxy. It had to do with an extortion letter that Edsel received, threatening his two sons. Similar to other extortion letters that Edsel had received over the years, it instructed Edsel to leave a large amount of cash at a specified location. This time, the money was to be dropped off at a church in Detroit.

To Backus' surprise, Harry Bennett called him to his office and asked him if he would pose as Edsel and deliver the ransom money. "Don't worry," Bennett told him, "we'll be watching your every move."

Backus could hardly refuse, and Bennett instructed him what to do.

Carrying a shoebox-size container of phony money, Backus was chauffeured to the church in Edsel's personal Lincoln. There he deposited the money as instructed.

Bennett's undercover agents were planted throughout the church and surrounding area, and they monitored Backus' every move.

Fortunately, the entire event went without incident. The perpetrator was found watching the events unfold through a window of a bar across the street. He was apprehended by one of Bennett's men and was turned over to the police.[50]

***

Initially, Ford Service was established to supervise personnel and offer plant protection at the great Rouge factory. By the 1930s, however, under Bennett's direction, it became an engine of repression and regimentation—and a force to keep the growing tide of unionism that was sweeping the country from entering the Rouge plant.

Since its inception at the turn of the century, employment in the automobile industry had always been helter-skelter. When sales were brisk, so was employment. When sales dropped, men were laid off. However, this cycle became especially devastating at the onslaught of the Great Depression. Between 1929 and 1932, automobile production had dropped a staggering 80 percent, and more than 60 percent fewer men were working in the industry.[51] Bennett and his Ford Servicemen used these dire economic times to place a stranglehold over Ford workers. With jobs scarce, Ford workers did whatever they could to hold onto their paychecks, and Bennett and his thugs preyed on this fact. Any of Bennett's hooligans could fire whomever he wanted, at any time, for any reason—there was no appeal.

Bennett developed within the Rouge facility a spy system that permeated every nook and cranny of the gargantuan plant.[52] His spies and stool pigeons reported to Bennett every action, every remark, and every expression of every worker in the plant. One misspoken word—

particularly if it related to unionism—meant the loss of that person's job. "Whenever an organization drive is announced," wrote one reporter, "the [Ford] Service Men become thicker than flies around a dead body. As many as eight men patrol 400 feet of production line, so that no worker can open his mouth without it being noticed."[53]

If this weren't enough to keep the men off balance, they were constantly shifted about so they could never learn whom they could trust. When a new man came into a section, he was viewed with suspicion and observed carefully. The other workers tried little stratagems to test him. Eventually, one of them would risk his job and ask him a question. If a look of genuine terror came over his face, he was all right. He was a Ford worker like the rest. However, if he answered and tried drawing his questioner away from his work, he was a stool pigeon.[54] No one who worked for Ford was safe from the spies—from the superintendents down to the poor creature who had to clean a certain number of toilets per hour. Because Ford employed men from all over the world, he had spies that could speak every language under the sun.[55]

Even away from the plant, Ford workers had to be cautious, for Bennett had his Ford Servicemen eavesdropping at grocery stores, taverns, restaurants, and hotel lobbies. They even joined the union to see which Ford employees were attending the union meetings.[56] If they could not find a legitimate reason to get rid of a worker, they would frame him, usually forcing the worker to do something against some obscure safety regulation.[57] It is hard to imagine, but between 1937 and 1941, Bennett's men fired more than 4,000 Ford workers because of actual or suspected union sympathies.[58]

This oppressive atmosphere was unbearable. Many men succumbed to what they called "Serviceman phobia"—a perennial feeling of insecurity.[59] So high was the tension at the Rouge plant, explained a reporter, that if someone jokingly fired a cap pistol, "35,000 heads would burst."[60]

The workers fought back as best they could, by devising ingenious methods of communication. They taped notes to parts moving along the assembly line, and they developed an art of covert speech known as the "Ford whisper." With this method, workers would talk to each other in muffled tones, never taking their eyes off their work. Others became virtual ventriloquists, being able to talk without moving their lips.[61] As the union became more entrenched, it began airing propaganda over a low-powered radio station in Detroit. Ford workers on final assembly would set the cars' radios to the union radio station and listen to the program as the cars went down the line.[62]

To aggravate these deplorable working conditions, pay at Ford had fallen woefully behind that at other automobile plants.[63] In 1914, Henry Ford had become a national hero when he doubled his workers' pay from $2.50 a day to $5 a day. From that point, however, Ford never increased salaries to keep up with the cost of living or with the pay offered by other automobile manufacturers. In fact, other automobile companies felt that as far as labor policy was concerned, Henry Ford was "still rattling along in the Model T era."[64] "I would be ashamed to have anyone tell me our conditions and pay are wrong," Henry told a reporter. "That part of my job was started 20 years ago. I have heard no complaints from our men."[65]

Little wonder. Every man was afraid to complain about anything, for fear of his job. "Ford and Bennett boast that anybody who wants to complain can always go to the Sociological Department," wrote a reporter about the climate within the Rouge plant. "And anybody who wearies of his life can also shoot himself!"[66]

At least Edsel understood the workers' plight. In the mid-1930s, he advocated increasing their wages and reducing their work week. "Pay a workman more than just enough to live on," he espoused. "He's got to buy a few things extra—otherwise, some industries will have to shut down. Give him a 40-hour week now—maybe a 30-hour week later, but that's in the future. A man's got to live as well as work."[67]

With this backdrop—Bennett's spy system, low wages, and Depression era uncertainty—worker restlessness grew at Ford Motor Company. The penal-like conditions at the Rouge plant kept the union out of Ford for years, but it also provided a fertile ground for the union to take root.

***

The first signs of labor discontent occurred in the spring of 1932. At that time, Ford was the only automobile company still working, although on a reduced schedule. The company was not only building Edsel's beautifully designed 1932 Fords, but it was also busy fulfilling a large tractor order from the Russian government.[68]

On Monday, March 7, 1932, a large crowd of approximately 3,000 disgruntled automobile workers gathered in downtown Detroit to protest arbitrary discharges, assembly line speed-ups, and general working conditions in all automobile factories, but particularly at Ford. To prevent a riot, Detroit mayor Frank Murphy was happy to give the marchers a police escort all the way to Dearborn. However, they did not receive such a cooperative welcome in Ford country. When the marchers arrived at the Dearborn city limits, they were met by a wall of police, who were in no mood to offer an escort. They told the marchers that if they crossed the Dearborn line, they would be dispersed by force. Undeterred, the marchers carried onward.

The Dearborn police shot tear gas into the crowd, but the wind shifted and blew the gas back into their faces. With the police temporarily stunned, the marchers quickly made their way to Gate 4 of the Rouge plant, one of the main entrances. After the police regrouped, they began spraying water into the crowd.

Bennett was in another part of the Rouge plant when the riot occurred. When he was informed of the ruckus, he got into his car and drove directly to the scene. He found the fire chief and told him to put the hoses down. "I'll talk to them," Bennett said.

"We want Bennett," yelled the crowd as Bennett's car pulled up to the plant. "And he's in that building!"

"No, you're wrong," yelled the little man as he got out of his car. "I'm Bennett!"

At that moment, a hail of slag descended upon him—a large piece hitting him in the head and knocking him to the ground. At that moment, the Dearborn police opened fire, killing five people, one of whom was Bennett's driver, and injuring many others. Bennett got up again but was hit in the head again, this time knocking him out.[69]

Henry, Edsel, and Sorensen were having lunch at the Round Table when the demonstration occurred. Russell Gnau, Sorensen's secretary, telephoned Sorensen and told him what had happened. Sorensen turned white. He returned to the table but did not want to mention anything to Henry in front of the other managers. Therefore, he scribbled a note on a pad and handed it to Edsel. Edsel read it, looked at Sorensen, and then handed the note to his father. Henry excused himself, and Edsel and Sorensen followed him. The three of them drove to the Rouge plant in Sorensen's car.[70]

The scene resembled something from World War I. Bodies were lying in pools of blood. Ambulances were everywhere. The three men could not believe their eyes. When Henry asked about Bennett, he was told that Bennett had been taken to the Rouge hospital, with other men injured during the ruckus.[71] When Henry saw Bennett lying there, blood oozing through the bandages wrapped around his head, he went into a state of shock. "Henry was white as a sheet—there were tears in his eyes," recalled one in attendance. "It was like Edsel was lying there."[72]

There is no question that Bennett was brave, but some people placed the deaths of his driver and the four protestors at his feet, for none of that would have occurred had Bennett not driven into the crowd and incited the protestors.

That did not bother Bennett. Henry Ford had witnessed his brashness, and that is all that mattered.

***

Labor strife laid low for the remainder of 1932, but it began to build steam again in early 1933. Fed up with low wages, harsh working conditions, and being paid only when the assembly line was running, workers at Briggs Manufacturing Company went on strike, although they were not organized. As a major body supplier to Ford and other automobile companies, the Briggs strike shut down assembly lines throughout Detroit.

The Briggs workers went on strike in haste, but their walkout lasted one week. They lost this battle, but the war was only beginning.

***

During these times, a number of labor organizations tried to gain a foothold in the automotive industry but to no avail. The first was the Automotive Industrial Association and then the Associated Automotive Workers. Then all workers' hopes were raised when Congress passed the Wagner Act in 1935, which guaranteed laborers the right to collective bargaining.

Eventually, the American Federation of Labor (AFL) tried its lot but also failed. Then the United Automobile Workers (UAW) was organized under the auspices of the Congress of Industrial Organization (CIO). With the strength of the CIO, the UAW gained a foothold and never let go.

The UAW knew that it had an immense undertaking. All of the automobile companies were started by free-spirited entrepreneurs who felt that management had its place and labor had its place, with full power resting in the management camp. Nowhere was this philosophy more ingrained than at Ford Motor Company. The UAW knew that Ford would be the most difficult to organize and would be the last to fall.

Therefore, the UAW started with General Motors—the largest but most susceptible motor company. On December 28, 1936, a sit-down strike occurred at the GM Fisher Body plant in Cleveland. Two days later, the strike spread to the Fisher plant in Flint and eventually throughout General Motors. Much to the surprise of the company, the strike dragged on through January and February 1937, idling 90 percent of its workforce.

Unable to sustain the losses it was incurring, General Motors capitulated on February 11, giving the UAW its first victory.

Flush with success, the UAW next went after Chrysler Corporation. Using the same tactics as had been used with General Motors, some 60,000 Chrysler workers struck and sat down in their factories early in March. Walter Chrysler, although somewhat sympathetic to his workmen, held out for as long as possible but eventually capitulated, one month after the strike had begun.

The next target of the UAW was Ford Motor Company. However, in this fight, they would not have it so easy.

***

From the outset, Edsel's advice to his father was to negotiate a favorable agreement with the union. General Motors and Chrysler had already signed agreements with the UAW, and Edsel knew it was only a matter of time before his company would have to do the same.[73] One time, Edsel deliberately stayed after an awards ceremony at which UAW officers were present. He pulled them aside and had a long discussion with them, telling them that he was in full accord with their mission. "There will be a better tomorrow," Edsel told them.[74]

This made a real hit with union officials. "I just hope he [Edsel] will continue to devote some time to labor relations," recalled a government official who worked in the plant, "because more than anyone else in the Ford Motor Company, he can help with the situation, as the Union has a lot of confidence in him."[75]

Edsel wrote to his parents, who were in Georgia on holiday, and told them about his meeting with the union. He summarized the situation very succinctly: "They don't trust any of our labor-relations men very much. And who does?"[76]

Henry's reaction to Edsel's letter was one of disdain. He ordered Edsel and Sorensen not to meet with union officials and not to discuss labor issues with anybody, particularly the press. If necessary, they were to leave town. Then Henry dropped a real bomb. "I've picked someone to talk with the unions," he said. "I want a strong, aggressive man who can take care of himself in an argument, and I've got him. He has my full confidence, and I want to be sure that you, Edsel, and you, Charlie, will support him. He's waiting in Charlie's office at the Rouge now, so let's go down there and see him."[77]

Sorensen drove Henry and Edsel to his office, and as they walked in, who was there but Harry Bennett. Both Sorensen and Edsel were dumbfounded. How could Henry put an ill-educated ex-pugilist into such a position of power? It was absurd to think about it.

"There was not much to say while we were together," recalled Sorensen. "Bennett was bubbling over with enthusiasm on how he could handle the union. He assured Henry Ford that he would take over the union. This was the beginning of Harry Bennett…"[78]

Almost immediately thereafter, Edsel and Bennett were at each other's throats. As long as Bennett had remained ensconced in his basement office, Edsel never paid much attention to him. Now, Henry had placed Bennett in a position of power that challenged Edsel as president. "The hurtful thing about all this is that Father takes Harry's word for all this and he won't believe mine!" Edsel confided to one of his managers. "Harry is nothing but a gob! What the hell does he know about running a company!"[79] Clara Ford asked the same question. "Who is this man who has so much control over my husband and is ruining my son's health?" she asked Sorensen, tears running down her face.[80] This man was Harry Bennett, and he had, in effect, become her husband's surrogate son.[81]

Confident that he had a man who was both daring and dutiful, Henry decided to fight the union. He told Bennett to stave off union demands indefinitely. "Arm everyone in the plant and use tear gas if necessary!" instructed Ford.[82] If that did not work, Henry gave Bennett "the power to close down the plant at any time."[83] Bennett, of course, was in full agreement with his boss' stance.

Edsel, however, was not, and he was alarmed at the plans his father and Bennett were ready to instigate.[84] Even at this early stage of the impending conflict, Edsel wanted to approach it systematically and intelligently. "If you want to fight the union," Edsel told his father, "then let's do it in the proper manner, and let's fight it within the law. Let's hire a good constitutional lawyer to assist Mr. Bennett, and let me sit in on the negotiations to offer my advice."[85] Similar to countless times in the past, however, Henry ignored his son's sound advice and told Bennett to prepare for battle. "Edsel was so worked up," recalled Sorensen, "that he was almost in tears."[86]

***

On May 26, 1937, the UAW obtained a permit from Dearborn City Hall to distribute handbills at the Rouge plant. When Bennett learned of this, he placed groups of thugs at all gates of the plant. Although Bennett told the press that his men would not instigate anything, one of his comrades, John Gillespie, said otherwise. He called his old friend, Ed Hall, who was now a union member, and warned him. "I don't want you to go out there today," he said. "Something is going to happen—it's going to be extremely unpleasant." Hall appreciated Gillespie's warning but quickly retorted, "I have a license to carry two revolvers," he said. "If any of your stooges out there lay a hand on me, I'll blow them so full of holes they won't make a good sieve."[87]

Walter Reuther, Richard Frankensteen, Richard Merriweather, and Ralph Dunham of the union, and Reverend Raymond Sanford, a minister from Chicago, drove to Gate 4 of the Rouge plant. At this location, an overpass crossed Miller Road, the main thoroughfare to the plant, connecting the parking lot to the entrance gate. They arrived an hour before shift change, where they met newspapermen and posed for photographs. No one present could have avoided noticing a number of Ford Servicemen milling around the area.

Without provocation, Bennett's men walked up to the union organizers. "This is Ford property!" they shouted. "Get the hell off of here!"

The union men, wanting to avoid a confrontation, began to leave but were blocked by other Ford Servicemen. As they attempted to pass, they were attacked.

Reuther was knocked down, kicked in the face and body, picked up, and knocked down again. The thugs pulled Frankensteen's coat up over his shoulders and then commenced kicking him in the head and groin. Merriweather's back was broken, and Dunham was seriously injured.

Unlike previous scuffles that Henry denied had happened, this one was caught on film by Detroit newspapermen. The "Battle of the Overpass" was front-page news throughout the country the next day.

Immediately, the National Labor Relations Board, the federal agency in charge of enforcing the Wagner Act, filed a complaint against the company. A short time later, the board found Ford Motor Company guilty of violating the Wagner Act and ordered Ford to stop interfering with union organizing.

Armed with the edict of the National Labor Relations Board, the UAW informed the City of Dearborn that they were again going to march on the Rouge plant and asked for protection. The mayor declined.

Nevertheless, the union marched on the Rouge plant again on August 11, and this time they were prepared. Instead of having only a handful of union members on hand, they produced 1,000 sympathizers to hand out fliers, with hundreds more awaiting call. Many outside supporters were also present, not only to support the union but to keep a watchful eye on Ford officials Service. The county prosecutor, newspapermen, state patrol officers, federal officials, and Cornelia Pinchot, wife of the former governor of Pennsylvania, were there to monitor the activities. Mrs. Pinchot was there, she said, to see if Dearborn was "still in America."[88]

With such a gathering of local, state, and national representatives watching, Bennett dared not do anything to thwart the legal activities of the union this time. Thousands of union fliers were handed out to Ford workmen as they entered and left the Rouge plant.

Henry Ford and Bennett were not about to give up so easily. Shortly after this march, Ford officials persuaded the Dearborn City Council to make it illegal to distribute paper material of any kind within the "congested areas" of the city. These restricted areas, of course, encompassed all of the entry points into the Rouge complex. Immediately, Dearborn police began arresting union organizers who tried to distribute union fliers. Surprisingly, this tactic kept the union out of Ford for three more years, until a local judge finally declared the former ruling unconstitutional on December 7, 1940.

Immediately after the judge's ruling, the union stormed the Rouge plant with thousands of fliers. After years of frustration, the union was finally free to solicit the union in Ford. However, the union did not have to do much persuasion. Ford workers were already wearing union badges, and pro-union signs began appearing on machinery and packing crates throughout the plant.[89]

Although bound by law to allow his men to organize and to form a union if they wished, Henry Ford knew that he had one final ace to play. If push came to shove, as it surely would, Henry determined that he would not *have* to reach an agreement with the union. Parroting Henry's attitude, Bennett said, "We will bargain with the union because the law says so. We'll bargain 'til Hell freezes over, but they won't get anything."[90] Knowing how Henry Ford and Bennett worked, it would not be unthinkable for them to do just that—bargain with the union for so long that it would either break the union or break Ford, or both.

Ironically, Bennett himself brought down the roof. As talks were about to begin, he ordered the dismissal of several shop stewards in the steel mill who were brandishing their union pins in the faces of workers who did not want the union.[91]

Riled by the firings, 3,000 Ford employees immediately sat down on the job—precipitating the first strike in company history. Spooked by the union show of solidarity, Bennett quickly rescinded his orders, and the mill was back at work within the hour. However, the precedent had been set, and the union realized that Bennett was indeed afraid of the workers going on strike. He was vulnerable. During the next couple of weeks, the union staged a number of isolated sit-down strikes in various departments within the Rouge plant. Each time, Bennett acquiesced and made concessions.[92]

***

In the middle of this labor unrest, Henry and Clara Ford celebrated their fiftieth wedding anniversary. Because of Edsel, their celebration was one to be remembered. The festivities were held at Edsel's Gaukler Pointe mansion, and Edsel had his friend, Walter Dorwin Teague, design the layout for the party. In his usual attention to detail, Edsel worked closely with his secretary, Albert Lepine, to plan the celebration and develop a guest list. "I have written in a few names on the typed list and am also including the doctors' list," he wrote to Lepine. "We wish to ask the department heads and wives as well."[93]

In the end, more than 500 guests were invited to the gala event—from Ford and Bryant family members to local and national dignitaries. However, the people whom Henry enjoyed seeing most were the many automotive elite, who could have filled a "Who's Who" of the automotive industry, if not one of automotive pioneers. In attendance were John Anderson, one of the original 12 Ford stockholders (he sold his 50 shares of stock that he bought for $5,000 to Henry Ford for $12 million 15 years later); Ransom Olds, founder of the Oldsmobile Motor Works and an early proponent of mass production; Charles Kettering, the General Motors wunderkind and inventor of the automobile engine self-starter; and C. Harold Wills, an engineer who assisted Henry in designing the Model T. Also present were Walter Chrysler, founder of the Chrysler Corporation; Bill Knudsen, former Ford employee and now president of General Motors; Walter Briggs, owner of Briggs Manufacturing; Clarence Avery, Edsel's one-time schoolteacher and contributor to Ford mass-production techniques; and Albert Kahn, architect of many of the mammoth Ford factories. Ford executives present were Charles Sorensen, Clifford Longley, William Mayo, Pete Martin, and Fred Black. Ernest and Josephine Kanzler were there, as were Dr. McClure and Dr. Sladen of the Henry Ford Hospital, John D. Rockefeller, Jr., William and Caecilie Valentiner, and Henry's longest held friend, now doctor, Edsel Ruddiman.[94] Most of Clara's family were invited, but surprisingly, not many Fords were. Although Henry abhorred alcohol, Edsel had champagne served.[95]

An eleven piece, all-girl ensemble was brought in on the *Detroiter* from New York at a cost of $1,000 to play at the celebration, and a New York harpist was flown in for the event. Most of the songs they played were popular during the Fords' courtship such as *Swannee River*, *Old Dog Tray*, and *Old Folks at Home*, plus a litany of Irish tunes such as *Wearing of the Green* and *Irish Washerwoman*. For the finale, *The Dargason* by Holst was performed, and Dearborn mayor John Carey presented Henry and Clara a scroll of congratulatory remarks—purported to be two miles long—gathered over the previous year from 700,000 people throughout the nation.[96]

Before the festivities began, however, Henry, Edsel, Clara, and Eleanor sat for a photograph, in front of large portraits of a young Henry and Clara at the time of their marriage. "What's the secret to a successful marriage?" asked one reporter.

"Stick with one model," Henry advised.[97]

As a climax to the evening, the orchestra played *The Wedding March*, and Henry, wearing a black tuxedo, and Clara, in a long burgundy gown, walked down the long hallway from the terrace to the gallery between the rows of guests to cut the wedding cake. When they arrived in the gallery, they were surprised to find their grandson, Henry Ford II, standing there. He had arrived from Yale just in time for the party.[98]

It was a marvelous event—executed with perfection as only Edsel could orchestrate. However, it was to be the last cordial encounter between Edsel and his father. From this point, the rift that had started more than a decade earlier and had intensified during the union conflict would soon reach a mortal conclusion.

***

The breaking point in the labor unrest occurred on April 2, 1941. While Henry was on holiday at his plantation in Georgia and Edsel was at his Hobe Sound retreat in Florida, Bennett fired the eight union organizers who had set off the various sit-down strikes during the previous weeks. When the union men approached him to work out a settlement, he refused to see them. With that, the union had no alternative but to wage a general strike against Ford Motor Company.

The call of "STRIKE!" swept throughout the plant, and all of the men quit and left the building. "They just stopped their machines and walked out," recalled one of the workers. "We did not try to stop the men from going because we could not—there were too many of them."[99]

Within hours, thousands of Ford workers and union sympathizers encircled the mighty River Rouge complex. Automobiles and trucks were parked bumper to bumper and blocked all streets leading to the Rouge. Even railroad tracks and waterways were barricaded. Workers who were unsympathetic to the union were met at the gates by throngs of men armed with two-by-fours and black pipes.

"Where d' yuh think yuh're goin', buddy?"

"To work."

"Naw. Yuh ain't goin' t' work. Henry Ford don't run this plant no more."[100]

Nothing like it had ever been seen at the Rouge complex. "By the end of the day," wrote one reporter, "about the only way one could get into the plant was by parachute."[101]

***

During these tumultuous days, Bennett too had to be cautious of his own well-being. He was under constant surveillance by union goons who would just as soon kill him as look at him. To get out of harm's way, Bennett would admit himself into the Henry Ford Hospital with an "acute sore throat" because the hospital was one of the few secure places where he could escape his attackers. Dr. Conrad Lam, chief surgeon and friend of the Ford family, was happy to oblige.

"How bad is it, doc?" Bennett would ask.

"Looks pretty bad," consoled Dr. Lam.

"How long do you think I will need to stay?"

"Looks like about five days."

"That sounds about right, doc," replied Bennett.[102]

***

When the union had control of the Rouge plant, the union officials issued three demands to Bennett. They wanted the eight dismissed men reinstated, they wanted an immediate government-sponsored union election, and, if the union won approval, they wanted the company to formally recognize the union and begin negotiations on a contract.

Feeling trapped, Bennett's first thought was to resort to violence. He seriously considered ordering a complete confrontation between his Ford Servicemen and the workers but quickly changed his mind when he realized that he and the Servicemen were now woefully outnumbered. Instead, Bennett asked the state to intervene. However, his request fell on deaf ears. For years, Henry and Bennett had refused government interference. Thus, when Bennett asked for help now, all city, state, and federal officials turned a deaf ear.

Meanwhile, Henry and Edsel had returned to Dearborn to address the labor issue. Again, Edsel tried to convince his father to negotiate with the union.[103] However, Henry ordered Edsel to keep out of Bennett's way. Regardless, Edsel could not simply sit back and let an all-out war erupt. He pleaded with his father and Bennett to drop their antagonism toward the union. Again, Edsel reminded them that General Motors and Chrysler had already signed agreements with the UAW and, from his perspective, it was only a matter of time before they would have to do the same—it was the right thing to do. "This is bad public relations, and it's costing us sales!" Edsel pressed his father. "It's costing us more money to fight than it would if we just settled with the union!"[104]

As usual, Henry was obstinate over the matter, but Edsel put up a good fight. Sorensen recalled that Edsel had long, heated arguments with his father over the union matter. "They quarreled so violently," he recalled, "that I avoided their meetings, or left as quickly as I could when the topic came up."[105]

Meanwhile, John Crawford, Edsel's right-hand man, had kept Edsel informed of the union negotiations. "Why, Bennett is giving away the plant!" he said.[106] Concerned, Edsel got word to the union leaders that as president of Ford Motor Company he would never accept any agreement in which Bennett was involved.[107] When Henry heard of this, he became incensed. "You stay away, Edsel! Let Bennett handle it!"[108]

Edsel became so frustrated by the way in which his father and Bennett were bungling the labor issue that he threatened to quit.[109] "Well, this is your baby, now!" he told them.[110] Fortunately, Sorensen talked Edsel out of quitting, saying that Henry would not let Bennett negotiate a settlement without his approval. "Please stay away and don't ever deal with the problem," Sorensen urged him. "It will be the end of you if you do."[111] Edsel promised to keep out of it, but he reiterated that he was not in accord with Bennett handling the union negotiations. No one saw Edsel for more than a week after that.[112]

That did not improve the situation. In fact, it made it worse. When Henry heard of Edsel's intentions, he simply badgered Edsel all the more. "The doctors tell me that Edsel goes home and throws himself on the bed and has fits of weeping, because he's ashamed of us, and we won't do the things he wants us to," Henry told Bennett. "Can you imagine that, Harry?"[113]

Sorensen could not do much either, but he did call to Henry's attention that his stance on the union and his unflagging support of Bennett was having a devastating affect on Edsel. "I told Henry Ford that trying to drive Edsel in line by using Bennett...was breaking down Edsel's health and his respect for him."[114]

Henry ignored Sorensen's warnings, believing that he was simply trying to protect Edsel. However, Sorensen was right—Edsel was not in good health, and Henry and Bennett's shenanigans were simply aggravating his condition.

* * *

As the strike wore on and the threat of bloody scenes recurred, Henry Ford thought long and hard about the current position of the company, the plight of his workers, and for what

Edsel had been fighting. Finally, he settled the question in his own mind and called Bennett. "Give'em everything, goddamn it!" Henry exclaimed.[115] Nine days after the strike on the Rouge plant had begun, Henry telephoned Sorensen and told him that the strike had been settled. "Though Mr. Ford and Edsel were far apart on this," recalled Bennett, "Mr. Ford gave in to Edsel's wishes. I don't think the union would have won out if it hadn't been for Edsel's attitude."[116]

On May 21, 1941, 83,000 Ford workers went to the voting booth. By the end of the day, 75 percent had voted for the UAW, and one-man rule had finally come to an end at Ford Motor Company. Short of closing down the entire company, Henry now had at least one entity in the company that had as much power as he had.

Formal negotiations began immediately and were held at the Dearborn Inn. Bennett was Ford's chief negotiator, but he received his orders directly from Henry Ford. Within a matter of days, a contract was worked out, and it was nothing less than astonishing. Where Ford had at one time denounced the union, he now was about to award them a contract unprecedented in union history. First, all plant security men were to wear badges and uniforms. Next, Ford would meet the highest wage in the industry for all labor classifications. Also, Ford offered a closed shop—where only union members could work—and offered a check-off system, whereby the company would deduct union dues directly from the workers' paycheck and deposit them into the union coffers. Finally, Ford would allow his product to be shipped with tags that said, "Union made."[117]

With the details thus worked out, Edsel made a public statement expressing his pleasure that an agreement had finally been reached between the company and its workers after nearly 10 years of turmoil. "The company has always considered every problem from the viewpoint of its workingmen," he said. "Through the years it has voluntarily instituted changes that were not only of immediate benefit to Ford workmen, but by example had a far reaching effect in improving labor conditions elsewhere. As the company viewed the measure, we felt that no half measures would be effective. So we have decided to go the whole way."[118]

All the contract needed now was Henry's formal approval, and Bennett hand-carried it to Henry for his signature. The next day, after spending most of the day reading the details, Henry called Edsel late in the afternoon for a conference in Sorensen's office. All were anxious to get the document signed so that this labor issue could be put to bed once and for all. However, none of them were prepared for what was about to happen.

"It was soon evident that Mr. Ford was in a bad frame of mind about this contract," recalled Sorensen. "The more he talked about it, the more bitter he became. He finalized it by saying, 'I'm not going to sign this contract! I want you and Edsel both to understand that as far as I'm concerned, the key is in the door. I'm going to throw it away, and I don't want anymore of this business. I will never accept this type of contract. Close the plant down if necessary. Let the union take over if it wishes!' "[119]

"You can't do that, Mr. Ford," exclaimed Sorensen, with Edsel's support. "If you do that, the government will take over the company!"

"Well, if that happens, they will be in the motorcar business, and we won't. Close the plant!"

It was almost 7:00 P.M. before they left.

The next day, as Sorensen was getting ready for work, he switched on the radio and could not believe what he was hearing: "Ford grants union shop and check-off."

Sorensen rushed to his office, where both Henry and Edsel were waiting. "Of course I expected Mr. Ford was going to confirm the stories that were being flashed... I did not bring up the question, I thought it would come from him. We sat there for about 15 minutes, just talking about generalities. No hint from him at all as to what had happened overnight. He stayed there for about another 15 minutes, and then left me and Edsel sitting there."

Looking directly at Sorensen, Edsel asked, "Well, what in the world has happened?"

"I thought perhaps you might know," replied Sorensen.

For the remainder of the morning and through lunch, Henry did not say a word about the contract. Therefore, Sorensen pulled Bennett aside and asked him what had happened. "Mr. Ford called me late last night and told me to sign the contract," explained Bennett. "He didn't say anything else."[120]

Why did Henry suddenly change his mind for the second time in as many days? Because of the pressure he felt from Edsel and, surprisingly, from his wife, Clara. The evening that Henry changed his mind, Edsel had called him on their private telephone line and pleaded his case one last time. The union was here to stay, Edsel said, and there was no point putting it off any longer. Plus, the comments that Sorensen had made about the government taking over the company were not idle threats. As we shall see later, the country was gearing up for war in Europe, and it needed the vast Ford manufacturing capacity to help in that effort. President Roosevelt would not allow an old, feeble autocrat to stand in the way of producing war materiel for the country.[121] Finally, Clara came to her son's aid. She had been aware of the predicament in which Henry had placed Edsel—she knew how distressed Edsel was over the situation, and it broke her heart. Clara had to do something to keep her husband and son together.[122] Therefore, when Henry came home and told her that he had ordered Edsel and Sorensen to shut down the plant, Clara could endure no more. For the first time in their 50-year marriage, Clara went against her husband's wishes. "Henry, if you don't sign that contract and shut the plant down," she remonstrated, "I will leave you!"[123]

"What could I do?" Ford explained to Sorensen several weeks later. "She became frantic about it. She insisted that I sign what she termed a peace agreement. I'm sure now she was right. The whole thing was not worth the trouble it would make. I'm glad that I did see it her way."[124]

It was all for naught, as Edsel had suggested it would be from the outset. Henry's bullheadedness and Bennett's determination notwithstanding, the whole sordid affair was a complete waste of money, resources, and prestige. "I don't think what Bennett did really helped in keeping the union out of the company," recalled a Ford worker. "Some of the moves he made just gave them more food to work with."[125]

***

After Bennett had lost his chance of ruling the Rouge complex without a union, he reverted to the days when he had gained favor with Henry by protecting Edsel and his family, explaining that the threat to them was as great as ever. "As late as October 1942, Edsel and I had a hot session with Harry Bennett," recalled Sorensen. "Bennett had been telling Henry Ford how his grandsons Henry II and Benson were in danger of being kidnapped. He wanted to surround them and Edsel's house with his Servicemen. Edsel blew up. He told Bennett to stop such talk and let his boys alone. He didn't want protection for either himself or his sons—and as for the kidnap story, he didn't believe it and was sure that it had been played up to Henry Ford."[126]

Bennett became incensed and put on one of his acts. He jumped from his chair, tore off his coat, and with fists clasped as though he were ready to strike, pressed his barrel chest against Edsel. Immediately, Sorensen jumped up and rushed between the two men, shouting for Edsel to leave.

"It is difficult to imagine such a scene" explained Sorensen, "but it shows what Edsel was up against. Mr. Ford could have put an end to it. Instead, he encouraged Bennett to keep at Edsel. Constant turmoil was Henry Ford's idea of harmony. By keeping things stirred up, no one else could swell with importance."[127]

Henry's attitude and Bennett's talebearing simply widened the breach between Henry and Edsel.[128]

***

This was all a ruse on Henry's part. His apparent unqualified support of this gob Bennett was simply another of Henry's demented ordeals for his son. In truth, Henry did not like Bennett any more than his son did, but he did admire Bennett's nerve. He told Sorensen that he kept Bennett around to aggravate Edsel, with the hope that Edsel would take a stand against the cocky "little guy"—to "make him more self-reliant and more courageous."[129] "If Edsel could have taken Bennett by the nape of the neck and thrown him out," recalled Sorensen, "he would have come up to his father's expectations. But Edsel couldn't even think of such a thing."[130]

Thus, Henry's plan backfired. Instead of standing up to Bennett, as Henry had hoped, Edsel acquiesced—believing that is what his father wanted. "Edsel knew of serious, blundering mistakes that I had made," recalled Bennett, "yet he never told his father about them."[131]

Behind the scenes, however, Edsel made it clear that not only did he dislike Bennett's tactics in the labor issue, he disliked Bennett as a person. "Edsel was always belittling me," Bennett said.[132] One time, Sorensen noted in his diary, Bennett had called, asking him "to help him with Edsel, who is on his neck."[133]

That was the scene: Bennett obsequiously adhering to Henry's every command. Henry putting up with it as a means of teaching Edsel. And Edsel enduring the ridicule simply to please his father. "There was really no sense to all the monkey business that Edsel went through," recalled Al Wibel. "You'd have to live through that thing to really grasp the situation."[134]

Charcoal still life done by Edsel in 1909.

Mexican artist Diego Rivera at work on the Detroit industry frescoes at the Detroit Institute of Arts, 1933. (Photo courtesy of the Detroit Institute of Arts)

Edsel (second from right) at the Museum of Modern Art in New York City, May 8, 1939. He is accompanied by his good friend, Nelson Rockefeller (fourth from right). The painting is Picasso's Les Demoiselles d'Avignon. *(The Museum of Modern Art Archives, A. Cogner Goodyear Papers, #52. Photo courtesy of The Museum of Modern Art, New York.)*

Except where noted, all images are from the Collections of Henry Ford Museum & Greenfield Village.

Airplane built by Edsel and his friends, circa 1909.

Left to right: Edsel Ford, Admiral Richard E. Byrd, and William Stout pose next to a Ford Tri-motor airplane, 1926.

Edsel at the Air Olympics in Dearborn, September 1927.

Except where noted, all images are from the Collections of Henry Ford Museum & Greenfield Village.

The Lelands and Fords pose outside the Lincoln plant on the day the Fords purchased it in February 1922. Left to right: Henry Leland, Mrs. Henry Leland, Wilfred Leland, Eleanor Ford, Edsel Ford, Clara Ford, and Henry Ford.

Edsel in his 1912 Mercer, which he shipped to California for his use during his honeymoon.

Edsel's 1922 Hispano-Suiza with custom-made Brunn body.

Except where noted, all images are from the Collections of Henry Ford Museum & Greenfield Village.

Edsel in one of his Model T speedsters, circa 1911.

Edsel sits behind the wheel of one of the Model T speedsters that he designed and built himself. Cars such as this helped Edsel develop his eye for automobile design, circa 1915.

Edsel at the wheel of his six-cylinder speedster, circa 1912.

Edsel driving a 1922 Lincoln sedan, body by Judkins.

Edsel's 1929 Type 35-A Bugatti.

Lincoln sedan with Brunn & Company body, circa 1924.

Edsel used many custom body builders to design and build bodies for his well-made Lincolns. This design was created by Brunn & Company of Buffalo, New York, circa 1924.

Brewster & Company of New York created this body design for Edsel, circa 1924.

Another Brewster & Company-designed custom body for Edsel Ford. Edsel would critique these blue-prints until the designers came up with a body design of his liking, circa 1924.

Edsel's personal 1923 Lincoln coupe, body by Locke of New York. Edsel approved the design of every Lincoln between 1922 and 1943.

Ford design department, circa 1939.

Ford executives meeting at the famous Round Table in the Ford Engineering Laboratory, circa 1936. Left to right: Edsel Ford, Henry Ford, Charles Sorensen, Pete Martin, and Albert Wibel.

Engineering and production executives of Ford Motor Company in 1941. Standing, left to right: Dale Roeder, truck engineering; Laurence Sheldrick, Ford engineering; William Pioch, tooling; Jack Wharam, Lincoln engineering; E.T. Gregorie, design chief; unknown; Ed Scott, body engineering; and Gene Farkas, design engineer–engines. Seated, left to right: Frank Johnson, Lincoln engineering; Peter Martin, vice president–production; Charles Sorensen, vice president–manufacturing; and Joe Galamb, body engineering. (Photo courtesy of E. T. Gregorie)

Edsel (left) and Henry, walking through the Ford Engineering Laboratory, circa 1935.

Edsel (left) and his father inspect a new 1932 Ford coming down the assembly line.

Edsel (right) and Henry watch as the one-millionth Ford V8 is assembled, June 20, 1934.

A 1936 Lincoln Zephyr four-door sedan.

A 1938 Ford full-size clay model at the Ford Design Department. This was the first Ford designed by E.T. Gregorie as chief designer.

A 1939 Lincoln Continental prototype. (Photo courtesy of Ford Motor Company)

Stunning profile of a 1939 Ford coupe. E.T. Gregorie conceived this design, as well as nearly all others during the 1930s and 1940s, through consultations with Edsel.

The 1939 Ford Motor Company product lineup (left to right):
Lincoln, Lincoln-Zephyr, Mercury, Ford Deluxe, and Ford Standard.

Henry (seated) and Edsel, circa 1940.
(Photo courtesy of Mike Skinner.)

Except where noted, all images are from the Collections of Henry Ford Museum & Greenfield Village.

The Edsel Fords (left) sit next to the Henry Fords, in front of portraits of Henry and Clara Ford taken around the time of their wedding in 1888. The event was Henry and Clara's fiftieth wedding anniversary, held at Edsel's Grosse Pointe mansion in 1938.

Left to right: Al Wibel, Harry Bennett, unknown, Edsel Ford, Charles Sorensen, and unknown, November 1942.

Edsel at the wheel of a Ford-built "Blitz Buggy" (Jeep), February 1941.

Left to right: Charles Sorensen, Lieutenant William Knudsen, and Edsel Ford, outside the mammoth Willow Run bomber plant in summer 1942. (Photo courtesy of Edsel & Eleanor Ford House)

Edsel and Eleanor arrive at the wedding of their son, Benson, in July 1941.

Except where noted, all images are from the Collections of Henry Ford Museum & Greenfield Village.

Clay models of the Ford Motor Company postwar lineup, June 1945.

Meeting of the Automotive Council for War Production, 1942. Left to right, around the table: George Mason (with his back to the camera), B.E. Hutchinson, George Romney, V.W. Fries, and Edsel Ford. (Photo courtesy of Detroit Public Library.)

Edsel (third from right) watches his son, William (third from left), receive a tennis award in 1943. Edsel has lost a lot of weight from the ravages of cancer and undulant fever. (Photo courtesy of Edsel & Eleanor Ford House)

Except where noted, all images are from the Collections of Henry Ford Museum & Greenfield Village.

Henry Ford II's wedding, July 1940. Standing, left to right: Mr. McDonnell, Reverend Fulton Sheen, Anne McDonnell Ford, Henry Ford II, and Edsel Ford. Mrs. McDonnell is seated at the left; Eleanor Ford is seated at the right.

Edsel and Eleanor arrive at the wedding of their daughter, Josephine, January 2, 1943. Edsel was gravely ill at this time.

This photograph, taken in 1943, is one of the last images of Edsel before his untimely death. Standing, left to right: Henry Ford II, Eleanor Ford, Katrina Kanzler (Edsel's niece), Anne Ford, Josephine Ford, and Edsel Ford. Seated, left to right: Charlotte Ford, Clara Ford, and Henry Ford. (Photo courtesy of Mike Skinner)

Pallbearers carry Edsel's coffin from Christ's Church located on Woodward Avenue in Detroit to an awaiting Lincoln hearse, May 28, 1943.

1. Walter Buhl Ford
2. Walter Buhl Ford III *(son of Mr. & Mrs. Walter Buhl Ford)*
3. Benson Ford
4. Eleanor Clay Ford *(Mrs. Edsel Ford)*
5. Edsel Ford II *(son of Mr. & Mrs. Henry Ford II)*
6. William Clay Ford
7. Henry Ford II
8. Sheila Ford *(daughter of Mr. & Mrs. William Clay Ford)*
9. Martha Ford *(Mrs. William Clay Ford)*
10. Josephine Ford *(Mrs. Walter Buhl Ford)*
11. Alfred Ford *(son of Mr. & Mrs. Walter Buhl Ford)*
12. Josephine Ford *(daughter of Mr. & Mrs. Walter Buhl Ford)*
13. Edith Ford *(Mrs. Benson Ford)*
14. Lynn Ford *(daughter of Mr. & Mrs. Benson Ford)*
15. Benson Ford, Jr. *(son of Mr. & Mrs. Benson Ford)*
16. Eleanor Ford *(daughter of Mr. & Mrs. Walter Buhl Ford)*
17. Anne Ford *(daughter of Mr. & Mrs. Henry Ford II)*
18. Martha Ford *(daughter of Mr. & Mrs. William Clay Ford)*
19. Anne Ford *(Mrs. Henry Ford II)*
20. Charlotte Ford *(daughter of Mr. & Mrs. Henry Ford II)*

Edsel's family. Edsel lived to see only two of his grandchildren born, but his three sons and one daughter expanded the family considerably, as evidenced in this 1953 family portrait.

Chapter 25

Willow Run

"The Willow Run project was the last great Ford enterprise that Edsel was involved in."[1]
Patricia Barbara Smith

Although Edsel was still knee-deep in labor problems at the Ford Motor Company, 1940 began as a prodigious year for the Edsel Fords. In June, their oldest son, Henry Ford II, married Anne McDonnell, a socialite from New York. Young Henry had met her on the *Normandie* during his return trip to America after spending the summer of 1936 in Europe.

The wedding was held at the McDonnell's summer home at Southampton, Long Island. Edsel and Eleanor sailed up from Florida on their yacht, *Onika*, and Henry and Clara were railed out in their private railroad car, *Fair Lane*.

All of the details of the wedding were well planned except for one—how were the bridesmaids to get from their homes to the McDonnell estate without wrinkling their beautiful gowns? Edsel, always a master for details, quickly suggested hiring a bus. That way the young ladies could ride standing up, holding onto the handrails for support.[2]

With that settled, the wedding went off without a hitch. Monsignor Fulton J. Sheen of New York presided over the ceremony, and, as such things go, it was finished in a matter of minutes. The Fords and the McDonnells then gathered on the lawn for photographs. In these photographs, Edsel appears to be very happy, but he is beginning to look old and haggard, more than his 46 years should indicate.

As his parents had done 25 years earlier, Henry II took his bride to Hawaii for their honeymoon.

Wedding gifts to the heirs of one of the richest fortunes in the world were more symbolic than necessities, and the gifts the newlyweds received from Edsel and Eleanor were truly an indication of their love for their son—a Georgian home in Grosse Pointe and 25,000 shares of Ford Motor Company stock worth approximately $2.5 million.[3] The house was located on Provencal Road, only three miles from Edsel's estate.[4] The stock was the significant gift, not so much for its value but because of its intent. "I'm giving you this stock," Edsel wrote to his son, "in recognition of the fact that you are finishing your college career this month, and after being married will join the Ford Motor Company as your future business..."[5]

A year later, Henry's brother, Benson, married Edith McNaughton. She and Benson had known each other practically all of their lives, meeting for the first time in kindergarten. After Benson had attended Princeton and Edith attended a private school in Manhattan, Benson proposed to Edith on the *Onika* in Hobe Sound. They were married on July 9, 1941 in Grosse Pointe.[6]

On April 3, 1941, Charlotte was born, the first child of Henry II and Anne. Less than two years later, a second baby girl, named Anne, was born into the young Ford family.

The Ford family was now growing more rapidly than ever, and Clara Ford was both pleased and pensive. To a long-time friend, she wrote: "The little boy you knew, is a grandfather, how time flies."[7]

Meanwhile, both Henry II and Benson were ordered to report to their draft boards. Instead of risking being drafted into the Army, Henry II jointed the Navy. Although he did not ask for a deferment, he was conveniently assigned to a cushy job at the Great Lakes Naval Training Center in Chicago. Benson, rejected twice because he was blind in one eye, sneaked to Kalamazoo and enlisted in the Army Air Corps.[8]

***

On October 15, 1940, William S. Knudsen, former employee of Ford Motor Company, former president of General Motors, and soon-to-be lieutenant general of the Army, was invited to attend a meeting of the Automobile Manufacturers Association at the Waldorf-Astoria Hotel in New York. His former colleagues and rivals in the automobile industry, were in New York for what was to be the last pre-war automobile show and had invited the venerable production man to express their appreciation of the work he had done during the past 30 years. The previous May, President Roosevelt had appointed Knudsen as Commissioner for Industrial Production (later he was made a lieutenant general) to supervise industrial production of defense rearmament as a member of the National Defense Advisory Commission.[9]

As Knudsen walked to the podium, the members expected him to tell a few stories about his exciting career in working for Henry Ford and Alfred Sloan. Instead, however, Knudsen launched into an hour-long dissertation of his grave concern about the state of the nation's unpreparedness for war, which was then threatening to engulf the world. Already, Hitler's Third Reich had conquered Poland and the Low Countries, the British had evacuated Dunkirk, and the German war machine was marching toward Paris. Understandably, Washington officials were stirred toward intensive defense measures. After his address, Knudsen converted the meeting into a round-table discussion of the most urgent problems he envisioned in organizing the disparate industrial enterprises of the country into one cooperative industry for defense.

Knudsen had come to the right forum. Many of the country's industrialists were there: Sloan from General Motors, K.T. Keller from Chrysler, and George Mason from Nash. Representatives from the tire industry were there, as well as people from the body manufacturing and tool and die industries. Although Ford Motor Company was not a member of the Automobile Manufacturers Association, Edsel Ford and Charles Sorensen were there.[10] Edsel was in New York for the automobile show, and when he heard about the Automobile Manufacturers Association meeting, he attended to honor his old friend, Knudsen.

Earlier, Knudsen had asked the automobile manufacturers to suspend any model changes requiring machine tool purchases to conserve the resources of this important part of the industry, in case they were needed for war production. Now, he wanted to obtain their cooperation in an even more crucial program. Pledging them to secrecy, he told them of the dangers confronting the nation—dangers that had been revealed to him as an official of the government. The United States, he told them, was precariously behind in the production of military aircraft, and Knudsen then asked for their help—as an industry—to supply hundreds of millions of dollars worth of airplane parts and components. General Henry "Hap" Arnold, chief of the U.S. Army Air Corps, had recently returned from England where he had seen firsthand the devastation of aerial bombardment. That trip convinced the general that the victor in this conflict would be the country that could produce the most medium and heavy-duty bombers. Knudsen also mentioned that his own inspection of aircraft facilities in the United States revealed that those facilities were woefully inadequate. They had already increased their production by 20 times, but their output was not enough. For these reasons, Knudsen was asking for the help of the automotive leaders, although most of them had never seen an airplane part, let alone made one.

"We need more bombers—more big bombers—than we can now hope to get," Knudsen pleaded. "Even the British now agree with General Arnold that they cannot win the war with fighter planes. It had been a bitter, bloody experience for them—this realization. Bombers, big bombers, are needed sooner than we dare hope to get them under present circumstances. We must build them at once. You've got to help!"[11]

"You're the Boss, Bill," replied someone from the audience.

"Give us your orders!" yelled another. "We want to help!"[12]

Ford Motor Company and Edsel Ford were uniquely qualified among the other automotive industries in helping the government. Edsel had been instrumental in providing war materiel 20 years earlier during World War I. With Edsel's expertise in the construction of airplanes during the 1920s and 1930s, he visualized his contributions to be great indeed.

Edsel was prepared to do anything for his country, and he gave orders to Sorensen and the other Ford executives that all war contracts were to be top priority. He and Sorensen would do all of the negotiations with the government.[13]

***

A few weeks after the meeting in New York, Edsel received the first of many phone calls from Knudsen. Knudsen invited Edsel and Sorensen to Washington to inspect an aircraft engine designed and manufactured by Rolls-Royce of Britain. The British were then using this engine—capable of producing well over 1,000 horsepower—in their fighter planes. Knudsen got right to the point. He wanted Ford Motor Company to produce 6,000 of these highly sophisticated Rolls-Royce engines.

Edsel and Sorensen were enamoured by the offer. The Rolls-Royce aircraft engine was a mastery of aeronautical engineering. Although it offered a significant test to Ford manufacturing, both Edsel and Sorensen fully believed that they could produce the engine in quantity, and they accepted the challenge.[14]

Later that day, Sorensen telephoned Henry Ford and informed him of the contract, and Henry "agreed over the phone that we should take it on."[15] Sorensen was expecting some resistance from Henry, for Henry had made it clear over the years that he was a pacifist and that he would not open his shops to manufacture war materiel for outside interests. If the United States government needed his help, he would give it to them, but not to "belligerent countries," Henry said. However, Henry accepted the Rolls-Royce contract because ostensibly it was with the United States government.

***

Sorensen had a Rolls-Royce engine shipped to Dearborn so that he and his production men could begin work. He had two large rooms in the Engineering Laboratory set up—one for engineering, and one for tool design. After the engine arrived, Sorensen had it disassembled right down to every last nut and bolt, and all of the parts were laid out on large tables. In this way, Sorensen and his production men could analyze each part and determine the best method of manufacturing it. Meanwhile, Rolls-Royce had sent Sorensen a complete set of engineering drawing on microfilm, and the Ford engineering department was in the process of making enlargements of these drawings. Because of the secrecy of this work, it was done in the company payroll department in the same room where the cash vault was located. Armed guards watched over this area 24 hours a day—it was the most secure area in the Engineering Laboratory.[16]

One afternoon after lunch, Edsel asked his father to look at the progress he and Sorensen were making on the Rolls-Royce project. Sorensen went over the multitude of parts with Henry and explained the changes he planned to make the complex engine a mass-production job. He already had plans for reducing the number of parts and casting the crankshaft instead of forging it.[17] Henry listened patiently but was not impressed. To him, the Rolls-Royce engine was much too complicated.[18]

Shortly thereafter, the press began playing up the fact that Henry Ford was going to help the British after all. Although in essence this was true, and Henry Ford knew it to be so, he did not want the public to believe that he was now in favor of the military conflict in Europe. To Henry, the war in Europe was nothing more than a "big profiteering scheme" dreamed up by greedy industrialists and bankers—he "refused to have anything to do with it."[19] With that, Henry summarily cancelled the contract. He called Edsel and Sorensen to his office to inform them of his change of heart. Unaware of what was about to happen, "Edsel came in looking quite happy," recalled Fred Black. "But as soon as his father told them that he wanted the contract cancelled, Edsel's face became drawn."[20] Edsel did his best to reason with his father. After all, Edsel had put his reputation on the line with the United States government. "We wouldn't have made these commitments if you hadn't expressed yourself in favor," Edsel told his father as strongly as he could. "Edsel's tone of voice with his father wasn't angry," explained Black. "He sounded more sorrowful than angry—that his father had placed him, personally, in a terrible situation. You could see that Edsel was terrifically unhappy about it."[21]

"I don't care," replied Henry. "Call up Knudsen and tell him we've changed our mind."

Edsel reached Knudsen in Washington. "Bill, we can't make those motors for the British," Edsel said.

"Why not?" asked Knudsen, now in a daze.

"Father won't do it."

"But you are president of the company."

"I know," replied Edsel, "but Father won't do it, and you know how he is."[22]

After his conversation with Knudsen, Edsel personally called all of the other people involved in the Rolls-Royce negotiations and apologized to them.[23] He did not have to do that—he could have delegated the task to his secretary or simply sent a form letter. However, due to the severity of the predicament, and the fact that it was Edsel who had made the commitment in the first place, he felt it proper that he call each of them personally.

Perhaps it was for the best. "That was a difficult engine to make," recalled one of the high-level engineers assigned to the project. "I was very happy when they kicked it out..."[24]

Edsel, of course, was heartbroken. However, Knudsen was well aware of Henry Ford's capriciousness, and soon he would give Edsel another, more important project.

***

During the summer of 1940, Ford engineers under the direction of Dale Roeder, Ford truck engineer, had been working on the final plans for a small four-wheel-drive vehicle, code name "GP," for "general purpose" vehicle. (Eventually, everyone began calling it jeep.) This tough, all-terrain vehicle was initially designed by American Bantam Car Company of Butler, Pennsylvania, which had been building motorcars only since 1937. The government projected needing at least one-quarter-million jeeps, a quantity much too large for Bantam; therefore, the Army turned first to Willys-Overland of Toledo, Ohio, and then to Ford Motor Company for additional production capacity.[25] Meanwhile, Willys-Overland had modified the Bantam design by adding its own 60-horsepower engine.

Laurence Sheldrick, Ford chief engineer, Edsel, and Roeder went to Washington to meet with General Gregory to discuss the matter. "Mr. Ford," said Gregory, "the Army wants to standardize this vehicle. You and your company can do the United States an immeasurable service in this war if you will agree to manufacture it according to Willys' design, with the Willys motor and every single part interchangeable."

The three Ford executives looked at each other in bewilderment. They were under the impression that the government was going to ask them to become involved in some highly technical project, but all the government wanted was for the company to build a simple vehicle from the plans of another manufacturer. Roeder had designed his version of the jeep around the Ford tractor engine and had incorporated other standard Ford parts into the design, but he could just as easily make that version or another version if he had exact blueprints from which to work. He and Edsel conversed for a moment, then Edsel said, "If that's what the Army wants, we would be glad to produce it."[26]

Willys sent the plans to the Rouge plant, and Sheldrick and Roeder began making production plans immediately. Within a few short months, the first production prototypes were complete and ready for testing.

The first test drive of the Ford-produced jeep was done on a piece of scrub between the Engineering Laboratory and the Rouge plant. Sheldrick told Edsel during lunch where the test would be performed and that if Edsel would stop by after lunch, he could see the jeep in operation.

Edsel, Sorensen, and Wibel drove to the area, parked their cars, and stood along the side of the road.

Unknown to Edsel, Sheldrick had arranged to have Henry II and Benson drive the prototype jeep to their father. "I had a prearranged signal set up with the boys," recalled Sheldrick. "They were in the jeep, way back in the woods, back of all this underbrush. At the signal, they came tearing through all this underbrush and high weeds. Their jeep just came out of nowhere up to their father.

"Edsel got one hell of a kick out of that. That was the one time that I saw Edsel when he was thoroughly enjoying himself. He was awfully proud of his boys."[27]

***

The challenging project that Edsel was expecting when he instead received the jeep assignment came a few months later. Bill Knudsen came to Dearborn to talk to Edsel and Sorensen about the possibility of building the newly designed Army M-4 tank. The Army needed the M-4 because its predecessor, the M-3, was woefully under-powered. The problem, however, was that the Army had no power plant to put in the larger design.

Fortunately, Ford engineers were still working on a water-cooled aircraft engine for a proposed pursuit plane that the government wanted them to build about a year earlier. The program was scrapped at an early date, but Henry Ford had his engineers continue working on the project because he was enthralled by it. He had already invested $2 million in the project, and by July 1941, a prototype engine was ready for testing. When the tank project arose, Edsel and Sorensen thought of Henry's new airplane engine immediately. As a twelve-cylinder engine, it was too long for a tank. However, Ford engineers said that by lopping off four cylinders, it could easily be adapted to the M-4.

Ford engineers also redesigned major components of the M-4 and improved the process of manufacturing armor plates, all in an effort to make the tanks simpler and easier to build.

With the technical aspects resolved, Edsel signed another contract with the government in October 1942 to produce M-4 tanks. It would become a monumental undertaking. The Army wanted 20 tanks per day, and General Gregory told Edsel to spare no cost in obtaining that rate of production. In the end, $250 million was spent to produce the new tank—from tools and equipment, to a new plant.[28] More could have been spent, but castings were done at the Rouge plant, with only final assembly being done at the new facility in Highland Park.

Edsel kept a close watch on the project—coming to the plant twice a week to check on production. By the end of June 1943, only eight months after accepting the M-4 project, 15 tanks per day were leaving the Ford plant.[29]

***

A few months later, Knudsen again approached Edsel with another proposal. This time, he wanted Ford to make 4,000 Pratt & Whitney aircraft engines. Similar to the Rolls-Royce engine, the Pratt & Whitney engine was already being produced, but not at a rate high enough for the needs of the government.

On the surface, it seemed as though the government had a short memory when it came to the reliability of the Ford Motor Company. However, the truth was that the government needed production capacity and was willing to go anywhere to obtain it—even to the company that had short-shrifted it only a few months earlier. Except for the Rolls-Royce debacle, Edsel and Ford

Motor Company had come through swimmingly on all subsequent requests from the government.

Plus, Edsel and Sorensen were not oblivious to their own predicament. They were concerned that if they and the Ford Motor Company did not cooperate—regardless if Henry Ford approved or not—Roosevelt would order the takeover of the company by government officials. Consequently, Edsel and Sorensen readily volunteered for projects.[30]

Edsel and Sorensen flew to Hartford, Connecticut, to inspect the engines.[31] Sorensen was amazed at the size of the power plant. "Just think, Edsel," said Sorensen, "the displacement of one of these cylinders is equal to the displacement of the entire Zephyr engine!"[32]

Again, the two agreed to take on the project. However, this time, Edsel and Sorensen worked quickly, signing contracts for building construction, tools and machinery, and assembling a production staff, so that if Henry Ford decided to cancel this contract, he would have to pay dearly for the privilege. Fortunately, that did not happen, and two months later, on November 1, 1940, Edsel signed a contract with the government to supply it with $30 million in aircraft engines.[33]

The new airplane engine factory was constructed on one of the few remaining open pieces of land of the Rouge plant. It was a two-story, air-conditioned building covering more than one million square feet. Installation of machinery began in April 1941, and by August the first engine was completed. Edsel had the honor of starting that first Ford-built airplane engine in the test facility at the Rouge facility. Through the observation window of multiple glass separating the test cell from the control panel, Edsel pushed a button and the 1,000-horsepower engine roared to life.[34]

Adjacent to this plant was an aircraft training school that Henry Ford built at his own expense. Here, more that 1,100 young men were trained in the making of aircraft and diesel engines. Edsel presided over the opening of the school.

***

When the Pratt & Whitney project was well underway, the government asked for Edsel's help again. Dr. George Mead, Knudsen's chief consultant in aviation, and Air Force major Jimmy Doolittle asked Edsel if the Ford Motor Company would be interested in building parts and assemblies for 1200 B-24 Liberator bombers. Three large aircraft manufacturers were already building large bombers for the government. First, North American Aviation in Kansas City, Kansas, was building the B-25, two-engine bomber. Second, the Glenn L. Martin Company in Omaha, Nebraska, was building the B-26, two-engine bomber. Third, Consolidated in San Diego, California, was building the B-24, four-engine bomber.[35] The government was planning that Ford, General Motors, and Chrysler would assume responsibility for the production and procurement of the parts and subassemblies for one complete bomber each.[36] These parts and subassemblies would then be shipped to two new assembly plants—one in Fort Worth, Texas, and the other in Tulsa, Oklahoma. Doolittle suggested that a logical first step would be for Ford personnel to visit the Consolidated plant.

Edsel agreed, but Henry thought the whole idea was preposterous. Henry was not against taking on the project, fortunately, but he did not think the planes would be needed. "Those planes will never be used for fighting," he told Sorensen optimistically. "Before you can build them, the war will be over."[37]

In January 1941, Knudsen ordered a bomber to fly Edsel and Sorensen, several Ford engineers, and Edsel's two oldest sons, Henry II and Benson, to San Diego to evaluate the aircraft plant.[38] They spent one whole day touring the facility.

Being high-production men, both Edsel and Sorensen were not impressed with what they saw in San Diego. Consolidated had one large factory in which it built its airplanes, and the final assembly was done outdoors. To Consolidated, building one bomber per day would be an achievement. However, both Edsel and Sorensen doubted that they could accomplish even that goal. From what they could determine, each plane was essentially custom-made, similarly to Edsel's custom-made motor cars. They were beautiful and functional machines but painstakingly slow to construct.

Consolidated "built their planes just like we build houses. They put on a piece of siding and nailed it on. They'd cut a piece of aluminum and put it up there and rivet it on. They'd cut it to fit with a pair of tin snips or a squaring shear if they had one available. All of their work was largely handwork. They had a very, very poor system."[39]

"Their major problem was to get their wings to line up with their fuselage and then drop them on," recalled one Ford engineer. "They used shims to make up the difference in their joints. They never made two wings alike. They never made two fuselages alike."[40] That, of course, is anathema to high-volume production. Tolerances in Ford plants were routinely held to one one-hundred-thousandth of an inch. Pull out any eight pistons of the millions made each year, and they would fit in any cylinder block.

At the end of the day, Edsel, Sorensen, and the other Ford personnel met with Doolittle and the management of Consolidated. Sorensen did most of the talking and did not quibble on words. If the government wanted thousands of B-24 bombers, making them one at a time was not the answer. Doolittle's obvious reply was, "How would you do it?" Sorensen said he would have to think about it but would have an answer in the morning.

Working all night in his room in the luxurious Coronado Hotel, Sorensen, with his decades of experience in high-volume manufacturing, developed a plan that would give the government not one B-24 Liberator per day, but one per *hour*. It would not be easy. (Sorensen said that compared with a Ford car, the bomber was like a skyscraper to a garage). However, he was confident that the theories of mass production—theories worked out by Sorensen himself more than 30 years earlier—could be applied here.[41] It was, after all, a matter of pride. After discussing it, Edsel and Sorensen thought it was ridiculous for the government to subordinate the mighty Ford Motor Company into the role of parts maker. Had the government forgotten that the company, under Edsel's direction, had mass-produced airplanes in the past? He and Sorensen wanted the government to assign a difficult problem to them.[42] Therefore, the two Ford executives assigned this seemingly impossible task to themselves.

At breakfast the next morning, Sorensen showed Edsel what he had developed during the night. The focal point of the discussion was a one-page sketch of a mammoth airplane manufacturing plant—it was to be one mile long and one-quarter mile wide, and would have 70 ancillary feeder assembly lines feeding a main assembly line. The number of men needed to operate the factory was 100,000. It would be a monumental undertaking, explained Sorensen, but with this facility Ford Motor Company could produce 18 bombers a day, 540 bombers a month, 6,500 bombers per year. "I knew we had the solution," wrote Sorensen years later, "and I was elated by the certainty that the Germans had neither the facilities nor the conception for greater bomber mass production."[43] If General Arnold were right—that victory in this war would go to the one that had the most bombers—then the United States would surely win because the Ford Motor Company would fill the skies with bombers.

Edsel looked at Sorensen's sketches and listened to his plan with his typical understanding. An hour later, Edsel literally signed off on Sorensen's plan by signing his bomber layout. They presented the $200 million plan to Major Doolittle and Major Reuben Fleet, president and general manager of Consolidated.

Naturally, Doolittle and Fleet were impressed by the Ford plan, but they also were highly skeptical. What Edsel and Sorensen were proposing had never been done. Fleet quickly countered by reiterating the desire of the government for Ford to supply only major sub-assemblies to Consolidated for final assembly.

"We're not interested in assemblies," Sorensen told Fleet. "We'll make the complete plane or nothing."[44]

Dr. Mead, sensing a standoff, intervened. "It's a marvelous plan," he said. "It's just what we were looking for."[45]

On February 25, 1941, Washington approved Edsel and Sorensen's plan for building complete bombers. From that moment, Ford Motor Company was no longer in the motorcar industry but was a subsidiary of the United States government and the largest airplane manufacturer in the world.

***

Henry Ford owned a large soybean farm near Ypsilanti, a small community 30 miles west of Dearborn. The ground was relatively flat, located away from large population areas, and through the middle of the field ran a stream that the locals called Willow Run. Over the past several years, Ford had set up a summer camp there for underprivileged boys. It was here that the Fords decided to build their new bomber plant.

Work on the great plant progressed at breathtaking speed. In March 1941, a throng of workers and machines descended upon the countryside, similar to what they had done years earlier during the construction of the Ford airport. Clearing and grading continued for three weeks, and then buildings were started.

The factory never made it to a mile in length, as Sorensen had envisioned, but at 3200 feet, it nearly did so. In total, the factory took up 3.5 million square feet, or slightly more than 80 acres. Adjacent to the factory were seven concrete runways, one of which was 6,250 feet long. Even then, Edsel did not think this would be adequate in the future—he had in mind runways two miles long.[46]

While work was progressing steadily, the government was hounding Sorensen about spending so much money. Edsel worried about being audited, but he never blocked Sorensen—he agreed with him that they could justify what was being done.[47] In fact, Edsel did everything he could to encourage everybody to extend their efforts.[48] In a speech he gave to one of the many visitors to Willow Run, Edsel made clear to everyone the magnitude of the great bomber plant and the urgency of getting it done as quickly as possible[49]:

There are two things I should like you to notice specially while you are here: first, these men and women who operate the machines of Willow Run. They have come out to greet you today as a pledge of their deadly serious interest in their war work. This bomber job is new to all of us. Practically every person you see before you—man or woman—had to be newly trained. We had everything to learn. And I want to say for these men and women that they learned fast.

The plant itself needs no praise—it speaks for itself. It is a fine plant, splendidly equipped, and eloquent of what can be accomplished by cooperation—cooperation between government and industry, management and labor, civilians and military men. Unfortunately, this plant has been the subject of some premature and inaccurate publicity. That, perhaps, was natural, for "Willow Run" has become a symbol of the might of the United States in the air. We are glad to report to you that the plant is practically on schedule—amazingly so, for so vast a project—and that in production we expect to exceed the schedule. Skeptics in and out of industry must get ready to acknowledge that our plant here will be more than equal to its gigantic task.

Every employee and every manager here pledges all out cooperation until the "V" of velocity in production, and valor in combat, becomes the "V" of final victory.

By March 1942, although the plant was not yet complete, the first set of parts made in the plant was shipped to Consolidated. By May, the first Ford-made B-24 bomber was completed.

"I am not overly impressed with the quality of this bomber," wrote Charles Lindbergh, who had been hired by Ford as a consultant. "The first article we have delivered is not only as bad but considerably worse than the aviation people said it would be—rivets missing, rivets badly put in, rivet holes started and forgotten, whole lines of rivets left entirely out... It was the worst piece of metal aircraft construction I have ever seen. I would certainly hate to be in a bomber of this type if a few pursuit planes caught up with it. Yet it was passed by both the Ford Company and the Army inspectors."[50]

All of the blame could not be placed on Ford Motor Company alone. Ford was indeed responsible for the quality of workmanship, but the Liberator was not a Ford design—it was the design of Consolidated and left something to be desired.[51] Additionally, the government kept making changes in the design and wanted those changes incorporated into all planes as soon as they were developed—a procedure anathema to mass production.

After production of the bombers began in earnest, Sorensen, full of himself for his accomplishment, ordered his craftsmen in the plant to make a model of two B-24's put together as a biplane! If that were not enough, Sorensen asked Lindbergh what he would think of building a plane with a 1,000-foot wingspan. "You know," boasted Sorensen, "Ford Company could do something like that."[52]

***

When the plant began producing bombers on a regular basis, President Roosevelt and his wife made a surprise visit. Because of security reasons, only the top executives at the plant knew of the plans, but the employees realized something was up when they started seeing many government officials setting up stations throughout the plant.

On the afternoon of September 18, the president's private railroad car pulled into a siding beside the plant. A chauffeur-driven Lincoln touring car, dispatched by Edsel, pulled up beside the railroad car, and a ramp was lowered from the rear of the railroad car to the door of the car. Roosevelt was helped down the ramp and settled down in the right rear seat. Henry Ford was asked to sit between the president and Mrs. Roosevelt, and Sorensen sat in a jump seat directly in front of the president. At this time, Edsel, who was running late, pulled up in his Lincoln Continental and ran over to meet the presidential party.[53] He took the jump seat in front of Mrs. Roosevelt.

As the Lincoln drove along a designated route through the enormous plant, Sorensen answered the president's questions, while Edsel replied to those of Mrs. Roosevelt. The tour lasted twice as long as anticipated and ended at the end of the assembly line, where the entourage stopped and watched completed bombers exit the plant.

"They entered the cars, Mr. Ford sitting in the rear seat between the President and Mrs. Roosevelt, and Edsel in one of the middle row along with Sorensen and Michigan governor Murray Van Wagoner. The procession headed into a big doorway leading to the assembly bays, passing the towering presses and the tall wing assembly fixtures."[54]

***

Not everything went smoothly at Willow Run. The dreadful urgency was daunting enough, but then there was the slow start-up of production, exacerbated by the lack of skilled manpower.

Willow Run was deliberately built in the remote countryside so as not to pick desperately needed workers from the plants in and around Detroit. All automotive factories were now producing war materiel, and it would do no good to siphon skilled workers from one war production job to another. Initially, both Ford and government officials felt that the small towns surrounding Willow Run could provide all the workers needed. This was true before the United States entered the war, but after the attack on Pearl Harbor, the need for men in the military increased beyond all expectations. This put a strain on available men to work in the factories.

The Ford Motor Company, similar to all automobile companies, received pleas from Washington to increase production. Ford and the others had already adopted a 24-hour, 7-day week. "We fully realize gravity of the situation," telegraphed Edsel to the request, "and importance of earliest completion and operation of our aircraft motor and bomber plants. A full shift was working tonight at the Pratt & Whitney aircraft factory, and construction crews were operating under floodlights at the bomber plant. We have taken this action in response to the government's declaration of an all-out effort in the war."[55]

Ford was doing its share in training new aircraft workers at its new training center at the Rouge facility. However, the government was drafting the workers as soon as they were trained.

Willow Run needed more than 100,000 workers, and the Ypsilanti area was simply not populous enough to supply that number. Many were recruited from the Detroit area, and when that was not enough, workers from all over the country were recruited, especially from the South, to man the plant. Within two months of the attack on Pearl Harbor, more than 5,000 men had left Willow Run—most were drafted. Of the remainder, most were single men within drafting age, and it was only a matter of time before they would be called to serve.[56]

Edsel was keenly aware of this manpower crisis. He met with Sorensen, Harry Bennett, Mead Bricker, and Roscoe Smith on a regular basis to discuss this problem.[57] He solicited the help of Congressman John Vorys of Michigan, providing him a plan to stem the tide. To keep a minimum of trained workers, Edsel suggested that a select number of men who were already employed at Willow Run be recruited into the Army Reserve. Those selected would not be permitted to transfer to other plants or quit their jobs—they could be drafted immediately—and the Army would not draft any of the workers.[58]

This helped stem the mass exodus, but it did not solve all of the personnel problems at the factory. Harry Bennett deliberately caused the rest. Sorensen, through Edsel, was in command of production and facilities at Willow Run, but Bennett was in charge of personnel and labor relations. With Henry Ford's support, Bennett wielded as much power as Sorensen. It was Bennett's idea, for example, to recruit workers from the South because he felt they were less sympathetic to unionization. Ironically, he hired more men than necessary, particularly unskilled men, bringing them into a community ill-equipped to house the influx of people. At an early time, Sorensen had predicted that the plant would require more than 100,000 workers. Although that amount never came to fruition, the total number of workers exceeded 40,000. Housing was so short that men had to live in trailers and makeshift dormitories.[59] At one point, Edsel asked his friend Walter Dorwin Teague to work up some plans for prefab housing.[60] Nevertheless, Bennett "over-hired,"[61] at a time when production was still slow. "Many of the men are loafing on their jobs. At first glance, they may appear to be working; but as one looks closer, they are only in the positions that make it look as though they are doing something..."[62] Because production work was slow in coming and there was a lack of seasoned managers, absenteeism at the plant ran higher than 10 percent, which was more than twice that of the Rouge plant.[63]

The lack of production, high turnover and absenteeism, and the lack of adequate housing gave the press a field day in reporting on the massive complex. Newspapers began calling the plant "Will It Run?" When Edsel heard this, he became enraged. He knew that the entire Ford organization—at least everyone save Bennett—was working diligently to get the plant into full production. He told Fred Black that nothing more should be said to the press about production at Willow Run until bombers were actually rolling off the line in quantity. "There have been too many leaks on this story."[64] Edsel put part of this blame on Sorensen, who "had made all kinds of production promises for months and listened to none of the warnings the aviation engineers and executives gave him."[65]

***

Eventually, all of the hard work began paying off. On May 15, 1942, the first bomber came off the line, only 14 months after construction crews began clearing the ground for the new plant. From that point, bomber production progressed steadily. By April 1943, five bombers per day were coming off the line. "Our war jobs are coming along fine, and we will be making a lot of stuff by August," Edsel wrote Knudsen. "The enclosed cards will give you a few high spots of two jobs, bombers and aircraft engines. All told, our employment will total 220,000 people against 120,000 on automobiles."[66] Edsel was right. By August, bomber production had nearly doubled. So prodigious was the rate of production that the press began calling the B-24 the "Flivver Bomber."[67]

That March, Edsel was invited to the Rouge plant to accept the Army-Navy "E" award on behalf of the workers who were making Pratt & Whitney engines. Each year, this award was given to the plant with the highest-quality output, and the Ford Rouge aircraft engine plant had won it this year.

Asked to make a few comments, Edsel was introduced to the workers by Colonel Alonzo Drake. As they shook hands, Colonel Drake presented Edsel with a blue flag with a large white letter "E" emblazoned on it to fly over the Rouge. Edsel's speech lasted about 15 minutes, and in part he said[68]:

> *All of us understand, I believe, that the presentation of this flag does not indicate that we have reached the top of our stride. We should regard it as challenging us to still larger production of the aircraft engines so badly needed at this moment. The engine we worked on this morning will be carrying American airmen over some distant theatre of war a few weeks from now. Let us make it a good engine for the safety of our airmen and let us produce it promptly that it may reach them in time and in ever increasing numbers.*
>
> *We like to hear of victories, but we are told that, as a people, we have a tendency to relax our efforts when victories are announced from the fighting fronts. Authorities say there is a perceptible letdown in the fighting spirit of American industry whenever the news is good. Perhaps we should think about that. It stands to reason that if we have news of a victory today, the materials that made the victory possible were produced weeks or months ago. A letdown over here now will mean a letdown over there later—and that is no way to help the men who are fighting this war—it is no way to shorten the war.*
>
> *Once more I congratulate you on earning this wartime honor, and I thank the War Department for conferring it. Your job is not finished. It will grow larger and larger as the weeks roll on. Your responsibilities will increase. Greater numbers of men and women will be employed here, and you who are experienced will have to imbue the newcomers with the same spirit that won this flag. What has been done here is only an indication of what will be done.*

After Edsel's speech, Jeanette MacDonald sang the National Anthem.

The award was well deserved, for the contribution by Ford to the war effort was remarkable: nearly 9,000 B-24 bombers, 60,000 aircraft engines, 278,000 jeeps, 93,000 military trucks, 27,000 tank engines, 4,000 gliders, 2,700 tanks, 13,000 amphibians, 12,500 armored cars, and countless other parts.[69]

After the ceremony, Edsel flew back to Detroit with Sorensen and a couple other Ford executives who had flown down in a B-24 to pick him up.[70] The pilot invited Edsel to sit in the co-pilot seat, but Edsel would not do that until the others had had an opportunity to sit up front first.[71] Besides, Edsel was not feeling well. He had a high fever and chills, and he wanted to be covered.[72] He had recently celebrated his forty-ninth birthday, but he appeared as if he were 80 years old. His face was gaunt and skinny, and he was wearing glasses, reflecting his failing health. It took all the strength he could muster to get through the hour-long ceremony.

***

While Sorensen was in complete control of the Willow Run plant, he reported to Edsel, and Edsel signed all of the contracts and approved all major decisions.[73] Since the time that Knudsen had asked for Edsel's help in the fall of 1940, Edsel had been working at a frantic pace—to Willow Run in the morning, to Highland Park in the afternoon, then on to Washington to meet with government officials.[74] He was a member of the Automotive Council for War Production—a committee composed of all of the automotive executives whose companies were making war materiel instead of motorcars. Headed by a young George Romney, he said that "Edsel attended all of the meetings and made significant contributions."[75]

Many believed that Edsel should wear a uniform, similar to Knudsen.[76] Instead, when Edsel entered the Willow Run plant, he donned white coveralls, the same as the other workers. He said he was "Working at the hangar."[77] "Spent all day yesterday at Willow Run," wrote Edsel to his parents, who were vacationing in Georgia.[78] He worked from 12 to 16 hours a day, day in and day out, in the war effort. "The war won't wait," he insisted. "We have no time to waste."[79]

For the most part, Henry was living the retired life at this time and would go to Willow Run only to escort dignitaries with Sorensen or Edsel.[80] Sorensen tried to discuss production problems with Henry, but "it just went in one ear and out the other."[81] When bomber production was in full swing, Henry and Bennett would go to the plant and watch bombers come off this line, much the same way that Henry and Edsel used watch Model T's or Model A's come off the assembly line.[82] Henry knew that the mighty Sorensen had collapsed twice under the workload, and he was aware that Edsel was not doing well either.[83] One day, Henry ran into Helen Gore, the little girl who used to come to the Ford house to play with Edsel when they were children. The first thing that Helen asked Henry was how her old friend Edsel was doing. "He is very sick," Henry confided to her. "You know, this war has dragged him down. He worries too much."[84]

***

Edsel was wearing himself out. "That last year he lived was strenuous far beyond his endurance," recalled Sorensen.[85] Between all the war work, Edsel also was helping Gregorie to develop an entirely new line of Ford, Mercury, and Lincoln vehicles. Although the war was only beginning, Edsel was concerned about what his company should be doing to prepare for when the war ended.[86]

Edsel and Gregorie did not have much with which to work, either. When America entered World War II in December 1941, many of Gregorie's design staff left the department. They were either drafted, joined the service on their own, or were reassigned to Willow Run. The only reason that Gregorie and a handful of other designers were able to remain in the Design Department was because the government gave them deferments to work on bomber turret designs, camouflage, and military vehicles. If the government learned that they also were working on civilian products, they all would have been drafted on the spot.[87]

***

From a styling perspective, Edsel and Gregorie had done a Herculean job until the start of World War II, but the designs they had created during that time had run their course by the early 1940s. "Those bodies went back to the Dark Ages," recalled Gregorie. "Edsel realized this better than anybody, and early in 1941 he sort of made up his mind that we needed some pretty drastic changes."[88]

Additionally, Edsel was all too aware that, mechanically, the Ford car was almost an entire decade behind its competition. The only mechanical advantage the Ford had over other makes was its V8 engine. However, at a time when economy overruled power, even this feature became a disadvantage. (Edsel was never enamored with the V8 engine.[89] By this time, his favorite power plant would have been an overhead cam, straight six.[90]) For almost the same price as a Ford, a customer could buy a Chevrolet, which was more economical because of its six-cylinder engine, easier riding because of its independent front suspension and dual elliptical rear springs, and had superior braking with its hydraulic brakes.

Henry Ford was also getting up in age—he was nearly 80 years old when the war started. Realizing that his father would not be around much longer—although not wanting the inevitable to happen any sooner than it would—Edsel had to assume the responsibility of preparing the company for the future.

For these reasons, Edsel felt the timing was ripe to pursue a new line of cars. He met with Bob Gregorie and Jack Davis and began talking to them about developing an entirely new line of Ford automobiles. These were not intended to be face-lifted models with the same old chassis and drive trains with which they had been strapped for years. Rather, they would be completely new cars, with new bodies, new chassis, and new drive trains. Gregorie also proffered the idea of adding a small car to the new Ford lineup.[91]

Edsel agreed and directed Gregorie to "squeeze in" design work on the Ford, Mercury, and Lincoln lines whenever Gregorie and his team were not doing work for the military. Edsel spent what time he could in directing future planning with Gregorie, but he was rapidly becoming overburdened with war production matters, particularly now that the Willow Run bomber plant was under construction.[92]

***

Over the years, the idea of producing a small-size Ford arose every now and then at the Round Table meetings of the company. The need for such a small car now seemed greater than ever. "Everybody felt pretty good about the idea of two sized Fords," explained Gregorie. "At that time, cars were increasing in size, weight, and price—the trend was generally up. In other words, we had pushed the basic Ford up to a higher level, and we wanted to fill in beneath it with a small Ford. That was the way the thinking went. We were certainly headed for a small car in some manner or form, and we felt we were on pretty safe ground by building a small car."[93] As early as 1941, Edsel mentioned to Joe Galamb that "we will have to build a cheaper car someday."[94]

Thus, the decision was made. A new small Ford would be designed around a 98-inch wheelbase, while the larger Ford would be built on a 118-inch wheelbase. According to Edsel's directions, Gregorie was to make the small Ford and the large Ford appear "as similar as possible."[95]

With the decision of the small Ford put to rest, Edsel, Davis, and Gregorie finalized the remainder of the lineup. In addition to the small Ford and the large Ford, there would be a new Mercury and two Lincolns—a small Lincoln and a large Lincoln. Each model was to have a variety of body styles. The two sized Fords would be available in a business coupe, a convertible, a two-door sedan, and a four-door sedan. The Mercury would be offered in a coupe, a convertible, and two- and four-door sedans. The small and large Lincolns—which were also to be similar in appearance—would be available in a convertible and two- and four-door sedans.

The big convertible Lincoln also was to be offered as a Continental by adding the trademark trunk-mounted spare tire. "Between the two of us," explained Gregorie, "Edsel and I decided on the proper shape of these new cars. We thought as one about what these cars should look like. In the meantime, engineering was supposed to be paralleling our work as far as the chassis, suspension, and various other mechanical features were concerned."

Al Esper, a mechanic and test driver, recalled a heated discussion among the Ford executives in 1941 about changing the chassis on the new cars. "Sorensen, Martin, Mr. Edsel Ford, and Mr. Henry Ford were all in the Engineering Lab," Esper explained. "Quite a battle was going on as to the design of these cars and front spring suspension. Mr. Edsel Ford was fighting the battle for the coil spring suspension. Sorensen and Martin were for the coil springs as well, but they didn't have enough gumption to back Edsel Ford up. So, naturally, he lost. They didn't want to be crossed up by Mr. Henry Ford."[96]

Nevertheless, Edsel instructed Sheldrick to surreptitiously work on various types of independent front suspension systems.[97] Gregorie recalls driving one Ford prototype that had independent torsion bar suspension. According to Sheldrick, Edsel was also pushing for the development of an automatic transmission and a new suspension.[98]

***

Gregorie seemed to have come full circle. His first assignment in 1931 was to design a body for a short chassis; now, a decade later, he was to do that again. However, this time, Gregorie was practically free of constraints of any kind—no chassis restrictions, no width restrictions, and as far as the body design was concerned, he had free reign over that as well. For the first time in his career, Gregorie did not have to contend with extraneous factors.

Gregorie began first by developing design ideas for the small Ford/large Ford combination. He made fastback designs, notchback designs, and convertibles for both sizes. For the small Ford, he toyed with both rear-wheel-drive designs and front-wheel-drive designs. Although the front-wheel-drive concept did not take hold in the United States until the late 1970s, Gregorie understood its space-saving capabilities for small cars in the early 1940s and had several models made to test its viability for his new small Ford. "For the front-wheel-drive car," explained Gregorie, "one of the possibilities we considered was to place the new engine Farkas was developing longitudinally into the chassis with the transmission along side of it with stub axles coming out each side. It made a nice compact picture. We also explored the possibility of standing the engine upright to save space. You know, placing the engine with the flywheel-end down. It would have saved all kinds of space. We built a wooden model of it, just to have it in position, something to look at."[99]

Gregorie and his men developed these front-wheel drive concepts in the Design Department. Working concurrently with him was Gene Farkas, who was working on his own ideas at the Rouge plant, away from Henry's view. Because Henry would visit occasionally and thus could cause grief for the designers and engineers, "the front wheel drive concept was done mostly on the Q.T., and we made them in a special area," explained Gregorie. "Gene Farkas worked on a small, vertical six-cylinder motor—it was only 24 inches long—but it was just a dummy. And we made a buck to check dimensions and placement of the motor and transmission."[100]

Laurence Sheldrick and the other chassis engineers were aggravated by Gregorie working on mechanical aspects for which they were responsible, "and I guess it would have aggravated us just as well if they started making body shapes," says Gregorie. "But I knew that Edsel was interested in mechanical innovations of various kinds, and we didn't attempt to go over anyone's head. We thought we might as well see what we could do. As a matter of fact, because of the war work the company was doing for the government, the chassis people weren't able to do development work, but we could."[101]

***

By November 1941, Gregorie and his design team had full-size clay models of the two proposed Ford sizes. Both were fastback designs. Because Gregorie was no longer concerned about putting bushel baskets in the trunk—a requirement upon which Henry had insisted to make cars more useful to farmers—the roof lines were lower and the sloping rear ends were not as extreme as they were on the 1940 and 1941 models. Edsel favored fastback designs, so that is what Gregorie developed. "Yeah, it was alright with me," said Gregorie. "Whatever pleased Edsel. As long as it looked reasonable, why, I didn't favor one or the other. Both of them basically fulfilled what we had to work with."

For years, Edsel and Gregorie had wanted to lower, widen, and reduce the overall height of their cars. With their new design program, those were the attributes toward which they worked. "We were looking for something that was lower and wider—within reason," explained the designer. "In other words, get the body profile as low as possible, consistent with adequate headroom." This led Gregorie to what he refers to as the slab-side look: side body panels brought out past the plane of the wheels, fenders melding into the body, lower roof height, and a more extensive "greenhouse" area.

"By this time," Gregorie continues, "we were losing the benefit of fender shapes, which are wonderful. They formed wonderful relief in those older cars. But when we got into this slab-sided thing, it's like a loaf of bread. You had to do things to relieve the surface, to give the car identity, and to relieve the ponderous bulk of it. Originally, all of my new designs had the front fenders blending into the doors. I remember getting my clay knife (I used to carry one with me all the time), and putting a dip in the door beltline. I was trying to bring back the feeling of a fender and to give the cars some identity. In other words, it was a pretty line. It tapered gracefully, and it dropped down—a kind of sparkly little drop. If you stripped this off—as was done initially—there was nothing. It looked just like a balloon."[102]

Overall, the cars were lower to the ground—something for which Gregorie had been arguing almost from the time he started working for the company. Also, the beltlines were lower, the roofs were thinner, and there was more glass area. Following Edsel's directions to the letter, both cars appeared almost identical, except for their size, and that shows Gregorie's genius. It is almost impossible to make a small car appear as good as a big car, but that is exactly what Gregorie achieved.

By the fall of 1942, Edsel had approved both versions for production and had given Gregorie the go-ahead to work on additional body styles based on the slab-sided design of these two Fords. "I designed the small Ford and the large Ford with fastbacks because Edsel was partial to fastbacks," explained Gregorie. "And once he had approved these two designs, the intent was to develop a business coupe and a convertible for each Ford size, as well as a bustle-back or trunk-back design for both sedans. Ever since we came out with the Continental with its trunk-back design, Edsel slowly took a liking to that concept in our other cars."[103]

Edsel, Davis, and Gregorie also settled on the design for the remainder of the new Ford line. Similar to the two Fords, Gregorie was instructed to develop two-door and four-door sedans, coupes, and convertibles for the Mercury and Lincoln lines. Gregorie and his men started all of these as soon as the basic design of the two Ford sedans had been approved. With this lineup of vehicles, they felt that Ford could compete successfully on any level with General Motors or Chrysler in the post-war market.

Chapter 26

Dark Angel

"So gracious and so fair was he,
So well his work was done,
He showed the world how fine can be
A famous father's son."
Edgar Guest[1]

At the time that Henry Ford ran into Helen Gore, he had not comprehended how gravely ill his son was. Edsel had complained about his stomach for years, but his doctors diagnosed his illness simply as ulcers—a malady that beset many men at Ford. They even had a word for it: *Forditis,* a stomach so upset that eating was impossible.[2] Edsel ate ice to deaden the pain.[3] However, the doctors did try to keep abreast of his condition, as the following letter sent to Edsel from Dr. John Mateer suggests[4]:

> *Dr. McClure and I were discussing the status of your health yesterday. We both felt that it would be a good idea at this time to obtain a progress X-ray examination of your stomach... As you know, an occasional periodic checking of one's health status is a worthwhile procedure, even though there are no symptoms to suggest anything fundamentally new.*
>
> *The above X-ray check, along with a routine blood and urine examination, would be apparently all that would be indicated, and we probably would not be bothering you again for another year or 18 months. There would be no indication for any tube swallowing this time, and the above several examinations could all be done at one visit.*
>
> *I realize how very busy you are at the present time. However, if the above idea appeals to you, I would suggest that you ask your secretary, Mr. Lepine, to telephone me the morning which would suit you best, and I shall send a small box of barium out to your office, with directions for taking it.*

To make matters worse, Edsel had contracted undulant fever.[5] Its formal name is *brucellosis*, caused by the *brucella* bacteria found in cows, sheep, and goats. Infection results from consuming dairy products such as milk or cheese that had not been pasteurized. It is a pernicious malady. The symptoms are similar to a severe case of the flu—high fever, chills, diarrhea, and weight loss. The fever can run as high as 104 degrees, accompanied with severe dehydration. The afflicted person often might soak his clothing with sweat during one of the bouts. The spells would come and go, similar to an ocean tide, striking the patient during the morning, for example, and then subsiding in the afternoon, only to repeat the devastating cycle the next day.

In the days before adequate antibiotics, any person who contracted the disease was doomed to live with it for the remainder of his life. Such was the case for Edsel. "Undulant fever is so debilitating" explained an expert on the disease, "that in the days before tetracycline—the only known cure for the disease—a person would eventually want to commit suicide."[6]

It is not clear how Edsel contracted the disease. One of the theories is that he contracted it by drinking non-pasteurized milk from his father's farm. Henry did not believe in pasteurization and said that if the milk were properly taken care of from the source, it did not need to be disinfected.[7] That may have been the case, for many people had drunk milk from Henry's farm and nobody else contracted the disease.[8] The other possibility was that Edsel contracted the bacteria from drinking milk from his own cows at Haven Hill or from goat's milk at his residence in Maine.[9]

Regardless of the source, Henry stopped the use of milk at Greenfield Village and ordered Ray Dahlinger, manager of the Ford farm, to get rid of all of the cows on his farm.[10]

***

In January 1942, while Edsel was heavily involved in war work, his stomach condition took a turn for the worse. About a month earlier, he began noticing that something was seriously wrong. However, when Edsel called his physician, Dr. John Mateer, about his symptoms, the doctor brushed it off, saying that Edsel "was suffering from some nervous depletion due to overwork."[11] Edsel's pains became worse, and Dr. Mateer finally asked him to come into the office for some tests. Edsel was diagnosed as having numerous and far-advanced ulcers of the stomach.[12]

That did not surprise Sorensen. "I had known of Edsel's troubles for some time," he later explained, "and I was frightened to see the pain he was in at times. On one of our trips to Washington, DC, he [Edsel] was suddenly taken ill after a seafood dinner. I got him back to our hotel rooms and called a doctor. Edsel was in such agony that I sat up all night, afraid that he was dying."[13]

On January 12, Edsel entered the Henry Ford Hospital. Five days later, he had an operation in which surgeons removed part of his stomach.[14] Until now, Henry Ford had dismissed Edsel's ailments, saying there was nothing wrong with Edsel that proper diet and proper living would not cure.[15] However, when Edsel went into the hospital for the operation, Henry was "plainly worried."[16]

Edsel remained in the hospital for the entire month of January. Sorensen wrote in his diary that he went to visit Edsel on January 26: "He looks better than I expected. Had a nice chat for about 15 minutes." Sorensen told Edsel that he would come again. "He held my hand when I said good-bye."[17]

After releasing Edsel from the hospital, the doctors sent him to his Florida retreat to rest, and they put him on a special diet of low-fat foods, water, and crackers. Edsel seemed to be recuperating well. Robert Rankin, Clara Ford's chauffeur and Edsel's friend, saw Edsel in Florida and said that he "looked wonderful," and that he had never seen "a man with such a fine suntan." However, Rankin soon learned that Edsel was not doing so well on the inside.

Edsel knew that Rankin had had a similar operation about a year earlier, except that Dr. Roy McClure, Henry Ford's personal physician, had operated on Rankin, rather than Dr. Mateer. Being curious, Edsel asked Rankin what he was eating. "Oh, anything!" exclaimed the Ford employee. "French-fried potatoes, steak, fish."

"That's the trouble," complained Edsel. "They don't give me anything to eat. They just feed me pills!"

"You can't go far on pills."

"No," chuckled Edsel. "I get hungry once in a while."

"Why don't you come down to Palm Beach and have dinner with us?"

"I wish I could, but I can't," Edsel replied. "Doctor's orders."[18]

Nevertheless, Edsel seemed to improve. "I am feeling much better than I did the first two weeks," he wrote to his mother. "My gastritis is practically cured up. But Dr. Mateer thinks I am wise to stay a little longer for a build up. I'm hoping I can put on about five pounds."[19]

***

When Edsel returned to Dearborn in March, he resumed his normal workload as if nothing were wrong. He went to his office in the morning, met with his father and other Ford officials for lunch, and visited Bob Gregorie in the Design Department in the afternoon. Albert Lepine, Edsel's long-time secretary, did not notice anything different about his boss. "I had no idea, seeing him almost every day at the office," he recalled, "that he was seriously ill."[20] However, Edsel was gravely ill, and in November, after celebrating his birthday, he suffered a relapse.

This time, Eleanor intervened and summoned Dr. Roscoe Graham, described as "one of the most noted abdominal surgeons of the Western Hemisphere," to come to Detroit from Toronto to operate on her husband.[21] In surgery, Edsel was opened up and then closed. Dr. Graham had feared cancer but had hoped that it was localized. What he found was indeed cancer, but it had already spread to other organs.[22] "I'm sorry," the doctor informed Eleanor, "but I can do nothing to help Mr. Ford."[23] The doctor gave Edsel only a few months to live.

Sorrowfully, Dr. Mateer had botched Edsel's treatments. Perhaps he was convinced that Edsel's problem was only ulcers. Then there was the incurable undulant fever, which would have complicated any proper diagnosis.[24] Even in the 1940s, stomach cancer was not necessarily fatal. It was common practice even then to remove the stomach and part of the intestines if necessary to effect a cure.[25] In Edsel's case, Dr. Mateer may have waited too long or simply misdiagnosed Edsel's new malady as recurring ulcers. "Dr. Mateer was somewhat of a quack," recalled Dr. Conrad Lam, a surgeon on the staff of the Henry Ford Hospital at the time. "He may have prescribed pills to Edsel when he should have prescribed another operation."[26] That is what happened to Robert Rankin, but in his case the doctors took the appropriate action, removing his stomach and part of his intestines. Under a strict diet, Rankin lived for another 30 years.[27]

What Dr. Mateer should have done from the outset was to convince Edsel that the root of his problems was the constant pressure of his job, and that the only hope for a recovery would be for Edsel to take a long rest.[28] They did discuss it, and Edsel promised that he would consider his health and become less active.[29] Edsel's family also wanted him to get away from work for awhile. Eleanor, in fact, wanted him to quit. She saw no likelihood of Edsel recovering as long as he was subjected to the pressures of the job, his father, and Harry Bennett.[30]

"Stress and strain is the worst thing for that ailment," recalled Sorensen. "I never saw Henry Ford have any sympathy for Edsel when he was ill. He kept preaching that he must learn to eat right and live right. If Edsel could have dropped out of sight for a while at this time and gone in for a cure, he would have recovered...[31] Another close associate put it more succinctly: "What he should have done...is to have taken his dough out of the company and quit."[32]

Unfortunately, Edsel never did slow down, not completely nor to the extent that he should have. There was simply too much to do—to assist in the war effort and to help Gregorie in designing the post-war models, not to mention keeping his father and Bennett in check. In fact, "that last year he lived was strenuous far beyond his endurance," recalled Sorensen.[33]

***

Edsel was bedridden in the hospital for about a month and then was taken to his Grosse Pointe mansion, where he remained through January 1943.

While at Gaukler Pointe, Edsel received an invitation to attend a testimonial dinner to be held on December 29 on behalf of his old friend, William Mayo. Although Mayo had been gone from Ford Motor Company for more than a decade, he wanted Edsel to be there so they could reminiscence about the days of plant expansion and airplane development. However, Edsel was too ill to attend, and he wrote Mayo of his predicament[34]:

> *I was not more disappointed than I can tell you at not being able to be at your dinner last night.*
>
> *I had planned to be there, but the damned Ford Hospital said otherwise.*
>
> *I have been partially laid up with undulant fever for the past two months and sometimes have to lay up. This week I have been home and could not go out.*
>
> *I hope you will understand because I <u>wanted</u> to be there.*

Still sick and exhausted, four days later Edsel mustered enough strength to attend the wedding of his daughter Josephine, and it was one of the highlights of his last year. His little girl had grown into a beautiful young woman, with long dark brown hair and full lips. As Edsel escorted Josephine down the isle, he could not help but compare her to Eleanor when she was a young woman.

Josephine had completed her schooling at Chatham Hall in Chatham, Virginia, and had met a handsome young industrial designer, Walter Buhl Ford (no relation), at parties in Grosse Pointe. They courted for a few years and decided to get married in the summer of 1943. However, due to Edsel's precarious health, the couple moved up their wedding plans to the first of the year. Friends and family were surprised when they received invitations in the middle of December, announcing that the couple was to be married a day after New Year's.[35]

Josephine and Walter were married at Christ Church in Grosse Pointe.[36] Eleanor was radiant as ever, but Edsel appeared gaunt and worn out. After the wedding, a reception and dinner was held at her parents' Gaukler Pointe estate.[37]

Later that month, on January 29, Anne Ford, young Henry's second child, was born. Edsel managed to travel to Chicago for her christening. He wrote a letter to his parents about the event, who were vacationing in Georgia[38]:

> *We have had below zero weather since you left, and now we have four inches of beautiful new snow and lots of drifts. I don't think summer will ever get here.*
>
> *Thursday noon, we left for Chicago... We went to the Drake and cleaned up, then to Henry's apartment and loaded up two cars for the church. I was godfather to baby Anne, and Mrs. McDonnell was godmother. The baby is a beauty and didn't make a murmur during the ceremony. But Charlotte made up for both—she didn't stop talking or walking all of the time.*

In February, Edsel went to his Florida home, as he had done in years past. He stayed there until mid-March, returning to Detroit via Philadelphia after receiving the "E" Award.

***

While Edsel was in Hobe Sound recuperating, Henry went on one of his rampages. Because of his ongoing belief that all of Edsel's problems were caused by the men with whom Edsel worked, Henry wanted to fire John Crawford, Edsel's long-time right-hand man, and Al Wibel, the company's expert purchasing director and Edsel's close associate. Henry also brought up Kanzler again, although Kanzler had not been with the company for nearly a decade. Henry telephoned Edsel and informed him of his intentions, which greatly agitated Edsel. When Sorensen called Edsel a little later, that was all Edsel could talk about. It took Sorensen all he could muster to calm Edsel.[39]

Edsel returned to work on April 14, 1943, and had lunch with his father and the other executives the next day. Henry was not sympathetic toward his son and criticized him severely whenever he complained about his stomach. Edsel was in no mood to extend an olive branch.[40]

That night, Henry Ford telephoned Sorensen at home—they talked about Edsel. Henry told Sorensen that he wanted him to see Edsel tomorrow morning and change his attitude on everything in general. "Some job!" Sorensen jotted in his diary. Sorensen also made a list of the things about which Henry wanted Edsel to change his attitude[41]:

> a. *Discord over handling labor unions.*
>
> b. *Wibel and his attitude on Bennett, says Wibel is through.*
>
> c. *Bennett in full accord with Henry Ford. Henry Ford will support Bennett against every obstacle. Seeing labor leaders.*
>
> d. *Bennett's job, no one else.*
>
> e. *Change relations with Bennett.*

f. Kanzler relationship—wants it broken up.

g. Regain health by cooperating with Henry Ford.

The next morning, Sorensen told Edsel about the telephone conversation he had had with Henry. Obviously, the situation was awkward—one that Sorensen had experienced in the past. Sorensen said he would gladly act as an intermediary to smooth over the situation, but the decision was Edsel's. He would go no further if that is what Edsel wanted.

Edsel was heartbroken and in tears. "The best thing for me to do is to resign," he told Sorensen. "My health won't let me go on."[42]

However, Sorensen talked Edsel out of resigning, saying that if Edsel left, he would go also.[43] That would have placed the company in a perilous condition, practically ensuring that the government would take over Ford Motor Company.[44]

After Edsel had regained his composure, Sorensen talked to him about Henry's message. Although Henry was more succinct in his thoughts, it was all the same complaints Henry had voiced for years—Edsel's friendship with Kanzler, his animosity toward Bennett, and his living his own life. The only new thing was Henry's desire now to get rid of Wibel, one of the most able executives in the company. That was surprising to Sorensen, but it became clear to him after he learned that Edsel and Wibel had gotten into a row with Bennett the previous day over some supplier Bennett wanted favored. "Keep your nose out of purchasing!" Edsel exclaimed.[45] Naturally, Bennett immediately called Henry and complained, using Edsel's health against him. "You know Edsel is sick," he told Henry, "and here's that Wibel, making him upset! You make up your mind whether you are going to lose Edsel or get rid of Wibel."[46] That immediately turned old Henry against Wibel.

Interestingly, when Henry Doss, the Lincoln sales manager, realized that Edsel was seriously ill, he suggested to Edsel that he appoint Al Wibel as interim president until Edsel was able to recuperate. "I think he's a top executive," explained Doss. "I am volunteering his name, and I realize it is none of my business, but I would like to see you relieved of a lot of these things."

"Well, Doss," replied Edsel, "he couldn't function if I put him in here. They [meaning Henry, Sorensen, and Bennett] wouldn't let him function."[47]

Sorensen had talked Edsel out of resigning—it was not that difficult for him to do. After all, Edsel had been with the company for 30 years, as its president for nearly 25 and its heir apparent. To give up now would have been a devastating defeat for Edsel and, in a twisted way, a victory for Henry. However, Eleanor and her three sons all thought that Edsel should resign. They hoped he would quit, or at least take an extended vacation, to recuperate from his ailment.[48] Edsel's doctors had recommended this to him in 1941 when his ulcers first became a problem.[49] Edsel might have done that, but with the outbreak of World War II and the strategic presence of the company in the war effort, Edsel felt that he had a duty to the company and to the country and could not leave at such a critical time.

Satisfied that another crisis had been averted, Sorensen left for a well-deserved vacation to Florida. Before leaving, however, he walked into Henry's office and made one more plea for him to bring Henry II into the company.

Five days later, however, Sorensen received a phone call from his office. It was not good news. Russell Gnau told him that Henry had fired Jack Crawford and that he now wanted

Wibel's resignation.[50] The recalcitrant Bennett had kept working on old Henry after the ruckus had ended. Realizing there was nothing he could do, Sorensen called Wibel and informed him of the situation. Wibel tendered his resignation immediately.[51]

***

For months, Gregorie had no contact with his patron, but he was still at work developing post-war designs for the Mercury and the Lincoln. "I had quite a lot of stuff ready for Edsel to approve," explained Gregorie, "and I went a couple of months trying to contact him." In the old days, Gregorie would mail design ideas to Edsel when he was at his Florida retreat, and Edsel would return them with his comments, thereby keeping Gregorie on schedule. This time, Gregorie did not know where Edsel was. "I called Mr. Lepine several times a week, asking when Edsel would be coming in, but he was very evasive. Finally, he told me that Mr. Ford had gone down to Hobe Sound and that he couldn't be disturbed."[52]

After returning in March, Edsel unexpectedly walked into Gregorie's office through the back door. "I could see that he appeared very ill," recalled Gregorie. "He was tanned, but he had lost a lot of weight." They talked for a while, Edsel asking in a general way how things were going, but they never addressed any specific design or project. Finally, in a moment of compassion, Gregorie offered some advice: "Why don't you go over to the eastern shore of Maryland and buy a nice, big farm over there, and just go away," he told Edsel. "Spend all the damn time you want there. This thing will hold together."

"I wish I could," Edsel solemnly replied. "I wish I could."[53]

"Edsel didn't see any of the additional models we had worked up for him," says Gregorie. "He just came in that one afternoon, sat down for about an hour, then left. I never saw him again."[54]

Edsel wanted to spend more time with Gregorie, working on the post-war designs, but war work was more pressing. He had tried returning to the same grueling schedule with which he was involved before his operations, but he simply could not do it.[55] While in his office at the Rouge plant, Edsel asked Lepine to bring him crackers and milk from the cafeteria.[56] As time passed, Edsel became weaker and weaker, and at the end of April, he went to Willow Run and took one last walk through the gigantic plant alone.[57] Edsel was proud of that plant and of the people there who had given so much blood and sweat in getting Ford Liberators into the air. After that, Edsel returned to his magnificent mansion on Lake St. Clair and collapsed.

***

Ford Motor Company was now in dire straits. Edsel was an invalid, and Henry was too old and feeble to take on responsibilities. Sorensen felt that neither Edsel nor Henry could be a real factor in the running of the company from this point, and both Dr. McClure and Clara Ford agreed with his assessment.[58] The only resolution, as far as Sorensen was concerned, was to bring Edsel's sons, Henry II and Benson, into the company.

Edsel had been thinking about bringing his two oldest sons into the company as early as 1938. At that time, however, both were too young and in college. However, in 1940, with Henry II fresh out of college and newly married, and Benson dropping out of school, the two came to work for the company of their father and their grandfather. Their first assignment was assisting their father and Sorensen in negotiating the Pratt & Whitney engine contract.[59]

Their first real job, however, was in the Engineering Department. Laurence Sheldrick put them to work in the engine dynamometer room, where they took their orders from Leonard Williams, a Negro foreman.

"Neither boy pulled any punches about getting their hands dirty, their faces greasy, and their clothes all messed up," recalled Sheldrick. "But about a week after they had been there, Henry Ford came through to see where his grandsons were working. Harry Bennett was with him. When Bennett saw that they were working for a colored man, he just hit the ceiling. He called me outside the room and called me everything on the map."

"That happened in the morning, and just before lunch, I told Edsel what had happened."

" 'Don't pay a bit of attention to Bennett,' he told me. 'That's good for the boys. Keep them right there, that's all right.' "[60]

However, the two Ford heirs were not there long. Edsel needed them to experience many departments quickly, and he met with them in his office on a regular basis to assess their development.[61]

At this early stage of their careers, it was impossible for Edsel to know how well his sons would take to the business. After all, he had grown up with the company and was always expected to become the eventual leader of the mammoth Ford enterprises. As for his sons, that remained an unknown. One thing was clear, however, and that was the fact that Edsel was not about to push his boys into the business if they were not interested. "I have three boys," he once explained to a reporter. "Perhaps one of them will prove able to manage the business, but if none of them does, then someone else will be found who can."[62]

***

Henry Ford resented Edsel's sons coming into the company. He did not get along with Henry II or Benson, particularly Benson. One time while old Henry was giving Benson "a talking to," the young Ford simply walked away.[63] Harking back to earlier days, Henry told Sorensen to get rid of the two young Fords—to send them to California. He wanted them as far away from the plant as possible. When Sorensen told Edsel of this, Edsel broke down in tears.[64]

Sorensen was disgusted. He grabbed Edsel, and the two of them marched into Henry's office. "I am opposed to getting rid of the boys," Sorensen told Henry, "and if you have any idea that I'll carry it out, forget it. I refuse. And, if you do this yourself, I'm through." With that, he and Edsel left. Nothing more was said. When the men met for lunch the next day, Henry acted as if nothing had happened.[65]

In a way, Henry got his wish a few months later when Henry II enlisted in the Navy and Benson enlisted in the Army Air Corps. Sorensen tried to convince the two young Fords to stay with the company, rendering a greater service to their country by helping their father produce war materiel. However, Sorensen's words fell on deaf ears. Both young men wanted to serve their country in the military, and Edsel, based on his experience during World War I, supported them in that decision.

Sorensen did not give up. He tried to convince Henry Ford to make Henry II an officer in the company, thereby forcing the government to assign him to the company. However, old Henry would not listen.[66] Therefore, Sorensen was left to manage a company that for all intents and purposes was rudderless.

***

For the entire month of May, Edsel slowly succumbed to his illness. At first, the family put him in the infirmary of the mansion, which was located in the attic. It was a well-equipped room with bedroom and restroom, similar to a hospital suite. A retinue of doctors was by his side at all times. Dr. Brock Brush, one of the doctors in attendance, said that on the rare days that Edsel felt up to it, he would take Edsel for a walk along the lake shore.[67] When Arthur Backus, one of Lepine's assistants, visited Edsel, he said that Edsel "was in constant pain and gulped gas continually."[68] However, "he didn't complain and he wasn't morose," explained the doctor. "He knew that the end was near and that he couldn't do anything about it."[69] As Sorensen recalled, "He took it all so very bravely."[70]

Henry and Clara visited their son regularly after he had gone home from his last operation, and they would stay most of the day after Edsel had become bedridden.[71] During these last days, Henry did not pay attention to anything else. "He was out there all the time," recalled a Fair Lane employee.[72] Even now, Henry could not accept the fact that his son had terminal cancer, and he continued to blame Edsel's ill-health on the "ear-pissers" with whom Edsel had surrounded himself, such as Ernest Kanzler and the Grosse Pointe set, and Edsel's "high living."[73] "If the doctors can't cure him," yelled Henry, "I'll get rid of the whole bunch."[74] If that did not work, Henry would have his chiropractor treat him.[75]

Clara, also in denial, refused to accept the graveness of Edsel's illness. "You know, some days he's so much better," she told her close friend, Camille Ruddiman.[76]

"Ernest Kanzler, in a gracious way, telephoned Edsel quite regularly," recalled Dr. Brush. "He was just trying to cheer him up, but Edsel didn't appreciate the calls after a while. He was tired and sick, and he didn't have the strength to talk for long, but Kanzler talked and talked, and Edsel was too polite to cut him off. He told me, 'Kanzler doesn't know what a telephone is for.' "[77]

Edsel fell into a coma the evening of May 25, and all of his doctors were called to his side. After examining him carefully, they informed Eleanor that "there is no chance that he will pass the night."[78] Eleanor called her children—Henry II, Josephine, and William—to his bedside. Benson, although notified at the military base, could not make it back to Edsel in time.[79]

Edsel died at 1:10 the following morning.[80] He was only 49 years old.

Of Edsel's children, Henry II took Edsel's death the worst. "He was a saint, just a saint," he said, tears flowing. "He didn't *have* to die. *They* killed him!"[81]

***

Edsel's death was a surprise to everyone, except the Ford family, of course.[82] Dr. McClure telephoned Sorensen to notify him, and later that day, after Sorensen had gone into work, Henry Ford called him. Sorensen said that Henry was well composed (the shock had not yet affected him), but Sorensen "cracked up."[83]

Everybody knew Edsel was ill, but they did not realize how serious his condition was. Ross Cousins, Gregorie's illustrator, had talked to Henry Ford the previous day and asked about Edsel's health. "He said he was on his way to visit him and thought he was improving," explained Cousins. "The next morning, as we drove through the gate (we were four in a car, sharing rides in those war days), we saw the flag at half-staff. We knew it was Edsel. We had lost our man. We all started crying."[84]

It was an odd situation. Either the Ford family had tried to keep Edsel's illness a secret, or nobody was willing to accept his grave condition.[85] William Ford was away at boarding school during his father's convalescence and was not even told that his father was ill. "I really thought he was going to be up for my graduation," recalled William, "then I got a handwritten letter from him, saying that he was laid up and wouldn't be able to make it." It was out of sheer luck that William happened to come home after the end of the semester and see Edsel before he died. "My father probably should have died about three weeks before my graduation from boarding school," continued William, "and I think he kind of willed himself alive until the day I got home. I walked into his bedroom, he was very coherent. I had my diploma and a trophy that I had won in the school's tennis championship. He was very happy to see those, and we had a fine discussion. About two days later, he died."[86]

Telegrams and letters of condolence were sent to the Ford family from all parts of the world—from England, Egypt, Australia, Finland, Mexico, South America, the West Indies, Canada, and every state of the union. They came from persons and groups in every walk of life—from kings, presidents, automobile executives, and friends, both rich and poor. One of the first telegrams Eleanor received was from President Roosevelt. Several of Edsel's old schoolteachers from more than 30 years earlier sent their sympathies.[87] Because only a handful of people knew of Edsel's illness, many quarters found it difficult to credit the news when it was first reported.[88]

Nobody was more shocked to learn of Edsel's death than was his old friend, William Valentiner. "An era of my life has come to an end," Valentiner wrote in his diary. "Everything I did in the museum was connected with him, the successes as well as the battles and the inhibitions. A tragic nature: he recognized the fate of capitalism, which he represented by heredity, not by his own will. He never had a chance to develop, he was the symbol of the generation that was skipped over. He recognized the dynamic forces of our time and was intelligent, judicious, and an immutable friend. One could write a beautiful book about him, about the necessity of his end. Everywhere in the country they will bemoan him, although only a few knew him."[89]

***

Edsel's death had come so quickly, so unexpectedly, that Eleanor had little time to make funeral arrangements or to arrange for a proper burial site. Whether or not Henry asked to have his son buried in the Ford family plot in Dearborn is unknown. However, even if he had, Eleanor most surely would have rejected his idea. She wanted Edsel buried close to her family—the Hudsons—at Woodlawn Cemetery north of Detroit along Woodward Avenue. She and Edsel used to drive by Woodlawn when Edsel took her courting in his Model T speedsters so many years earlier.

With the help of her daughter-in-law, Anne, they made arrangements for the viewing, ceremony, and funeral. Edsel's body was taken to the William R. Hamilton Funeral Home located at the corner of Cass and Alexandrine avenues in Detroit.[90] Eleanor selected a simple but elegant bronze casket weighing 1,400 pounds.[91]

To allow the public to pay their respects, Eleanor allowed viewing for most of the day on Thursday. She was probably not prepared for the turnout. Beginning early in the morning, thousands of people lined up along Alexandrine Avenue, wanting a chance to say good-bye to the felled son of Henry Ford. They entered the parlor through the Alexandrine Avenue door, passed Edsel's body which lay in state in a small room, and then passed out to Cass Avenue. Simplicity dominated the scene, with the plain casket flanked by two clusters of red roses and a blanket of palms.

There was hardly an end to the line of viewers—men in overalls going to work or coming home from the factories, mothers and children, and the aristocracy of Detroit, pulling up in chauffeur-driven limousines.[92] Eleanor had the doors closed at noon for a private family viewing and then ordered the doors opened again to the public at one o'clock. People continued to come all afternoon and well into dusk.[93] By the end of the day, more than 7,000 people had filed past Edsel's bier.[94]

***

That night, Henry Ford called Sorensen and told him that he was going to take over Edsel's position as president. Sorensen was flabbergasted. Henry was much too old to take over the company, especially now that it was so involved in war work. However, Henry was serious and said that it would end all speculation as to whom would inherit the job that Edsel had held for a quarter of a century.[95] Bennett could probably have had Edsel's job for the asking, but even he realized that he was not qualified for the position. Likewise, Bennett liked working in the background. Some thought that Sorensen would be the logical choice, and he probably would have been. However, Sorensen said, "I never had any such ambition."[96] He may have had greater ambitions a few years earlier, but now Sorensen was looking toward retirement. He suggested bringing in Henry Ford II.[97]

The next day, funeral services were held at Christ Church in Grosse Pointe. People started arriving at 2:00 P.M., a half-hour before the service was supposed to start. Before long, the church, which held 500 persons, was nearly full. The Ford women—Clara, Anne, Edith, and, of course, Eleanor—all wore black mourning garb and veils. As they entered the front pews, Henry Ford, taut with the vigor he had carried through his 80 years, touched their elbows and helped them.[98]

Sitting behind the immediate Ford family were other family members, close friends, and company associates—the Kanzlers, McNaughtons, McDonalds, Firestones, Sorensens, Craigs, and Roy Bryant and his wife. General William Knudsen was there, as was Governor Charles Edison, the son of Thomas Edison, who had been Edsel's friend as their fathers had been friends in earlier times.[99] Also present were Fred Searle, Edsel's former teacher at the Detroit University School, and the automotive elite of Detroit—Carl Breer, Fred Zeder, Owen Skelton, K.T. Keller of Chrysler, and Charles Wilson of General Motors.[100]

Two pastors officiated at the services, the Right Reverend Frank Creighton, bishop of the diocese, and Reverend Francis Creamer, rector of the church. The formal service started at 2:30 P.M. Eastern War Time.[101] The chapel bell tolled, the organ played softly, and the choir in white surplices entered the church. "Let not you be troubled; ye believe in God, believe also in Me," read the bishop.

At the orders of Henry Ford, all Ford factories were to observe five minutes of silence in memory of Edsel, the president of the company and the friend of many. The observance was to take place at precisely the same time as the funeral service. Telegrams were sent to all Ford dealers, asking that they too pay tribute in a similar manner. Pupils in all of the public schools in Detroit were asked to observe a moment of silence at the same time, with a pupil in each school to pay tribute before the student body to Edsel's life and work.[102] For five minutes, hundreds of thousands of men and women and boys and girls stood silent in memory of Edsel Ford. "At the Rouge and Willow Run plants, mightiest of them all," wrote one reporter, "the silence fell like a heavy weight of bombs, shocking senses attuned to the constant din of production."[103] Mary-Alice Merchant, a worker in the Ford Highland Park plant, was one of those in observance. "When the plant shut down for Edsel's observance," she recalled, "the motors of the machine tools stopped, the pounding of the massive metal presses quit, and the piercing rhythm of production ceased. The silence was deafening. As all of us stood there in silence, an Irish tenor began singing the *Lord's Prayer*, his tender voice echoing throughout the plant. We all broke down. It was terribly moving."[104]

It was, indeed. However, many observers on this day were shocked to learn that old Henry had ordered only five minutes of observance in memory of his only son, but he had ordered the closing of all of his plants for a half day when his friend Thomas Edison had died.[105]

From the church, Edsel's casket was loaded into a Lincoln hearse and taken to Woodlawn Cemetery. Behind the funeral car was a procession of Lincoln-Zephyr sedans, carrying Henry and Clara, Eleanor and William, Henry II and Anne, Benson and Lynn, and Josephine and Walter. They pulled up to the Hudson family mausoleum and gathered inside a canopy, where Reverend Creamer said a few final words. Only immediate family members were allowed at the cemetery, except for Charles Sorensen and his wife.[106] Henry Ford had the audacity to ask Harry Bennett to attend the funeral, but Bennett refused, admitting that it would be hypocritical of him to attend the funeral of a man whom he had treated so poorly. "I knew that Edsel despised me," he said.[107] Hundreds of others had come to watch the ceremony, but they were kept back by police.[108]

There was no mausoleum for the Edsel Ford family to receive Edsel's remains. Temporarily, his casket was placed in the Hudson family vault. Within a few months, however, Eleanor had a new vault built specifically for her husband—a simple vault made of black granite with Edsel's name engraved on it.

For a moment, the family stood together in silence. Then Eleanor, on the arm of her young son William, walked away slowly, followed by Henry, holding Clara's arm, and the remainder of the Ford family.[109]

All of the cars moved away quickly, except for Henry and Clara's car. Their car remained still, and there they sat, quietly waiting until the other cars and all of the spectators had gone. At that moment, Henry and Clara returned by themselves and stood by the mausoleum in a "final, silent farewell to their only son."[110]

***

Eleanor was distressed over Edsel's death. He was so young. He had given so much to the family and to the company. And for what? To die at such a young age, only to be forgotten by the family and to history because of his self-effacement? The night after the funeral, Eleanor decided to join her husband. Slipping into Lake St. Clair, she began swimming out to her demise. Fortunately, one of the groundskeepers saw her and summoned help.[111] Eleanor regained her composure during the next few days, and, to help her cope with her loss, she wrote a letter to Henry and Clara: "Somehow I have a rather difficult time sleeping at night, so I am writing you a line to thank you for giving me the most wonderful husband and father God ever made. I don't mean to be emotional, but I realize the formative years were yours, and your son and my husband is our greatest treasure in the years to come..."[112] A few months later, Eleanor and William took a trip to Hot Springs, Virginia. Still hurting from her loss, Eleanor again wrote to Clara: "I do miss him so terribly that at times it seems impossible to go on, and yet I feel so very close to him that that gives me comfort. I brought down so many pictures so that everywhere I look I can see him."[113]

Henry had not fared much better. Throughout the funeral services, he maintained his composure "like a piece of stone."[114] He helped Clara to and from the church and to and from the cemetery. Throughout that night, Henry sat quietly in his den, saying hardly a word. However, the next day, when Henry went to work, as he normally did on Saturdays, "he cried like a little kid," recalled one of the engineers.[115] After that, Henry fell into a state of melancholy. He went on a hunger strike and ate nothing but cracked wheat.[116] Relying on his belief in reincarnation, Henry "felt sure that in some way, he would be with Edsel after he had shed his human body," recalled Catherine Ruddiman. "He felt they would be more of the same mind when they were united again."[117] Henry really did not believe that Edsel had died until Harry Bennett told him so.[118] "I just can't get over it," confided Henry to a long-time Ford employee. "I've got this lump right here. The Misses sits down and cries and gets over it, feels a little better. I just cannot do it. I just have a lump here, and there is nothing I can do about it."[119]

But Henry tried. In retaliation, he fired everyone whom he thought had caused Edsel to worry. He had already gotten rid of Wibel. Laurence Sheldrick was the next to go. Then Henry got rid of his long-time secretary, Ernest Liebold.[120]

Henry could blame whomever he wanted—Sorensen, Bennett, or even Edsel for not being more of a man. However, Henry had nobody to blame but himself.[121] Said Fred Black, "Henry Ford killed his son just as dead as though he had shot him with a pistol."[122]

***

Meanwhile, Eleanor was not only mourning the loss of her husband, but she was beginning a long drawn-out battle with the federal government over inheritance taxes. At the time of Edsel's death, the value of Ford Motor Company was estimated at $718 million. How much would Eleanor have to pay? $100 million? $200 million? $300 million? Stockbrokers on Wall Street were itching for the chance to capitalize on her predicament.

As one of the richest men in the world, Edsel did not have that many worldly possessions—most of his wealth was tied up in Ford Motor Company stock. He had his magnificent properties in Grosse Pointe, central Michigan, Maine, and Florida, and rare and expensive pieces of artwork. In his closet were hundreds of suits, ties, and shoes, and he had a small amount of jewelry. In his garage at Gaukler Pointe were only seven cars: the little speedster that Gregorie had made for Edsel back in 1934, four 1941 Fords and Mercurys, a 1942 Mercury coupe, and one 1941 Lincoln Continental cabriolet, his favorite car.[123]

Before anybody could get anything, however, a value had to be placed on the stock because no share had ever been sold on the open market. Government officials placed a value of $195 per share, but Eleanor's attorney, Clifford Longley, who was there to protect the Ford family as usual, said that $58 was ample. The two factions finally agreed on a per-share price of $135. At that price, the value of the Ford Motor Company was estimated at $718 million. With

Edsel's 42 percent ownership in the company, that placed Eleanor's inheritance at slightly more than $300 million.[124]

Wall Street soon was disappointed when it learned that Eleanor would not have to sell one measly share of Ford stock to pay her inheritance taxes. Edsel left most of his possessions to her. She received all of the properties, works of art, automobiles, and personal articles. However, Edsel gave his stock in Ford Motor Company to the Ford Foundation and to his family. The Ford Foundation received all of his Class A shares except "such as it may be necessary to sell to pay the obligations of my estate." Of Edsel's Class B shares, he left equal amounts to Eleanor and to his children—20 percent to his wife immediately and 20 percent to each child when he or she reached 25 years of age.[125] Because of the Ford Foundation that Edsel had founded only seven years earlier, Eleanor's tax liability came to only $42 million, which, of course, was paid by the Ford Foundation.[126]

The provisions of the Ford Foundation that Edsel had so meticulously arranged worked perfectly. If it were not for the Ford Foundation, the Ford family would have had to sell Edsel's Ford Motor Company stock to pay inheritance taxes.

It was indeed ironic, however, that the provisions of the Ford Foundation were tested through Edsel's death rather than the death of his father, as was naturally anticipated.

***

When Henry Ford died in 1947 and Clara died in 1950, they both bequeathed most of their estates to the Ford Foundation. The Ford Foundation now owned nearly 90 percent of the Ford Motor Company, and the dividends that it received on a yearly basis amounted to more than $100 million. By the end of 1955, the Foundation had given away nearly $900 million in grants and endowments.[127]

As time passed, the directors of the Ford Foundation found their ownership in the Ford Motor Company untenable. The fortunes of the Foundation rose and fell with the income of the Ford Motor Company. Because the Foundation did not have any voice in the management of the company, it was concerned that Ford management could at any time not declare a dividend and plow all of its earning back into the business. The directors of the Foundation wanted to diversify their holdings. To do that, they decided to sell the Foundation-owned Ford stock on the open market. To do this, another Ford stock reclassification was required.

Henry Ford II, ever vigilant of his family's fortune, wanted assurances that his family would continue to remain in control of the company, regardless of the number of shares sold on the open market. On the surface, this seemed to be an impossible request because the New York Stock Exchange required that all stock offered must carry voting privileges. However, through negotiations among the Exchange, the Ford Foundation, and the Ford family, a special arrangement was established through another reclassification of Ford stock. This time, the Ford stock was split into three classes: Class A, Class B, and Common, with the result that two-fifths of the voting power would be held by the Ford family, and three-fifths would be scattered among the general public. Finally, any stock retained by the Ford Foundation would still carry no voting rights. If this were not enough, the family holdings would continue to have two-fifths of the voting rights—no matter how many shares of the Foundation-owned stock were converted into Common stock—as long as the family owned at least 2.7 million shares of Class B stock.[128]

With the ownership question now out of the way, the Ford Foundation finally was able to pursue its plans of selling a large part of its Ford Motor Company stock. On January 26, 1956, 10,200,000 shares of Ford Motor Company stock went on the open market at $63 per share. At the time, it was the largest public offering in history. Within a few weeks, the Ford Foundation pocketed more than $600 million, and the Ford Motor Company for the first time in its history was now a publicly owned company.[129]

***

A little more than a decade earlier, Henry, Clara, and Edsel had owned the mighty Ford Motor Company in its entirety. Now, the mammoth industrial enterprise was owned by 300,000 shareholders—some buying only one share so they could say that they owned a piece of one of the greatest industrial enterprises in the world. The number of shareholders was significantly increased in 1959 when the Foundation sold another four million shares, and again in 1961 when it placed another 2.75 million shares on the block. At the end of that year, the Ford Foundation held 28 million Class A shares, the Ford family owned 6 million Class B shares, and the public held 21 million Common shares. All three classes shared equally in company dividends, providing the public, the Foundation, and the Ford family with a steady stream of income.

During the past 50 years, the Ford Foundation has distributed more than $9 billion, making it the largest philanthropic organization in the world.[130] The individuals, communities, and countries that received these funds over the years should thank Edsel Ford for his prudence and benevolence.[131]

Chapter 27

Epilogue

"Where's Harry?"
Henry Ford[1]

On Monday, May 31, five days after Edsel's death, Henry Ford called a meeting in his office at the Rouge plant. Present were Charles Sorensen, Bert Craig, Eleanor, Henry II, and Benson. After a few minutes of small talk, Henry Ford, owner of 51 percent of Ford Motor Company, elected himself president. He appointed Sorensen as executive vice-president and Craig as treasurer.[2] Eleanor, Benson, and Bennett were elected to the board.[3]

Eleanor and Henry II were enraged. Edsel's widow stood up and scathingly denounced Harry Bennett and demanded his removal on the spot.[4] Young Henry made it clear to his grandfather that not only was it inconsiderate to elect a new president so soon after his father's death, but that he was strongly against old Henry becoming president. Old Henry paid no attention to either one.

As everyone stood up to leave the meeting, old Henry grabbed Sorensen's arm so that he could not talk to young Henry and took him to his car. Henry tried to explain to Sorensen how things were going to be like the good old days again, but Sorensen interrupted him. "Things will never be the same," Sorensen told the old man, "and you can't run the company now. The best thing that you can do now is to make young Henry president and Benson vice president. Better give that some thought."

Old Henry said that he would do that, but Sorensen did not believe him.[5]

That evening, Clara Ford called Sorensen at home, saying that Eleanor wanted to see him immediately. Clara told Sorensen that Eleanor was quarrelling with Mr. Ford. Eleanor had called him to voice her disapproval of the new management structure, and when the old man began defending his actions, Eleanor became incensed. "You, Mr. Ford. You're the one who killed your son!"[6]

When Sorensen arrived at the Ford estate at Gaukler Pointe, young Henry and Benson were with their mother. Eleanor, similar to her two sons, resented what had been done at the so-called board meeting that morning. The four of them talked for quite a while, Sorensen explaining to Eleanor that the only solution would be for young Henry to assume the presidency of the company, and he asked for her help to make that happen.[7]

***

Before returning to his military duties, young Henry walked through the Rouge plant. He talked to managers, superintendents, and workers on the line, and then went to visit Bob Gregorie. Prior to this, Henry II had never seemed much interested in the automobile business, but apparently his father had tutored him more than anyone had realized. In particular, Edsel had explained to him the importance of automobile design in the continued success of Ford Motor Company. He walked into the chief designer's office, shook hands, and said, "Father told me to start here."[8]

"That's amazing, when you think about it," recalled Gregorie. "Here Edsel was on his deathbed, and one of the things he wanted to impress upon his son was the importance of the Design Department to the future of the company. It shows just how much he loved the design of automobiles. It also says a lot about what Edsel thought about the working relationship we had developed over the years. I was moved by the thought."[9]

Young Henry and Gregorie had quite a discussion, the chief designer bringing the young man up to date on Edsel's post-war plans. They also drove to the Lincoln plant, where they met with Robby Robinson, who explained to Henry II the problems he was having in getting the tooling made for the new Lincolns. With Henry II showing such interest, Gregorie thought that perhaps the young man would pick up where his father had left off. Gregorie hoped he would.

Gregorie had known Henry II since he was a little boy. Young Henry and his younger brother Benson had come to the Design Department with Edsel on many occasions, roller skating among the clay models. With such an intimate relationship, there was no reason why Gregorie and the new Ford boss could not build a relationship similar to what he and Edsel had developed over the years.

Unfortunately, after that initial meeting, Henry II never showed much interest in car design. "He wasn't even conversant with it," explained Gregorie. "He acted bored with it." Whenever Gregorie tried to get Henry II involved in a design decision, he would nonchalantly say, "Yeah, go ahead with it," without really understanding—or really caring about—what he had just approved.[10]

***

Young Henry returned to the Navy. Meanwhile, Gregorie's old nemeses came knocking at his door. Old Henry Ford's men had never forgotten how Gregorie, with Edsel's support, had nullified their power over the years, especially Joe Galamb. Galamb had originally worked with Edsel on the design of new cars, and Galamb's input into their design had been all but eliminated when Edsel appointed Gregorie as design chief. Galamb had always resented Gregorie for that. As long as Gregorie had Edsel to protect him, Galamb stayed clear. However, now that Edsel was gone, Galamb returned with a fury. "Joe gave me a bad time after Edsel died," Gregorie says. "He thought, 'Oh, boy, I'm really going to fix this guy!' " Galamb started coming to the Design Department almost every day, telling Gregorie what to do, which was precisely what Gregorie had feared. Gregorie and Edsel had fought hard to separate design from engineering, and now they were back at Gregorie's doorstep. "I didn't want to get involved with it," says Gregorie. "That was one thing I fought with Sheldrick and the other engineers about when I was a kid. I'd finally gotten design split apart from them. I put design on more or less of a separate department under Edsel and myself. And, man, I wasn't about to step into *that* again."[11] Gregorie did all he could to keep Galamb at bay, but it eventually became a lost cause. Galamb wanted revenge. He wanted Gregorie out of the company.

Until this time, Gregorie had held his own against the Napoleonic Galamb. Now that Edsel was gone, Gregorie had lost the one man in the company who could protect him, and he was open to attack. The final blow came one Friday afternoon in June while Henry Ford II was in the Navy. In the months following Edsel's death, Galamb not only began taunting Gregorie, he began looking for someone on the board of directors to get rid of Gregorie. He found his ally in Frank Campsall, old Henry's confidant. Campsall called Gregorie to his office in the Ford Administration Building and fired him.

Gregorie was not too surprised. "At that time," he said, "firing of officials was popular—it was a trend. They got rid of Sheldrick and Sorensen at about that same time. I got notice from Frank Campsall—that's all I know about it. I don't recall ever having any contact with Henry Ford II, or I think he would have intervened."[12]

During this same period, Gregorie had no one to direct him or to approve his designs. "So I was left at a standstill, and it was awkward for me," he explained. "There were no other company officials that would assume any responsibility. The place hadn't been reorganized on any basis—they hadn't any committees or anything like that. That came along later. But I'm afraid it came along in such a way that it wouldn't appeal to me at all. I liked the simplicity of working with Edsel Ford. But now, it was an entirely different situation." From Gregorie's meetings with Henry II, it was clear that young Henry simply did not take to the discussion of automobile design. It became apparent that Gregorie soon would be forced back under the direction of engineering, by the sheer vacuum left by Edsel's death. "I accepted my fate as it came," said Gregorie. "I couldn't get anything done anyway. It was all right with me."[13]

***

A couple months later, Henry II informed Sorensen that he was ready to come into the company. He had resisted for a long time, thinking he was not yet ready to assume such monumental responsibilities. Plus, he was doing his duty for his country. However, both his mother and grandmother had impressed upon the young Ford that it was his destiny to run the family business. Although the timing may not have been right, the company and the family needed him now. Eleanor and Clara would support him totally and completely against old Henry Ford and Harry Bennett.[14]

Delighted with the news, Sorensen immediately called his contacts in Washington to obtain young Henry's release from the Navy. By the end of July, Secretary of War Frank Knox released Henry Ford II from any further military service, and on August 10, 1943, young Henry took over his father's office in the Ford Administration Building. "Henry hated to leave the Navy," recalled William Bonbright, his close friend. "He wanted sea duty or any kind of duty that would have kept him from going back to the plant and getting into that situation with Bennett. But his mother and grandmother wanted him in the company, and that was that. With Henry, his family's wishes always came before anything else."[15]

At 23, Henry Ford II was not only young, he looked young. His mother's fleshy features did not help his overweight appearance. Many were concerned that he would not act forcefully and decisively, and old Henry, not wanting him there anyway, used Bennett to harass the young man.

However, young Henry knew what he was facing. He had looked through his father's papers, and the two of them had had long talks about the condition of the company, which was not good. The organization was poor, the engineering was out of date, and it was losing money, regardless of lucrative government contracts. Above all else, young Henry had to watch out for Harry Bennett.[16] "Edsel would always compromise to avoid an unpleasant scene," recalled Jack Davis, "and Bennett thought young Henry would be the same way. But he found out that when he ran into young Henry, he ran into a brick wall."[17]

Sorensen discovered that also. "One morning, I was with him [Henry II] when he got a telephone call from Bennett," recalled Sorensen. "He was getting an earful and could hardly get a word in. Not wishing to listen, I stepped into the outer office and went back in when Henry II put down the receiver. Not a word was said about the phone call. He was under fire—I could see that—but young Henry was composed and resumed his talk with me as though nothing had happened. The boy can take it, I said to myself happily—everything will turn out all right."[18]

Nevertheless, it took Henry II months to gain a foothold. Initially, men were afraid to be seen with him, fearing reprisal from Bennett.[19] However, Sorensen took the young Ford under his wing, and Henry II stayed away from Bennett as much as possible.[20]

***

Meanwhile, Henry II was building around him a group of executives that would be loyal to him. From Willow Run he had Mead Bricker and Logan Miller, two manufacturing men with long experience and untainted by Bennett's machinations. Young Henry also brought back Jack Davis, Edsel's good friend and former sales manager of the company. Davis had been in exile in California since Edsel and Bennett got into an argument over him. Loyal to the company and especially to Edsel, Davis waited patiently until called back. He was glad to hear from Edsel's son. Henry II also called on the services of John Bugas, former director of the Federal Bureau of Investigation in Detroit.

Ironically, Bennett instigated his own downfall. Sometime in late 1942, Bennett hired John Bugas as his assistant.[21] Bugas had been director of the FBI Detroit office since 1938. Around the beginning of the war, Edsel contacted Bugas and asked him to investigate thefts that were occurring at both Willow Run and the Rouge facilities. After Bugas had made significant progress, Edsel then asked him to investigate Bennett.[22] Before long, Bugas established his own information network in the plants to counter those of Bennett, and he kept Edsel informed of what he learned.[23]

When Bennett learned about Bugas' clandestine investigations, he protested to Edsel. "I'm sick and tired of having my men called in by the FBI," he protested. However, Edsel did not flinch nor did he back down. "We've got to do something," Edsel retorted. "They're carrying the goddamn plant away!"[24]

Not much is known about what dirt Bugas may have found on Bennett, but it would not be unreasonable to assume that he had uncovered plenty, for he became "a source of irritation" to

Bennett. "Finally I decided that the best way to cope with him was to hire him into the plant," recalled Bennett.[25] It was the only sure way that Bennett could keep an eye on Bugas.[26] "John's the smartest man they've got," Bennett said. "I'm going to hire him. I can give him twice what he's getting with the FBI, and he's worth it. He can be a big shot in the company, and that's what he likes. He doesn't have to cater to any of the big shots we got. He'll work for me."[27]

"I drove out to see Bennett in a ramshackle old car that I had," recalled Bugas. "When I left his office and walked out to the parking lot, my car was gone and a new Lincoln was standing in its place. That was how Bennett operated. I had a terrible time getting my old car back from him."[28]

Bennett had Bugas work on industrial relations, hoping he could turn this G-man into one of his stellar assistants. However, Bugas was more astute than Bennett's other men, and he quickly realized how Bennett operated.[29] Patiently, Bugas observed the political goings-on soon after Henry Ford II came to work and, assessing the situation, decided to support the new regime rather than the old. When Henry II began asking questions about labor relations, Bugas readily supplied him with accurate and detailed data.

After that, Eleanor and Clara made their next move. Again, using Eleanor's stock as leverage, they insisted that Henry II be made executive vice president of the company—placing him one step away from the presidency.

Regardless, old Henry Ford did not want Edsel's boys around the company.[30] He felt that if young Henry came into the company now, he would be too influenced by Ernest Kanzler, a thought that sent shivers up old Henry's spine.[31] However, that soon changed. For the past 55 years, Clara had done what her husband had wanted, bowing to his every need. Not this time. Clara now sided with Eleanor, and, with their control of large blocks of voting stock, the two women insisted that young Henry immediately take an important and active interest in the company.[32] "If you don't," threatened Eleanor, "I will sell my stock."[33]

To make matters worse, the family had learned that old Henry had set up a codicil to his will which, in effect, would establish a board of trustees to run the company after his death and until young Henry and Benson were older and more experienced, and which would negate the Ford Foundation by establishing a trust.[34] The board of trustees would be composed of Bennett and Bennett's friends. Old Henry Ford thought that this trust arrangement was a better way of transferring power than his son's foundation plan. However, as in so many of old Henry's previous decisions that were detrimental to the company, this one, if implemented, would have destroyed Ford Motor Company. Old Henry would not have to worry about who ran the company, because there would be no company to run, and both Henry II and Benson believed that.[35]

Henry II asked Bugas to see Bennett and to find out if there were any truth to the existence of a codicil. When Bugas asked Bennett about the codicil, he did not deny that it existed. "Come back here tomorrow," he told Bugas, "and we'll straighten the whole thing out."

The next day, a strange scene took place in Bennett's office. Bennett showed Bugas the original typewritten addendum to Henry Ford's will and a carbon copy of it. Thus, it was true. Old Henry had succumbed to his warped sense of reality and had literally signed away his company.

However, Bennett did not put up a struggle. He knew he had met his match with Bugas. Bennett placed the original and the copy of the codicil on the floor and lit a match to them. The two men watched the papers curl up in flames. When the fire smoldered out, Bennett swept the ashes into an envelope and handed them to Bugas.

"Take this back to Henry," Bennett said.[36]

If Bennett had the piece of paper that would assure him a power base within the company long after old Henry was gone, why did he destroy it? "It wasn't any good," Bennett explained to the company attorney when he asked him the same question. "Mr. Ford had carried the instrument around in his pocket for a long time and had made a lot of scribblings on it, including verses from the Bible." Some of the names of the trustees were misspelled. The attorney came to the conclusion that Henry Ford had never signed the codicil. It was merely a ploy on Bennett's part.[37]

Fortunately for the Ford family, there was no bona fide codicil, but the threat of its existence finally prompted the family to act. It was time to take over power from old Henry and pass it on to his grandson. Eleanor and Clara had had long discussions about this matter and had kept Henry II informed of their discussions. He was in favor of their direction. "It's about time we did something to clean the place up," he said.[38]

Clara had done her part. She finally convinced her husband that the time had come to give up the presidency to his grandson.[39] Eleanor set up the meeting at Fair Lane. "My grandfather told me he [old Henry] was planning to step down and let me be president of the company," Henry recalled. "I told him I'd take it only if I had a completely free hand to make changes I wanted to make. We argued about that—but he didn't withdraw the offer."[40]

From that meeting, Henry II went to the Administration Building and instructed Frank Campsall, old Henry's secretary, to draft a letter of resignation for his grandfather's signature. He also instructed Campsall to set up a meeting of the board of directors for the next day so that they could elect Henry II president.

On that day, September 21, 1945, the reigns of power were finally transferred from old Henry and his crony, Harry Bennett, to Henry Ford II. Present were Henry Ford, Henry Ford II, Harry Bennett, Mead Bricker, Bert Craig, Ray Rausch, and Eleanor Ford. Campsall had old Henry's resignation letter ready and read it aloud to the board members. Immediately thereafter, the board elected Henry Ford II president of Ford Motor Company.[41] Young Henry was only 28 years old, two years older than Edsel was when he became president.

***

Henry Ford II's first official duty as the new president was to fire Harry Bennett. As everyone was leaving the board meeting, young Henry pulled Bennett aside and told him that he had plans for reorganizing the company and that Bennett was not part of them. "I told him that John Bugas was taking over his job as industrial-relations director as of the next day," Henry II recalled years later. "As a matter of fact, although I fired Bennett immediately from his job and from his position of authority, I told him he could stay with the company if he wanted to. But he saw how silly that would be and left the next day. We kept him on the payroll for sixteen more months so that he'd be eligible for a pension."[42]

Young Henry and Bugas followed Bennett back to his office to discuss the transfer of responsibilities from Bennett to Bugas. "Bennett seemed to be in a state of shock," recalled Bugas. He said that he wanted to show young Henry and Bugas several things that old Henry had asked him to do around the plant, and, in Bennett's car, the three men made a round of the Rouge facilities.

"I don't know what was going through Bennett's mind," Bugas recalled. "He didn't have anything to show us. He'd point to a pile of lumber or some discarded machinery and mutter something about having it cleaned up. It was ridiculous, the three of us trying to make casual conversation as if nothing had happened. We came back to his office, and Henry left us. Bennett and I sat there for a few minutes, without saying anything. Then he turned on me suddenly and said, 'You son-of-a-bitch, you did this to me!' and pulled a .45 automatic out of his desk.' " Anticipating such a confrontation, Bugas pulled a .38 out of his belt. "Don't make the mistake of pulling the trigger, because I'll kill you," Bugas said calmly. "I won't miss. I'll put one right through your heart, Harry."[43]

Bennett immediately backed down and set his gun on his desk. Bugas turned and walked toward the door, wondering if he would be shot in the back.

"All that afternoon, Bennett's office was filled with smoke," recalled Bugas. "He was burning papers in his wastebasket. That night, he went away, and we never saw him again."[44]

With Bennett now gone, Henry II went to Fair Lane to inform his grandfather. "I went to him with my guard up," young Henry recalled. "I was sure he'd blow my head off."

However, old Henry did not even raise an eyebrow. The old man sat patiently while his grandson informed him of the new state of affairs at the company. All he said was, "Well, now Harry is back where he started from."[45]

Although Bennett was gone from the company, Henry Ford longed for his old companion. During the last years of old Henry's life, he did not spend much time at the factory. He simply laid around his mansion, followed Clara in the same way as would a little puppy, and asked repeatedly for Bennett.[46] "Where's Harry? Hasn't Harry been here today? Why doesn't Harry come?"[47] Bennett probably would have visited his old boss had Henry invited him, but Clara had barred Bennett from the house.[48]

* * *

Henry Ford was now an old and lonely man, left with only his memories of his son and of building an industrial empire. He missed Edsel terribly. Henry's barber, Joe Zaroski, recalled the time when Henry came into his shop shortly after Edsel's death: "There was no one in the shop, and I was ready to go home. He [Henry] was the last customer. He pulled out his wallet, which had loose leaflets in it with pictures of Edsel and the grandchildren. He paid me and said, 'Joe, look. This is Edsel when he was a little boy, and I was very proud of him. My son had everything to live for, but he worked too hard and now he is gone.' I looked at his face, and he had tears in his eyes. He was a lonely old gentleman."[49]

Shortly thereafter, Henry Ford unexpectedly walked into Bob Gregorie's office early one morning, which was surprising to the designer. Old Henry rarely came to the Design Department when Edsel was alive, and Gregorie thought it odd that the old man should show up now. "Let's take a ride," said the automobile magnate.

Gregorie did not know why Henry wanted to go for a ride with him, but they went to the service garage behind the Engineering Laboratory and climbed into a new Ford. "Where would you like to go?" asked Gregorie.

"Let's go down to the test track," directed Henry.

As they drove to the entrance of the track, Gregorie began to slow down to check in at the gate, but Henry instructed him to drive right through it. "They're all asleep anyhow," Henry growled. Then, pointing to the left, Henry said, "Come on, let's go on around there."

"But, Mr. Ford, that's the wrong direction!" yelled Gregorie.

"That's all right. That's all right," replied Henry defiantly.

Gregorie steered the Ford into oncoming traffic, and test cars immediately began swerving and skidding in an effort to avoid hitting the errant Ford approaching them. A number of cars almost collided with Gregorie's car. "Fortunately," says Gregorie in relief, "one of gatemen saw Mr. Ford in the car with me, and he passed the word very quickly what was going on. Danger lights began flashing, and all of the traffic came to a standstill. We could have been the victims of a horrible thing. It would have been the two of us if one of those test cars had hit us. It came that close."

As Gregorie drove back to the Engineering Laboratory, old Henry acted as if nothing had happened and tried making small talk with Gregorie, asking about current events and how the car was running. Gregorie was cautious about saying too much, for fear that he might mention something he would regret. "I wasn't quite sure what the old gentleman wanted, because he had never talked to me before," explained Gregorie. "I just told Mr. Ford that everything was going smoothly in the design department and that we'd appreciate him stopping by whenever he felt like it—we'd show him what was going on."

"In all the years that I worked at Ford," continued Gregorie, "that was the first and only meeting I had ever had with Henry Ford, and he was the one who instigated it. I never prompted it. I never made any attempt to get in contact with Henry Ford—to offer my condolences to him—after Edsel had died. I thought the more I laid low, the better off I'd be. So I let him do all the talking. But I had a feeling that he wanted to talk to me about Edsel. He knew that Edsel would come to see me after he had lunch with him and the other Ford executives, and I think he wanted to find out how much influence I had had over Edsel. He always thought that I was the one who put newfangled ideas into Edsel's head about design and new models and so on. I assumed that the old man figured that I might want to have a personal discussion about Edsel and waited for me to bring him up. But I didn't bring up Edsel. If he wanted to bring up Edsel, I preferred that it came from him first. But he didn't talk about Edsel, either. For some unknown reason, he clammed up at the last minute, which suited me just as well. I didn't want to get into a discussion pertaining to Edsel."

When Gregorie pulled the car into the garage, Henry got out of the car and walked away. "It was a peculiar instance," continued Gregorie. "It left me with a strange feeling. I didn't know what was going through his mind. (Neither did a lot of other people.) I didn't see the old man much after that."

"Oh, the old man was crazy!" added Gregorie. "He was still in a daze over Edsel's death. He had put Edsel through the ringer for years. Edsel would come into my office after lunch and just shake his head in frustration. He never told me any details of the discussions that took place with his father, but he didn't have to. I could tell that they weren't pleasant, just by the expression on his face and the tone in his voice."[50]

Perhaps the reason old Henry had gone to see Gregorie was to fire him. Since Edsel's death, old Henry was having visions of returning the company to its glory days of the Model T.

He had never gotten over losing his beloved Model T nor had he understood why Edsel was so obsessed with getting rid of it and proliferating the company with more and more products and redesigning them so often. Gregorie was a constant remainder to him of his late son and of Edsel's desire to constantly change the product. Old Henry was probably thinking about getting rid of Gregorie and the entire Design Department. As he was walking through the Engineering Building one day, old Henry ran into Joe Galamb, the body engineer who had helped him design the Model T in 1907. After making small talk, old Henry said, "Joe, you know we have to go back to the old days, the Model T days. We've got to build only one car. There won't be any Mercury, no Lincoln, and no other car, just the one Ford." [51]

***

Clara was not faring any better. With her husband in a stupor, she had one of her sisters move into Fair Lane to keep her company.[52] Once lively and witty, "the spark in Clara's eyes was gone."[53] She and Henry would talk about everything except their son.[54] The wound was too deep and too painful.

While Bennett was still at Ford, the two men would take long drives together, and Henry could not keep from talking about Edsel. "Now we're not going to talk about it anymore," he'd tell Bennett, but then he would return to the subject again and again.

"Harry," old Henry once asked Bennett, "do you honestly think I was ever cruel to Edsel?"

"Well, if that had been me you'd treated that way, it wouldn't have been cruelty," replied Bennett cautiously.

"Why don't you give me an honest answer?"

"Well, cruel, no; but unfair, yes. If that had been me, I'd have got mad."

"That's what I wanted him to do," Henry replied, exposing at last his *modus operandi*, "to get mad."[55]

A long-time Ford executive confirms this. "I think Henry Ford would have been tickled to death if Edsel would have jumped up and down and said, 'For Christ's sake, Dad, you've made me president of this company, and I'm going to run it! If you don't like it, get somebody else!" I think Henry Ford would have gone home and grinned to his wife and said, 'Well, I guess that boy is getting to become a man.' "[56]

Many criticized Sorensen, and rightfully so, for contributing to Edsel's complications. "I have always blamed Sorensen," recalled Joe Galamb, "because he backed Henry Ford up rather than doing what was right."[57] On many occasions over the years—the changeover from Model T to Model A is a good example—Sorensen could have intervened on Edsel's behalf and acted as a stalwart against Henry Ford.[58] However, Sorensen usually did not, for reasons that cannot be explained, other than fear for his own position within the company. That too is a weak argument, for every time Sorensen stood up to Henry Ford, the old man backed down. "My God, Charlie, someday I'm going to show you who's running this place!" Henry Ford would scold Sorensen each time the two got into a row.[59] However, Henry never followed through with his threats. He liked Sorensen's audacity and truculence and found those traits admirable in an executive.

In the end, Henry fell back on his belief on reincarnation. "Well, Harry, you know my belief—Edsel isn't dead."[60]

***

When Henry Ford II was released from his Navy obligations in August 1943 and began working at Ford on a full-time basis, he found that Gregorie had been fired. "Tom Hibbard came up in the vacuum that was left when I left," explains Gregorie. "He was down at the Rouge plant, and Joe Galamb brought him up to do my job. But I don't think he came over right away. I didn't designate anyone in particular when I left. It was kind of a dead period, anyway."[61]

For four months, Henry II had worked in the company with no official title and no particular duties. Throughout most of that time, he simply wandered through the Rouge plant and asked questions. As soon as he was appointed vice president, however, Henry II quickly began assuming responsibility. Although young and inexperienced, he learned the business quickly. Plus, he seemed perfectly suited for his new position. "Young Henry was ambitious, aggressive, fearless," recalled Jack Davis. "He was physically rugged, and he would win if it killed him."[62]

From the few experienced and loyal managers that had survived his grandfather's purges, Henry II established a policy committee to help him run the beleaguered company. This was only an interim step because Henry II wanted to weed out the old regime and recruit a group of young, college-educated men to become his executives. Until that happened, he ran Ford Motor Company through this new arrangement.

Henry II and the policy committee had to address many problems. Months had passed before they began to evaluate the Design Department. Henry II was well aware of the product plans that his father and Gregorie had devised a few years earlier, but Jack Davis was the only one of the original team who was still there to implement the small Ford/large Ford program. Davis could sell the program to the dealers and to the public, but he could do little to complete the design of the cars that Edsel and Gregorie had conceptualized. Clearly, Henry II needed Gregorie in the company.

"Henry Ford II called me, and he wanted me to come up to his office," recalled Gregorie. "He said, 'Now, Bob, we'll get things going again.' "[63] Gregorie was reluctant to go at first, but he could not refuse a personal invitation from Edsel's son. He went the next morning.

During their meeting, Gregorie explained to Henry II the obstruction he started encountering from engineering as soon as Edsel died. He described how it took him and Edsel years to insulate their work from the petty and unnecessary interference of engineering in the past, and he stated that he would not care to work under those old conditions again. Henry II was receptive because he knew that at General Motors, design worked independently from engineering. Gregorie said, "I'll come back on the condition that I won't have any interference from certain people that I know will create problems for me and the department."

"Like who?" asked the young Ford.

"Well, Joe Galamb, for one."

"I'll get rid of Joe right now!" exclaimed Henry II.

"We don't have to do it that way, Mr. Ford," appealed Gregorie. "Can't you give him something else to do, or tell him to stay the hell out of my area?"

"Nah," replied Ford, "I'll get rid of him."

"That's the way he was," explains Gregorie of the new boss. "He was impulsive as hell. He grabbed the phone, called somebody, and said, 'Get rid of Joe Galamb. Get rid of him. I don't care what you do. Get rid of him!' "[64]

It was a sign of changing times within the company. Joe Galamb had been old Henry's close associate since 1905 and had helped develop the Model N and the Model T. Now, he was out, fired by Henry Ford's grandson. Regardless, Ford had paid him well over the years. It was said that Galamb was the richest Hungarian in Detroit.

With a salary twice that of what he had been making prior to his termination, and with engineering removed from his way (or so he thought), Gregorie closed his design firm and returned to Ford in April 1944, regaining his position as chief designer. Hibbard was rather surprised when he was told that Gregorie was returning to take over the department again, but he did not put up any argument. When Gregorie arrived, the two men introduced themselves to each other, and Hibbard asked, "What would you like me to do?" Gregorie was well aware of Hibbard's background in the custom-body business, and instead of returning him to the Rouge or relegating him to a drafting table, he made Hibbard his assistant.[65]

Most of the men who were there when Gregorie left were still there when he returned. Therefore, the entire team, again headed by Gregorie, continued from where they had left off. Because the small Ford and large Ford sedans were already in the tooling stage, Gregorie and his men began a flurry of activity to complete the Mercury and Lincoln lines that they had started more than a year earlier. By June 1945, full-size clay models of Gregorie's new Ford, Mercury, and Lincoln lines were ready to be shown to company officials and plant managers throughout the country. All of these cars should have been complete working prototypes by this time, but the Engineering Department had dragged its feet in developing new suspension systems and chassis for all of these new models. When the first prototype of the large Ford was completed, it still had the old buggy suspension under it. "We were way ahead of engineering," said Gregorie. "We had the appearance of the new cars for three years before we got new chassis to go along with them. Nevertheless, we had to keep sufficiently flexible to accommodate any areas that were questionable, such as the floor pan, the tunnel for the drive shaft, and other heavy mechanical aspects of these cars. As it turned out, the engineering for these cars was not started until the summer of 1946."[66]

Henry II was glad to have Gregorie in the company again, but he had concerns other than design about which to worry. By the time World War II ended, Ford Motor Company was losing $10 million a month, and Henry II realized that he needed expert help to stop the bleeding and save the company. Many of the old Ford executives were nearing retirement, and Henry II was not confident that their subordinates had the wherewithal to undertake the monumental task that confronted the company. Therefore, he looked outside the company, particularly at General Motors, for new talent.

***

For years, Edsel Ford had tried to emulate the great General Motors, not only from a marketing standpoint—trying to copy its "car for every purse and purpose" strategy—but also its highly successful management techniques. Edsel must have emphasized this with his son over the years, because Henry II also thought highly of how General Motors was run. If he could recruit some General Motors executives to Ford, he thought, they could help him reorganize Ford Motor Company into the image of General Motors.

After months of negotiations, Henry II finally coaxed one of the brightest young executives from General Motors to help him run Ford. Ernest Breech, a self-taught accountant from Arkansas, had risen in only two years to the rank of general manager of the General Motors Bendix Division. Bendix had been losing $3 million a year when Breech took over the division. Two years later, it was making a profit of $5 million. By the end of the following year, 1945, Bendix posted a net profit of more than $55 million. It was commonly felt that Breech's next move would be to head Chevrolet and eventually become president of General Motors. From a managerial standpoint, Breech was a logical choice for Henry Ford II.

Partially through young Henry's persuasive manner and partially through the challenge that Ford Motor Company offered him, Breech quit General Motors and joined Ford in July 1946. He brought with him three other men who were experts in finance, engineering, and manufacturing. They, too, had all achieved high positions within General Motors and were willing to take on the challenge at Ford that Breech offered to them.

First, Lewis Crusoe was a talented finance man who had been assistant controller of General Motors. He was Breech's assistant at Bendix when he was asked to join Ford. Second, Harold T. Youngren had been an engineer at Oldsmobile and was instrumental in developing the Hydra-matic transmission. He was chief engineer of Borg-Warner when he got the call from Breech. Third, Delmar Harder was a manufacturing expert. He had served with Crusoe for years at Fisher Body, and when Breech asked for his help, he was president of the E.W. Bliss Company.

Each of these men immediately attained high positions in Ford when they joined Breech. Crusoe became the chief financial officer at Ford, reporting directly to Breech. Youngren was the new chief engineer at Ford, replacing Sheldrick, who had been fired around the time Gregorie had been let go. (Sheldrick went to work for the General Motors Detroit Diesel Division.). Harder became vice president of operations at Ford.

As for Breech, Henry Ford II made him executive vice president and gave him a seat on the board of directors. From that point, Ernest Breech called the shots at Ford Motor Company, and he began making swift and monumental changes. "A new broom sweeps clean," is the way Gregorie put it.

At first, Gregorie welcomed the changes. As far as he was concerned, before Breech and his men arrived, conditions within the company were not much better than they had been after Edsel's death. Although he no longer had to contend with petty obstructions from engineering, at least in the old days there had been some semblance of an organization. Gregorie said that in 1944, "The whole place was a nuthouse." It was not much better in 1946, before Breech joined the company. During meetings of the policy committee, of which Gregorie was a member, he could see that most of the people on the management team were not worth their salt. Gregorie had his hands full in the Design Department, so he could not do much to assist Henry II but did advise him to start seeking new talent to help him run the company. "I put a bug in young Ford's ear one Saturday morning," explained Gregorie. "I told him we were badly in need of a top management rearrangement, that we needed some new department heads, and we needed some new thinking in the company itself. And I think as a result of my prodding, he went out and got Breech. That's a fact! I told him things were in a bad state, and he couldn't be expected to come up with the ideas to re-establish the company himself. He was just a kid that came in there, he had to listen to everybody's story, and he was so dammed confused he didn't know whether he was coming or going. He needed a man just like Breech. I didn't like the bastard, but he got things done. He knew how to stimulate things. Even though they didn't go my way, I have to give him credit for getting the wheels going. He was a good businessman."[67]

***

One of Breech's first tasks was to review the Ford product lineup for the post-war market, and he was not impressed. He could not see the need for the small Ford, and he thought the large Ford was too big. Youngren had told Henry II and Breech that General Motors was making every effort to keep weight down, and that is what Ford should concentrate on as well. Youngren also suggested to the two executives that "We should not place all our hopes on Mr. Gregorie."[68] Breech suggested that the company hire an outside styling consultant to review Gregorie's designs.

George W. Walker was an industrial designer whose only experience in the automobile industry was designing hardware: door handles, window cranks, and the like. His greatest assets were his gift of gab and his friendship with Ernest Breech.

Breech drove to Walker's office in Detroit and told him of the situation. "We want you to come out to Ford and take a look at our products and give us your opinion on them," Breech said. When Walker came over, he and Breech concentrated on the large Ford. "Gees, that's terrible," said Walker. "You'll go broke doing that."

"Well, we're going to put that one out," said Henry II.

"Don't do that," replied Walker. "It's heavy thing. It's all out of shape. It looks like a pregnant Buick."

"Could you do any better?" Breech asked.

"I could do better with my eyes closed," replied the cocky designer. He wanted the account desperately, but he was already under contract with Nash-Kelvinator and International Harvester, which were each paying him $75,000 a year.

"Well, think it over," requested Breech.

A few nights later, Walker met with Breech at the Bloomfield Country Club.

"Well, have you made up your mind?" Breech asked. "It's a great opportunity. I can't tell you what's going to happen, but you can make yourself millions of dollars if you come with us."

Walker was ready for the offer. He had already spoken with both Fowler McCormick at International Harvester and George Mason at Nash-Kelvinator, and both agreed to release him from his contracts. When Breech asked again, Walker replied, "Well, fine, I'll do it."[69]

By the time Walker joined Ford, tooling had all but been completed on the small and large Fords. If Walker had had his way, he would have scrapped both cars and started all over. However, the company had already spent millions of dollars tooling these two models and was running out of time to develop a new line of cars for the post-war market. It needed to find a way to save some of Gregorie's work.

Then Breech had an idea. At the moment, he did not know what to do with the small Ford, but he suggested to Henry Ford and to Walker that they should make the large Ford the new Mercury and start afresh on a new Ford design.

"Yes, that's what you should do," concurred Walker.

However, if the small Ford were too little and the large Ford too big, what size should the new post-war Ford be? Youngren settled that question as old Henry Ford and his chassis people used to settle it for Gregorie in the old days—except that Youngren set many more parameters for the car, not only the length and width of the chassis, as Henry Ford had done. "Youngren and his boys bought two 1947 Studebakers and dissected them from every angle, and they arrived at what they felt was the proper hip room and headroom, length, width, and height," explained Gregorie. "So, what we were given to work with was that formula, or package, all based on that little, double-ended Studebaker."[70]

At Breech's directive, two design groups were set up—a sort of design competition—one headed by Gregorie and the other by Walker. Each group would develop a new Ford design based on Youngren's dimensions. The policy committee, headed by Breech, would determine which of the new designs would become the new post-war Ford.

Gregorie had the resources of his Design Department at his disposal, but he worked primarily with Tom Hibbard to develop a design. Walker, working from his office in the New Center Building across the street from the General Motors headquarters in Detroit, delegated the design efforts to Joe Oros, Elwood Engel, and Holden Koto. As usual, Gregorie did much of the design work himself, drawing on Hibbard for ideas and feedback. Walker, on the other hand, never lifted a pencil in developing his design.

By now, both design teams had completed quarter-size models of their design proposals, and Gregorie and Walker were instructed to begin work on full-size models. Because Walker did not have the facilities or manpower to construct a full-size model, Breech allowed him to set up an area in the north end of the Engineering Laboratory and to borrow some of Gregorie's clay modelers to construct his model. According to Walker, "There was supposed to be a guard there all morning, noon, and night to make sure nobody entered the area that we had set up. But Youngren allowed Gregorie and John Najjar to go in one evening and make sketches of what we had." His implication was that Gregorie was stealing his design.

Gregorie recalls this time much differently and is adamant that neither he nor any of his other designers were banned from the area. Nor were Walker's men forbidden from entering Gregorie's work area. In fact, there was no secrecy whatsoever between the two work groups. From the outset, Gregorie knew what Walker was working on, and Walker knew what Gregorie was working on. "We were a party to it," explains Gregorie. "My men set up the partitions. My men did the modeling. It was the damnedest thing the way it worked out."

By November 22, 1946, both teams had clay models ready, and on December 11, Henry II, Breech, Youngren, and other members of the policy committee examined the two models. After a brief discussion, the members chose Walker's design over Gregorie's design. Nobody was supposed to know who designed which car—the choice was to be totally objective. However, Gregorie knew better. "Breech ran the policy committee," he exclaims. "Christ, they were all beholden to him. They're not going to say, 'Gee, let's take Gregorie's car.' No, that would have been impossible. I realized that."[71]

So much similarity existed between the two cars that the policy committee did not have to be too concerned about choosing the wrong one. "I couldn't see a hell of a lot of difference," explains Gregorie. "They were both designed to a formula: the hood was to be so long, and the body was to be so wide, the tread was fixed on the wheel size. We came up with practically the same sheet metal design. You couldn't change it very much. The grille was different, and maybe the hubcap design was different, but the car—you put one alongside the other—I couldn't tell one from the other."

Gregorie was right. Both cars were of the slab-sided design, and both had a similar greenhouse. Gregorie's design was a little more rounded than Walker's design, the grille was different, and Gregorie's design had more definition in the side panels. "There was a little more shape in my car," explains Gregorie. "It tucks under a little, and it has a sheer line running the entire length of the body to give it a little form, whereas Walker's design has a raw edge—the side panels were completely smooth. The grille on my car, of course, was subject to change. We had several versions ready to try, but never got a chance to evaluate them on the model."[72]

What distinguished the two cars from one another were their paint jobs. Gregorie's was painted a conservative bluish-green—a color carried over from the Edsel days—whereas Walker's was painted bright yellow.

Shortly after the policy committee had approved Walker's design, Henry Ford II brought his grandfather to the Design Department to show him the twenty-eighth-generation Ford model. Old Henry was not impressed. Reminiscent of the garage scene in the 1920s, he walked around the model a few times. It was too low, he thought, and he could not see how a couple of milk cans could fit into that small trunk. When he tried to open one of the doors, the mock handle fell off in his hand. "That doesn't work," he said. Young Henry laughed and then escorted the founder back to his 1942 Ford sedan to be chauffeured home.[73] (This car, old Henry's last, had been retrofitted with mechanical brakes. Although the company had been making cars with hydraulic brakes since 1939, Henry still did not believe in them.[74])

Gregorie reflected on this tumultuous period with humor. "It's better than crying," he said. "I realized it was a hopeless situation. Breech wanted to put on a show for Henry Ford II and the board of directors. There was never any doubt that their version of the car would be selected. It was foolish to even try to compete with it. We didn't have enough leeway; we were both heading in the same direction. I stopped the thing halfway through; it was senseless to duplicate their efforts. I didn't make any effort to finish up the car we were working on; it was foolish. It was just a gesture on our part."[75]

* * *

When the design of the basic Ford was settled, Gregorie and his design staff completed the design of the large Ford (now the Mercury) and the large Lincolns. One of the first things they did was to revamp the front end of the Mercury. Initially, all of Gregorie's post-war models had conventional grilles with horizontal bars similar to the 1946 Fords. The sheet metal on these new models, however, had a tucked-in appearance in the grille opening from the outset. At the last minute, Gregorie eliminated the horizontal grille bars and replaced them with a long horizontal grille piece made with a multitude of fine, vertical, convex-shaped grille bars. This design mimicked the rolled sheet metal opening and "let the grille sort of float in there," describes the designer. Gregorie had made a similar grille on one of his Lincoln clay models, and it looked so good there that he decided to adapt it to the smaller car as well.

The post-war Lincoln designs went through a similar metamorphosis at the end. Gregorie started with massive horizontal grilles that appeared almost solid. However, he decided on a more conventional design that had two large horizontal bars. The most dramatic change occurred with the headlights. The original intent was to have pop-up headlights on the Lincolns—a design cue not new to the industry but one that would have been unique in the post-war market. Gregorie designed the front end to accommodate such a feature. "The Lincoln engineers were to work up the pop-up mechanism," explains Gregorie, "but they weren't successful at getting the covers to open right. So, at the last minute, in a desperate attempt to get the cars into production, they designed a stamped-metal insert to fit into the headlight holes as a substitute for the pop-up headlamps. That's why the Lincolns look a little strange in that area with the headlights recessed into the sheet metal."

In addition to their new forms, all of the new Ford models had new chassis. Gone at last was old Henry's favorite buggy suspension, which he had used on his cars for half a century. New coil-spring, independent front suspension on the front and semi-elliptic springs on the back finally replaced old Henry's favorite transverse-spring and wishbone suspension.

***

Shortly before Gregorie threw the Mercury and Lincoln designs over the wall to engineering, he decided to leave the company. "I couldn't work with Walker there," he says. Of course, Breech had sold Henry II on the idea that there should be a whole new setup, and my connection with him had dulled with Walker coming in. It was useless. The only thing for me to do was to make up my mind to get out of there. That's when I decided to go to Florida."[76]

To Gregorie's credit, he decided not to make an untenable situation worse by fighting it out with Walker. "I knew damn well that I wouldn't stand a chance if it came to a knock-down-drag-out contest between the two of us," he says. "So rather than be a part of it, I just decided to leave on friendly terms. I said to myself, 'If that's the way they want it, why, it's okay with me.' I looked forward to being relieved of it." Gregorie quit four days after his new Ford design was rejected.

***

Gregorie had analyzed the situation correctly. If he had any doubt in his mind as to whether he had made the right move, Breech's remarks at a press conference introducing the new 1949 Ford set him straight. "This car," declared Breech, "was styled by George Walker and his staff, and don't you forget it. I don't want anybody to forget about that."[77]

Regarding the large Ford that went into production as the Mercury, Gregorie is nostalgic. "The '49 Mercury was a very nice-looking car," he said. "It was one of the nicer designs we did. Edsel Ford had approved the clay model of this car. It was the last design that he and I did together."[78]

As for the small Ford/large Ford concept that Gregorie and Edsel had conceived in the early 1940s—the lineup that did not impress Breech—that same approach came to fruition more than a decade later, long after Gregorie had left Ford Motor Company. In 1960, Ford introduced the compact Falcon to thwart the increasing number of small imports selling in the country and to offer an inexpensive and fuel-efficient alternative against the big cars that the entire industry was building. If Breech had thought that Gregorie's and Edsel's small car appeared too small at the time, he must have been shocked at the sight of the Falcon because it was almost 33 inches shorter than the standard Ford sedan. However, the car sold well—in excess of 400,000 units the first year alone—and ultimately became one of the most successful models produced by Ford.

Although he had been gone from Ford Motor Company for years, Gregorie did not miss the significance or irony when he saw the little Falcon in dealer showrooms. "Edsel and I had that same plan 20 years earlier!" he exclaimed. "If it wasn't for that damn Breech coming in and scrapping our plans, why, we could have come out with a similar product lineup about 15 years earlier and caused a great stir in the industry. I think the success of the Falcon proves that our thinking of coming out with a small Ford after the war was both revolutionary and accurate. But as soon as Breech walked in the door of Ford Motor Company, he began poisoning Henry Ford II on all the things that Henry Ford, Edsel, and the rest of Ford management had implemented over the years. He wanted to erase any memory of them altogether. Surprisingly, and unfortunately, Number Two Ford went along with him. Coming from General Motors, Breech's one purpose in life was to 'GM-ize' Ford Motor Company, and he immediately gave the orders to hire ex-GM hacks and stuff the company with them. Eventually, he was able to throw a blanket over any and all policies that had been put in place by Henry and Edsel Ford. No matter if their plans, processes, or methods were logical and reasonable, Number Two Ford and his men scrapped them all, at Breech's recommendation, in order to implement their own, even if they were untried. I never understood why Henry Ford II went along with Breech to the extent that he did. I guess he just wanted to disassociate himself from the old days and do things his own way. I suppose he didn't want to go down in history as being accused of not having a backbone, like his father."[79]

***

Since Edsel's death, the people of Detroit and the Ford Motor Company have tried their best to recognize their forgotten son. After the war, General Knudsen recommended to the Detroit City Council that the proposed East-West freeway through the city be named the Edsel Ford Freeway, and the recommendation was approved. Signs along the road indicate that it is the Edsel Ford Freeway, but today it is better known as I-94. In 1956, with contributions from the Ford family, Ford dealers, and the City of Detroit, an auditorium for the performing arts was constructed beside the Detroit River in downtown Detroit and was named after Henry and Edsel Ford. About 10 years later, the Edsel Ford High School was built in Dearborn.[80]

As for Ford Motor Company, it acknowledged Edsel in the only way it knew how—it named a car after him. On August 28, 1957, among much fanfare and publicity, Ford Motor Company launched its first new model since Edsel introduced the Mercury almost two decades earlier. Designed by one of the new breed of designers recruited by the company after Gregorie had left, the new model was glitzy, gaudy, and overly dramatic—the antithesis of the designs Edsel and Gregorie loved. It was long, loaded with chrome and gadgets, and had a very unusual oval-shaped front-end treatment. Trying to be kind, when some critics first saw the new car, they referred to the unique front end as a horse collar or a toilet seat. More critical observers said it resembled "an Olds sucking on a lemon" or "the female sexual organ."[81]

The new model was designed at the zenith of albatrosses coming from Detroit during the 1950s, and the Ford Motor Company spent months trying to find an appropriate name for its new entry. After the marketing staff of the company failed to provide a suitable name, even after compiling a list of 6,000 names from which the executives could choose, it asked for assistance from the renowned American poet Marianne Moore. Her suggestions—Resilient Bullet, Arcencial, and Aeroterra—were, to say the least, creative but unacceptable for the name of an automobile, no matter how bizarre the vehicle appeared. "May I submit," she wrote the company in one last attempt, "Utopian Turtletop?"[82]

After months of wrangling, Ernest Breech settled on the one name that had not been on the exhaustive list but had been bantered around informally since the idea for a new line was first suggested, and he submitted it to the Ford family for final approval. The name he wanted was Edsel.[83] Breech felt that this name would be a fitting tribute to the man who had been president of Ford Motor Company for 25 years and who was instrumental in dragging the company into the modern era, particularly through styling.

Breech anticipated a warm reception from the Ford family, but instead he received only sighs of disbelief. William Clay was adamantly against it. He said that other post-war cars named after individuals, such as the Henry J, had a poor record of success, and he did not want a similar fate to befall a car named after his father.[84] Benson Ford's reaction was, "Over my dead body."[85] When a Ford representative went to see Eleanor Ford to inform her of the decision, she slammed the door in his face.[86] Henry Ford II did not think much of the choice either, wondering if his father would approve of having his name stamped on a million hubcaps.[87] However, to break the logjam that had developed over naming the new car, Henry II reluctantly agreed. It was a decision he would regret for the remainder of his life.

The Edsel entered the market with a fizzle. Sales during the first few months were well below the break-even point, and they declined from there. Aside from its grotesque styling, the Edsel was rushed into production in an economy that had started falling into recession two years earlier. Problems related to quality soon began to surface, which caused the few buyers in the market to avoid the new model. By the time the economy started to rebound in 1961, Edsel sales had been so dismal during the preceding three years that the company decided to drop the line entirely in November 1959.

In the end, Ford Motor Company lost $350 million in its Edsel car fiasco, and the name Edsel has been associated with disaster since then. The humiliation was not so much for the monetary losses that the Edsel sustained, but because it tarnished forever the name of Edsel Ford. "You're kind of sensitive," bemoaned William Clay, "when your father's name becomes a synonym for failure."[88]

Bob Gregorie had been long gone from Ford Motor Company when he heard the news that it planned to produce a new car and name it after his patron. "It was a travesty on Edsel's name!" Gregorie exclaimed. "The name Edsel is not a particularly eloquent name, and it certainly wasn't appropriate for the name of an automobile. Maybe the company officials wanted to recognize Edsel's contributions, but the public did not make the connection."[89]

* * *

It was the end of the middle era of Ford Motor Company, the time between when Henry Ford founded the company 40 years earlier and when Henry Ford II moved the company into the modern post-war era. The company could not have survived without the genius of Henry Ford in those early years, nor could it have survived without the dedication, tenacity, and perseverance of Henry Ford II after the war. However, there is little doubt that Ford Motor Company would not have survived without the energy, vision, and devotion of Edsel Ford. There were so many times over the years that Edsel had placed himself in jeopardy by fighting his father for what he knew was best for the company—changing from the Model T to the Model A, taking hold of the styling of Ford products, and establishing the Ford Foundation—to name only a few. There were many more instances, almost on a daily basis. Failure in any one of these attempts would have spelled the demise of one of the greatest industrial enterprises the world has ever known.

To his credit, Edsel did not fail. In each instance, and against great odds and with great physical sacrifice, Edsel pushed through his ideas, ensuring the continuation of the company. "The various plans that Edsel Ford wanted to have done were very necessary to the survival of the company," recalled a high executive. Said another, "We have to give him credit for having made a brave effort."[90]

Why did Edsel remain for all those years and endure his father's abuse for so long? First and foremost, he was a loyal son, and he loved his father. "I just think that he adored his father and what he stood for," recalled Al Wibel, "and nothing that he was going to do was going to react and disturb Henry Ford's way of thinking or have him have a suspicion that his son was contradictory to the way he wanted things to go."[91] Edsel did not like the way in which his father treated him, but he overlooked that treatment to serve his father and the company. Second, Edsel was a Ford—a fighter, a builder of motorcars, and heir apparent to the greatest family-owned industrial enterprise in the world. Suppose Edsel had quit the company, as he had threatened on a number of occasions. He could have quit, of course. He had plenty of money. He did not have to work another day in his life. However, Edsel could not fathom the inevitable headlines: EDSEL FORD QUITS HIS FATHER'S COMPANY. He could not damage the company in that way.[92]

No, Edsel held tough, and so did his son, Henry II. Similar to Edsel, Henry Ford II did not have to become involved with the company after his father's death. He could have hired competent managers to run the company for the family. However, even at a young age, Henry II wanted to get revenge on his grandfather. He wanted to prove to himself and to the world at large that a Ford—other than old Henry himself—could effectively manage the mammoth Ford industries. To prove that point, Henry II ran the company following his father's democratic system, rather than by his grandfather's autocratic methods. As young Henry became immersed in the heat of managing the beleaguered company and was asked to make a critical decision on some important matter or the other, he would ask himself, "Now, what would Father have done?"[93]

Acknowledgments

In any endeavor such as writing a biography, there are many people who gladly give their time, assistance, and encouragement. No one did this more for me than the late Dave Crippen. As chief archivist at the Ford Archives for many years, Dave voiced his wholehearted support when I mentioned that I wanted to write Edsel Ford's biography. Unfortunately, it took me too long to write it, and Dave never saw the final work. That is something I will always regret. If it were not for Dave's early encouragement, this book may never have been started.

Also during the early years of researching this book, Liz Jordan was the administrative assistant in the Ford Archives. She is a wonderful lady and a hard worker. At that time, she was moonlighting at the Archives, transcribing the interviews that Dave Crippen was conducting with many old-time automobile designers into Design Oral Histories. Normally, the Ford Archives is closed during the weekends, but Dave arranged with Liz to allow me to do much of my research on Saturday and Sunday afternoons, working while she worked. There were many Sunday afternoons when we were simply glad to leave the Archives, only to return the next weekend.

To Dave and Liz, I extend my utmost thanks.

Dave and Liz were early supporters of my research, but all the other members of the Ford Archives staff were also supportive. My thanks go to Judith Endleman, Luke Gillium-Swetmen, Catherine Latendresse, Steven Hamp, Charles Hanson, Jeanine Head, Denise Moline, Cynthia Read-Miller, Linda Skolarus, Alene Soloway, Melissa Haddock, and John Tobin.

The staff at the Edsel & Eleanor Ford House was also helpful. I believe I was the first outsider to conduct research there, and I came away with a bounty of information. I especially want to thank Lynn Ford Alandt, James Bridenstine, Maureen Devine, Josephine Shea, and Donna Oliver.

I want to thank all of the people who shared their memories of Edsel Ford with me. These people are listed in the bibliography. However, I especially want to thank Bob Gregorie, Edsel Ford's chief designer during the 1930s and 1940s, for all of the time he spent with me. Bob told me some wonderful stories, most of which are included in this book, and provided a rare glimpse into the artistic mastery of Edsel. Bob is a wonderful, witty, articulate man and, in my opinion, the best automobile designer who ever lived. I must also thank Bob's wonderful wife,

Evelyn. She was a most gracious hostess and provided Bob and me with plenty of food and drink while we talked about the old days. I will always cherish the time I spent with the Gregories.

However, I would not have had a chance to see Bob Gregorie without the help of Mike Lamm. Mike has known Gregorie for 30 years, and when I asked Mike to introduce me to Bob, he was more than willing. Within five minutes of Mike's phone call, I was talking to the great designer.

Then there is Dick Folsom. Dick is the consummate Fordophile, and, over the years, he was fortunate to meet with and interview many people who knew the Henry and Edsel Fords personally. Most of those people are gone now, but through Dick's connections, I was able to talk to them as well, specifically about Edsel.

Mike Skinner is also a Fordophile. For example, he has been conducting tours of the Henry Ford estate for the past 20 years. Mike has a considerable collection of Ford material, and he allowed me to comb through it for information about Edsel. Mike also did research for me. No matter what odd question I asked, Mike always came through with an answer.

I would also like to thank my publisher's staff, especially Ed Manns, Martha Swiss, Linda Snyder, Terry Wilson, and Cindy Brack. They were a pleasure to work with, and I appreciate their hard work and dedication to this project.

I'm sure I missed others who helped me, and to them I apologize for my sloppiness. I should have kept better track of those with whom I spoke.

There is one person whom I cannot forget to thank, and that is my wife, Pat DeHerrera. She put up with me while I was ensconced in my office for untold nights and weekends writing, and while I was traipsing off to Michigan, Florida, or wherever my research took me. I am sure that if she never hears the word Edsel again, it will probably be too soon.

Selected Bibliography

Books

Adams, G. Eugene. *40 Years in Product Design and Engineering at Ford Motor Company*. Adams Publishing Company, Green Valley, AZ, 1992.

Appel, Joseph H. *The Business Biography of John Wanamaker, Founder and Builder*. The Macmillan Company, New York, 1930.

Armi, C. Edson. *The Art of American Automobile Design*. The Pennsylvania State University Press, University Park, PA, 1988.

Arnold, Marvin. *Lincoln and Continental, Classic Motorcars: The Early Years*. Taylor Publishing Company, Dallas, TX, 1989.

Automobiles of America. The Motor Vehicle Manufacturers Association. Wayne State University Press, Detroit, 1974.

Barnard, Harry. *Independent Man: The Life of Senator James Couzens*. Charles Scribner's Sons, New York, 1958.

Bennett, Harry. *We Never Called Him Henry*. Gold Metal Books, Fawcett Publications, New York, 1951.

Bingay, Malcolm W. *Detroit Is My Own Hometown*. The Bobbs-Merrill Company, New York, 1946.

Bonsall, Thomas E. *The Lincoln Motorcar: The Complete History of an American Classic*. Stony Run Press, Baltimore, MD, 1992.

Borgeson, Griffith and Eugene Jaderquist. *Sports and Classic Cars*. Prentice-Hall, New York, 1955.

Borth, Christy. *Masters of Mass Production*. The Bobbs-Merrill Company, New York, 1945.

Brauer, Norman. *There to Breathe the Beauty*. Norman Brauer Publications, Dalton, PA, 1995.

Bridenstine, James A. *Edsel and Eleanor Ford House*. Wolk Press, Woodlawn, MD, 1988.

Brough, James A. *The Ford Dynasty: An American Story*. Doubleday and Company, New York, 1977.

Bryan, Ford. *Beyond the Model T*. Wayne State University Press, Detroit, 1989.

Bryan, Ford. *Clara: Mrs. Henry Ford*. Ford Books, Dearborn, 2001.

Bryan, Ford. *The Fords of Dearborn*. Harlo Press, Detroit, 1989.

Bryan, Ford. *Henry's Attic: Some Fascinating Gifts to Henry Ford and His Museum*. Ford Books, Dearborn, 1995.

Bryan, Ford. *Henry's Lieutenants*. Wayne State University Press, Detroit, 1993.

Burlingame, Roger. *Henry Ford*. Alfred A. Knopf, New York, 1954.

Bush, Donald J. *The Streamlined Decade*. George Braziller, New York, 1975.

Bushnell, Sarah T. *The Truth About Henry Ford*. The Reilly and Lee Company, Chicago, 1922.

Byrd, Richard E. *Alone*. G.P. Putnam's Sons, New York, 1938.

Byrd, Richard E. *Little America*. G.P. Putnam's Sons, New York, 1930.

Byrd, Richard E. *Skyward*. G.P. Putnam's Sons, New York, 1928.

Byrne, John A. *The Whiz Kids*. Currency-Doubleday, New York, 1993.

Chambers, M.M. *Charters of Philanthropies.* The Merrymount Press, New York, 1948.
Clancy, Louise and Florence Davies. *The Believer: The Life Story of Mrs. Henry Ford.* Van Rees Press, New York, 1960.
Collier, Peter and David Horowitz. *The Fords: An American Epic.* Summit Books, New York, 1987.
Conot, Robert. *American Odyssey.* William Morrow Company, New York, 1974.
Crabb, Richard. *Birth of a Giant: The Men and Incidences That Gave America the Motorcar.* Chilton Book Company, New York, 1969.
Dahlinger, John Cote. *The Secret Life of Henry Ford.* Bobbs-Merrill Company, Indianapolis, 1978.
Dammon, George H. *Illustrated History of Ford.* Crestline Publishing Company, Sarasota, FL, 1970.
Detroit Style: Automotive Forms 1925–1950. The Detroit Institute of Arts, Detroit, 1985.
Eight Automobiles. The Museum of Modern Art, New York, 1951.
Ferdinand, Prince Louis. *The Rebel Prince: Memoirs of Prince Louis Ferdinand.* Henry Regnery Company, 1952.
Fine, Sidney. *The Automobile Under the Blue Eagle.* University of Michigan Press, Ann Arbor, 1963.
Firestone, Harvey S. *Men and Rubber.* Doubleday, Page and Company, New York, 1926.
The Ford Book of Styling: A History and Interpretation of Automobile Design. Public Relations, Styling Office, Ford Motor Company, Dearborn, 1963.
Ford, Henry. *Edison as I Know Him.* Cosmopolitan Book Corporation, New York, 1930.
Ford, Henry. *My Life and Work.* Doubleday, Page and Company, Garden City, NJ, 1926.
Ford, Henry. *Today and Tomorrow.* Doubleday, Page and Company, Garden City, NJ, 1926.
Forden, Lesley. *The Ford Air Tours: 1925–1931.* The Nottingham Press, Alameda, CA, 1973.
Freedom's Arsenal: The Story of the Automobile Council for War Production. Automobile Manufacturers Association, Detroit, 1950.
Garrett, Garet. *The Wild Wheel.* Pantheon Books, New York, 1952.
Gelderman, Carol. *Henry Ford: The Wayward Capitalist.* The Dial Press, New York, 1981.
General Motors: The First Seventy-Five Years of Transportation Products. Automobile Quarterly, Princeton, NJ, and General Motors Corporation, Detroit, 1983.
Greenleaf, William. *From These Beginnings: The Early Philanthropies of Henry and Edsel Ford, 1911–1936.* Wayne State University Press, Detroit, 1964.
Greenleaf, William. *Monopoly on Wheels: Henry Ford and the Selden Automobile Patent.* Wayne State University Press, Detroit, 1961.
Halberstam, David. *The Reckoning.* William Morrow & Company, New York, 1986.
Hays, Walter. *Henry: A Life of Henry Ford II.* Grove Weidenfeld, New York, 1990.
Henry, Leslie R. *Henry's Fabulous Model A Ford.* Floyd Clymer Publications, 1959.
Herndon, Booton. *Ford: An Unconventional Biography of the Men and Their Times.* Weybright and Talley, New York, 1969.
Hickerson, J. Mel. *Ernie Breech: The Story of His Remarkable Career at General Motors, Ford, and TWA.* Meredith Press, New York, 1968.
Holden, Henry M. *The Fabulous Ford Tri-Motors: The Saga of the Tin Goose.* TAB Books, Blue Ridge Summit, PA, 1992.

Hoyt, Edwin P. *The Last Explorer: The Adventures of Admiral Byrd*. The John Day Company, New York, 1968.

Jardim, Anne. *The First Henry Ford: A Study in Personality and Business Leadership*. The Colonial Press, 1970.

Jones, Jesse H. with Edward Angly. *Fifty Billion Dollars: My Thirteen Years with the RFC*. The Macmillan Company, New York, 1951.

Lacey, Robert. *Ford: The Men and the Machine*. Little, Brown and Company, Boston, 1986.

Lamm, Michael and Dave Holls. *A Century of Automotive Styling: 100 Years of American Car Design*. Lamm-Morada Publishing Company, Stockton, CA, 1996.

Lasky, Victor. *Never Complain, Never Explain: The Story of Henry Ford II*. Richard Marek Publishers, New York, 1981.

Leonard, Jonathan Norton. *The Tragedy of Henry Ford*. G.P. Putnam's Sons, New York, 1932.

Lewis, David L. *The Public Image of Henry Ford*. Wayne State University Press, Detroit, 1976.

Lewis, Eugene W. *Motor Memories*. Alved Publishers, Detroit, 1947.

The Lincoln Highway. By the Lincoln Highway Association. Dodd, Mead and Company, New York, 1935.

Lindbergh, Charles A. *The Wartime Journals of Charles A. Lindbergh*. Harcourt, Brace, Jovanovich, New York, 1970.

Lochner, Louis P. *Henry Ford: America's Don Quixote*. International Publishers, New York, 1925.

Long, J.C. *Roy D. Chapin*. Private printing, 1945.

MacDonald, Dwight. *The Ford Foundation: The Men and the Millions*. Reynal and Co., New York, 1956.

Marquis, Samuel S. *Henry Ford: An Interpretation*. Little, Brown and Company, Boston, 1923.

Nevins, Allan. *Ford: The Times, The Man, The Company*. Charles Scribner's Sons, New York, 1954.

Nevins, Allan and Frank Ernest Hill. *Ford: Expansion and Challenge: 1915–1933*. Charles Scribner's Sons, New York, 1957.

Nevins, Allan and Frank Ernest Hill. *Ford: Decline and Rebirth 1933–1962*. Charles Scribner's Sons, New York, 1962.

Neyhart, Louise Albright. *Henry Ford, Engineer*. Houghton Mifflin Company, Boston, 1950.

O'Callaghan, Timothy J. *Henry Ford's Airport and Other Aviation Interests: 1909–1954*. Proctor Publications, Ann Arbor, 1993.

Olson, Sidney. *Young Henry Ford: A Pictorial History of the First Forty Years*. Wayne State University Press, Detroit, 1963.

Peck, William H. *The Detroit Institute of Arts*. The Founders Society, The Detroit Institute of Arts, Detroit, 1991.

Pfau, Hugo. *The Custom Body Era*. Castle Books, New York, 1970.

Plumb, Mildred. *History of Tau Beta*. Evans-Winter-Hebb, Detroit, 1938.

Porter, Darwin and Danforth Prince. *England from $60 a Day*. Macmillan Travel, New York, 1998.

Quaife, M.M. *The Henry Ford Family*. Burton Historical Collection, Detroit Public Library, Detroit, MI.

Ray, John B. *Henry Ford: Great Lives Observed*. Prentice-Hall, New Jersey, 1969.

Richards, William C. *The Last Billionaire*. Charles Scribner's Sons, New York, 1948.
Ritch, Ocee. *The Lincoln Continental*. Motorbooks International Publishers & Wholesalers, Osceola, WI, 1974.
Rivera, Diego. *My Art, My Life*. Citadel Press, New York, 1960.
Seltzer, Lawrence H., Ph.D. *A Financial History of the American Automobile Industry*. Houghton Mifflin Company, New York, and The Riverside Press, Cambridge, 1928.
Simonds, William A. *Henry Ford: His Life, His Work, His Genius*. The Bobbs-Merrill Company, New York, 1943.
Sinsabaugh, Chris. *Who, Me?* Arnold-Powers, Detroit, 1940.
Sloan, Alfred P., Jr. *My Years with General Motors*. Doubleday and Company, New York, 1964.
Sorensen, Charles E. *My Forty Years with Ford*. W.W. Norton & Company, New York, 1956.
Sterne, Margaret. *The Passionate Eye: The Life of William R. Valentiner*. Wayne State University Press, Detroit, 1980.
Stoddard, Gloria May. *Henry Leland: The Story of the Vermonter Who Created Cadillac and Lincoln*. The New England Press, Shelburne, VT, 1986.
Stout, Richard H. *Make 'em Shout Hooray!* Vantage Press, New York, 1988.
Stout, William Bushnell. *So Away I Went!* The Bobbs-Merrill Company, New York, 1951.
Sward, Keith. *The Legend of Henry Ford*. Rinehart & Company, New York, 1948.
Thomas, Bob. *Confessions of an Automobile Stylist*. Self-published, 1984.
Thomas, Gordon and Max Morgan-Witts. *The Day the Bubble Burst*. Doubleday and Company, New York, 1979.
Weiss, David Ansel. *The Saga of the Tin Goose*. Crown Publishers, New York, 1971.
Wik, Reynold M. *Henry Ford and Grass-Roots America*. University of Michigan Press, Ann Arbor, 1972.
Wilkins, Mira and Frank Ernest Hill. *American Business Abroad*. Wayne State University Press, Detroit, 1964.
Wolfe, Bertram D. *Diego Rivera: His Life and Times*. Alfred A. Knopf, New York, 1939.
Wolfe, Bertram D. *The Fabulous Life of Diego Rivera*. Alfred A. Knopf, New York, 1951.

Periodicals

Allen, George H. "Edsel Ford Selects a Cotswold House." *House & Garden*, February 1932.
American Magazine of Art, May 1933.
"Bill Stout." *Fortune*, January 1941.
"Body by Briggs." *Special-Interest Autos*, November–December, 1973.
Burck, Gilbert. "Henry Ford II." *Life*, October 1, 1945.
Byrd, Richard E. "Fly On! Fly On! Edsel Ford Talks About Aviation to Commander Richard E. Byrd." *Collier's*, May 12, 1928.
Cameron, William J. "In Memory." *The Herald*, Vol. X, No. 9, June 4, 1943.
Connelly, Mary. " 'The Duce' Reflects on Career, Colleagues." *Automotive News*, September 26, 1994.
Couzens, James. "What I Learned About Business from Henry Ford." *System*, September 1921.
Crowther, Samuel. "Edsel B. Ford." *Saturday Evening Post*, December 25, 1926, pp. 23, 77–79.

Crowther, Samuel. "Henry Ford Talks About Edsel and Edsel Talks About His Father." *World's Work*, August 1928.
"Detroit Da Vinci." *Saturday Evening Post*, December 7, 1940.
Dietrich, Raymond H. "Custom Designs Off the Assembly Line." *Wards Quarterly*, Summer 1965.
Downs, Linda. "Diego Rivera's Portrait of Edsel Ford." *Bulletin of The Detroit Institute of Arts*, Vol. 57, No. 1, 1979.
"Dynamic Detroit." *Creative Art*, April 1933.
"Edsel Ford: 1893–1943." *Newsweek*, June 7, 1943.
"Edsel Ford's Biography." News Department Release of the Ford Motor Company, August 27, 1957.
"Edsel Ford's Hot Rods." *Special-Interest Autos*, December 1970.
"Edsel, Future Master of Great Enterprise, Already Is Called the Boss by His Father." Ford Archives, clipbook, December 31, 1939.
"Edsel's Transcontinental Trip." *Ford Times*, October 1915.
"Facts About Edsel Ford." *Pipps Weekly*, May 15, 1920.
"Fantastic Ford Finds." *Special-Interest Autos*, December 1970.
"Father of the Automobile Dies." *Life*, April 21, 1947.
Flynn, John T. "Michigan Magic: The Detroit Bank Scandal." *Harper's*, December 1933.
"Ford Becomes a Matriarchy." *Business Week*, April 26, 1947.
Ford, Edsel. "A New Era Dawns in Industry." *Nation's Business*, April 1929.
Ford, Edsel. "Why We Have Faith in Aircraft." *Nation's Business*, May 1, 1925.
Ford, Edsel B. "The Motor Car of the Future." *Commercial Art and Industry*, November 1935.
Ford, Henry. "I've Been Helped by Everybody." *True Story*, April 1935.
"The Ford Heritage." *Fortune*, June 1944.
Ford, Sheila F. "Edsel Ford: Artistic Industrialist." *The Herald*, Vol. 8, No. 2, Spring 1979.
Ford Times, Henry Ford Centennial Issue, July 1963.
"The Ford Way of Doing Business." *New Republic*, May 20 and May 27, 1940.
"Ford's Grandson Marries Anne McDonnell in Major 1940 Wedding." *Life*, July 29, 1940.
Gillette, H.P. "Ford Business Philosophy." *Engineering & Contracting*, March 1928, pp. 137–140.
Gillmore, Dan. "Ford's Fascism: Proof Part 2." *Friday*, Vol. 2, No. 5, January 31, 1941, pp. 3–7.
Graham, J. Erle. "Edsel Ford's Classic Sports Discovered." *Car Classics*, December 1975.
Grigg, Ramona. "Diego Rivera: Former Assistants Remember the Man, the Genius." *The Eccentric*, Thursday, March 20, 1986.
"Henry and Edsel." *Printer's Ink*, November 10, 1938.
"Henry Ford and the Believer." *Ladies' Home Journal*, September 1923.
"Henry Ford's Commander-in-Chief." *Forum*, February 1938.
"Henry Ford's Death Will Leave His Empire Intact." *Business Week*, April 12, 1947.
"History's Biggest Stock Issue." *Life*, January 30, 1956.
Jenkins, Glenn F. "A Man for Mr. Ford's Shoes." *Christian Science Monitor Weekly Magazine*, May 6, 1935.
"The Judgement of Detroit." *Art News*, Vol. 31, No. 8, May 1933.
Kanzler, Ernest. "He Was Interested in Everything." *Wards Quarterly*, Summer 1965.

Kellogg, Paul U. "Henry Ford's Hired Men." *Survey Graphic*, All of 1928, and January and February 1929.
King, Lawrence G. "A New Era Dawns in Industry." *Nation's Business*, April 1929.
Laas, William. "The Meaning of a Name." *Your Edsel Marketer*, Vol. 1, No. 2, publication of the Ford Motor Company, no date.
Lamm, Michael and David L. Lewis. "The First Mercury and How It Came to Be." *Special-Interest Autos*, July–August 1974.
Lewis, David L. "Edsel Ford, The Gentle Businessman." *Michigan History Magazine*, May–June 1994.
Lewis, David L. "Lincoln Cosmopolitan: The Gleam in Edsel Ford's Eye." *Car Classics*, April 1973.
MacDonald, Dwight. "Foundation." *New Yorker*, a series of articles: November 26, 1955; December 3, 1955; December 10, 1955; December 17, 1955; and February 11, 1956.
McCarten, John. "The Little Man in Mr. Ford's Basement." *American Mercury*, May and June 1940.
McCarthy, Joe. "The Ford Family." *Holiday*, June, July, August, and September 1957.
McCarthy, Joe. "From Henry to Edsel to Henry." *Reader's Digest*, April 1958.
Montgomery, James. "Henry Ford & Son, Partners." *Personality Magazine*, December 1927.
"Mr. Ford Doesn't Care." *Fortune*, December 1933.
Museum News. Publication of The Detroit Institute of Arts, June 15, 1943.
Neville, Howard Ralph. "The Detroit Banking Collapse of 1933." Occasional Paper No. 2, Bureau of Business and Economic Research, Michigan State University, East Lansing, MI, 1960.
"Personality Profile: John Najjar: Life with 'Mother Ford.' " *Collectible Automobile*, April 1995.
Pierrot, George F. "The Gentle Gentleman of the Industry." *Ward's Quarterly*, Summer 1965.
Pipp, E.G. "Ford and That Leland Deal." *Pipps Weekly*, February 18, 1922.
Pipp, E.G. "The Ford-Lincoln Deal." *Pipps Weekly*, February 10, 1923.
Pipp, E.G. "Why." *Pipps Weekly*, July 21, 1923.
Pound, Aurther. "The Ford Myth." *Atlantic Monthly*, January 1924.
"River Rouge." *Life*, August 19, 1940.
Rivera, Diego. "Dynamic Detroit." *Creative Art*, April 1933.
Sayers, Michael. "Conspiracy of Silence: The Case Against Henry Ford." *Friday*, Vol. 1, No. 35, pp. 13–17, November 8, 1940.
Sorensen, Lorin. "Fishleigh Fords." *Special-Interest Autos*, March–April 1977.
Sprague, Jesse R. "Confessions of a Ford Dealer." *Harper's*, June 1927.
Taylor, Marion. "Reliability Proved by the Air Tour." *The Wind Cone*, October 1929.
Teague, Walter Dorwin. "Edsel Ford—Designer." *Lincoln-Mercury Times*, May–June 1953.
Teague, Walter Dorwin. "The Motor Car of the Future—Designer." *Commercial Art and Industry*, November 1935.
Thomas, Cullen. "Lincoln Continental." *Automobile Quarterly*, Vol. 6, No. 4, 1968.
Tjaarda, John. "How the Lincoln-Zephyr Was Born." *Motor Trend*, February 1954.
Tjaarda, Tom. "I Remember My Father." *Special-Interest Autos*, April–May 1972.
"Touring to the Expedition." *Ford Times*, October 1915.

Washburn, Robert Collyer. "Peach Pits, Plaster Cupids, Gold Trim: Many Roads Lead to Motor Car Designing." *Ford Times*, April 1946.
Welliver, Judson. "Henry Ford, Dreamer and Worker." *American Review of Reviews*, November 1921.

Newspapers

Bingay, Malcolm W. "Good Morning." *Detroit Free Press*, May 28, 1943.
"Calls Edsel Ford Presidential Pet." *Grand Rapids Press*, July 16, 1919.
"Concerning the Exemption of Edsel Ford." *Detroit Saturday Night, Detroit Free Press*, November 2, 1918.
"Draft Board Denies Edsel Ford's Claim." *The Detroit News*, October 5, 1917.
"Edsel Ford Buys Out Father's Associates." *The Detroit News*, July 11, 1919.
"Edsel Ford for Years Real Head of the Great Industry." *The Detroit News*, January 6, 1927.
"Edsel Offers 'Ford Plan' as Pattern for Recovery." *The Detroit News*, September 4, 1935.
"Edsel Seeks to Offset His Father's Speeches." *Detroit Saturday Night, Detroit Free Press*, March 29, 1919.
"Elder Ford Trained Son to Take Over." *Detroit Free Press*, May 27, 1943.
Ford, Edsel B. "Europe Loses Auto Position." *Detroit Free Press*, January 20, 1924.
Jenkins, Glenn. "The Rise of Edsel Ford." *Detroit Saturday Night, Detroit Free Press*, July 11, 1936.
Kaempffert, Waldemar. "The Mussolini of Highland Park." *The New York Times*, January 8, 1928.
McCutcheon, John T. "A Sad Ford Story: How Father Got Exempted from the Senatorial Race." *Huntington Herald Dispatch*, Huntington, WV, August 29, 1918.
"Meet the Real Harry Bennett." *Detroit Free Press*, February 8, 1942.
Nathan, Jacob. "Edsel Ford Seeks to Offset the Effects of His Father's Interviews." *Detroit Saturday Night, Detroit Free Press*, March 29, 1919.
"New Plan May Exempt Ford." *The Detroit News*, October 24, 1917.
Pooler, James D. "Edsel Ford Rites to Be Held Friday." *Detroit Free Press*, May 27, 1943.
Smith, Patricia Barbara. "Edsel Ford." *Detroit Magazine, The Detroit Free Press*, October 17 and October 24, 1965.
"What About Edsel Ford?" *The Detroit News*, May 10, 1924.
Williams, Elmer. "Plot to Kidnap Edsel Fizzles." *The Detroit News*, December 15, 1960.

Government Documents

Macomb County Probate Court, Microfilm #23836, I–IX, Mount Clemens, MI.
U.S. Congressional Committee on Banking & Currency, Stock Exchange Practices, Hearing 73rd Congress, 2nd Session, Parts 9–13.

Interviews

Bannon, Kenneth.
Barbera, Placid.
Bonbright, William.
Booth, Wayne.
Brush, Dr. Brock.
Chapin, Roy Jr.
Collins, Eleanor.
Cousins, Ross.
Dimitroff, Lucienne Bloch.
Dimitroff, Stephen Pope.
Fauster, Reverend Arthur.
Gregorie, E.T.
Hadda, Joe.
Hay, John.
ImOberstag, Francis.
Jackson, Leigh.
Lackenaw, Floyd.
Lam, Dr. Conrad.
Longley, James.
Madawick, Tucker.
Mallast, Roberta Rankin.
Merchant, Mary-Alice.
Miller, Rhys.
Najjar, John.
O'Rear, Emmett.
Prunk, Grace.
Reuther, Victor.
Roberts, Ralph.
Romney, Governor George.
Schermerhorn, Jane.
Smith, Robert.
Thomas, Bob.
Willemse, Jan.

Notes

Listed in this section are all of the references used in this book. The books and magazine articles are self-explanatory, as are the interviews that were conducted. The materials culled from the Ford Archives of the Henry Ford Museum and Greenfield Village in Dearborn, Michigan, however, warrant some explanation.

The Ford Archives, arguably the most significant collection of materials compiled on a corporation, was established in 1950 after the death of Mrs. Henry Ford. When the curators of the archives began sorting the vast amount of materials left by Henry, Clara, and Edsel Ford, they divided it into what they called Accessions. An Accession can consist of as little as one box of material to numerous boxes. Accession 6, for example, consists of 464 boxes of letters, photographs, and other materials of Edsel Ford, mostly in his capacity as president of Ford Motor Company. Many of the references used in the compilation of this biography came from these Accessions.

Another significant collection of the Ford Archives is the Oral History Collection, called Reminiscences. While the Ford Motor Company was preparing for its fiftieth anniversary in 1953, it began interviewing Ford family relatives and long-time Ford employees to record their experiences for posterity. The Reminiscences are a wealth of material to the Ford researcher.

Other collections consulted were: The Bentley Historical Library of the University of Michigan, Ann Arbor, Michigan; The Burton Historical Collection and the National Automotive History Collection of the Detroit Public Library, Detroit, Michigan; The Detroit Institute of Arts Archives, Detroit, Michigan; The Henry Ford Hospital Archives of the Henry Ford Hospital, Detroit, Michigan; The Edsel & Eleanor Ford House Archives reposited at Edsel and Eleanor's mansion in Grosse Pointe, Michigan; and The Library of Congress, Washington, DC.

Chapter 1: A Man Called Edsel

1. E.T. Gregorie, various interviews between 1992 and 1998.
2. *Reminiscences*, Fred Black, p. 94.
3. Crowther, Samuel. "Edsel B. Ford." *Saturday Evening Post*, December 25, 1926, p. 78.
4. *Reminiscences*, Fred Black, p. 94.
5. Accession 6, Box 270.
6. Scrapbook, The Edsel & Eleanor Ford House Archives.
7. Richards, William C. *The Last Billionaire*. Charles Scribner's Sons, New York, 1948, p. 200.
8. Collier, Peter and David Horowitz. *The Fords: An American Epic*. Summit Books, New York, 1989, pp. 136–137.
9. Clancy, Louise and Florence Davies. *The Believer: The Life of Mrs. Henry Ford.* Van Rees Press, New York, 1960, p. 197.
10. *Reminiscences*, Fred Black, p. 94.
11. *Reminiscences*, Mr. and Mrs. Louis C. Scott, p. 31; also Accession 65.

12. Dr. Conrad Lam, interview, September 21, 1989.
13. Lacey, Robert. *Ford: The Men and the Machine*. Little, George Brown and Company, Boston, 1986, p. 38.
14. Lacey, Robert. *Ford: The Men and the Machine*. Little, George Brown and Company, Boston, 1986, p. 39; also Lamm, Michael and David Lewis. "The Mercury and How It Came to Be." *Special-Interest Autos*, July–August 1974, p. 16.
15. *Reminiscences*, Dr. Edsel Ruddiman, p. 2.
16. Simonds, William A. *Henry Ford: His Life, His Work, His Genius*. The Bobbs-Merrill Company, New York, 1943, p. 29.
17. Herndon, Booton. *Ford: An Unconventional Biography of the Men and Their Times*. Weybright and Talley, New York, 1969, p. 225.
18. Bryan, Ford. *Clara: Mrs. Henry Ford.* Wayne State University Press, Detroit, 2001, p. 81.
19. Accession 6, Box 260.
20. Accession 6, Box 262.
21. Herndon, Booton. *Ford: An Unconventional Biography of the Men and Their Times*. Weybright and Talley, New York, 1969, p. 143.
22. *Reminiscences*, A.A. Lepine, p. 68.
23. *St. Louis Star*, April 1, 1920.
24. Brough, James. *The Ford Dynasty: An American Story*. Doubleday and Company, New York, 1977, p. 29.
25. Catherine Ruddiman, interview, October 11, 1992; also *Reminiscences*, A.A. Lepine, p. 87.
26. Accession 6, Box 30; other accounts claim Edsel was 5 feet 7-1/2 inches tall; also *Ford Life*, January–February 1973, p. 38.
27. Bushnell, Sarah T. *The Truth About Henry Ford.* The Reilly and Lee Company, Chicago, 1922, pp. 169–170.
28. *Reminiscences*, Clara Snow, p. 59.
29. *Reminiscences*, Mr. and Mrs. Louis C. Scott, p. 35.
30. *Ward's Quarterly*, Summer 1965, Vol. 1, No. 3, p. 68.
31. *Reminiscences*, Irving Bacon, p. 59.
32. *Reminiscences*, Irving Bacon, p. 246.
33. Smith, Patricia Barbara. "Edsel Ford." *Detroit Magazine, The Detroit Free Press*, October 17 and October 24, 1965, p. 14.
34. *Reminiscences*, Clara Snow, p. 59.
35. *Reminiscences*, A.A. Lepine, p. 80.
36. Accession 6, Box 260.
37. Ibid.
38. Accession 1689.
39. Scrapbook, The Edsel & Eleanor Ford House Archives.
40. Lacey, Robert. *Ford: The Men and the Machine*. Little, George Brown and Company, Boston, 1986, p. 258.
41. *The Detroit News*, November 29, 1935.
42. *Reminiscences*, A.A. Lepine, p. 62.
43. Ibid.
44. *Reminiscences*, Ernest Liebold, p. 1303.

45. *Reminiscences*, A.A. Lepine, pp. 61–62.
46. *Fortune*, June 1944.
47. *Reminiscences*, Theodore Gehle, p. 101.
48. *Reminiscences*, A.A. Lepine, p. 62.
49. Ralph Roberts, various interviews between 1992 and 1997.
50. *Reminiscences*, Mr. and Mrs. Clarence Davis, p. 975.
51. *Reminiscences*, Margaret Ruddiman, p. 98.
52. *Reminiscences*, Dr. David O'Donnell, p. 2.
53. *Reminiscences*, Ernest Liebold, p. 1321.
54. Bushnell, Sarah T. *The Truth About Henry Ford.* The Reilly and Lee Company, Chicago, 1922, p. 167.
55. Dahlinger, John Cote. *The Secret Life of Henry Ford.* The Bobbs-Merrill Company, Indianapolis/New York, 1978, p. 200.
56. Radebold, Klaus, M.D., Ph.D., *eMedicine Journal*, September 5, 2001, Vol. 2, No. 9.
57. Jan Willemse, interview, February 10, 1992.
58. Ford Archives clip book, July 28, 1921.
59. Accession 975.
60. E.T. Gregorie, various interviews between 1992 and 1998.
61. *Reminiscences*, Charles Carroll, p. 21.
62. *Ward's Quarterly*, Summer 1965, Vol. 1, No. 3, p. 68.
63. E.T. Gregorie, various interviews between 1992 and 1998.
64. Laas, William. "The Meaning of a Name." *Your Edsel Marketer*, Vol. 1, No. 2, publication of Ford Motor Company, no date.

Chapter 2: A Son for Henry

1. Burton Historical Collection, Detroit Public Library.
2. *Detroit Free Press*, November 6, 1893.
3. Bryan, Ford. *Clara: Mrs. Henry Ford.* Ford Books, Dearborn, 2001, p. 333.
4. Quaife, M.M. "The Ford Family." Burton Historical Collection, The Detroit Public Library, p. 5.
5. Bryan, Ford. *Clara: Mrs. Henry Ford.* Ford Books, Dearborn, 2001, p. 57.
6. Olson, Sidney. *Young Henry Ford.* Wayne State University Press, Detroit, 1963, p. 60.
7. Olson, Sidney. *Young Henry Ford.* Wayne State University Press, Detroit, 1963, pp. 60 and 65.
8. Smith, Patricia Barbara. "Edsel Ford." *Detroit Magazine, The Detroit Free Press*, October 17 and October 24, 1965, p. 13.
9. Olson, Sidney. *Young Henry Ford.* Wayne State University Press, Detroit, 1963, p. 60.
10. *Reminiscences*, Margaret Ruddiman, p. 78.
11. Baut, Donald, Winfield Arneson, and Edward Cushman. *Interview with E. Roy Bryant.* Dearborn Historical Museum, Dearborn, MI, February 23, 1968.
12. Ford Motor Company press release.
13. *Ford Times*, July 1963, p. 4.
14. Nevins, Allan. *Ford: The Times, The Man, The Company.* Charles Scribner's Sons, New York, 1954, p. 147.

15. *Reminiscences*, Frederick Strauss, p. 87.
16. Simonds, William A. *Henry Ford: His Life, His Work, His Genius*. The Bobbs-Merrill Company, New York, 1943, p. 55.
17. Olson, Sidney. *Young Henry Ford*. Wayne State University Press, Detroit, 1963, p. 81.
18. Bryan, Ford. *Clara: Mrs. Henry Ford*. Ford Books, Dearborn, 2001, p. 42.
19. Simonds, William A. *Henry Ford: His Life, His Work, His Genius*. The Bobbs-Merrill Company, New York, 1943, p. 54.
20. *Reminiscences*, Margaret Ruddiman, pp. 66–67.
21. *Reminiscences*, Margaret Ruddiman, p. 1.
22. Crowther, Samuel. "Henry Ford Talks About Edsel and Edsel Talks About His Father." *World's Work*, LVI, p. 390; also Burton Collection, Detroit Public Library; also Crowther, Samuel. "Edsel B. Ford." *Saturday Evening Post*, December 25, 1926, p. 78.

Chapter 3: The Early Years

1. Crowther, Samuel. "Edsel B. Ford." *Saturday Evening Post*, December 25, 1926, p. 78.
2. Nevins, Allan. *Ford: The Times, The Man, The Company*. Charles Scribner's Sons, New York, 1954, p. 166.
3. Ford, Henry. "I've Been Helped by Everybody." *True Story*, April 1935, p. 84.
4. Ford, Henry. "I've Been Helped by Everybody." *True Story*, April 1935, p. 85.
5. Nevins, Allan. *Ford: The Times, The Man, The Company*. Charles Scribner's Sons, New York, 1954, p. 167.
6. Olson, Sidney. *Young Henry Ford*. Wayne State University Press, Detroit, 1963, p. 104.
7. Olson, Sidney. *Young Henry Ford*. Wayne State University Press, Detroit, 1963, p. 129.
8. Nevins, Allan. *Ford: The Times, The Man, The Company*. Charles Scribner's Sons, New York, 1954, pp. 169–191.
9. Olson, Sidney. *Young Henry Ford*. Wayne State University Press, Detroit, 1963, p. 127.
10. Nevins, Allan. *Ford: The Times, The Man, The Company*. Charles Scribner's Sons, New York, 1954, pp. 202–205.
11. Levine, Leo. *Ford: The Dust and the Glory*. The Macmillan Company, New York, 1968, p. 5.
12. Olson, Sidney. *Young Henry Ford*. Wayne State University Press, Detroit, 1963, p. 130.
13. Olson, Sidney. *Young Henry Ford*. Wayne State University Press, Detroit, 1963, pp. 132–133.
14. Olson, Sidney. *Young Henry Ford*. Wayne State University Press, Detroit, 1963, p. 134.
15. Olson, Sidney. *Young Henry Ford*. Wayne State University Press, Detroit, 1963, p. 149.
16. Olson, Sidney. *Young Henry Ford*. Wayne State University Press, Detroit, 1963, p. 150.
17. Accession 65.
18. Ibid.
19. Ibid.
20. Accession 65; also Catherine Ruddiman, various interviews between 1989 and 1992.
21. Accession 65.
22. Ibid.
23. Ibid.

24. *Reminiscences*, Faye Beebe, pp. 3–4.
25. Ibid.
26. Ibid.
27. *Reminiscences*, Faye Beebe, p. 10.
28. Nevins, Allan. *Ford: The Times, The Man, The Company.* Charles Scribner's Sons, New York, 1954, p.206.
29. Olson, Sidney. *Young Henry Ford.* Wayne State University Press, Detroit, 1963, p. 147.
30. Olson, Sidney. *Young Henry Ford.* Wayne State University Press, Detroit, 1963, p. 148.
31. Olson, Sidney. *Young Henry Ford.* Wayne State University Press, Detroit, 1963, p. 156.
32. Accession 1, Box 27.
33. Nevins, Allan. *Ford: The Times, The Man, The Company.* Charles Scribner's Sons, New York, 1954, pp. 260–261.
34. Clancy, Louise Florence Davies. *The Believer: The Life of Mrs. Henry Ford.* Van Rees Press, New York, 1960, p. 28.
35. Bryan, Ford. *Henry's Attic.* Ford Books, Dearborn, 1995, p. 333.
36. Crowther, Samuel. "Edsel B. Ford." *Saturday Evening Post*, December 25, 1926, p. 78.

Chapter 4: From Bicycles to Bugattis

1. Richards, William C. *The Last Billionaire.* Charles Scribner's Sons, New York, 1948, p. 205.
2. *Reminiscences*, Max Wollering, p. 13.
3. Crowther, Samuel. "Edsel B. Ford." *Saturday Evening Post*, December 25, 1926, p. 78.
4. *Reminiscences*, C.J. Smith, p. 24.
5. *Reminiscences*, Frank Hadas, p. 76.
6. *Reminiscences*, Richard Kroll, p. 19.
7. *Reminiscences*, Ernest Grimshaw, p. 9; also Sinsabaugh, Chris. *Who, Me?* Arnold-Powers, Detroit, 1940, p. 122.
8. *Reminiscences*, Charles Sorensen, p. 303.
9. *Reminiscences*, Eugene Farkas, p. 54.
10. *Reminiscences*, C.J. Smith, p. 24.
11. *Reminiscences*, George Brown, p. 170.
12. *Reminiscences*, Irving Bacon, p. 7; also *Reminiscences*, William Klann, p. 39.
13. *Reminiscences*, Charles Niepoth, pp. 51–52; also Accession 6, Box 376.
14. Ford Archives clip book, July 28, 1921.
15. Bushnell, Sarah T. *The Truth About Henry Ford.* The Reilly and Lee Company, Chicago, 1922, p. 177; also Accession 65.
16. Tour of the Edison Avenue house by owner Jerald Mitchell.
17. Accession 65.
18. Catherine Ruddiman, various interviews between 1989 and 1992.
19. Accession 65.
20. Catherine Ruddiman, various interviews between 1989 and 1992.
21. *Reminiscences*, Margaret Ruddiman, p. 132.
22. *Ford Times*, October 1963.

23. Grace Prunk, interview, July 12, 1989.
24. Ibid.
25. *Reminiscences*, Mr. and Mrs. Louis C. Scott, p. 35.
26. *Reminiscences*, Margaret Ford Ruddiman, p. 122.
27. *Reminiscences*, John Wandersee, p. 53.
28. *Detroit Free Press*, May 28, 1943.
29. *Reminiscences*, Catherine Ruddiman, p. 89; also *Reminiscences*, Robert Rankin, p. 31.
30. Grace Prunk, interview, July 12, 1989.
31. Cameron, William J. "In Memory." *The Herald*, Vol. X, No. 9, 1943, p. 3.
32. *Reminiscences*, Edward Cutler, p. 91.
33. Simonds, William A. *Henry Ford: His Life, His Work, His Genius*. The Bobbs-Merrill Company, New York, 1943, p. 110.
34. Catherine Ruddiman, various interviews between 1989 and 1992.
35. Accession 65; also Catherine Ruddiman, various interviews between 1989 and 1992.
36. Simonds, William A. *Henry Ford: His Life, His Work, His Genius*. The Bobbs-Merrill Company, New York, 1943, p. 110.
37. Simonds, William A. *Henry Ford: His Life, His Work, His Genius*. The Bobbs-Merrill Company, New York, 1943, p. 30.
38. *Reminiscences*, J.L. McCloud, p. 384.
39. Catherine Ruddiman, various interviews between 1989 and 1992.
40. *Reminiscences*, Margaret Ruddiman, p. 123.
41. Neyhart, Louise Albright. *Henry Ford, Engineer*. Houghton Mifflin Company, Boston, 1950, p. 102.
42. Accession 65; also *Reminiscences*, Ernest Liebold, p. 256.
43. *Reminiscences*, Conrad Menge, p. 2.
44. *Reminiscences*, Charles Sorensen, pp. 197–198.
45. Levine, Leo. *Ford: The Dust and the Glory*. The Macmillan Company, New York, 1968, p. 32.
46. Levine, Leo. *Ford: The Dust and the Glory*. The Macmillan Company, New York, 1968, pp. 32–34.
47. Levine, Leo. *Ford: The Dust and the Glory*. The Macmillan Company, New York, 1968, p. 37.
48. *Reminiscences*, William Hughson, p. 15.
49. Levine, Leo. *Ford: The Dust and the Glory*. The Macmillan Company, New York, 1968, p. 39.
50. Nevins, Allan. *Ford: The Times, The Man, The Company*. Charles Scribner's Sons, New York, 1954, p. 406.
51. *Reminiscences*, Charles Sorensen, p. 251.
52. *Reminiscences*, John Wandersee, p. 53.
53. Olson, Sidney. *Young Henry Ford*. Wayne State University Press, Detroit, 1963, p. 99.
54. Crowther, Samuel. "Edsel B. Ford." *Saturday Evening Post*, December 25, 1926, p. 78.
55. Ibid.
56. Ibid.
57. Olson, Sidney. *Young Henry Ford*. Wayne State University Press, Detroit, 1963, p. 158.

58. *Reminiscences*, Mr. and Mrs. Louis C. Scott, p. 34.
59. *Reminiscences*, C.J. Smith, pp. 8–9.
60. Jenkins, Glenn F. "A Man for Mr. Ford's Shoes." *Christian Science Monitor Weekly Magazine*, May 6, 1935, p. 6.
61. Greenleaf, William. *From These Beginnings: The Early Philanthropies of Henry and Edsel Ford, 1911–1936*. Wayne State University Press, Detroit, 1964, p. 162; also Accession 6, Box 258; also *Reminiscences*, Harold Hicks, p. 152; also *Reminiscences*, A.A. Lepine, p. 28.
62. Accession 6, Box 258.
63. Jenkins, Glenn F. "A Man for Mr. Ford's Shoes." *Christian Science Monitor Weekly Magazine*, May 6, 1935, p. 6.
64. Catherine Ruddiman, various interviews between 1989 and 1992.
65. Ibid.
66. Ibid.
67. Baut, Donald, Winfield Arneson, and Edward Cushman. *Interview with E. Roy Bryant*. Dearborn Historical Museum, Dearborn, MI, February 23, 1968.
68. Accession 872.
69. Ibid.
70. Baut, Donald, Winfield Arneson, and Edward Cushman. *Interview with E. Roy Bryant*. Dearborn Historical Museum, Dearborn, MI, February 23, 1968.
71. Ibid.
72. Accession 1, Box 22.
73. Baut, Donald, Winfield Arneson, and Edward Cushman. *Interview with E. Roy Bryant*. Dearborn Historical Museum, Dearborn, MI, February 23, 1968.
74. Wilkins, Mira and Frank Ernest Hill. *American Business Abroad*. Wayne State University Press, Detroit, p. 48.
75. Baut, Donald, Winfield Arneson, and Edward Cushman. *Interview with E. Roy Bryant*. Dearborn Historical Museum, Dearborn, MI, February 23, 1968.
76. Accession 82.
77. Richards, William C. *The Last Billionaire*. Charles Scribner's Sons, New York, 1948, p. 202.
78. Smith, Patricia Barbara. "Edsel Ford." *Detroit Magazine, The Detroit Free Press*, October 17 and October 24, 1965, p. 16.
79. Marquis, Samuel S. *Henry Ford: An Interpretation*. Little, George Brown and Company, Boston, 1923, p. 181.

Chapter 5: College Is a Waste of Time

1. Detroit University School Yearbook, "Helicon" 1912, Grosse Pointe Historical Society, no page number.
2. Brough, James. *The Ford Dynasty: An American Story*. Doubleday and Company, New York, 1977, p. 49.
3. Accession 1.
4. Lindbergh, Charles A. *The Wartime Journals of Charles A. Lindbergh*. Harcourt, Brace, Jovanovich, New York, 1970, p. 630.

5. Catherine Ruddiman, various interviews between 1989 and 1992.
6. Olson, Sidney. *Young Henry Ford.* Wayne State University Press, Detroit, 1963, p. 131.
7. Olson, Sidney. *Young Henry Ford.* Wayne State University Press, Detroit, 1963, p. 134.
8. *Reminiscences*, Frederick Searle, pp. 1–2.
9. Accession 1.
10. *Reminiscences*, Frederick Searle, p. 2; also Accession 1.
11. Detroit Public Library, Burton Historical Collection.
12. Accession 1.
13. Ibid.
14. Ibid.
15. Accession 6, Box 345.
16. *Ward's Quarterly*, Summer 1965, Vol. 1, No. 3, pp. 68–69.
17. *Ward's Quarterly*, Summer 1965, Vol. 1, No. 3, p. 129.
18. *Reminiscences*, Mr. and Mrs. Louis C. Scott, p. 13.
19. Smith, Patricia Barbara. "Edsel Ford." *Detroit Magazine, The Detroit Free Press,* October 17 and October 24, 1965, p. 14.
20. Smith, Patricia Barbara. "Edsel Ford." *Detroit Magazine, The Detroit Free Press,* October 17 and October 24, 1965, p. 14.
21. Mrs. Evelyn Book, various interviews in 1993.
22. Accession 1.
23. Wormer, Anne Getz. *Century: The History of University Liggett School, 1878–1978.* University Liggett School, Grosse Pointe Woods, MI, 1981, p. 64.
24. Accession 1.
25. Ibid.
26. University Liggett School Archives.
27. Accession 1.
28. Accession 526, Box 3.
29. There are no records of Edsel graduating, nor is his photo in the school yearbook for 1912.
30. Richards, William C. *The Last Billionaire.* Charles Scribner's Sons, New York, 1948, p. 205.
31. *Reminiscences*, Charles Sorensen, pp. 27–28.
32. Accession 1660.
33. Accession 6, Box 329.
34. Accession 6, Box 308.
35. Wormer, Anne Getz. *Century: The History of University Liggett School, 1878–1978.* University Liggett School, Grosse Pointe Woods, MI, 1981, p. 123.
36. Brough, James. *The Ford Dynasty: An American Story.* Doubleday and Company, New York, 1977, p. 64.

Chapter 6: San Francisco Sojourn

1. Simonds, William A. *Henry Ford: His Life, His Work, His Genius.* The Bobbs-Merrill Company, New York, 1943, p. 125.
2. *Reminiscences*, Joseph Galamb, p. 63.

3. Simonds, William A. *Henry Ford: His Life, His Work, His Genius*. The Bobbs-Merrill Company, New York, 1943, p. 125.
4. Simonds, William A. *Henry Ford: His Life, His Work, His Genius*. The Bobbs-Merrill Company, New York, 1943, p. 125.
5. *Ford Times*, October 1963.
6. Ibid.
7. Mrs. Evelyn Book, various interviews in 1993.
8. Edsel Ford Diary, Edsel & Eleanor Ford House Archives.
9. *Ford Times*, October 1963.
10. Ford Archives clip book, July 4, 1915.
11. Accession 1, Box 27.
12. Accession 1, Box 103.
13. Ibid.
14. Ibid.
15. Ford Archives clip book, January 23, 1915.
16. Ibid.
17. *Ford Times*, October 1963.
18. Mrs. Evelyn Book, various interviews in 1993.
19. *Reminiscences*, Frank Vivian, pp. 18–19.
20. Ibid.
21. Ibid.
22. *Reminiscences*, Frank Vivian, p. 42.
23. All quotations of Edsel's cross-country trip, except where noted, are from his personal logbook of his trip, which is reposited at the Edsel & Eleanor Ford House Archives.
24. Accession 1, Box 103.
25. Brauer, Norman. *There to Breathe the Beauty*. Norman Brauer Publications, Dalton, PA, 1995, p. 27.
26. Brauer, Norman. *There to Breathe the Beauty*. Norman Brauer Publications, Dalton, PA, 1995, pp. 27–28.
27. *Reminiscences*, Frank Vivian, pp. 28–29.

Chapter 7: Courtship and Marriage

1. Conot, Robert. *An American Odyssey*. William Morrow Company, New York, 1974, p. 185.
2. Accession 1.
3. Nevins, Allan. *Ford: Expansion and Challenge: 1915–1933*. Charles Scribner's Sons, New York, 1954, p. 87.
4. Mrs. Evelyn Book, various interviews in 1993.
5. Lacey, Robert. *Ford: The Men and the Machine*. Little, George Brown and Company, Boston, 1986, p. 153.
6. Mrs. Evelyn Book, various interviews in 1993.
7. Simonds, William A. *Henry Ford: His Life, His Work, His Genius*. The Bobbs-Merrill Company, New York, 1943, p. 167.
8. *The Detroit News*, June 6, 1971.
9. Mrs. Evelyn Book, various interviews in 1993.

10. The Liggett School Yearbook, "Rivista" 1915, Grosse Pointe Historical Society, no page number.
11. Wormer, Anne Getz. *Century: The History of University Liggett School, 1878–1978*. University Liggett School, Grosse Pointe Woods, MI, 1981, p. 33.
12. Mrs. Evelyn Book, various interviews in 1993.
13. Ibid.
14. Accession 1, Box 2.
15. Lacey, Robert. *Ford: The Men and the Machine*. Little, George Brown and Company, Boston, 1986, p. 153.
16. *Reminiscences*, Frederick Searle, p. 2.
17. Mrs. Evelyn Book, various interviews in 1993.
18. Accession 7, Box 2.
19. *Detroit Saturday Night, The Detroit Free Press*, November 4, 1916.
20. Wedding scrapbook, The Edsel & Eleanor Ford House Archives.
21. Ibid.
22. *Reminiscences*, John Williams, pp. 25–26.
23. *Reminiscences*, Ernest Liebold, p. 27.
24. Lacey, Robert. *Ford: The Men and the Machine*. Little, George Brown and Company, Boston, 1986, p. 149.
25. *Reminiscences*, Mr. and Mrs. Lewis Simpson, p. 9; also *Reminiscences*, Catherine Ruddiman, p. 12.
26. *Reminiscences*, C.J. Smith, p. 26.
27. Accession 7, Box 2.
28. *Detroit Free Press*, November 2, 1916.
29. Lacey, Robert. *Ford: The Men and the Machine*. Little, George Brown and Company, Boston, 1986, p. 154.
30. *Detroit Free Press*, November 2, 1916.
31. *Reminiscences*, Eugene Farkas, p. 353–354.
32. Smith, Patricia Barbara. "Edsel Ford." *Detroit Magazine, The Detroit Free Press*, October 17 and October 24, 1965, p. 16.
33. *Detroit Free Press*, November 2, 1916.
34. *Detroit Saturday Night, The Detroit Free Press*, November 4, 1916.
35. Ibid.
36. *Detroit Free Press*, November 2, 1916.
37. Ibid.
38. Exhibit at the Edsel & Eleanor Ford House.
39. Wedding scrapbook, The Edsel & Eleanor Ford House Archives.
40. Accession 1, Box 27.
41. Ibid.
42. Accession 6, Box 108.
43. *Reminiscences*, William Hughson, p. 61.
44. *Reminiscences*, Frank Vivian, pp. 35–36.
45. Accession 1, Box 27.

46. Ibid.
47. Wedding scrapbook, The Edsel & Eleanor Ford House Archives.

Chapter 8: Live Very Simply

1. Accession 7, Box 2.
2. Lacey, Robert. *Ford: The Men and the Machine*. Little, George Brown and Company, Boston, 1986, p. 256.
3. *Reminiscences*, A.A. Lepine, p. 72.
4. *Reminiscences*, Joseph Beh, p. 17.
5. *Reminiscences*, A.A. Lepine, p. 71.
6. Collier, Peter and David Horowitz. *The Fords: An American Epic*. Summit Books, New York, 1987, p. 137.
7. *Reminiscences*, Fred Loskowske, p. 44.
8. Porter, Darwin and Danforth Prince. *England from $60 a Day*. Macmillan Travel, New York, 1998, p. 453.
9. *The Detroit News*, March 20, 1927.
10. *Reminiscences*, A.A. Lepine, p. 73.
11. *The Detroit News*, March 20, 1927.
12. *Detroit Free Press*, August 2, 1927.
13. The Edsel & Eleanor Ford House Archives.
14. Allen, George H. "Edsel Ford Selects a Cotswold House." *House and Garden*, February 1932.
15. *Reminiscences*, A.A. Lepine, p. 23.
16. Evelyn Collins, interview, July 9, 1994.
17. McCarthy, Joe. "The Ford Family." *Holiday*, September 1957, p. 133.
18. McCarthy, Joe. "The Ford Family." *Holiday*, June 1957, pp. 69–70.
19. *Reminiscences*, A.A. Lepine, p. 94; also Accession 6, Box 345.
20. *Reminiscences*, Arthur Backus, p. 11.
21. *Time*, June 7, 1943.
22. Smith, Patricia Barbara. "Edsel Ford." *Detroit Magazine, The Detroit Free Press*, October 17 and October 24, 1965, p. 16.
23. Collier, Peter and David Horowitz. *The Fords: An American Epic*. Summit Books, New York, 1987, p. 137.
24. *Detroit Free Press*, September 15, 1931.
25. *Reminiscences*, Arthur Backus, pp. 11–12.
26. Ibid.
27. Nevins, Allan. *Ford: Expansion and Challenge: 1915–1933*. Charles Scribner's Sons, New York, 1954, p. 489.
28. *Reminiscences*, Charles Sorensen, p. 250.
29. Lewis, David. L. "Edsel Ford's Farm Environmental Center." *Cars and Parts*, February 1975.
30. *Reminiscences*, Arthur Backus, p. 12.
31. Chris Arble, interview, July 29, 2001.
32. Floyd Lackanow, interview, July 12, 1989.

33. Ibid.
34. Ibid.
35. *Detroit Free Press*, April 16, 1944.
36. *Grosse Pointe News*, November 4, 1976.
37. *Reminiscences*, Catherine Ruddiman, pp. 66–67.
38. *Reminiscences*, Catherine Ruddiman, p. 92.
39. Josephine Goman Papers, University of Michigan, p. 1
40. *Reminiscences*, Charles Sorensen, p. 123.
41. Josephine Goman Papers, University of Michigan, p. 2.

Chapter 9: Accusations of Evasion

1. Wik, Reynold M. *Henry Ford and Grass-Roots America*. University of Michigan Press, Ann Arbor, 1972, p. 171.
2. *Pipp's Weekly*, May 15, 1920, p. 6.
3. *Detroit Free Press*, May 27, 1943.
4. Accession 6, Box 31.
5. Accession 6, Box 456.
6. Ibid.
7. *The Detroit News*, November 19, 1917.
8. Ford Archives clip book, October 13, 1917.
9. Accession 6, Box 456.
10. Ibid.
11. *Pipp's Weekly*, May 10, 1924.
12. Accession 6, Box 456.
13. Ibid.
14. Smith, Patricia Barbara. "Edsel Ford." *Detroit Magazine*, *Detroit Free Press*, October 17 and October 24, 1965, p. 16.
15. *Pipp's Weekly*, May 15, 1917.
16. Accession 6, Box 34.
17. Marquis, Samuel S. *Henry Ford: An Interpretation*. Little, George Brown and Company, Boston, 1923, pp. 184–185.
18. Accession 6, Box 34.
19. Ibid.
20. Ford Archives clip book, October 11, 1917.
21. Accession 1.
22. Burton Collection, Detroit Pubic Library.
23. Ford Archives clip book, October 11, 1917.
24. *Grand Rapids Press*, July 16, 1919.
25. *Port Huron Herald*, July 17, 1919.
26. Sward, Keith. *The Legend of Henry Ford*. Rinehart and Company, New York, 1948, p. 95.
27. Wik, Reynold M. *Henry Ford and Grass-Roots America*. University of Michigan Press, Ann Arbor, 1972, p. 171; also *Grand Rapids Press*, July 16, 1919.
28. *Huntington Herald Dispatch*, August 29, 1918, p. 1.
29. National Archives, Enoch H. Crowder memo, September 7, 1918.

30. *The Detroit News*, October 24, 1917.
31. *Detroit Saturday Night*, *The Detroit Free Press*, October 27, 1917.
32. Ibid.
33. *State Journal*, Lansing, Michigan, September 7, 1918.
34. National Archives, Enoch H. Crowder memo, September 7, 1918.
35. *Reminiscences*, Ernest Liebold, p. 353.
36. *Reminiscences*, Charles Sorensen, p. 378.
37. *Reminiscences*, Charles Sorensen, p. 306.
38. *Reminiscences*, Charles Sorensen, pp. 305–306.

Chapter 10: Buy-Out

1. Lewis, Eugene W. *Motor Memories*. Alved Publishers, Detroit, 1947, p. 115.
2. *The Detroit News*, January 2, 1919.
3. *Reminiscences*, Harry Hanson, p. 175.
4. Lacey, Robert. *Ford: The Men and the Machine*. Little, George Brown and Company, Boston, 1986, p. 172.
5. Simonds, William A. *Henry Ford: His Life, His Work, His Genius*. The Bobbs-Merrill Company, New York, 1943, p. 180.
6. Accession 7, Vol. 18, p. 53.
7. *Reminiscences*, William N. Mayo, p. 24.
8. *Reminiscences*, Ernest Liebold, pp. 183–184.
9. Barnard, Harry. *Independent Man: The Life of Senator James Couzens*. Charles Scribner's Sons, New York, 1958, p. 130.
10. *Reminiscences*, Ernest Liebold, pp. 185–186.
11. *Reminiscences*, William N. Mayo, p. 24.
12. Accession 1.
13. *The Detroit News*, July 11, 1919.
14. Lacey, Robert. *Ford: The Men and the Machine*. Little, George Brown and Company, Boston, 1986, p. 187.

Chapter 11: The Administrator

1. *Reminiscences*, Charles Thomas, p. 424.
2. Crowther, Samuel. "Edsel B. Ford." *Saturday Evening Post*, December 25, 1926, p. 78.
3. *Reminiscences*, A.A. Lepine, pp. 6 and 70.
4. Simonds, William A. *Henry Ford: His Life, His Work, His Genius*. The Bobbs-Merrill Company, New York, 1943, p. 131.
5. *Reminiscences*, A.A. Lepine, p. 7.
6. *Reminiscences*, Arthur Backus, p. 18.
7. *Washington Post*, May 30, 1943; also *Reminiscences*, A.A. Lepine, p. 79.
8. Robert Smith, interview, May 18, 1989.
9. Ibid.
10. *Reminiscences*, A.A. Lepine, pp. 28.
11. Accession 6, Box 259.

12. *Reminiscences*, Arthur Backus, pp. 17–18.
13. Ralph Roberts, various interviews between 1992 and 1997.
14. Crowther, Samuel. "Edsel B. Ford." *Saturday Evening Post*, December 25, 1926, p. 78.
15. Lacey, Robert. *Ford: The Men and the Machine*. Little, George Brown and Company, Boston, 1986, p. 255.
16. *Reminiscences*, George Ferghete, p. 28.
17. *Reminiscences*, A.A. Lepine, p. 26.
18. E.T. Gregorie, various interviews between 1992 and 1998.
19. *Reminiscences*, A.A. Lepine, pp. 3–5.
20. Accession 6.
21. *Ward's Quarterly*, Summer 1965, Vol. 1, No. 3, pp. 68–69.
22. Accession 6, Box 284.
23. Accession 6, various memos; also *Reminiscences*, A.A. Lepine, p. 19.
24. *Ward's Quarterly*, Summer 1965, Vol. 1, No. 3, p. 129.
25. *Reminiscences* , A.A. Lepine, p. 80.
26. *Reminiscences*, A.A. Lepine, p. 18.
27. Accession 7, Box 9.
28. *Reminiscences*, Carmen Hueston, p. 75.
29. *Reminiscences*, A.A. Lepine, p. 32.
30. *Reminiscences*, A.A. Lepine, pp. 5–6.
31. *Reminiscences*, Liebold, pp. 1126–1127; *Reminiscences*, Cowling, p. 58.
32. *The Detroit News*, November 6, 1940.
33. Byrd, Commander Richard E. *Skyward*. G.P. Putnam's Sons, New York, 1928, p. 331.
34. *Reminiscences*, Charles Sorensen, p. 311.
35. *Reminiscences*, C. Bernhardt, pp. 7–8.
36. *Reminiscences*, Henry Doss, p. 78.
37. *Reminiscences*, Henry Doss, pp. 78–79.
38. Accession 6, Box 258.
39. Ibid.
40. *Reminiscences*, Henry Doss, p. 87.
41. *Reminiscences*, Henry Doss, p. 76.
42. *Reminiscences*, Henry Doss, p. 87.
43. Accession 6, Box 452.
44. Marquis, Samuel S. *Henry Ford: An Interpretation*. Little, George Brown and Company, Boston, 1923, pp. 186–187.
45. *Design Oral History*, E.T. Gregorie, pp. 27–28.
46. *Reminiscences*, Harold Hicks, p. 211.
47. Accession 6, Box 259.
48. *Reminiscences*, A.A. Lepine, p. 65.
49. *Reminiscences*, Norman Ahrens, p. 23.
50. *Reminiscences*, Al Esper, p. 12.
51. *Reminiscences*, Leslie McDonnell, p. 28.
52. *Reminiscences*, Harold Hicks, p. 48.
53. *Reminiscences*, Oscar Olsen, p. 113.

54. *Reminiscences*, Norman Ahrens, p. 23.
55. *Reminiscences*, Ernest Liebold, p. 1492.
56. Accession 6.
57. Accession 6, Box 375.
58. Accession 6, Box 336.
59. Accession 975.
60. *Reminiscences*, Albert Wibel, p. 109.
61. Jan Willemse, interview, February 10, 1992.
62. Ibid.
63. Accession 975.
64. *Reminiscences*, Henry Doss, pp. 50–51, and Joseph Anderson, p. 4.
65. *Reminiscences*, Henry Doss, p. 80.
66. *Reminiscences*, Herman Moekle, p. 193.
67. *Reminiscences*, Howard Simpson, p. 102.
68. *Reminiscences*, Ernest Liebold, p. 1330.
69. *Reminiscences*, William Cameron, pp. 209–210.
70. *Reminiscences*, Laurence Sheldrick, p. 393.
71. *Reminiscences*, Howard Simpson, p. 103.
72. *Ward's Quarterly*, Summer 1965, Vol. 1, No. 3, p. 70.
73. *Reminiscences*, Emil Zoerlein, p. 73.

Chapter 12: Differences in Direction

1. *Reminiscences*, Irving Bacon, p. 7.
2. *Reminiscences*, George Brown, p. 171.
3. Ford Archives clip book, July 28, 1921.
4. Ibid.
5. *Reminiscences*, William Hughson, p. 35.
6. *Reminiscences*, Fred Black, p. 95.
7. *Reminiscences*, Roscoe Smith, pp. 77–78.
8. *Reminiscences*, Ernest Liebold, p. 1485.
9. *Reminiscences*, Herman Moekle, p. 100.
10. *Reminiscences*, George Holley, pp. 70–71.
11. Dahlinger, John Cote. *The Secret Life of Henry Ford.* Bobbs-Merrill Company, New York, p. 201.
12. Lacey, Robert. *Ford: The Men and the Machine*. Little, George Brown and Company, Boston, 1986, p. 262.
13. *Reminiscences*, Catherine Ruddiman, pp. 89–90.
14. Ibid.
15. *Reminiscences*, Emil Zoerlein, pp. 51–52.
16. E.T. Gregorie, various interviews between 1992 and 1998.
17. *Reminiscences*, Frank Hadas, p. 287.
18. *Reminiscences*, Ernest Liebold, p. 1145.
19. *Reminiscences*, George Holly, p. 69; also *Reminiscences*, Theodore Gehle, p. 99.

20. *Reminiscences*, Frederick Searle, p. 38.
21. *Reminiscences*, E. Roy Bryant.
22. *Reminiscences*, Fred Black, p. 50.
23. *Reminiscences*, Frederick Searle, p. 38.
24. *Reminiscences*, Charles Sorensen, pp. 709–710.
25. *Reminiscences*, Walter Doehler, p. 14.
26. *Reminiscences*, A.G. Wolfe, p. 171.
27. *Reminiscences*, Fred Black, p. 87.
28. *Reminiscences*, Harold Hicks, p. 173.
29. *Reminiscences*, L.E. Briggs, pp. 58–59.
30. Floyd Lackanow, interview, July 12, 1989.
31. *Reminiscences*, Frank Hadas, pp. 240–242.
32. Accession 1512.
33. Ibid.
34. Ibid.
35. *Reminiscences*, Harold Wasson, p. 42.
36. *Reminiscences*, Fred Black, p. 83.
37. *Reminiscences*, Joseph Galamb, p. 34.
38. *Reminiscences*, Harold Hicks, p. 29.
39. *Reminiscences*, Eugene Farkas, p. 353.
40. Accession 1512.
41. E.T. Gregorie, various interviews between 1992 and 1998.
42. *Design Oral History*, E.T. Gregorie, p. 61.
43. *Reminiscences*, Carmen Hueston, pp. 103–105.
44. *Reminiscences*, Fred Black, p. 95.
45. Dahlinger, John Cote. *The Secret Life of Henry Ford.* The Bobbs-Merrill Company, New York, p. 202.
46. Nevins, Allan. *Ford: Expansion and Challenge: 1915–1933.* Charles Scribner's Sons, New York, 1954, p. 272.
47. *Reminiscences*, Fred Black, p. 50.
48. *Reminiscences*, A.A. Lepine, p. 35.
49. *Reminiscences*, Albert Wibel, pp. 284–285.
50. *Reminiscences*, Charles Sorensen, no page number.
51. *Reminiscences*, Irving Bacon, p. 129.
52. *Reminiscences*, Charles Sorensen, p. 200.
53. *Reminiscences*, Ernest Liebold, pp. 1143–1144.
54. *Reminiscences*, Charles Sorensen, p. 301.
55. *Reminiscences*, Ernest Liebold, p. 1493.
56. *Reminiscences*, Emil Zoerlein, p. 240.
57. *Reminiscences*, Theodore Gehle, pp. 150–151.
58. Accession 1512.
59. *Reminiscences*, Charles Sorensen, p. 302.
60. *Reminiscences*, Charles Voorhess, p. 14.
61. Crowther, Samuel. "Edsel B. Ford." *Saturday Evening Post*, December 25, 1926, p. 78.

62. *Reminiscences*, Emil Zoerlein, pp. 72–73.
63. Nevins, Alan and Frank Ernest Hill. *Ford: Decline and Rebirth 1933–1962.* Charles Scribner's Sons, New York, 1962, p. 115.
64. Bennett, Harry. *We Never Called Him Henry*. Gold Metal Books, Fawcett Publications, New York, 1987, p. 17.
65. *Reminiscences*, William Hughson, p. 50.
66. *Reminiscences*, William Hughson, pp. 48–50.
67. *Reminiscences*, Charles Sorensen, p. 214.
68. *Time*, June 7, 1943, p. 81.
69. *Reminiscences*, Ernest Liebold, p. 1115.
70. *Fortune*, December 1933, p. 132.
71. *Reminiscences*, Howard Simpson, p. 103.
72. *Reminiscences*, A.A. Lepine, p. 66.
73. *Design Oral History*, E.T. Gregorie, p. 27.
74. *Reminiscences*, Fred Black, pp. 86–87.
75. Accession 1512.

Chapter 13: From Model T to Model A

1. Nevins, Allan. *Ford: Expansion and Challenge: 1915–1933*. Charles Scribner's Sons, New York, 1954, p. 428.
2. *Reminiscences*, George Brown, pp. 108–115.
3. *Reminiscences*, Al Esper, p. 96; also *Reminiscences*, George Holley, p. 70; also *Reminiscences*, Joseph Galamb, p. 141.
4. *Reminiscences*, Charles Sorensen, p. 615.
5. Lewis, David L. *The Public Image of Henry Ford.* Wayne State University Press, Detroit, 1976, p. 192.
6. Neyhart, Louise Albright. *Henry Ford, Engineer*. Houghton Mifflin Company, Boston, 1950, p. 149.
7. *Reminiscences*, Charles Sorensen, 219.
8. *Reminiscences*, Charles Sorensen, p. 309.
9. *Reminiscences*, Charles Sorensen, no page number.
10. Nevins, Allan. *Ford: Expansion and Challenge: 1915–1933*. Charles Scribner's Sons, New York, 1954, pp. 61–62.
11. Ibid.
12. *Reminiscences*, William Cameron, p. 98.
13. *Reminiscences*, Charles Carroll, p. 19.
14. Nevins, Allan. *Ford: Expansion and Challenge: 1915–1933*. Charles Scribner's Sons, New York, 1954, p. 155.
15. *Reminiscences*, Theodore Gehle, p. 24.
16. *Reminiscences*, Walter Nelson, pp. 103–104; also *Reminiscences*, Arthur Backus, p. 15.
17. *Reminiscences*, Arthur Backus, p. 12.
18. *Detroit News*, November 6, 1940.
19. Nevins, Allan. *Ford: Expansion and Challenge: 1915–1933*. Charles Scribner's Sons, New York, 1954, pp. 271–272; also *Reminiscences*, Theodore Gehle, p. 56.

20. *Reminiscences*, Laurence Sheldrick, pp. 46–47.
21. *Reminiscences*, Frank Hadas, p. 288.
22. *Reminiscences*, Theodore Gehle, p. 56.
23. *Reminiscences*, Ernest Liebold, p. 1116.
24. *Reminiscences*, Charles Sorensen, no page number.
25. *Reminiscences*, Charles Niepoth, p. 25.
26. E.T. Gregorie, various interviews between 1992 and 1998.
27. *Reminiscences*, Charles Sorensen, pp. 15–16.
28. *Reminiscences*, Ernest Liebold, p. 1123.
29. *Reminiscences*, Albert Wibel, p. 210.
30. *Reminiscences*, N.H.F. Olsen, p. 40; also Accession 6, Box 274; also *Reminiscences*, Laurence Sheldrick, p. 29.
31. *Reminiscences*, Eugene Farkas, pp. 219–220.
32. *Reminiscences*, Fred Black, p. 92.
33. *Reminiscences*, Albert Wibel, p. 251.
34. *Reminiscences*, William Cameron, p. 99.
35. *Reminiscences*, Charles Sorensen, p. 17.
36. *Reminiscences*, Charles Sorensen, no page number.
37. *Reminiscences*, Laurence Sheldrick, p. 37.
38. Accession 1512.
39. *Reminiscences*, Charles Sorensen, no page number.
40. Ibid.
41. *Reminiscences*, Eugene Farkas, p. 218.
42. *Reminiscences*, Laurence Sheldrick, p. 48.
43. *Reminiscences*, Eugene Farkas, p. 215.
44. *Reminiscences*, Charles Voorhess, pp. 13–14.
45. *Reminiscences*, Albert Wibel, pp. 207–208.
46. *Reminiscences*, Herman Moekle, pp. 90–91.
47. *Reminiscences*, Charles Carroll, p. 22.
48. *Reminiscences*, Ernest Liebold, p. 1487.
49. *Reminiscences*, Charles Sorensen, no page number.
50. Collier, Peter and David Horowitz. *The Fords: An American Epic*. Summit Books, New York, 1987, pp. 150–151.
51. *Reminiscences*, Charles Sorensen, p. 308.
52. Accession 1512.
53. Accession 1, Box 180.
54. *Reminiscences*, Charles Voorhess, pp. 13–14.
55. *Reminiscences*, Laurence Sheldrick, pp. 47–48.
56. *Reminiscences*, Walter Doehler, p. 19.
57. *Reminiscences*, Mrs. Georgia Boyer, p. 13.
58. *Reminiscences*, Fred Black, p. 93.
59. Accession 1512.
60. Nevins, Allan. *Ford: Expansion and Challenge: 1915–1933*. Charles Scribner's Sons, New York, 1954, p. 428.

61. *Reminiscences*, Henry Doss, p. 22.
62. Nevins, Allan. *Ford: Expansion and Challenge: 1915–1933*. Charles Scribner's Sons, New York, 1954, p. 439.
63. Accession 6, Box 244.
64. *Reminiscences*, William Klann, p. 242.
65. E.T. Gregorie, various interviews between 1992 and 1998.
66. *Reminiscences*, Laurence Sheldrick, p. 30.
67. *Reminiscences*, Joseph Galamb, p. 138.
68. *Reminiscences*, Eugene Farkas, p. 182.
69. Accession 6, Box 274.
70. *Reminiscences*, Richard Kroll, pp. 36–37.
71. *Reminiscences*, Harold Hicks, p. 140.
72. *Reminiscences*, Harold Hicks, p. 148.
73. *Reminiscences*, William Klann, pp. 241–245.
74. *Reminiscences*, Harold Hicks, p. 146.
75. Accession 975.
76. *Reminiscences*, Theodore Gehle, p. 101.
77. *Reminiscences*, Frank Hadas, p. 288.
78. *Reminiscences*, Charles Sorensen, pp. 15–16.
79. *Reminiscences*, Frank Hadas, p. 288.
80. *Reminiscences*, Charles Sorensen, pp. 309–310.
81. *Reminiscences*, Harold Wasson, p. 42.
82. Accession 6, Box 273.
83. Josephine Goman Papers, University of Michigan.
84. *Detroit Free Press*, November 30, 1927.
85. Accession 6, Box 257, File "By-De."
86. *Reminiscences*, William Klann, p. 254.
87. *Reminiscences*, Fred Black, p. 184.
88. Accession 6, Box 274.
89. Ibid.
90. Accession 6.
91. Laas, William. "The Meaning of a Name." *Your Edsel Marketer*, Vol. 1, No. 2, publication of Ford Motor Company, no date.
92. Simonds, William A. *Henry Ford: His Life, His Work, His Genius*. The Bobbs-Merrill Company, New York, 1943, p. 214.
93. *Reminiscences*, Charles Sorensen, p. 177; also *Reminiscences*, Carmen Hueston, p. 101.
94. Nevins, Allan. *Ford: Expansion and Challenge: 1915–1933*. Charles Scribner's Sons, New York, 1954, p. 428.
95. *Reminiscences*, A.G. Wolfe, p. 172.
96. Sorensen, Charles. *My Forty Years with Ford*. W.W. Horton and Company, New York, 1956, p. 223.
97. *Reminiscences*, Charles Voorhess, p. 19.

Chapter 14: Bank Holiday

1. Lacey, Robert. *Ford: The Men and the Machine*. Little, George Brown and Company, Boston, 1986, p. 328.
2. *Reminiscences*, Ernest Liebold, p. 988.
3. *Reminiscences*, Ernest Liebold, p. 997.
4. U.S. Congressional Committee on Banking and Currency, Stock Exchange Practices, Hearing 73rd Congress, 2nd Session, Parts 9–13, p. 4600.
5. Accession 6, Box 262.
6. Accession 6, Box 276.
7. Ibid.
8. Ibid.
9. Accession 6, Box 273.
10. *Reminiscences*, Ernest Liebold, p. 997.
11. Nevins, Allan. *Ford: Expansion and Challenge: 1915–1933*. Charles Scribner's Sons, New York, 1954, p. 465.
12. *Reminiscences*, Ernest Liebold, p. 1485; also Bennett, Harry. *We Never Called Him Henry*. Gold Metal Books, Fawcett Publications, New York, 1987, p. 147.
13. Nevins, Allan. *Ford: Expansion and Challenge: 1915–1933*. Charles Scribner's Sons, New York, 1954, p. 477.
14. Accession 975.
15. U.S. Congressional Committee on Banking and Currency, Stock Exchange Practices, Hearing 73rd Congress, 2nd Session, Parts 9–13, p. 4544.
16. Neville, Howard Ralph. "The Detroit Banking Collapse of 1933." Occasional Paper No. 2, Bureau of Business and Economic Research, Michigan State University, East Lansing, MI, 1960, p. 18.
17. Flynn, John T. "Michigan Magic: The Detroit Bank Scandal." *Harper's*, December 1933, p. 7.
18. U.S. Congressional Committee on Banking and Currency, Stock Exchange Practices, Hearing 73rd Congress, 2nd Session, Parts 9–13, p. 4598.
19. Neville, Howard Ralph. "The Detroit Banking Collapse of 1933." Occasional Paper No. 2, Bureau of Business and Economic Research, Michigan State University, East Lansing, MI, 1960, p. 14.
20. Conot, Robert. *An American Odyssey*. William Morrow Company, New York, 1974, p. 302.
21. U.S. Congressional Committee on Banking and Currency, Stock Exchange Practices, Hearing 73rd Congress, 2nd Session, Parts 9–13, p. 4661.
22. Accession 6, Box 318.
23. Conot, Robert. *An American Odyssey*. William Morrow Company, New York, 1974, p. 302.
24. Flynn, John T. "Michigan Magic: The Detroit Bank Scandal." *Harper's*, December 1933, p. 5.
25. Accession 6, Box 293.
26. Accession 6, Box 276.
27. U.S. Congressional Committee on Banking and Currency, Stock Exchange Practices, Hearing 73rd Congress, 2nd Session, Parts 9–13, p. 4547.

28. "The Guardian Building." Brochure published by the Michigan Consolidated Gas Company, no date.
29. James Longley, interview, May 10, 1989.
30. *Reminiscences*, William Cameron, p. 147.
31. Neville, Howard Ralph. "The Detroit Banking Collapse of 1933." Occasional Paper No. 2, Bureau of Business and Economic Research, Michigan State University, East Lansing, MI, p. 9.
32. Sward, Keith. *The Legend of Henry Ford.* Rinehart and Company, New York, 1948, p. 225.
33. U.S. Congressional Committee on Banking and Currency, Stock Exchange Practices, Hearing 73rd Congress, 2nd Session, Parts 9–13, p. 4656.
34. U.S. Congressional Committee on Banking and Currency, Stock Exchange Practices, Hearing 73rd Congress, 2nd Session, Parts 9–13, p. 4595.
35. Flynn, John T. "Michigan Magic: The Detroit Bank Scandal." *Harper's*, December 1933, p. 8.
36. Ibid.
37. Conot, Robert. *An American Odyssey.* William Morrow Company, New York, 1974, p. 303.
38. U.S. Congressional Committee on Banking and Currency, Stock Exchange Practices, Hearing 73rd Congress, 2nd Session, Parts 9–13, p. 4602.
39. U.S. Congressional Committee on Banking and Currency, Stock Exchange Practices, Hearing 73rd Congress, 2nd Session, Parts 9–13, p. 4605.
40. Conot, Robert. *An American Odyssey.* William Morrow Company, New York, 1974, pp. 302–303.
41. Ibid.
42. U.S. Congressional Committee on Banking and Currency, Stock Exchange Practices, Hearing 73rd Congress, 2nd Session, Parts 9–13, pp. 4664, 4679–4680.
43. U.S. Congressional Committee on Banking and Currency, Stock Exchange Practices, Hearing 73rd Congress, 2nd Session, Parts 9–13, p. 4570.
44. U.S. Congressional Committee on Banking and Currency, Stock Exchange Practices, Hearing 73rd Congress, 2nd Session, Parts 9–13, p. 4682.
45. U.S. Congressional Committee on Banking and Currency, Stock Exchange Practices, Hearing 73rd Congress, 2nd Session, Parts 9–13, p. 4687.
46. U.S. Congressional Committee on Banking and Currency, Stock Exchange Practices, Hearing 73rd Congress, 2nd Session, Parts 9–13, p. 4667.
47. U.S. Congressional Committee on Banking and Currency, Stock Exchange Practices, Hearing 73rd Congress, 2nd Session, Parts 9–13, p. 4668.
48. *Reminiscences*, Charles Sorensen, p. 5.
49. *Reminiscences*, Arthur Backus, p. 11.
50. *Reminiscences*, Ernest Liebold, pp. 1029–1030.
51. *Reminiscences*, Irving Bacon, p. 224.
52. *Reminiscences*, Arthur Backus, p. 11.
53. *Reminiscences*, Joseph Zaroski, pp. 25–26.
54. Accession 6, Box 302.
55. U.S. Congressional Committee on Banking and Currency, Stock Exchange Practices, Hearing 73rd Congress, 2nd Session, Parts 9–13, p. 4563.

56. U.S. Congressional Committee on Banking and Currency, Stock Exchange Practices, Hearing 73rd Congress, 2nd Session, Parts 9–13, p. 4848.
57. U.S. Congressional Committee on Banking and Currency, Stock Exchange Practices, Hearing 73rd Congress, 2nd Session, Parts 9–13, p. 4593.
58. U.S. Congressional Committee on Banking and Currency, Stock Exchange Practices, Hearing 73rd Congress, 2nd Session, Parts 9–13, p. 4626.
59. U.S. Congressional Committee on Banking and Currency, Stock Exchange Practices, Hearing 73rd Congress, 2nd Session, Parts 9–13, p. 4687.
60. U.S. Congressional Committee on Banking and Currency, Stock Exchange Practices, Hearing 73rd Congress, 2nd Session, Parts 9–13, p. 4546.
61. U.S. Congressional Committee on Banking and Currency, Stock Exchange Practices, Hearing 73rd Congress, 2nd Session, Parts 9–13, pp. 4548–4549.
62. Conot, Robert. *An American Odyssey.* William Morrow Company, New York, 1974, p. 303.
63. U.S. Congressional Committee on Banking and Currency, Stock Exchange Practices, Hearing 73rd Congress, 2nd Session, Parts 9–13, pp. 4550–4551.
64. U.S. Congressional Committee on Banking and Currency, Stock Exchange Practices, Hearing 73rd Congress, 2nd Session, Parts 9–13, p. 4565.
65. U.S. Congressional Committee on Banking and Currency, Stock Exchange Practices, Hearing 73rd Congress, 2nd Session, Parts 9–13, p. 4627.
66. Barnard, Harry. *Independent Man: The Life of Senator James Couzens.* Charles Scribner's Sons, New York, 1958, p. 218.
67. U.S. Congressional Committee on Banking and Currency, Stock Exchange Practices, Hearing 73rd Congress, 2nd Session, Parts 9–13, p. 4691.
68. Barnard, Harry. *Independent Man: The Life of Senator James Couzens.* Charles Scribner's Sons, New York, 1958, p. 226.
69. *Reminiscences*, Ernest Liebold, p. 1010.
70. U.S. Congressional Committee on Banking and Currency, Stock Exchange Practices, Hearing 73rd Congress, 2nd Session, Parts 9–13, p. 4692.
71. U.S. Congressional Committee on Banking and Currency, Stock Exchange Practices, Hearing 73rd Congress, 2nd Session, Parts 9–13, p. 4693.
72. Accession 1512.
73. U.S. Congressional Committee on Banking and Currency, Stock Exchange Practices, Hearing 73rd Congress, 2nd Session, Parts 9–13, pp. 4626–4631.
74. U.S. Congressional Committee on Banking and Currency, Stock Exchange Practices, Hearing 73rd Congress, 2nd Session, Parts 9–13, pp. 4628–4629.
75. Conot, Robert. *An American Odyssey.* William Morrow Company, New York, 1974, p. 306.
76. Ibid.
77. *Reminiscences*, Ernest Liebold, pp. 1041–1042.
78. Conot, Robert. *An American Odyssey.* William Morrow Company, New York, 1974, p. 306.
79. Neville, Howard Ralph. "The Detroit Banking Collapse of 1933." Occasional Paper No. 2, Bureau of Business and Economic Research, Michigan State University, East Lansing, MI, 1960, p. 57.
80. Sward, Keith. *The Legend of Henry Ford.* Rinehart and Company, New York, 1948, p. 252.
81. Ibid.

82. U.S. Congressional Committee on Banking and Currency, Stock Exchange Practices, Hearing 73rd Congress, 2nd Session, Parts 9–13, pp. 4548–4575.
83. U.S. Congressional Committee on Banking and Currency, Stock Exchange Practices, Hearing 73rd Congress, 2nd Session, Parts 9–13, pp. 4655–4697.
84. *Reminiscences*, Irving Bacon, p. 223; also Bennett, Harry. *We Never Called Him Henry*. Gold Metal Books, Fawcett Publications, New York, 1987, p. 148.
85. Accession 1512.
86. *Reminiscences*, Ernest Kanzler, p. 975.
87. Accession 6, Box 275.
88. Accession 6, Box 291.
89. Ibid.
90. Accession 6, Box 275.
91. Accession 6, Box 155.
92. Accession 6, Box 156.
93. Ibid.

Chapter 15: Patron of the Arts

1. Sterne, Margaret. *The Passionate Eye: The Life of William Valentiner*. Wayne State University Press, Detroit, 1980, p. 268.
2. Accession 6, Box 320.
3. Ibid.
4. *Reminiscences*, Ernest Kanzler; also Crowther, Samuel. "Edsel B. Ford." *Saturday Evening Post*, December 25, 1926, p. 78.
5. *The Detroit News*, March 8, 1928.
6. *Reminiscences*, Fred Black, p. 48.
7. Brough, James. *The Ford Dynasty: An American Story*. Doubleday and Company, New York, 1977, p. 135.
8. *Reminiscences*, Irving Bacon, p. 108.
9. Ford, Sheila F. "Edsel Ford: Artistic Industrialist." *The Herald*, Vol. 8, No. 2, p. 35.
10. Accession 1; also Accession 6, Boxes 191–193.
11. Accession 1, Box 19–1.
12. *Ward's Quarterly*, Summer 1965, Vol. 1, No. 3, p. 128.
13. *Reminiscences*, J.L. McCloud, p. 40.
14. *Reminiscences*, Irving Bacon, p. 7.
15. Grace Prunk, interview, July 12, 1989.
16. Peck, William H. *The Detroit Institute of Arts*. The Founders Society, The Detroit Institute of Arts, Detroit, 1991, p. 71.
17. Sterne, Margaret. *The Passionate Eye: The Life of William Valentiner*. Wayne State University Press, Detroit, 1980, p. 173.
18. Accession 7, Box 9.
19. Accession 6, Box 334.
20. Sterne, Margaret. *The Passionate Eye: The Life of William Valentiner*. Wayne State University Press, Detroit, 1980, p. 161.
21. Detroit Institute of Arts, file G6-31.

22. Ibid.
23. *Reminiscences*, Arthur Backus, p. 6.
24. *Reminiscences*, Arthur Backus, p. 5.
25. *Reminiscences*, Arthur Backus, p. 7.
26. Sterne, Margaret. *The Passionate Eye: The Life of William Valentiner*. Wayne State University Press, Detroit, 1980, pp. 161–162.
27. *Reminiscences*, Arthur Backus, p. 6.
28. Sterne, Margaret. *The Passionate Eye: The Life of William Valentiner*. Wayne State University Press, Detroit, 1980, p. 162.
29. Sterne, Margaret. *The Passionate Eye: The Life of William Valentiner*. Wayne State University Press, Detroit, 1980, p. 161.
30. Crowther, Samuel. "Edsel B. Ford." *Saturday Evening Post*, December 25, 1926, p. 78.
31. The Detroit Institute of Arts Archives, William Valentiner File 8/8.
32. Sterne, Margaret. *The Passionate Eye: The Life of William Valentiner*. Wayne State University Press, Detroit, 1980, p. 162; also *Ward's Quarterly*, Summer 1965, Vol. 1, No. 3, p. 128.
33. Greenleaf, William. *From These Beginnings: The Early Philanthropies of Henry and Edsel Ford, 1911–1936*. Wayne State University Press, Detroit, 1964, p. 163.
34. The Detroit Institute of Arts Archives, William Valentiner File 29/15.
35. The Detroit Institute of Arts Archives, File G6-31.
36. *Detroit Free Press*, January 31, 1935.
37. Richards, William C. *The Last Billionaire*. Charles Scribner's Sons, New York, 1948, pp. 211–212.
38. The Detroit Institute of Arts Archives, William Valentiner File 12/4.
39. Accession 6, Box 336.
40. Ibid.
41. Ibid.
42. Ibid.
43. Ibid.
44. Ibid.
45. Sterne, Margaret. *The Passionate Eye: The Life of William Valentiner*. Wayne State University Press, Detroit, 1980, p. 186.
46. Ibid.
47. *Reminiscences*, Rosa Buhler and J.D. Thompson, p. 20.
48. Sterne, Margaret. *The Passionate Eye: The Life of William Valentiner*. Wayne State University Press, Detroit, 1980, p. 186.
49. Sterne, Margaret. *The Passionate Eye: The Life of William Valentiner*. Wayne State University Press, Detroit, 1980, p. 188.
50. Sterne, Margaret. *The Passionate Eye: The Life of William Valentiner*. Wayne State University Press, Detroit, 1980, p. 189.
51. Ibid.
52. *Reminiscences*, A.G. Wolfe, p. 340.
53. *The Detroit News*, February 15, 1931.
54. *The Detroit News*, March 21, 1933.

55. Biography 1948, p. 41.
56. *Reminiscences*, A.G. Wolfe, p. 340.
57. Rivera, Diego. *My Life, My Art*. Citadel Press, New York, 1960, p. 182.
58. Ibid.
59. *The Detroit News*, April 22, 1932.
60. *The Detroit News*, April 22, 1932.
61. Sterne, Margaret. *The Passionate Eye: The Life of William Valentiner*. Wayne State University Press, Detroit, 1980, p. 193.
62. Rivera, Diego. *My Life, My Art*. Citadel Press, New York, 1960, p. 190.
63. Rivera, Diego. *My Life, My Art*. Citadel Press, New York, 1960, p. 183.
64. Rivera, Diego. *Portrait of America*. Covici, Friede, New York, 1934, p. 18.
65. Mrs. Lucienne Bloch Dimitroff, interview, October 22, 1990.
66. *Detroit Free Press*, March 14, 1933.
67. Sterne, Margaret. *The Passionate Eye: The Life of William Valentiner*. Wayne State University Press, Detroit, 1980, p. 194.
68. Ibid.
69. Ibid.
70. *The Detroit News*, October 4, 1932.
71. *Reminiscences*, A.G. Wolfe, p. 345.
72. Sterne, Margaret. *The Passionate Eye: The Life of William Valentiner*. Wayne State University Press, Detroit, 1980, p. 201.
73. Ibid.
74. Ibid.
75. Mrs. Lucienne Bloch Dimitroff, interview, October 22, 1990.
76. Rivera, Diego. *My Life, My Art*. Citadel Press, New York, 1960, p. 186; also The Detroit Institute of Arts Archives, Clyde Burroughs File 48/9.
77. Accession 6, Box 308.
78. Sterne, Margaret. *The Passionate Eye: The Life of William Valentiner*. Wayne State University Press, Detroit, 1980, p. 195.
79. Downs, Linda. "Diego Rivera's Portrait of Edsel Ford." *Bulletin of The Detroit Institute of Arts*, Vol. 57, No. 1, p. 48.
80. Accession 6, Box 375.
81. Mrs. Lucienne Bloch Dimitroff, interview, October 22, 1990.
82. Ibid.
83. Sorensen, Lorin. "The Fishleigh Fords." *Special-Interest Autos*, March–April 1977, p. 55.
84. Downs, Linda. "Diego Rivera's Portrait of Edsel Ford." *Bulletin of The Detroit Institute of Arts*, Vol. 57, No. 1, p. 47.
85. Mrs. Lucienne Bloch Dimitroff, October 22, 1990.
86. Accession 6, Box 157.
87. *Reminiscences*, Fred Black, p. 116.
88. Downs, Linda. "Diego Rivera's Portrait of Edsel Ford." *Bulletin of The Detroit Institute of Arts*, Vol. 57, No. 1, 1979, p. 52.
89. Rivera, Diego. *My Art, My Life*. Citadel Press, New York, 1960, pp. 196–197.
90. *Detroit Free Press*, March 26, 1933.

91. Peck, William H. *The Detroit Institute of Arts*. The Founders Society, The Detroit Institute of Arts, Detroit, 1991, pp. 83 and 88.
92. Accession 6, Box 360.
93. *Reminiscences*, Fred Black, p. 117.
94. Sterne, Margaret. *The Passionate Eye: The Life of William Valentiner*. Wayne State University Press, Detroit, 1980, p. 211.
95. Greenleaf, William. *From These Beginnings: The Early Philanthropies of Henry and Edsel Ford, 1911–1936*. Wayne State University Press, Detroit, 1964, p. 163.
96. The Detroit Institute of Arts Archives, File G6-31.
97. Sterne, Margaret. *The Passionate Eye: The Life of William Valentiner*. Wayne State University Press, Detroit, 1980, p. 197.
98. Greenleaf, William. *From These Beginnings: The Early Philanthropies of Henry and Edsel Ford, 1911–1936*. Wayne State University Press, Detroit, 1964, p. 163.
99. *Detroit Free Press*, May 27, 1943.
100. The Detroit Institute of Arts, file G6-31; also Accession 6, Box 360.
101. *Reminiscences*, Fred Black, pp. 116–117.
102. Ibid.
103. Ibid.
104. *Ward's Quarterly*, Summer 1965, Vol. 1, No. 3, pp. 129–130.
105. *Reminiscences*, Fred Black, pp. 117–118.
106. The Detroit Institute of Arts Archives.
107. Sterne, Margaret. *The Passionate Eye: The Life of William Valentiner*. Wayne State University Press, Detroit, 1980, p. 92.
108. *Ward's Quarterly*, Summer 1965, Vol. 1, No. 3, pp. 128–129.
109. Sterne, Margaret. *The Passionate Eye: The Life of William Valentiner*. Wayne State University Press, Detroit, 1980, p. 197.
110. *Reminiscences*, Fred Black, pp. 118–120.
111. Ibid.
112. *The Art Digest*, April 15, 1933, p. 6.
113. Sterne, Margaret. *The Passionate Eye: The Life of William Valentiner*. Wayne State University Press, Detroit, 1980, p. 201.
114. *Reminiscences*, Fred Black, p. 122.
115. The Detroit Institute of Arts Archives, Clyde Burroughs, Box 48-10.
116. Sterne, Margaret. *The Passionate Eye: The Life of William Valentiner*. Wayne State University Press, Detroit, 1980, p. 189.
117. *Detroit Free Press*, March 21, 1933.
118. *Detroit Free Press*, March 23, 1933.
119. *Detroit Free Press*, March 17, 1933.
120. *The Detroit News*, March 21, 1933.
121. *Detroit Times*, March 18, 1933.
122. Rivera, Diego. *Portrait of America*. Covici, Friede, New York, 1934, p. 54.
123. Accession 6.
124. *Detroit Free Press*, March 22, 1933.
125. *The Detroit News*, March 26, 1933.

126. *Detroit Free Press*, March 23, 1933.
127. *The Detroit News*, March 22, 1933.
128. *The Detroit News*, March 21, 1933.
129. *The Detroit News*, March 19, 1933.
130. Ibid.
131. Ibid.
132. *Detroit Free Press*, April 12, 1933.
133. Ibid.
134. *Detroit Free Press*, March 29, 1933.
135. *The Detroit News*, March 21, 1933.
136. *American Magazine of Art*, May 1933, p. 254.
137. Mrs. Lucienne Bloch Dimitroff, interview, October 22, 1990.

Chapter 16: Aviation Pioneer

1. Accession 7, Box 9.
2. Byrd, Richard E. "Fly On! Fly On!" *Collier's*, May 12, 1928, p. 8.
3. Crowther, Samuel. "Edsel B. Ford." *Saturday Evening Post*, December 25, 1926, p. 79.
4. Accession 6, Box 252.
5. Stout, William Bushnell. *So Away I Went!* The Bobbs-Merrill Company, New York, 1951, p. 174.
6. Accession 6, Box 252.
7. Ibid.
8. Ibid.
9. Ibid.
10. Ibid.
11. Stout, William Bushnell. *So Away I Went!* The Bobbs-Merrill Company, New York, 1951, p. 74.
12. Accession 6, Box 252.
13. O'Callaghan, Timothy J. *Henry Ford's Airport and Other Aviation Interests: 1909–1954.* Proctor Publications, Ann Arbor, 1993, p. 12.
14. Nevins, Allan. *Ford: Expansion and Challenge: 1915–1933.* Charles Scribner's Sons, New York, 1954, pp. 66–67.
15. Stout, William Bushnell. *So Away I Went!* The Bobbs-Merrill Company, New York, 1951, p. 120.
16. Ibid.
17. Holden, Henry M. *The Fabulous Ford Tri-Motors: The Saga of the Tin Goose.* TAB Books, Blue Ridge Summit, PA, 1992, p. 19.
18. Stout, William Bushnell. *So Away I Went!* The Bobbs-Merrill Company, New York, 1951, p. 137.
19. Stout, William Bushnell. *So Away I Went!* The Bobbs-Merrill Company, New York, 1951, p. 156.
20. Stout, William Bushnell. *So Away I Went!* The Bobbs-Merrill Company, New York, 1951, p. 159.

21. Stout, William Bushnell. *So Away I Went!* The Bobbs-Merrill Company, New York, 1951, p. 160.
22. Stout, William Bushnell. *So Away I Went!* The Bobbs-Merrill Company, New York, 1951, p. 162.
23. *Reminiscences*, Ernest Liebold, p. 762.
24. Stout, William Bushnell. *So Away I Went!* The Bobbs-Merrill Company, New York, 1951, p. 69.
25. Accession 6, Box 271.
26. Stout, William Bushnell. *So Away I Went!* The Bobbs-Merrill Company, New York, 1951, p. 170.
27. Mrs. Evelyn Book, various interviews in 1993.
28. Accession 6, Box 271.
29. *Reminiscences*, Glenn Hoppin, p. 66.
30. Accession 6, Box 271.
31. Ibid.
32. Ibid.
33. Stout, William Bushnell. *So Away I Went!* The Bobbs-Merrill Company, New York, 1951, p. 108.
34. *Reminiscences*, William N. Mayo, p. 34.
35. Ford, Henry. *Today and Tomorrow*. Doubleday, Page and Company, Garden City, NJ, 1926, p. 205.
36. Stout, William Bushnell. *So Away I Went!* The Bobbs-Merrill Company, New York, 1951, p. 173.
37. Stout, William Bushnell. *So Away I Went!* The Bobbs-Merrill Company, New York, 1951, p. 175.
38. Stout, William Bushnell. *So Away I Went!* The Bobbs-Merrill Company, New York, 1951, p. 198.
39. Accession 6, Box 271.
40. Ibid.
41. *Reminiscences*, Ernest Liebold, pp. 772–773.
42. Stout, William Bushnell. *So Away I Went!* The Bobbs-Merrill Company, New York, 1951, p. 182.
43. *Reminiscences*, Ernest Liebold, p. 761.
44. Stout, William Bushnell. *So Away I Went!* The Bobbs-Merrill Company, New York, 1951, p. 186.
45. *Reminiscences*, Fred Black, p. 188.
46. Accession 6, Box 271.
47. Nevins, Allan. *Ford: Expansion and Challenge: 1915–1933*. Charles Scribner's Sons, New York, 1954, p. 242.
48. Accession 6, Box 271.
49. Ibid.
50. Accession 6, Box 270.
51. Stout, William Bushnell. *So Away I Went!* The Bobbs-Merrill Company, New York, 1951, p. 167.

52. *Reminiscences*, Ernest Liebold, p. 766.
53. Accession 6, Box 258.
54. Accession 6, Box 259.
55. Accession 6, Box 279.
56. *Reminiscences*, Al Esper, p. 62.
57. *Reminiscences*, Harold Hicks, p. 87.
58. *Reminiscences*, Al Esper, p. 62.
59. *Reminiscences*, Ernest Liebold, p. 769.
60. Ibid.
61. *Aviation*, April 1925.
62. *Reminiscences*, Ernest Liebold, pp. 770–771.
63. *Reminiscences*, Emil Zoerlein, p. 145.
64. Accession 7, Box 9.
65. Byrd, Richard E. "Fly On! Fly On!" *Collier's*, May 12, 1928, p. 8.
66. *Reminiscences*, Al Esper, p. 61.
67. Byrd, Richard E. "Fly On! Fly On!" *Collier's*, May 12, 1928, p. 8.
68. Byrd, Richard E. "Fly On! Fly On!" *Collier's*, May 12, 1928, pp. 8–9 and 55.
69. King, Lawrence G. "A New Era Dawns in Industry." *Nation's Business*, April 1929, pp. 15–16.
70. Ford, Edsel. "Why We Have Faith in Aircraft." *Nation's Business*, May 1, 1925.
71. *Reminiscences*, Glenn Hoppin, p. 197.
72. Accession 6, Box 398.
73. Ibid.
74. Forden, Lesley. *The Ford Air Tours: 1925–1931*. The Nottingham Press, Alameda, CA, 1973, pp. 1–2.
75. Accession 6, Box 398.
76. *Reminiscences*, Irving Bacon, p. 53.
77. Taylor, Marion. "Reliability Proved by the Air Tour." *The Wind Cone*, October 1929.
78. O'Callaghan, Timothy J. *Henry Ford's Airport and Other Aviation Interests: 1909–1954*. Proctor Publications, Ann Arbor, 1993, pp. 31–32.
79. Accession 6, Box 271.
80. Stout, William Bushnell. *So Away I Went!* The Bobbs-Merrill Company, New York, 1951, p. 203.
81. Holden, Henry M. *The Fabulous Ford Tri-Motors: The Saga of the Tin Goose*. TAB Books, Blue Ridge Summit, PA, 1992, p. 53.
82. *Reminiscences*, Harold Hicks, pp. 61–62.
83. *Reminiscences*, Harold Hicks, p. 68.
84. *Reminiscences*, Ernest Liebold, p. 762.
85. Holden, Henry M. *The Fabulous Ford Tri-Motors: The Saga of the Tin Goose*. TAB Books, Blue Ridge Summit, PA, 1992, p. 55.
86. Nevins, Allan. *Ford: Expansion and Challenge: 1915–1933*. Charles Scribner's Sons, New York, 1954, p. 244.
87. Appel, Joseph H. *The Business Biography of John Wanamaker, Founder and Builder*. The Macmillan Company, New York, 1930, p. 167.

88. Nevins, Allan. *Ford: Expansion and Challenge: 1915–1933.* Charles Scribner's Sons, New York, 1954, p. 245.
89. *Reminiscences*, Ernest Liebold, 760.
90. Ibid.
91. Stout, William Bushnell. *So Away I Went!* The Bobbs-Merrill Company, New York, 1951, p. 75.
92. *Reminiscences*, Ernest Liebold, p. 765.
93. O'Callaghan, Timothy J. *Henry Ford's Airport and Other Aviation Interests: 1909–1954.* Proctor Publications, Ann Arbor, 1993, pp. 98–102.
94. Lindbergh, Charles A. *The Wartime Journals of Charles A. Lindbergh.* Harcourt, Brace, Jovanovich, New York, 1970, p. 628.
95. *Reminiscences*, Ernest Liebold, pp. 767–768.
96. *Reminiscences*, Ernest Liebold, p. 782.
97. *Reminiscences*, William N. Mayo, p. 38–39; also O'Callaghan, Timothy J. *Henry Ford's Airport and Other Aviation Interests: 1909–1954.* Proctor Publications, Ann Arbor, 1993, p. 101.
98. *Reminiscences*, Harold Hicks, p. 202.
99. Accession 1512.
100. Holden, Henry M. *The Fabulous Ford Tri-Motors: The Saga of the Tin Goose.* TAB Books, Blue Ridge Summit, PA, 1992, pp. 121–122.
101. *Reminiscences*, William N. Mayo, p. 39.
102. Lindbergh, Charles A. *The Wartime Journals of Charles A. Lindbergh.* Harcourt, Brace, Jovanovich, New York, 1970, pp. 628–629.
103. Stout, William Bushnell. *So Away I Went!* The Bobbs-Merrill Company, New York, 1951, p. 189.

Chapter 17: The Byrd Expeditions

1. Accession 6, Box 328.
2. Accession 6, Box 273.
3. Hoyt, Edwin P. *The Last Explorer: The Adventures of Admiral Byrd.* The John Day Company, New York, 1968, p. 70.
4. Ibid.
5. Byrd, Commander Richard E. *Skyward.* G.P. Putnam's Sons, New York, 1928, p. 328.
6. Greenleaf, William. *From These Beginnings: The Early Philanthropies of Henry and Edsel Ford, 1911–1936.* Wayne State University Press, Detroit, 1964, p. 158.
7. Byrd, Commander Richard E. *Skyward.* G.P. Putnam's Sons, New York, 1928, p. 332.
8. Accession 6, File " '21-'27."
9. Accession 6, Box 257, File "By-De."
10. Ibid.
11. Byrd, Commander Richard E. *Skyward.* G.P. Putnam's Sons, New York, 1928, p. 330; also Greenleaf, William. *From These Beginnings: The Early Philanthropies of Henry and Edsel Ford, 1911–1936.* Wayne State University Press, Detroit, 1964, pp. 159–160.
12. Accession 6, Box 257, File "By-De."
13. Accession 6, File " '21-'27."

14. Accession 6, Box 257, File "By-De."
15. Ibid.
16. Ibid.
17. Byrd, Commander Richard E. *Skyward.* G.P. Putnam's Sons, New York, 1928, p. 328.
18. Ibid.
19. Accession 6, Box 257, File "By-De."
20. Ibid.
21. Ibid.
22. Ibid.
23. O'Callaghan, Timothy J. *Henry Ford's Airport and Other Aviation Interests: 1909–1954.* Proctor Publications, Ann Arbor, 1993, p. 80.
24. Byrd, Commander Richard E. *Skyward.* G.P. Putnam's Sons, New York, 1928, pp. 196–199.
25. Accession 6, Box 257, File "By-De."
26. Byrd, Commander Richard E. *Skyward.* G.P. Putnam's Sons, New York, 1928, p. 333.
27. Accession 6, Box 257, File "By-De."
28. Ibid.
29. Ibid.
30. Ibid.
31. Ibid.
32. Byrd, Commander Richard E. *Skyward.* G.P. Putnam's Sons, 1928, pp. 327–328.
33. *Detroit Free Press*, August 2, 1927.
34. Holden, Henry M. *The Fabulous Ford Tri-Motors: The Saga of the Tin Goose.* TAB Books, Blue Ridge Summit, PA, 1992, p. 81.
35. Accession 6, File " '21-'27."
36. Accession 6, Box 273.
37. Accession 6, File " '21-'27."
38. Accession 6, File " '21-'27"; also Accession 6, Box 257, File "By-De."
39. *Reminiscences*, Harold Hicks, pp. 90–91.
40. Hoyt, Edwin P. *The Last Explorer: The Adventures of Admiral Byrd.* The John Day Company, New York, 1968, p. 158.
41. Accession 6, Box 273.
42. Ibid.
43. "Byrd's Polar Flight Backed by Edsel Ford." *Detroit Free Press*, August 2, 1927.
44. Ibid.
45. Accession 6, Box 281.
46. Ibid.
47. Hoyt, Edwin P. *The Last Explorer: The Adventures of Admiral Byrd.* The John Day Company, New York, 1968, p. 253.
48. Hoyt, Edwin P. *The Last Explorer: The Adventures of Admiral Byrd.* The John Day Company, New York, 1968, p. 254.
49. Accession 6, Box 257, File "By-De."
50. Accession 6, Box 328.
51. Ibid.

52. Hoyt, Edwin P. *The Last Explorer: The Adventures of Admiral Byrd.* The John Day Company, New York, 1968, p. 260.
53. Accession 6, Box 328.
54. Hoyt, Edwin P. *The Last Explorer: The Adventures of Admiral Byrd.* The John Day Company, New York, 1968, p. 267.
55. Accession 6, Box 328.
56. Ibid.
57. Ibid.
58. Ibid.
59. Ibid.
60. Hoyt, Edwin P. *The Last Explorer: The Adventures of Admiral Byrd.* The John Day Company, New York, 1968, pp. 261–262.
61. Hoyt, Edwin P. *The Last Explorer: The Adventures of Admiral Byrd.* The John Day Company, New York, 1968, p. 298.
62. Ibid.
63. Accession 6, Box 352.
64. Accession 6, Box 333.

Chapter 18: The Lincoln Motor Car Company

1. Nevins, Allan. *Ford: Expansion and Challenge: 1915–1933.* Charles Scribner's Sons, New York, 1954, p. 172.
2. Stoddard, Gloria May. *Henry Leland.* The New England Press, Shelburne, VT, 1986, pp. 45–46.
3. Stoddard, Gloria May. *Henry Leland.* The New England Press, Shelburne, VT, 1986, p. 50.
4. Nevins, Allan. *Ford: Expansion and Challenge: 1915–1933.* Charles Scribner's Sons, New York, 1954, p. 174.
5. Stoddard, Gloria May. *Henry Leland.* The New England Press, Shelburne, VT, 1986, pp. 19 and 100.
6. Nevins, Allan. *Ford: Expansion and Challenge: 1915–1933.* Charles Scribner's Sons, New York, 1954, p. 175.
7. Nevins, Allan. *Ford: Expansion and Challenge: 1915–1933.* Charles Scribner's Sons, New York, 1954, p. 177.
8. Nevins, Allan. *Ford: Expansion and Challenge: 1915–1933.* Charles Scribner's Sons, New York, 1954, p. 171.
9. Accession 6, Box 276.
10. Nevins, Allan. *Ford: Expansion and Challenge: 1915–1933.* Charles Scribner's Sons, New York, 1954, p. 172.
11. Accession 6, Box 276.
12. Ibid.
13. *Reminiscences*, Harold Robinson, pp. 71–72.
14. Ford Archives film collection.
15. Accession 1512.
16. Accession 6, Box 265.
17. Ibid.

18. Ibid.
19. Accession 6, Box 252.
20. Accession 6, Box 276.
21. Ibid.
22. Accession 6, Box 276; also *Design Oral History*, E.T. Gregorie, p. 17.
23. Nevins, Allan. *Ford: Expansion and Challenge: 1915–1933*. Charles Scribner's Sons, New York, 1954, pp. 190–191.
24. Ibid.
25. Accession 6, Box 276.
26. Ibid.
27. Accession 6, Box 265.
28. Ibid.

Chapter 19: Carrossiers

1. Ford, Edsel B. "The Motor Car of the Future." *Commercial Art and Industry*, November 1935, p. 170.
2. Borgeson, Griffith and Eugene Jaderquist. *Sports and Classic Cars*. Prentice-Hall, New York, 1955, p. 222.
3. Ralph Roberts, various interviews between 1989 and 1997.
4. Ibid.
5. Borgeson, Griffith and Eugene Jaderquist. *Sports and Classic Cars*. Prentice-Hall, New York, 1955, pp. 222–223.
6. *Reminiscences*, Harold Robinson, p. 74.
7. Accession 6, Box 255.
8. Accession 6, Box 274.
9. *Ward's Quarterly*, Summer 1965, Vol. 1, No. 3, p. 70.
10. *Design Oral History*, E.T. Gregorie, p. 24.
11. Mandel, Leon. *American Cars*. Stewart, Tabori and Chang, New York, 1982, p. 172.
12. *Automobile and Culture*, p. 218.
13. Ibid.
14. Ralph Roberts, various interviews between 1989 and 1997.
15. *Design Oral History*, E.T. Gregorie, p. 17.
16. Ralph Roberts, various interviews between 1989 and 1997.
17. *Reminiscences*, J.M. Waggoner, p. 83.
18. Accession 6, Box 252.
19. *Reminiscences*, A.A. Lepine, p. 28.
20. Laas, William. "The Meaning of a Name." *Your Edsel Marketer*, Vol. 1, No. 2, publication of Ford Motor Company, no date; also Ralph Roberts, various interviews between 1989 and 1997.
21. Accession 6, Box 252.
22. Accession 6, Box 255.
23. Ibid.
24. Ibid.

25. Borgeson, Griffith and Eugene Jaderquist. *Sports and Classic Cars*. Prentice-Hall, New York, 1955, p. 159.
26. *Detroit Style: Automotive Form: 1925–1950*. The Detroit Institute of Arts, Detroit, 1985, p. 107.
27. Ralph Roberts, various interviews between 1989 and 1997.
28. Ibid.
29. Ibid.
30. Ibid.
31. Ibid.
32. Ibid.
33. Pfau, Hugo. *The Custom Body Era*. Castle Books, New York, 1970, p. 24.
34. Ralph Roberts, various interviews between 1989 and 1997.
35. Ross Cousins, various interviews between 1990 and 1996.
36. Borgeson, Griffith and Eugene Jaderquist. *Sports and Classic Cars*. Prentice-Hall, New York, 1955, p. 159.
37. Ralph Roberts, various interviews between 1989 and 1997.
38. Ibid.
39. Ibid.
40. Pfau, Hugo. *The Custom Body Era*. Castle Books, New York, 1970, p. 85.
41. Ralph Roberts, various interviews between 1989 and 1997.
42. Ibid.
43. Ibid.
44. *Reminiscences*, R.C. Getsinger, p. 26.
45. Accession 6, Box 252; also Accession 6, Box 279.
46. Accession 6, Box 252.
47. Ibid.
48. Ibid.
49. Accession 6, Box 259.
50. *Ward's Quarterly*, Summer 1965, Vol. 1, No. 3, p. 70.
51. Accession 6, Box 255.
52. Ibid.
53. Mandel, Leon. *American Cars*. Stewart, Tabori and Chang, New York, 1982, p. 172.
54. *Design Oral History*, E.T. Gregorie, p. 14.
55. Accession 6, Box 33; also *Design Oral History*, E.T. Gregorie, p. 29.
56. Accession 6, Box 30.
57. Accession 6, Box 333.
58. Mandel, Leon. *American Cars*. Stewart, Tabori and Chang, New York, 1982, pp. 174 and 194.
59. Accession 6, Box 258.
60. *Reminiscences*, J.L. McCloud, p. 415.
61. *Reminiscences*, Carmen Hueston, pp. 100–101.
62. *Reminiscences*, Laurence Sheldrick, p. 113.
63. Ford, Edsel B. "The Motor Car of the Future." *Commercial Art and Industry*, November 1935, p. 172.

64. Accession 6, Box 407.
65. Ralph Roberts, various interviews between 1989 and 1997.
66. Ibid.
67. Ibid.
68. Ibid.
69. Ibid.
70. *Ward's Quarterly*, Summer 1965, Vol. 1, No. 3, p. 70.
71. Ralph Roberts, various interviews between 1989 and 1997.
72. Ibid.
73. Ibid.
74. Ibid.
75. Mandel, Leon. *American Cars*. Stewart, Tabori and Chang, New York, 1982, p. 172.
76. Ralph Roberts, various interviews between 1989 and 1997.
77. Ibid.
78. Ibid.
79. Pfau, Hugo. *The Custom Body Era*. Castle Books, New York, 1970, p. 77.
80. Ralph Roberts, various interviews between 1989 and 1997.
81. Ibid.
82. *Ward's Quarterly*, Summer 1965, Vol. 1, No. 3, p. 70.
83. Accession 6, Box 265.
84. *Design Oral History*, E.T. Gregorie, p. 13.
85. Accession 6, Box 252.
86. Ralph Roberts, various interviews between 1989 and 1997.
87. Ibid.
88. Ibid.
89. Ibid.
90. *Ward's Quarterly*, Summer 1965, Vol. 1, No. 3, p. 127.
91. Langworth, Richard M. "LeBaron: Thoroughbred of Custom Coachbuilders." *Automobile Quarterly*, Vol. XII, No. 3, p. 234.
92. Ralph Roberts, various interviews between 1989 and 1997.
93. Ibid.
94. Ibid.
95. Langworth, Richard M. "LeBaron: Thoroughbred of Custom Coachbuilders." *Automobile Quarterly*, Vol. XII, No. 3, p. 234.
96. Ralph Roberts, various interviews between 1992 and 1997.
97. Rhys Miller, interview, November 21, 1995.
98. Ralph Roberts, various interviews between 1992 and 1997.
99. Ibid.
100. *Design Oral History*, John Najjar, p. 38.
101. *Detroit Free Press*, January 20, 1924.
102. Ritch, Ocee. *The Lincoln Continental*. Motorbooks International Publishers and Wholesalers, Osceola, WI, p. 25.

Chapter 20: Felicitous Eye

1. McNarin, William and Marjorie. *Quotations from the Unusual Henry Ford.* Quotamus Press, Redondo Beach, CA, 1978, p. 89.
2. Ford, Edsel B. "The Motor Car of the Future." *Commercial Art and Industry*, November 1935, p. 169.
3. Nevins, Allan. *Ford: Expansion and Challenge: 1915–1933.* Charles Scribner's Sons, New York, 1954, p. 444.
4. *Reminiscences*, Joseph Galamb, p. 34.
5. Ford, Edsel B. "The Motor Car of the Future." *Commercial Art and Industry*, November 1935, p. 169.
6. "Edsel Ford's Hotrods." *Special-Interest Autos*, December 1970, p. 37.
7. Nevins, Allan. *Ford: Expansion and Challenge: 1915–1933.* Charles Scribner's Sons, New York, 1954, p. 395.
8. Cameron, William J. "In Memory." *The Herald*, Vol. X, No. 9, 1943, p. 3.
9. Teague, Walter Dorwin. "Edsel Ford: Designer." *Lincoln-Mercury Times*, May–June 1953, p. 23.
10. Laas, William. "The Meaning of a Name." *Your Edsel Marketer*, Vol. 1, No. 2, publication of Ford Motor Company, no date.
11. Ibid.
12. *Ward's Quarterly*, Summer 1965, Vol. 1, No. 3, p. 128.
13. *Reminiscences*, A.G. Wolfe, p. 38.
14. Accession 6.
15. *Reminiscences*, W. Ernest Grimshaw, pp. 24–25.
16. *Ward's Quarterly*, Summer 1965, Vol. 1, No. 3, p. 68.
17. *Design Oral History*, William Clay Ford, p. 2.
18. *Ward's Quarterly*, Summer 1965, Vol. 1, No. 3, p. 68.
19. Accession 6, Box 252.
20. "Edsel Ford's Hotrods." *Special-Interest Autos*, December 1970, p. 37.
21. *Design Oral History*, E.T. Gregorie, p. 14.
22. *The Detroit News*, December 26, 1929.
23. *Ward's Quarterly*, Summer 1965, Vol. 1, No. 3, p. 70.
24. Ritch, Ocee. *The Lincoln Continental*. Motorbooks International Publishers and Wholesalers, Osceola, WI, p. 45.
25. E.T. Gregorie, various interviews between 1992 and 1998.
26. *Design Oral History*, E.T. Gregorie, pp. 16–17.
27. *Design Oral History*, E.T. Gregorie, pp. 1–14.
28. *Design Oral History*, E.T. Gregorie, p. 14.
29. *Reminiscences*, Laurence Sheldrick, p. 171.
30. *Reminiscences*, Al Esper, p. 92.
31. Laas, William. "The Meaning of a Name." *Your Edsel Marketer*, Vol. 1, No. 2, publication of Ford Motor Company, no date.
32. E.T. Gregorie, various interviews between 1992 and 1998.
33. *Reminiscences*, Laurence Sheldrick, p. 171.

34. E.T. Gregorie, various interviews between 1992 and 1998.
35. *Design Oral History*, E.T. Gregorie, p. 96.
36. Tucker Madawick, various interviews in 1996; also John Najjar, various interviews between 1990 and 1998.
37. E. T. Gregorie, various interviews between 1992 and 1998.
38. Laas, William. "The Meaning of a Name." *Your Edsel Marketer*, Vol. 1, No. 2, publication of Ford Motor Company, no date.
39. Armi, C. Edson. *The Art of American Automobile Design*. The Pennsylvania State University Press, University Park, PA, 1988, p. 35.
40. Armi, C. Edson. *The Art of American Automobile Design*. The Pennsylvania State University Press, University Park, PA, 1988, pp. 35 and 236.
41. *Design Oral History*, Franklin Q. Hershey, p. 88.
42. Thomas, Bob. *Confessions of an Automotive Stylist*. Self-published, 1984, pp. 18–19.
43. *Ward's Quarterly*, Summer 1965, Vol. 1, No. 3, p. 70.
44. Armi, C. Edson. *The Art of American Automobile Design*. The Pennsylvania State University Press, University Park, PA, 1988, pp. 24–25.
45. Ibid.
46. Armi, C. Edson. *The Art of American Automobile Design*. The Pennsylvania State University Press, University Park, PA, 1988, p. 237.
47. *Design Oral History*, Franklin Q. Hershey, p. 88; also *Design Oral History*, William Mitchell, p. 2.
48. Armi, C. Edson. *The Art of American Automobile Design*. The Pennsylvania State University Press, University Park, PA, 1988, p. 20.
49. *Design Oral History*, William Mitchell, p. 2; also Armi, C. Edson. *The Art of American Automobile Design*. The Pennsylvania State University Press, University Park, PA, 1988, p. 20.
50. Armi, C. Edson. *The Art of American Automobile Design*. The Pennsylvania State University Press, University Park, PA, 1988, p. 25.
51. Armi, C. Edson. *The Art of American Automobile Design*. The Pennsylvania State University Press, University Park, PA, 1988, p. 25; also Franklin Q. Hershey, various interviews between 1992 and 1997.
52. Armi, C. Edson. *The Art of American Automobile Design*. The Pennsylvania State University Press, University Park, 1988, p. 25.
53. *Design Oral History*, E.T. Gregorie, pp. 59 and 100.
54. *Design Oral History*, E.T. Gregorie, pp. 101 and 20.
55. Ralph Roberts, various interviews between 1992 and 1997; also E.T. Gregorie, various interviews between 1992 and 1998.
56. *Design Oral History*, E.T. Gregorie, p. 53.
57. Teague, Walter Dorwin. "Edsel Ford: Designer." *Lincoln-Mercury Times*, May–June 1953, p. 23.
58. E.T. Gregorie, various interviews between 1992 and 1998.
59. Armi, C. Edson. *The Art of American Automobile Design*. The Pennsylvania State University Press, University Park, PA, 1988, p. 35; also E.T. Gregorie, various interviews between 1992 and 1998.

60. *Design Oral History*, E.T. Gregorie, p. 65.
61. Armi, C. Edson. *The Art of American Automobile Design*. The Pennsylvania State University Press, University Park, PA, 1988, p. 235.
62. E.T. Gregorie, various interviews between 1992 and 1998.
63. *Reminiscences*, Laurence Sheldrick, p. 132.
64. *Reminiscences*, Joseph Galamb, p. 138.
65. Ralph Roberts, various interviews between 1992 and 1997.
66. *Design Oral History*, E.T. Gregorie, p. 97.
67. E.T. Gregorie, various interviews between 1992 and 1998.
68. *Reminiscences*, Henry Doss, p. 76.

Chapter 21: Edsel's Sanctuary

1. E.T. Gregorie, various interviews between 1992 and 1998.
2. *Reminiscences*, Harold Hicks, p. 46.
3. *Design Oral History*, Alex Tremulis, p. 14.
4. The Model B became more popular as the Model 18, equipped with the new Ford V8 engine.
5. *Reminiscences*, Laurence Sheldrick, p. 91.
6. Tjaarda, Tom. "I Remember My Father." *Special-Interest Autos*, April–May, 1972, p. 51.
7. "Body by L.E. Briggs." *Special-Interest Autos*, November–December 1973, pp. 26–27.
8. Ibid.
9. Ralph Roberts, various interviews between 1992 and 1997.
10. Ibid.
11. Ibid.
12. Ibid.
13. Ibid.
14. *Reminiscences*, William Pioch, p. 56.
15. Nevins, Allan. *Ford: Expansion and Challenge: 1915–1933*. Charles Scribner's Sons, New York, 1954, p. 564.
16. Nevins, Allan. *Ford: Expansion and Challenge: 1915–1933*. Charles Scribner's Sons, New York, 1954, p. 553.
17. Nevins, Allan. *Ford: Expansion and Challenge: 1915–1933*. Charles Scribner's Sons, New York, 1954, p. 565.
18. *Reminiscences*, Eugene Farkas, pp. 248–249.
19. Ralph Roberts, various interviews between 1992 and 1997.
20. Armi, C. Edson. *The Art of American Automobile Design*. The Pennsylvania State University Press, University Park, PA, 1988, p. 72.
21. *Reminiscences*, Eugene Farkas, p. 253.
22. E.T. Gregorie, various interviews between 1992 and 1998.
23. *Reminiscences*, Eugene Farkas, p. 253.
24. Emmett O'Rear, various interviews between 1995 and 1996.
25. E.T. Gregorie, various interviews between 1992 and 1998.
26. Ibid.
27. Ibid.

28. Ibid.
29. Ibid.
30. Ralph Roberts, various interviews between 1992 and 1997.
31. Accession 6, Box 258.
32. E.T. Gregorie, various interviews between 1992 and 1998.
33. *Reminiscences*, Laurence Sheldrick, pp. 190–191.
34. *Reminiscences*, Charles Sorensen, p. 12.
35. Accession 1512.
36. *Reminiscences*, Frank Hadas, p. 288.
37. *Reminiscences*, Eugene Farkas, pp. 201–202.
38. *Reminiscences*, Catherine Ruddiman, p. 52.
39. E.T. Gregorie, various interviews between 1992 and 1998.
40. John Hay, interview, July 28, 1996.
41. Ibid.
42. *Reminiscences*, George Ferghete, p. 28.
43. Emmett O'Rear, various interviews between 1995 and 1996.
44. Placid Barbera, interview, July 22, 1996.
45. Emmett O'Rear, various interviews between 1995 and 1996.
46. Tucker Madawick, various interviews in 1996.
47. E.T. Gregorie, various interviews between 1992 and 1998.
48. Ibid.
49. Ibid.
50. *Design Oral History*, E.T. Gregorie, p. 80.
51. *Design Oral History*, G.E. Adams, p. 58.
52. E.T. Gregorie, various interviews between 1992 and 1998.
53. *Design Oral History*, E.T. Gregorie, p. 20.
54. Ibid.
55. *Design Oral History*, John Najjar, p. 35.
56. *Design Oral History*, E.T. Gregorie, p. 68.
57. Emmett O'Rear, various interviews between 1995 and 1996.
58. John Najjar, various interviews between 1990 and 1998.
59. Emmett O'Rear, various interviews between 1995 and 1996.
60. Ibid.
61. E.T. Gregorie, various interviews between 1992 and 1998.
62. Ibid.
63. Ibid.
64. Ibid.
65. *Design Oral History*, G.E. Adams, p. 58.
66. *Design Oral History*, John Najjar, p. 34.
67. *Design Oral History*, G.E. Adams, pp. 58–59.
68. Ralph Roberts, various interviews between 1992 and 1997.
69. E.T. Gregorie, various interviews between 1992 and 1998.
70. Armi, C. Edson. *The Art of American Automobile Design*. The Pennsylvania State University Press, University Park, 1988, p. 18.

71. Laas, William. "The Meaning of a Name." *Your Edsel Marketer*, Vol. 1, No. 2, publication of Ford Motor Company, no date.
72. Armi, C. Edson. *The Art of American Automobile Design*. The Pennsylvania State University Press, University Park, PA, 1988, p. 239.
73. E.T. Gregorie, various interviews between 1992 and 1998.
74. Ibid.
75. Ibid.
76. Ibid.
77. Ibid.
78. Ibid.
79. Ibid.
80. Ford Archives clip book, December 31, 1939.
81. E.T. Gregorie, various interviews between 1992 and 1998.
82. Ibid.
83. E.T. Gregorie, various interviews between 1992 and 1998; also *Design Oral History*, E.T. Gregorie, pp. 73–74.
84. E.T. Gregorie, various interviews between 1992 and 1998.
85. *Design Oral History*, E.T. Gregorie, p. 123.
86. Armi, C. Edson. *The Art of American Automobile Design*. The Pennsylvania State University Press, University Park, PA, 1988, p. 35.
87. *Reminiscences*, Charles Sorensen, pp. 634–635.
88. Lamm, Michael and David Lewis. "The Mercury and How It Came to Be." *Special-Interest Autos*, July–August, p. 16.
89. Collier, Peter and David Horowitz. *The Fords: An American Epic*. Summit Books, New York, 1987, p. 159.

Chapter 22: Design Synergy

1. Placid Barbera, various interviews between 1996 and 1998.
2. Ford, Edsel B. "The Motor Car of the Future." *Commercial Art and Industry*, November 1935, p. 171.
3. *Reminiscences*, Al Esper, p. 66.
4. *Reminiscences*, Al Esper, p. 65.
5. *Reminiscences*, Emil Zoerlein, pp. 81–82.
6. Ibid.
7. *Design Oral History*, E.T. Gregorie, p. 62.
8. *Reminiscences*, Laurence Sheldrick, p. 132.
9. *Reminiscences*, William Pioch, p. 56.
10. *The Detroit News*, January 2, 1919.
11. *Eight Automobiles*, publication of the Museum of Modern Art, New York, 1951, p. 4.
12. Nevins, Allan. *Ford: Expansion and Challenge: 1915–1933*. Charles Scribner's Sons, New York, 1954, p. 69.
13. Tjaarda, John, p. 30.
14. *Reminiscences*, Charles Sorensen, Sections 40–41, pp. 1–2.

15. Tjaarda, John. "How the Lincoln-Zephyr Was Born." *Motor Trend*, February 1954, p. 30.
16. Ralph Roberts, various interviews between 1989 and 1997.
17. Ibid.
18. Ibid.
19. Ibid.
20. Tjaarda, Tom. "I Remember My Father." *Special-Interest Autos*, April–May 1972, p. 51.
21. Ralph Roberts, various interviews between 1989 and 1997.
22. *The Detroit News*, September 4, 1935.
23. Mandel, Leon. *American Cars*. Stewart, Tabori and Chang, New York, 1982, p. 128.
24. *Design Oral History*, Alex Tremulis, p. 4.
25. Mandel, Leon. *American Cars*. Stewart, Tabori and Chang, New York, 1982, p. 128.
26. *Design Oral History*, Tom Tjaarda, p. 30.
27. Ralph Roberts, various interviews between 1989 and 1997.
28. *Reminiscences*, Laurence Sheldrick, p. 179.
29. *Reminiscences*, Eugene Farkas, p. 255.
30. Ford, Edsel B. "The Motor Car of the Future." *Commercial Art and Industry*, November 1935, p. 171.
31. Tjaarda, John. "How the Lincoln-Zephyr Was Born." *Motor Trend*, February 1954, p. 31.
32. *Reminiscences*, Laurence Sheldrick, pp. 170–171.
33. *Reminiscences*, Laurence Sheldrick, p. 171.
34. *Design Oral History*, E.T. Gregorie, p. 42.
35. *Design Oral History*, E.T. Gregorie, p. 18.
36. Nevins, Alan and Frank Ernest Hill. *Ford: Decline and Rebirth 1933–1962*. Charles Scribner's Sons, New York, 1962, pp. 67–69.
37. Accession 6, Box 430.
38. Accession 6, Box 348.
39. Accession 6, Box 348.
40. Tjaarda, Tom. "I Remember My Father." *Special-Interest Autos*, April–May, p. 63.
41. Borgeson, Griffith and Eugene Jaderquist. *Sports and Classic Cars*. Prentice-Hall, New York, 1955, p. 224; Teague, Walter Dorwin. "Edsel Ford—Designer." *Lincoln-Mercury Times*, May–June 1953.
42. *Design Oral History*, E.T. Gregorie, p. 43.
43. *Design Oral History*, E.T. Gregorie, p. 40.
44. Franklin Q. Hershey, various interviews between 1992 and 1997.
45. E.T. Gregorie, various interviews between 1992 and 1998.
46. Ibid.
47. Accession 624, Box 9.
48. E.T. Gregorie, various interviews between 1992 and 1998.
49. *Design Oral History*, E.T. Gregorie, pp. 35–36.
50. Thomas, Cullen. "Lincoln Continental." *Automobile Quarterly*, Vol. 6, No. 4, 1968, p. 425.
51. *Design Oral History*, G.E. Adams, pp. 25–26.
52. E.T. Gregorie, various interviews between 1992 and 1998.
53. Accession 624, Box 9.
54. E.T. Gregorie, various interviews between 1992 and 1998.

55. *Design Oral History*, G.E. Adams, p. 25C.
56. *Design Oral History*, p. 37.
57. Armi, C. Edson. *The Art of American Automobile Design*. The Pennsylvania State University Press, University Park, PA, 1988, p. 242; also *Design Oral History*, E.T. Gregorie, various interviews between 1992 and 1998, p. 62.
58. *Design Oral History*, E.T. Gregorie, pp. 39–40.
59. Ross Cousins, various interviews between 1990 and 1996.
60. *Design Oral History*, William Clay Ford, p. 7.
61. Thomas, Bob. *Confessions of an Automobile Stylist*. Self-published, 1984, p. 19.
62. *Reminiscences*, Harold Robinson, p. 116.
63. *Reminiscences*, Harold Robinson, p. 117.
64. *Design Oral History*, E.T. Gregorie, p. 38.
65. *Reminiscences*, Harold Robinson, p. 118.
66. Thomas, Cullen. "Lincoln Continental." *Automobile Quarterly*, Vol. 6, No. 4, 1968, p. 426.
67. Ibid.
68. *Reminiscences*, J.M. Waggoner, p. 102.
69. *Reminiscences*, J.M. Waggoner, p. 102.
70. E.T. Gregorie, various interviews between 1992 and 1998.
71. E.T. Gregorie, various interviews between 1992 and 1998; also Armi, C. Edson. *The Art of American Automobile Design*. The Pennsylvania State University Press, University Park, PA, 1988, p. 74.
72. E.T. Gregorie, various interviews between 1992 and 1998.
73. *Automobile Quarterly*, Vol. 1, No. 1, p. 30.
74. *Reminiscences*, Ralph Harrison, p. 42.
75. Accession 6, Box 2387.
76. Ibid.
77. E.T. Gregorie, various interviews between 1992 and 1998.
78. *Reminiscences*, Laurence Sheldrick, pp. 164–165.
79. E.T. Gregorie, various interviews between 1992 and 1998.
80. Ibid.
81. Ibid.
82. Ibid.
83. Ibid.
84. Ibid.
85. Tucker Madawick, various interviews in 1996.
86. Lamm, Michael and David Lewis. "The First Mercury and How It Came to Be." *Special-Interest Autos*, July–August 1974, p. 19.
87. E.T. Gregorie, various interviews between 1992 and 1998.
88. Ibid.

Chapter 23: The Ford Foundation

1. Wik, Reynold M. *Henry Ford and Grass-Roots America*. University of Michigan Press, Ann Arbor, 1972, p. 5.

2. Greenleaf, William. *From These Beginnings: The Early Philanthropies of Henry and Edsel Ford, 1911–1936.* Wayne State University Press, Detroit, 1964, p. 155.
3. Accession 1189.
4. Crowther, Samuel. "Edsel B. Ford." *Saturday Evening Post*, December 25, 1926, p. 78.
5. *Reminiscences*, A.A. Lepine, p. 61.
6. Ibid.
7. Accession 6, Box 259.
8. *Reminiscences*, A.A. Lepine, p. 58.
9. *Automotive News*, May 31, 1943.
10. Accession 6, Box 258.
11. *Detroit Free Press*, May 27, 1943; also Accession 6, Box 259; also James Longley, interview, May 10, 1989.
12. Greenleaf, William. *From These Beginnings: The Early Philanthropies of Henry and Edsel Ford, 1911–1936.* Wayne State University Press, Detroit, 1964, p. 18; also *Reminiscences*, A.A. Lepine, pp. 20–21.
13. Robert Smith, interview, May 18, 1989.
14. *Reminiscences*, William Cameron, p. 165.
15. *Reminiscences*, William Cameron, pp. 228–229.
16. Greenleaf, William. *From These Beginnings: The Early Philanthropies of Henry and Edsel Ford, 1911–1936.* Wayne State University Press, Detroit, 1964, p. 65.
17. *Reminiscences*, William Cameron, p. 235.
18. Greenleaf, William. *From These Beginnings: The Early Philanthropies of Henry and Edsel Ford, 1911–1936.* Wayne State University Press, Detroit, 1964, p. 69.
19. Greenleaf, William. *From These Beginnings: The Early Philanthropies of Henry and Edsel Ford, 1911–1936.* Wayne State University Press, Detroit, 1964, p. 66.
20. Accession 6, Box 348.
21. Ibid.
22. Accession 6, Box 348; also Accession 6, Box 380.
23. Greenleaf, William. *From These Beginnings: The Early Philanthropies of Henry and Edsel Ford, 1911–1936.* Wayne State University Press, Detroit, 1964, p. 157.
24. Accession 6, Box 265.
25. Sinsabaugh, Chris. *Who, Me?* Arnold-Powers, Detroit, 1940, pp. 147–150.
26. Sinsabaugh, Chris. *Who, Me?* Arnold-Powers, Detroit, 1940, p. 152; also Accession 6, Box 265.
27. Accession 6, Box 265.
28. Ibid.
29. Ibid.
30. Greenleaf, William. *From These Beginnings: The Early Philanthropies of Henry and Edsel Ford, 1911–1936.* Wayne State University Press, Detroit, 1964, p. 155.
31. Ibid.
32. Ibid.
33. *Reminiscences*, William Cameron, p. 100.
34. *Detroit Motor News*, April 1930.

35. Greenleaf, William. *From These Beginnings: The Early Philanthropies of Henry and Edsel Ford, 1911–1936.* Wayne State University Press, Detroit, 1964, pp. 25–26.
36. Accession 1189.
37. Nevins, Alan and Frank Ernest Hill. *Ford: Decline and Rebirth 1933–1962.* Charles Scribner's Sons, New York, 1962, pp. 409–410.
38. *The New York Times*, August 11, 1935.
39. *The Detroit News*, December 26, 1926.
40. Richards, William C. *The Last Billionaire.* Charles Scribner's Sons, New York, 1948, p. 356.
41. MacDonald, Dwight. *The Ford Foundation: The Men and the Millions.* Reynal and Company, New York, 1956, p. 131.
42. *Reminiscences*, Ernest Liebold, p. 1451.
43. Greenleaf, William. *From These Beginnings: The Early Philanthropies of Henry and Edsel Ford, 1911–1936.* Wayne State University Press, Detroit, 1964, p. 187.
44. MacDonald, Dwight. *The Ford Foundation: The Men and the Millions.* Reynal and Company, New York, 1956, p. 132.
45. Accession 1189.
46. MacDonald, Dwight. *The Ford Foundation: The Men and the Millions.* Reynal and Company, New York, 1956, p. 43.
47. *Reminiscences*, Charles Sorensen, p. 68.
48. Accession 1740, Box 1.
49. Accession 1189.
50. Nevins, Alan and Frank Ernest Hill. *Ford: Decline and Rebirth 1933–1962.* Charles Scribner's Sons, New York, 1962, p. 411.
51. Accession 1189.
52. *Reminiscences*, Charles Sorensen, p. 68.
53. Accession 1512.
54. Simonds, William A. *Henry Ford: His Life, His Work, His Genius.* The Bobbs-Merrill Company, New York, 1943, p. 340.
55. *Reminiscences*, Charles Sorensen, p. 67.
56. Cameron, William J. "In Memory." *The Herald*, Vol. X, No. 9, 1943, p. 217.
57. James Longley, interview, May 10, 1989.
58. Accession 1189.
59. Ibid.
60. Ibid.
61. Chambers, M.M. *Charters of Philanthropies.* The Merrymount Press, New York, 1948, pp. 134–135.
62. Chambers, M.M. *Charters of Philanthropies.* The Merrymount Press, New York, 1948, p. 134; also Greenleaf, William. *From These Beginnings: The Early Philanthropies of Henry and Edsel Ford, 1911–1936.* Wayne State University Press, Detroit, 1964, p. 188.
63. Greenleaf, William. *From These Beginnings: The Early Philanthropies of Henry and Edsel Ford, 1911–1936.* Wayne State University Press, Detroit, 1964, p. 188.
64. MacDonald, Dwight. *The Ford Foundation: The Men and the Millions.* Reynal and Company, New York, 1956, p. 131.

65. Accession 1189.
66. Ibid.
67. Ibid.

Chapter 24: Bennett Motor Company

1. *Reminiscences*, William Cameron, p. 109.
2. *Reminiscences*, Albert Wibel, p. 353.
3. Sward, Keith. *The Legend of Henry Ford.* Rinehart and Company, New York, 1948, p. 338.
4. E.T. Gregorie, various interviews between 1992 and 1998.
5. Bennett, Harry. *We Never Called Him Henry.* Gold Metal Books, Fawcett Publications, New York, 1987, p. 7.
6. Bennett, Harry. *We Never Called Him Henry.* Gold Metal Books, Fawcett Publications, New York, 1987, pp. 7–8.
7. Bennett, Harry. *We Never Called Him Henry.* Gold Metal Books, Fawcett Publications, New York, 1987, p. 11.
8. Lewis, David. *The Public Image of Henry Ford.* Wayne State University Press, Detroit, 1976, p. 161.
9. Bennett, Harry. *We Never Called Him Henry.* Gold Metal Books, Fawcett Publications, New York, 1987, p. 34.
10. Bennett, Harry. *We Never Called Him Henry.* Gold Metal Books, Fawcett Publications, New York, 1987, p. 17.
11. Bennett, Harry. *We Never Called Him Henry.* Gold Metal Books, Fawcett Publications, New York, 1987, p. 21.
12. Lacey, Robert. *Ford: The Men and the Machine.* Little, Brown and Company, Boston, 1986, p. 363.
13. E.T. Gregorie, various interviews between 1992 and 1998.
14. Bennett, Harry. *We Never Called Him Henry.* Gold Metal Books, Fawcett Publications, New York, 1987, p. 39.
15. Bennett, Harry. *We Never Called Him Henry.* Gold Metal Books, Fawcett Publications, New York, 1987, p. 38.
16. Bushnell, Sarah T. *The Truth About Henry Ford.* The Reilly and Lee Company, Chicago, 1922, p. 188.
17. *Reminiscences*, Leslie McDonnell, p. 28.
18. Bennett, Harry. *We Never Called Him Henry.* Gold Metal Books, Fawcett Publications, New York, 1987, pp. 26–27.
19. Bennett, Harry. *We Never Called Him Henry.* Gold Metal Books, Fawcett Publications, New York, 1987, p. 31.
20. Josephine Goman Papers, University of Michigan, p. 49.
21. McCarten, John. "The Little Man in Mr. Ford's Basement." *American Mercury*, May and June 1940, p. 10.
22. Sward, Keith. *The Legend of Henry Ford.* Rinehart and Company, New York, 1948, p. 292.
23. *Reminiscences*, Irving Bacon, p. 210.

24. Bennett, Harry. *We Never Called Him Henry.* Gold Metal Books, Fawcett Publications, New York, 1987, pp. 63–64.
25. E.T. Gregorie, various interviews between 1992 and 1998.
26. E.T. Gregorie, various interviews between 1992 and 1998; also Sward, Keith. *The Legend of Henry Ford.* Rinehart and Company, New York, 1948, p. 336.
27. *Reminiscences*, Fred Black, p. 136.
28. *Reminiscences*, Ernest Liebold, pp. 1212–1213.
29. Sward, Keith. *The Legend of Henry Ford.* Rinehart and Company, New York, 1948, pp. 294 and 299.
30. *Reminiscences*, Irving Bacon, p. 248.
31. Bennett, Harry. *We Never Called Him Henry.* Gold Metal Books, Fawcett Publications, New York, 1987, p. 65.
32. Sward, Keith. *The Legend of Henry Ford.* Rinehart and Company, 1948, p. 297.
33. McCarten, John. "The Little Man in Mr. Ford's Basement." *American Mercury*, May and June 1940, p. 10.
34. Bennett, Harry. *We Never Called Him Henry.* Gold Metal Books, Fawcett Publications, New York, 1987, p. 68.
35. *Reminiscences*, Walter Doehler, p. 34.
36. Accession 1512.
37. Accession 6, Box 156.
38. Accession 6, Box 353.
39. *Reminiscences*, William Rinehart, pp. 12–13; also Evelyn Collins, interview, July 9, 1989.
40. Accession 6, Box 259.
41. Accession 6, Box 270.
42. Brough, James. *The Ford Dynasty: An American Story.* Doubleday and Company, New York, 1977, p. 148.
43. *Detroit Free Press*, February 11, 1933.
44. *The Detroit News*, January 26, 1962.
45. Sward, Keith. *The Legend of Henry Ford.* Rinehart and Company, New York, 1948, p. 239.
46. Bennett, Harry. *We Never Called Him Henry.* Gold Metal Books, Fawcett Publications, New York, 1987, p. 42.
47. *Reminiscences*, William Rinehart, p. 13.
48. Lacey, Robert. *Ford: The Men and the Machine.* Little, George Brown and Company, Boston, 1986, p. 384.
49. *Reminiscences*, Arthur Backus, p. 5.
50. *Reminiscences*, Arthur Backus, p. 4.
51. *Collectible Vehicle*, 1997 edition, p. 269; also Nevins, Alan and Frank Ernest Hill. *Ford: Decline and Rebirth 1933–1962.* Charles Scribner's Sons, New York, 1962, p. 31.
52. Leonard, Jonathan Norton. *The Tragedy of Henry Ford.* G.P. Putnam's Sons, New York, 1932, p. 232.
53. Gillmore, Dan. "Ford's Fascism: Proof Part 2." *Friday Weekly*, Vol. 2, No. 5, January 31, 1941, p. 5.
54. Leonard, Jonathan Norton. *The Tragedy of Henry Ford.* G.P. Putnam's Sons, New York, 1932, p. 236.

55. Ibid.
56. Sward, Keith. *The Legend of Henry Ford.* Rinehart and Company, 1948, p. 374–375.
57. *Reminiscences*, J.M. Waggoner, p. 26.
58. Sward, Keith. *The Legend of Henry Ford.* Rinehart and Company, New York, 1948, p. 377.
59. Sward, Keith. *The Legend of Henry Ford.* Rinehart and Company, New York, 1948, p. 361.
60. Sward, Keith. *The Legend of Henry Ford.* Rinehart and Company, New York, 1948, p. 369.
61. Collier, Peter and David Horowitz. *The Fords: An American Epic.* Summit Books, New York, 1987, p. 161; also Sward, Keith. *The Legend of Henry Ford.* Rinehart and Company, New York, 1948, p. 312.
62. Bennett, Harry. *We Never Called Him Henry.* Gold Metal Books, Fawcett Publications, New York, 1987, p. 117.
63. Sward, Keith. *The Legend of Henry Ford.* Rinehart and Company, 1948, pp. 347–348.
64. *Time*, March 17, 1941, p. 18.
65. Simonds, William A. *Henry Ford: His Life, His Work, His Genius.* The Bobbs-Merrill Company, New York, 1943, p. 267.
66. Gillmore, Dan. "Ford's Fascism: Proof Part 2." *Friday Weekly*, Vol. 2, No. 5, January 31, 1941, p. 5.
67. *The Detroit News*, September 4, 1935.
68. Bennett, Harry. *We Never Called Him Henry.* Gold Metal Books, Fawcett Publications, New York, 1987, p. 91.
69. Bennett, Harry. *We Never Called Him Henry.* Gold Metal Books, Fawcett Publications, New York, 1987, pp. 90–94.
70. Sorensen, Charles. *My Forty Years with Ford.* W.W. Horton and Company, New York, 1956, p. 255.
71. *Reminiscences*, Charles Sorensen, no page number.
72. Collier, Peter and David Horowitz. *The Fords: An American Epic.* Summit Books, New York, 1987, p. 157.
73. *Reminiscences*, Herman Moekle, p. 159.
74. Ken Bannon, interview, July 31, 1989.
75. Accession 1, Box 174, File 11.
76. Accession 1, Box 27.
77. Sorensen, Charles. *My Forty Years with Ford.* W.W. Horton and Company, New York, 1956, p. 260.
78. *Reminiscences*, Charles Sorensen, no page number.
79. *Reminiscences*, Albert Wibel, p. 254.
80. Sorensen, Charles. *My Forty Years with Ford.* W.W. Horton and Company, New York, 1956, p. 257.
81. Bennett, Harry. *We Never Called Him Henry.* Gold Metal Books, Fawcett Publications, New York, 1987, p. 63.
82. Sorensen, Charles. *My Forty Years with Ford.* W.W. Horton and Company, New York, 1956, p. 261; also Bennett, Harry. *We Never Called Him Henry.* Gold Metal Books, Fawcett Publications, New York, 1987, p. 136.
83. *Reminiscences*, Charles Sorensen, no page number.

84. Bennett, Harry. *We Never Called Him Henry.* Gold Metal Books, Fawcett Publications, New York, 1987, p. 136.
85. *Reminiscences*, Charles Sorensen, no page number; also Bennett, Harry. *We Never Called Him Henry.* Gold Metal Books, Fawcett Publications, New York, 1987, p. 118; also Sorensen, Charles. *My Forty Years with Ford.* W.W. Horton and Company, New York, 1956, p. 259.
86. *Reminiscences*, Charles Sorensen, no page number.
87. Nevins, Alan and Frank Ernest Hill. *Ford: Decline and Rebirth 1933–1962.* Charles Scribner's Sons, New York, 1962, p. 140.
88. Sward, Keith. *The Legend of Henry Ford.* Rinehart and Company, New York, 1948, p. 397.
89. Sward, Keith. *The Legend of Henry Ford.* Rinehart and Company, New York, 1948, p. 402.
90. *Time*, March 17, 1941, p. 17.
91. Simonds, William A. *Henry Ford: His Life, His Work, His Genius.* The Bobbs-Merrill Company, New York, 1943, p. 308.
92. Sward, Keith. *The Legend of Henry Ford.* Rinehart and Company, New York, 1948, p. 405.
93. Accession 7, Box 25.
94. Ibid.
95. Francis ImOberstag, interview, June 22, 1989.
96. Clancy, Louise and Florence Davies. *The Believer: The Life Story of Mrs. Henry Ford.* Van Rees Press, New York, 1960, p. 192; also Simonds, William A. *Henry Ford: His Life, His Work, His Genius.* The Bobbs-Merrill Company, New York, 1943, p. 275.
97. Simonds, William A. *Henry Ford: His Life, His Work, His Genius.* The Bobbs-Merrill Company, New York, 1943, p. 275.
98. Accession 7, Box 25.
99. *Reminiscences*, Joseph Lawry, p. 66.
100. Simonds, William A. *Henry Ford: His Life, His Work, His Genius.* The Bobbs-Merrill Company, New York, 1943, p. 309.
101. Simonds, William A. *Henry Ford: His Life, His Work, His Genius.* The Bobbs-Merrill Company, New York, 1943, p. 308.
102. Dr. Conrad Lam, interview, September 21, 1989.
103. Accession 1512.
104. *Reminiscences*, Fred Black, p. 95.
105. Sorensen, Charles. *My Forty Years with Ford.* W.W. Horton and Company, New York, 1956, p. 260.
106. Bennett, Harry. *We Never Called Him Henry.* Gold Metal Books, Fawcett Publications, New York, 1987, pp. 138–140.
107. Sorensen, Charles. *My Forty Years with Ford.* W.W. Horton and Company, New York, 1956, p. 261.
108. Sorensen, Charles. *My Forty Years with Ford.* W.W. Horton and Company, New York, 1956, p. 262.
109. Ibid.
110. Accession 1512.
111. Sorensen, Charles. *My Forty Years with Ford.* W.W. Horton and Company, 1956, p. 262.
112. *Reminiscences*, Charles Sorensen, no page number.

113. Accession 1512.
114. *Reminiscences*, Charles Sorensen, p. 123.
115. Accession 1512.
116. Nevins, Alan and Frank Ernest Hill. *Ford: Decline and Rebirth 1933–1962.* Charles Scribner's Sons, New York, 1962, p. 164.
117. Sward, Keith. *The Legend of Henry Ford.* Rinehart and Company, New York, 1948, pp. 419–420.
118. Nevins, Alan and Frank Ernest Hill. *Ford: Decline and Rebirth 1933–1962.* Charles Scribner's Sons, New York, 1962, p. 166; also Simonds, William A. *Henry Ford: His Life, His Work, His Genius.* The Bobbs-Merrill Company, New York, 1943, p. 316.
119. *Reminiscences*, Charles Sorensen, p. 36.
120. *Reminiscences*, Charles Sorensen, p. 431; also Sorensen, Charles. *My Forty Years with Ford.* W.W. Horton and Company, New York, 1956, p. 269.
121. *Reminiscences*, Charles Sorensen, p. 36.
122. *Reminiscences*, William Hughson, p. 34.
123. Nevins, Alan and Frank Ernest Hill. *Ford: Decline and Rebirth 1933–1962.* Charles Scribner's Sons, New York, 1962, p. 167.
124. Sorensen, Charles. *My Forty Years with Ford.* W.W. Horton and Company, New York, 1956, p. 271.
125. *Reminiscences*, Joseph Lawry, p. 65.
126. Sorensen, Charles. *My Forty Years with Ford.* W.W. Horton and Company, New York, 1956, pp. 257–258.
127. *Reminiscences*, Charles Sorensen, p. 84.
128. Sorensen, Charles. *My Forty Years with Ford.* W.W. Horton and Company, New York, 1956, pp. 257–258.
129. Josephine Goman Papers, University of Michigan, p. 52; also *Reminiscences*, Charles Sorensen, p. 85.
130. *Reminiscences*, Charles Sorensen, p. 85.
131. Accession 1512.
132. Bennett, Harry. *We Never Called Him Henry.* Gold Metal Books, Fawcett Publications, New York, 1987, pp. 149–150.
133. *Reminiscences*, Charles Sorensen, p. 130.
134. *Reminiscences*, Albert Wibel, p. 254.

Chapter 25: Willow Run

1. Smith, Patricia Barbara. "Edsel Ford." *Detroit Magazine, The Detroit Free Press*, October 17 and October 24, 1965, p. 23.
2. McCarthy, Joe. "The Ford Family." *Holiday*, June 1957, p. 121.
3. Floyd Lackanow, interview, July 12, 1989.
4. Mike Skinner, interview, August 15, 2001.
5. Nevins, Alan and Frank Ernest Hill. *Ford: Decline and Rebirth 1933–1962.* Charles Scribner's Sons, New York, 1962, p. 184.
6. Lasky, Victor. *Never Complain, Never Explain.* Richard Marek Publishers, New York, p. 57.

7. Accession 141.
8. *Detroit Times*, May 26, 1943.
9. Borth, Christy. Unpublished manuscript, Detroit Public Library.
10. *Freedom's Arsenal: The Story of the Automotive Council for War Production*. Automobile Manufacturers Association, Detroit, 1950, pp. 47–48.
11. *Freedom's Arsenal: The Story of the Automotive Council for War Production*. Automobile Manufacturers Association, Detroit, 1950, p. 3.
12. Borth, Christy. Unpublished manuscript, Detroit Public Library.
13. *Reminiscences*, Joseph Lawry, p. 77; also *Reminiscences*, Charles Sorensen, p. 14.
14. *Reminiscences*, Ernest Liebold, p. 1180.
15. Accession 65, Box 69.
16. *Reminiscences*, Laurence Sheldrick, p. 319.
17. Lindbergh, Charles A. *The Wartime Journals of Charles A. Lindbergh*. Harcourt, Brace, Jovanovich, New York, 1970, pp. 363–364.
18. *Reminiscences*, Rufus Wilson, p. 38.
19. *Reminiscences*, Ernest Liebold, p. 1182.
20. *Reminiscences*, Fred Black, p. 90.
21. *Reminiscences*, Fred Black, p. 91.
22. Lacey, Robert. *Ford: The Men and the Machine*. Little, George Brown and Company, Boston, 1986, p. 388.
23. *Reminiscences*, Fred Black, p. 90.
24. *Reminiscences*, William Pioch, p. 58.
25. *Reminiscences*, Dale Roeder, p. 108.
26. Borth, Christy. Unpublished manuscript, Detroit Public Library.
27. *Reminiscences*, Laurence Sheldrick, pp. 304–305.
28. *Reminiscences*, Joseph Lawry, p. 74.
29. *Reminiscences*, Joseph Lawry, pp. 74–75.
30. Sorensen, Charles. *My Forty Years with Ford*. W.W. Horton and Company, New York, 1956, p. 278.
31. Borth, Christy. Unpublished manuscript, Detroit Public Library.
32. *Reminiscences*, Laurence Sheldrick, p. 321.
33. Sorensen, Charles. *My Forty Years with Ford*. W.W. Horton and Company, New York, 1956, pp. 276–277.
34. Burton Collection, Detroit Public Library.
35. *Freedom's Arsenal: The Story of the Automotive Council for War Production*. Automobile Manufacturers Association, Detroit, 1950, pp. 43–44.
36. *Freedom's Arsenal: The Story of the Automotive Council for War Production*. Automobile Manufacturers Association, Detroit, 1950, p. 37.
37. Sorensen, Charles. *My Forty Years with Ford*. W.W. Horton and Company, New York, 1956, p. 279.
38. Borth, Christy. Unpublished manuscript, Detroit Public Library.
39. *Reminiscences*, Roscoe Smith, p. 58.
40. *Reminiscences*, William Pioch, p. 64.

41. Sorensen, Charles. *My Forty Years with Ford.* W.W. Horton and Company, New York, 1956, p. 281.
42. Borth, Christy. Unpublished manuscript, Detroit Public Library.
43. Sorensen, Charles. *My Forty Years with Ford.* W.W. Horton and Company, New York, 1956, p. 282.
44. Sorensen, Charles. *My Forty Years with Ford.* W.W. Horton and Company, New York, 1956, p. 284.
45. Ibid.
46. *The Washington Post*, May 30, 1943.
47. *Reminiscences*, Charles Sorensen, p. 157.
48. *Reminiscences*, Charles Sorensen, p. 139.
49. Ford Archives, Accession 6.
50. Lindbergh, Charles A. *The Wartime Journals of Charles A. Lindbergh.* Harcourt, Brace, Jovanovich, New York, 1970, pp. 613 and 644–645.
51. Nevins, Alan and Frank Ernest Hill. *Ford: Decline and Rebirth 1933–1962.* Charles Scribner's Sons, New York, 1962, p. 192.
52. Lindbergh, Charles A. *The Wartime Journals of Charles A. Lindbergh.* Harcourt, Brace, Jovanovich, New York, 1970, pp. 680 and 690.
53. *Design Oral History*, Frederick Hoadley.
54. Simonds, William A. *Henry Ford: His Life, His Work, His Genius.* The Bobbs-Merrill Company, New York, 1943, p. 333.
55. Nevins, Alan and Frank Ernest Hill. *Ford: Decline and Rebirth 1933–1962.* Charles Scribner's Sons, New York, 1962, p. 198.
56. Accession 6, Box 381.
57. Lindbergh, Charles A. *The Wartime Journals of Charles A. Lindbergh.* Harcourt, Brace, Jovanovich, New York, 1970, pp. 737–738.
58. Accession 6, Box 381.
59. Nevins, Alan and Frank Ernest Hill. *Ford: Decline and Rebirth 1933–1962.* Charles Scribner's Sons, New York, 1962, pp. 215–216.
60. Accession 6, Box 373.
61. *Reminiscences*, H. Ablewhite, p. 28.
62. Lindbergh, Charles A. *The Wartime Journals of Charles A. Lindbergh.* Harcourt, Brace, Jovanovich, New York, 1970, p. 677.
63. Accession 6, Box 381.
64. Lewis, David. *The Public Image of Henry Ford.* Wayne State University Press, Detroit, 1976, p. 353.
65. Lindbergh, Charles A. *The Wartime Journals of Charles A. Lindbergh.* Harcourt, Brace, Jovanovich, New York, 1970, p. 737.
66. Accession 6.
67. Ford Archives, clip book, 1941.
68. Accession 6, Box 174.
69. Nevins, Alan and Frank Ernest Hill. *Ford: Decline and Rebirth 1933–1962.* Charles Scribner's Sons, New York, 1962, p. 226.
70. *Reminiscences*, Charles Sorensen, p. 154.

71. *Reminiscences*, M.L. Wiesmyer, p. 120.
72. *Reminiscences*, Herman Moekle, p. 184.
73. *Reminiscences*, M.L. Wiesmyer, p. 143.
74. *Reminiscences*, Charles Sorensen, no page number.
75. Governor George Romney, interview, August 1, 1989.
76. *Automotive News*, May 26, 1943.
77. Don Fair, interview, May 2, 1989.
78. Accession 1, Box 27.
79. Clancy, Louise and Florence Davies. *The Believer: The Life Story of Mrs. Henry Ford.* Van Rees Press, New York, 1960, p. 199.
80. *Reminiscences*, Charles Sorensen, p. 69.
81. Sorensen, Charles. *My Forty Years with Ford.* W.W. Horton and Company, New York, 1956, pp. 103–105.
82. *Reminiscences*, Irving Bacon, p. 201.
83. Borth, Christy. Unpublished manuscript, Detroit Public Library.
84. *Reminiscences*, Helen Gore Doremus, pp. 35–36.
85. *Reminiscences*, Charles Sorensen, p. 211.
86. *Reminiscences*, Laurence Sheldrick, p. 374.
87. Lewis, David L. "Lincoln Cosmopolitan: The Gleam in Edsel Ford's Eye." *Car Classics*, April 1973, p. 53.
88. E.T. Gregorie, various interviews between 1992 and 1998.
89. *Reminiscences*, Charles Sorensen, no page number.
90. *Reminiscences*, Laurence Sheldrick, p. 268.
91. E.T. Gregorie, various interviews between 1992 and 1998.
92. "Fantastic Ford Finds." *Special-Interest Autos*, November–December 1970, p. 9.
93. E.T. Gregorie, various interviews between 1992 and 1998.
94. *Reminiscences*, Joseph Galamb, p. 113.
95. E.T. Gregorie, various interviews between 1992 and 1998.
96. *Reminiscences*, Al Esper, p. 70.
97. *Reminiscences*, V.Y. Tallberg, p. 111.
98. *Reminiscences*, Laurence Sheldrick, p. 266.
99. E.T. Gregorie, various interviews between 1992 and 1998.
100. Ibid.
101. Ibid.
102. Ibid.
103. Ibid.

Chapter 26: Dark Angel

1. Accession 536, Box 39.
2. Brough, James. *The Ford Dynasty: An American Story*. Doubleday and Company, New York, 1977, p. 165; also Accession 975.
3. Robert Smith, interview, May 18, 1989.
4. Accession 6, Box 320.

5. *Reminiscences*, Charles Sorensen, p. 178.
6. Dr. Neal Chamberlain, interview, May 12, 2001.
7. *Reminiscences*, Burt Watt, p. 4.
8. Robert Smith, interview, May 18, 1989.
9. Ibid.
10. Bennett, Harry. *We Never Called Him Henry*. Gold Metal Books, Fawcett Publications, New York, 1987, p. 164.
11. *The Detroit News*, May 26, 1943,
12. Ibid.
13. *Reminiscences*, Charles Sorensen, p. 317.
14. *The Detroit News*, January 17, 1943.
15. *Reminiscences*, Charles Sorensen, p. 67.
16. Sorensen, Charles. *My Forty Years with Ford*. W.W. Horton and Company, New York, 1956, p. 318.
17. *Reminiscences*, Charles Sorensen, p. 99.
18. *Reminiscences*, Robert Rankin, pp. 12–13.
19. Accession 1, Box 27.
20. *Reminiscences*, A.A. Lepine, p. 63.
21. *The Detroit News*, January 17, 1942.
22. Edsel Ford's death certificate, State of Michigan.
23. *Reminiscences*, Irving Bacon, p. 176.
24. Dr. Conrad Lam, interview, August 21, 1989.
25. Ibid.
26. Ibid.
27. Roberta Rankin Mallast, interview, June 21, 1989.
28. Dr. Conrad Lam, interview, September 21, 1989.
29. *Reminiscences*, Charles Sorensen, no page number.
30. *Reminiscences*, Charles Sorensen, no page number; also Sorensen, Charles. *My Forty Years with Ford*. W.W. Horton and Company, New York, 1956, p. 320.
31. *Reminiscences*, Charles Sorensen, p. 67.
32. *Reminiscences*, Arthur Backus, p. 15.
33. *Reminiscences*, Charles Sorensen, p. 211.
34. Accession 819.
35. *Detroit Free Press*, December 27, 1942.
36. *Detroit Free Press*, January 3, 1943.
37. *Reminiscences*, Charles Sorensen, p. 183.
38. Accession 1, Box 27.
39. *Reminiscences*, Charles Sorensen, pp. 112–113.
40. *Reminiscences*, Charles Sorensen, pp. 174–175.
41. Sorensen, Charles. *My Forty Years with Ford*. W.W. Horton and Company, New York, 1956, p. 320.
42. Sorensen, Charles. *My Forty Years with Ford*. W.W. Horton and Company, New York, 1956, p. 321.
43. Ibid, p. 321.

44. Nevins, Alan and Frank Ernest Hill. *Ford: Decline and Rebirth 1933–1962.* Charles Scribner's Sons, New York, 1962, p. 260.
45. Sorensen, Charles. *My Forty Years with Ford.* W.W. Horton and Company, New York, 1956, p. 322.
46. *Reminiscences*, Albert Wibel, p. 351.
47. *Reminiscences*, Henry Doss, pp. 104–105.
48. *Reminiscences*, Charles Sorensen, p. 173.
49. *Reminiscences*, Charles Sorensen, no page number.
50. *Reminiscences*, Albert Wibel, p. 352.
51. Sorensen, Charles. *My Forty Years with Ford.* W.W. Horton and Company, New York, 1956, pp. 322–323.
52. E.T. Gregorie, various interviews between 1992 and 1998.
53. *Design Oral History*, E.T. Gregorie, p. 61.
54. E.T. Gregorie, various interviews between 1992 and 1998.
55. Cameron, William J. "In Memory." *The Herald*, Vol. X, No. 9, 1943, p. 2.
56. *Reminiscences*, Arthur Backus, p. 18.
57. *Reminiscences*, Charles Sorensen, pp. 322–323.
58. *Reminiscences*, Charles Sorensen, p. 133.
59. *Reminiscences*, Charles Sorensen, p. 14.
60. *Reminiscences*, Laurence Sheldrick, pp. 303–304.
61. Accession 975.
62. Crowther, Samuel. "Edsel B. Ford." *Saturday Evening Post*, December 25, 1926, p. 79.
63. *Reminiscences*, Charles Sorensen, p. 113.
64. Sorensen, Charles. *My Forty Years with Ford.* W.W. Horton and Company, New York, 1956, pp. 312–315.
65. Sorensen, Charles. *My Forty Years with Ford.* W.W. Horton and Company, New York, 1956, p. 315.
66. Sorensen, Charles. *My Forty Years with Ford.* W.W. Horton and Company, 1956, p. 318–319.
67. Dr. Brock Brush, interview, March 14, 1993.
68. *Reminiscences*, Arthur Backus, p. 19.
69. Dr. Brock Brush, interview, March 14, 1993.
70. *Reminiscences*, Charles Sorensen, p. 96.
71. *Reminiscences*, Rosa Buhler and J D. Thompson, p. 17.
72. *Reminiscences*, A.G. Wolfe, p. 209.
73. Accession 1512.
74. *Reminiscences*, Charles Sorensen, p. 14.
75. *Reminiscences*, Charles Sorensen, p. 18.
76. *Reminiscences*, Camille Ruddiman, p. 96.
77. Dr. Brock Brush, interview, March 14, 1993.
78. *Reminiscences*, Mrs. Frank Campsall, p. 17.
79. *The Detroit News*, May 26, 1943.
80. Sorensen, Charles. *My Forty Years with Ford.* W.W. Horton and Company, New York, 1956, p. 324.

81. Collier, Peter and David Horowitz. *The Fords: An American Epic.* Summit Books, NY, 1987, p. 195.
82. Cameron, William J. "In Memory." *The Herald*, Vol. X, No. 9, 1943, p. 2.
83. Sorensen, Charles. *My Forty Years with Ford.* W.W. Horton and Company, New York, 1956, p. 324.
84. Ross Cousins, various interviews between 1990 and 1996.
85. Baut, Donald, Winfield Arneson, and Edward Cushman. *Interview with E. Roy Bryant.* Dearborn Historical Museum, Dearborn, MI, February 23, 1968.
86. *Design Oral History*, William Clay Ford, p. 11–12.
87. Cameron, William J. "In Memory." *The Herald*, Vol. X, No. 9, 1943, p. 2.
88. Ibid.
89. Sterne, Margaret. *The Passionate Eye: The Life of William Valentiner.* Wayne State University Press, Detroit, 1980, p. 276.
90. *Reminiscences*, J.E. Bossardet, p. 85.
91. *Detroit Times*, May 29, 1943.
92. *Detroit Free Press*, May 28, 1943.
93. The Edsel & Eleanor Ford House Archives.
94. Smith, Patricia Barbara. "Edsel Ford." *Detroit Magazine, The Detroit Free Press*, October 17 and October 24, 1965, p. 30.
95. Sorensen, Charles. *My Forty Years with Ford.* W.W. Horton and Company, New York, 1956, p. 324–325.
96. Ibid.
97. Ibid.
98. *Detroit Free Press*, May 29, 1943.
99. Ibid.
100. Detroit *Times*, May 29, 1943.
101. Accession 536, Box 39.
102. *Detroit Free Press*, May 28, 1943.
103. *The Detroit News*, May 29, 1943.
104. Mary-Alice Merchant, interview, June 13, 1993.
105. Gelderman, Carol. *Henry Ford: The Wayward Capitalist.* The Dial Press, New York, 1981, p. 213.
106. *Reminiscences*, Charles Sorensen, pp. 211–212.
107. Nevins, Alan and Frank Ernest Hill. *Ford: Decline and Rebirth 1933–1962.* Charles Scribner's Sons, New York, 1962, p. 248.
108. *Detroit Times*, May 29, 1943.
109. Ibid.
110. Ibid.
111. Collier, Peter and David Horowitz. *The Fords: An American Epic.* Summit Books, New York, 1987, p. 191.
112. Accession 1, Box 32.
113. Ibid.
114. *Reminiscences*, William Cameron, p. 110.
115. *Reminiscences*, Fred Thoms, p. 113.
116. Conot, Robert. *An American Odyssey.* William Morrow Company, New York, 1974, p. 397.

117. *Reminiscences*, Catherine Ruddiman, p. 85.
118. *Reminiscences*, Charles Sorensen, p. 208.
119. *Reminiscences*, Gus Munchow, p. 60.
120. Nevins, Alan and Frank Ernest Hill. *Ford: Decline and Rebirth 1933–1962*. Charles Scribner's Sons, New York, 1962, p. 258.
121. *Reminiscences*, Ernest Liebold, p. 1176; also *Reminiscences*, William Cowling, p. 57.
122. *Reminiscences*, Theodore Gehle, p. 153.
123. Edsel's will, reposited at the Macomb County government offices.
124. Nevins, Alan and Frank Ernest Hill. *Ford: Decline and Rebirth 1933–1962*. Charles Scribner's Sons, New York, 1962, pp. 412–413; also Accession 1189.
125. Edsel Ford's will, reposited at the Macomb County government offices.
126. MacDonald, Dwight. *The Ford Foundation: The Men and the Millions*. Reynal and Company, New York, 1956, p. 133.
127. Nevins, Alan and Frank Ernest Hill. *Ford: Decline and Rebirth 1933–1962*. Charles Scribner's Sons, New York, 1962, p. 417.
128. Nevins, Alan and Frank Ernest Hill. *Ford: Decline and Rebirth 1933–1962*. Charles Scribner's Sons, New York, 1962, pp. 420–421.
129. Nevins, Alan and Frank Ernest Hill. *Ford: Decline and Rebirth 1933–1962*. Charles Scribner's Sons, New York, 1962, pp. 422–424.
130. Ford Foundation Annual Report 1998, p. 1.
131. Nevins, Alan and Frank Ernest Hill. *Ford: Decline and Rebirth 1933–1962*. Charles Scribner's Sons, New York, 1962, pp. 424–426.

Chapter 27: Epilogue

1. Josephine Goman Papers, University of Michigan, p. 122.
2. Sorensen, Charles. *My Forty Years with Ford.* W.W. Horton and Company, New York, 1956, pp. 325–326.
3. Nevins, Alan and Frank Ernest Hill. *Ford: Decline and Rebirth 1933–1962*. Charles Scribner's Sons, New York, 1962, p. 250.
4. *Reminiscences*, Irving Bacon, p. 211.
5. Sorensen, Charles. *My Forty Years with Ford.* W.W. Horton and Company, New York, 1956, p. 327.
6. *Reminiscences*, Charles Sorensen, p. 214; also Accession 1512.
7. Sorensen, Charles. *My Forty Years with Ford.* W.W. Horton and Company, New York, 1956, pp. 217–218.
8. E.T. Gregorie, various interviews between 1992 and 1998.
9. Ibid.
10. Ibid.
11. Ibid.
12. Ibid.
13. Ibid.
14. Sorensen, Charles. *My Forty Years with Ford.* W.W. Horton and Company, New York, 1956, pp. 327–328; also Nevins, Alan and Frank Ernest Hill. *Ford: Decline and Rebirth 1933–1962*. Charles Scribner's Sons, New York, 1962, p. 250.

15. McCarthy, Joe. "The Ford Family." *Holiday*, August 1957, p. 68.
16. Accession 975.
17. McCarthy, Joe. "The Ford Family." *Holiday*, August 1957, p. 64.
18. Sorensen, Charles. *My Forty Years with Ford.* W.W. Horton and Company, New York, 1956, pp. 328–329.
19. McCarthy, Joe. "The Ford Family." *Holiday*, August 1957, p. 68.
20. Nevins, Alan and Frank Ernest Hill. *Ford: Decline and Rebirth 1933–1962.* Charles Scribner's Sons, New York, 1962, p. 254.
21. *Reminiscences*, Mead Bricker, p. 63.
22. *Reminiscences*, Fred Black, p. 64.
23. Conot, Robert. *An American Odyssey.* William Morrow Company, New York, 1974, p. 397.
24. Accession 1512.
25. Bennett, Harry. *We Never Called Him Henry.* Gold Metal Books, Fawcett Publications, New York, 1987, p. 175.
26. Lasky, Victor. *Never Complain, Never Explain.* Richard Marek Publishers, New York, 1981, p. 63.
27. Josephine Goman Papers, University of Michigan, p. 49.
28. McCarthy, Joe. "The Ford Family." *Holiday*, August 1957, p. 68.
29. Hayes, Walter. *Henry: A Life of Henry Ford II.* Grove Weidenfeld, New York, 1990, p. 60.
30. *Reminiscences*, Charles Sorensen, p. 242.
31. Nevins, Alan and Frank Ernest Hill. *Ford: Decline and Rebirth 1933–1962.* Charles Scribner's Sons, New York, 1962, p. 249.
32. Sorensen, Charles. *My Forty Years with Ford.* W.W. Horton and Company, New York, 1956, pp. 327–328; also Nevins, Alan and Frank Ernest Hill. *Ford: Decline and Rebirth 1933–1962.* Charles Scribner's Sons, New York, 1962, p. 264.
33. Nevins, Alan and Frank Ernest Hill. *Ford: Decline and Rebirth 1933–1962.* Charles Scribner's Sons, New York, 1962, p. 268.
34. *Reminiscences*, Ernest Liebold, p. 1172.
35. *Reminiscences*, Charles Sorensen, p. 234.
36. McCarthy, Joe. "The Ford Family." *Holiday*, August 1957, p. 78.
37. Nevins, Alan and Frank Ernest Hill. *Ford: Decline and Rebirth 1933–1962.* Charles Scribner's Sons, New York, 1962, p. 266.
38. Connelly, Mary. "The Duce Reflects on Career, Colleagues." *Automotive News*, September 26, 1994, p. 34.
39. Ibid.
40. McCarthy, Joe. "The Ford Family." *Holiday*, August 1957, p. 78.
41. Nevins, Alan and Frank Ernest Hill. *Ford: Decline and Rebirth 1933–1962.* Charles Scribner's Sons, New York, 1962, pp. 268–269.
42. McCarthy, Joe. "The Ford Family." *Holiday*, August 1957, p. 79.
43. *Reminiscences*, Irving Bacon, p. 211.
44. McCarthy, Joe. "The Ford Family." *Holiday*, August 1957, p. 79.
45. Ibid.
46. Bryan, Ford. *Clara: Mrs. Henry Ford.* Wayne State University Press, Detroit, 2001, p. 291.
47. Josephine Goman Papers, University of Michigan, p. 122.

48. Ibid.
49. *Reminiscences*, Joseph Zaroski, p. 23.
50. E.T. Gregorie, various interviews between 1992 and 1998.
51. *Reminiscences*, Joseph Galamb, p. 30.
52. *Reminiscences*, Catherine Ruddiman, p. 10.
53. *Reminiscences*, Catherine Ruddiman, p. 112.
54. Clancy, Louise and Florence Davies. *The Believer: The Life of Mrs. Henry Ford.* Van Rees Press, New York, 1960, p. 199.
55. Bennett, Harry. *We Never Called Him Henry.* Gold Metal Books, Fawcett Publications, New York, 1987, p. 168.
56. *Reminiscences*, Albert Wibel, pp. 129 and 212.
57. *Reminiscences*, Joseph Galamb, p. 31.
58. *Reminiscences*, Fred Black, p. 57.
59. *Reminiscences*, Charles Niepoth, p. 28.
60. Bennett, Harry. *We Never Called Him Henry.* Gold Metal Books, Fawcett Publications, New York, 1987, p. 165.
61. E.T. Gregorie, various interviews between 1992 and 1998.
62. Accession 975.
63. E.T. Gregorie, various interviews between 1992 and 1998.
64. *Design Oral History*, E.T. Gregorie, pp. 84–85.
65. E.T. Gregorie, various interviews between 1992 and 1998.
66. Ibid.
67. Ibid.
68. Nevins, Alan and Frank Ernest Hill. *Ford: Decline and Rebirth 1933–1962.* Charles Scribner's Sons, New York, 1962, p. 324.
69. *Design Oral History*, George Walker, p. 29.
70. E.T. Gregorie, various interviews between 1992 and 1998.
71. Ibid.
72. Ibid.
73. *Design Oral History*, George Walker, p. 34.
74. *Reminiscences*, Eugene Farkas, p. 194.
75. E.T. Gregorie, various interviews between 1992 and 1998.
76. Ibid.
77. *Design Oral History*, George Walker, p. 40.
78. E.T. Gregorie, various interviews between 1992 and 1998.
79. Ibid.
80. Lewis, David. *The Public Image of Henry Ford.* Wayne State University Press, Detroit, 1976, p. 541.
81. Byrne, John A. *The Whiz Kids.* Currency-Doubleday, New York, 1993, p. 336; also Collier, Peter and David Horowitz. *The Fords: An American Epic.* Summit Books, New York, 1987, p. 276.
82. Hayes, Walter. *Henry: A Life of Henry Ford II.* Grove Weidenfeld, New York, 1990, p. 72.
83. Herndon, Booton. *Ford: An Unconventional Biography of the Men and Their Times.* Weybright and Talley, New York, 1969, p. 226.

84. Collier, Peter and David Horowitz. *The Fords: An American Epic*. Summit Books, New York, 1987, p. 266.
85. Herndon, Booton. *Ford: An Unconventional Biography of the Men and Their Times*. Weybright and Talley, New York, 1969, p. 225.
86. Lacey, Robert. *Ford: The Men and the Machine*. Little, George Brown and Company, Boston, 1986, p. 484.
87. Smith, Patricia Barbara. "Edsel Ford." *Detroit Magazine*, *The Detroit Free Press*, October 17 and October 24, 1965, p. 12.
88. *Design Oral History*, William Clay Ford, p. 12.
89. E.T. Gregorie, various interviews between 1992 and 1998.
90. *Reminiscences*, V.Y. Tallberg, p. 113; also *Reminiscences*, Theodore Gehle, p. 56.
91. *Reminiscences*, Albert Wibel, p. 128.
92. Herndon, Booton. *Ford: An Unconventional Biography of the Men and Their Times*. Weybright and Talley, New York, 1969, p. 141.
93. Burck, Gilbert. "Henry Ford II." *Life*, October 1, 1945, p. 120.

Index

Abbreviations are used before the page number to indicate a photo (*p*). Page numbers in the two photo sections are numbered consecutively *p*1-1 through *p*1-16 for the first section of photos, and *p*2-1 through *p*2-16 for the second section of photos.

About the Author

As a quintessential Fordophile, Henry Dominguez has studied Ford Motor Company history for more than a quarter of a century. However, he is more than a student of Ford history—he relives it through the many old Fords that he has owned over the years. As a teenager, he purchased his first Ford—a 1946 business coupe—for $56 at an estate sale. Since then, he has owned about a dozen others, including a 1927 Model T coupe, a 1940 Ford deluxe coupe, and a 1951 Ford station wagon. His present old Ford product is a rare 1939 Mercury sedan coupe.

When not working on his old Fords, Mr. Dominguez writes about the neglected but important aspects of Ford history. *The Ford Agency*, his first work, chronicled the development of the Ford dealership network. His latest work, *Edsel Ford & E. T. Gregorie*, is the first in-depth study of how the classic Fords of the 1930s and 1940s were designed. The latter book received the prestigious Cugnot Award of Distinction from the Society of Automotive Historians in 2000.

Mr. Dominguez holds a B.S. degree from Weber State University and an M.B.A. from the University of New Mexico. He is a member of the Society of Automotive Engineers, the Society of Automotive Historians, the Henry Ford Heritage Association, and the Early Ford V8 Club of America.

Although a Ford aficionado, Mr. Dominguez has spent most of his professional life working for General Motors. Initially located in southeastern Michigan, he spent much of his leisure time in the Ford Archives, researching his books. He still works for General Motors but now lives in Northern California with his wife, Pat DeHerrera, and son, Samuel.

12

www.ingramcontent.com/pod-product-compliance
Ingram Content Group UK Ltd.
Pitfield, Milton Keynes, MK11 3LW, UK
UKHW051451290926
15471UKWH00008B/1137

9 781468 607192